Second Edition

ELECTRICAL MACHINES with MATLAB®

Second Edition

ELECTRICAL MACHINES with MATLAB®

TURAN GÖNEN

CRC Press
Taylor & Francis Group
Boca Raton London New York

CRC Press is an imprint of the
Taylor & Francis Group, an **informa** business

CRC Press
Taylor & Francis Group
6000 Broken Sound Parkway NW, Suite 300
Boca Raton, FL 33487-2742

© 2012 by Taylor & Francis Group, LLC
CRC Press is an imprint of Taylor & Francis Group, an Informa business

No claim to original U.S. Government works

Printed in the United States of America on acid-free paper
Version Date: 20111111

International Standard Book Number: 978-1-4398-7799-9 (Hardback)

Library of Congress Cataloging-in-Publication Data

Gönen, Turan.
    Electrical machines with MATLAB / Turan Gönen. -- 2nd ed.
      p. cm.
    Rev. ed. of: Electrical machines / Turan Gönen. 1998.
    Includes bibliographical references and index.
    ISBN 978-1-4398-7799-9 (hardback)
    1. Electric machinery. 2. MATLAB. I. Gönen, Turan. Electrical machines. II. Title.

TK2000.G66 2012
621.31'042--dc23   2011041800

Visit the Taylor & Francis Web site at
http://www.taylorandfrancis.com

and the CRC Press Web site at
http://www.crcpress.com

Contents

Preface to the First Edition

As electrical engineering programs have become overloaded with various new courses, many universities have started to offer only one course in electrical machinery. Therefore, the primary purpose of writing this book is to provide a meaningful and easily readable textbook for a three-semester-hour introductory-level electrical machinery course. Clearly, the purpose of this book is not an introduction to the design of electrical machinery but is intended for students in electrical and other engineering disciplines who may want to teach themselves. It is assumed that the students have already completed an electrical circuits analysis course and are familiar with electromagnetic fields.

This book has evolved from the content of courses that I have presented at California State University, Sacramento; the University of Missouri at Columbia; the University of Oklahoma; Florida International University; and Iowa State University. It has been written for junior-level undergraduate students as well as practicing engineers in the industry. The book is suitable for both electrical and nonelectrical engineering students or practicing engineers who may want to teach themselves.

Basic material has been explained carefully and in detail with numerous examples. Special features of the book include ample numerical examples and problems designed to use the information presented in each chapter. Each new term is clearly defined when it is first introduced in the text and a special effort has been made to familiarize the reader with the vocabulary and symbols used by the industry. Also, consistent with modern practice, the *International System* (*SI*) of units has been used throughout the book.

It is important for every electrical engineering student, regardless of his or her particular specialization area, to be familiar with the fundamental concepts involving three-phase circuits, power, and power measurement in ac circuits. However, based on my experience and observation throughout my long teaching career, such topics are often reviewed either inadequately or too quickly toward the end of a basic circuit analysis course, or not at all in some cases due to lack of time. Therefore, as a remedy, a brief review of these topics is included in the first two chapters of this book. Of course, instructors will decide for themselves whether or not to review them in detail, depending on the need.

Furthermore, it is a fact that most students who take a basic electrical machines course do not specialize in electrical power engineering. Certainly they will not be designing any electrical machines during their professional careers; even those students specializing in electrical power engineering will probably not need to do so. Therefore, based on such reasoning as well as the Power Engineering Society of the IEEE, a more general approach to electrical machines has been used throughout this book.

In addition, new modern topics have been introduced in the additional five chapters to keep the reader up-to-date with new developments in the area of electrical machine applications and electric power systems, including renewable energy, wind energy and wind energy conversion systems, solar energy systems, energy storage systems, and the smart grid.

Several new appendices have been included in this edition. Also included is an extensive glossary to define the terminologies used in electrical machines.

It is expected that a typical student has a working knowledge of complex algebra, sinusoidal analysis, phasor diagrams, phasor analysis, and other basic concepts. However, it is a good idea to review the first two chapters briefly to establish a common background before reading further chapters. It is recommended that the minimum amount of material covered in a one-semester course should include the chapters on magnetic circuits, transformers, induction machines, synchronous machines, and direct-current machines, depending on the purpose of the course. A complete solutions manual is also available for instructors from the publisher.

Preface to the Second Edition

In this edition, the whole book has been reexamined thoroughly to ensure that it is clean of any errors. For example, all the examples and end-of-chapter problems have been recalculated by hand as well as by using MATLAB® program to ensure that it is free of any possible errors.

In addition, new modern topics have been introduced in the additional five chapters to keep the reader up-to-date with new developments in the area of electrical machine applications and electric power systems.

For MATLAB® and Simulink® product information, please contact:

The MathWorks, Inc.
3 Apple Hill Drive
Natick, MA, 01760-2098 USA
Tel: 508-647-7000
Fax: 508-647-7001
E-mail: info@mathworks.com
Web: www.mathworks.com

Turan Gönen
Sacramento, California

Acknowledgments

I would like to express my appreciation to Dr. David D. Robb of D. D. Robb and Associates for his kind encouragement and invaluable suggestions and friendship over the years. I would also like to express my sincere appreciation to Dr. Paul M. Anderson of Power Math Associates for his continual encouragement and suggestions.

I am most grateful to numerous colleagues, particularly Dr. Herbert Hess of the University of Idaho and Dr. Juan C. Balda of the University of Arkansas for pointing out the errors in the first edition; Dr. Anjan Bose of Washington State University; Dr. Thomas H. Ortmeyer of Clarkson University; Dr. Yahya Baghzouz of the University of Nevada, Las Vegas; Dr. Alexander E. Emanuel of Worcester Polytechnic Institute; late Dr. Adly A. Girgis and Dr. Elham B. Makram of Clemson University; Dr. Alvin Day of Iowa State University; Dr. G. T. Heydt of Arizona State University; Dr. Charles Slivinsky, Dr. Richard G. Hoft, Dr. Cyrus O. Harbourt, and Dr. James R. Tudor of the University of Missouri at Columbia; Late Professor John Pavlat of Iowa State University; Dr. Max Anderson and Dr. Earl F. Richards of the University of Missouri at Rolla; Dr. James Story of Florida International University; and my friends late Dr. Don Koval of the University of Alberta and Dr. B. P. Lathi of California State University, Sacramento, for their interest, encouragement, and invaluable suggestions. Also acknowledged is Dr. Salah Yousif of California State University, Sacramento, for his support and encouragement.

A special thank you is extended to Gerhard W. Juette and Klaus Habur of Siemens A.G.; John R. Stoutland, Ron Stevens, and Darlene Heare of General Electric Company; Gary B. Lister of Canadian General Electric Company; Bill Petruska of ABB, Inc.; Andy Carpenter and Marilyn Muscenti of Reliance Electric, Inc.; Judy Chaves of North American Transformer, Inc.; and Robert Murray and Otto Stoll of MagneTex, Inc.

I am also indebted to numerous students who studied portions of the book at California State University, Sacramento; the University of Missouri at Columbia; the University of Oklahoma; and Florida International University and made countless contributions and valuable suggestions for improvements.

I am most grateful to Lynne Onitsuka for her computer assistance and expertise during times of crises.

Finally, I am indebted to my MS student, Alan Escoriza, for resolving all the examples and problems in the book as well as solving the MATLAB® examples.

Author

Turan Gönen is professor of electrical engineering and director of the Electrical Power Educational Institute at California State University, Sacramento (CSUS). Previously, he was professor of electrical engineering and director of the Energy Systems and Resources Program at the University of Missouri–Columbia. Professor Gönen also held teaching positions at the University of Missouri–Rolla, the University of Oklahoma, Iowa State University, Florida International University, and Ankara Technical College. He has taught electrical machines and electric power engineering for over 38 years.

Professor Gönen also has a strong background in the power industry. He worked as a design engineer in numerous companies for eight years, both in the United States and abroad. He has served as a consultant for the United Nations Industrial Development Organization (UNIDO), Aramco, Black & Veatch Consultant Engineers, and the public utility industry. He has written over 100 technical papers as well as 4 books: *Electric Power Distribution System Engineering*; *Modern Power System Analysis*; *Electric Power Transmission System Engineering: Analysis and Design*; and *Engineering Economy for Engineering Managers*.

Professor Gönen is a life fellow of the Institute of Electrical and Electronics Engineers and the Institute of Industrial Engineers. He served on several committees and working groups of the IEEE Power Engineering Society and is a member of numerous honor societies, including Sigma Xi, Phi Kappa Phi, Eta Kappa Nu, and Tau Alpha Pi.

Professor Gönen received his BS and MS in electrical engineering from Istanbul Technical College in 1964 and 1966, respectively, and his PhD in electrical engineering from Iowa State University in 1975. He also received his MS in industrial engineering in 1973 and his PhD co-major in industrial engineering in 1978 from Iowa State University, and did his MBA from the University of Oklahoma (1980). Professor Gönen received the Outstanding Teacher Award twice at CSUS in 1997 and 2009.

1 Basic Concepts

I can resist everything except temptation.

Oscar Wilde, Lady Windermere's Fan, 1892

I wouldn't belong to any club that would have me as a member.

Groucho Marx

If you are not puzzled yet, you are just not listening to me.

Author Unknown

1.1 INTRODUCTION

In the United States, the use of electrical energy increased quickly after 1882, and power plants mushroomed across the entire country.* The main reasons for such a rapid increase in demand for electrical energy are several: Electrical energy is, in many ways, the most convenient energy form. It can be sent by power lines over great distances to the consumption point and easily transformed into mechanical work, radiant energy, heat, light, or other forms. Electrical energy cannot be stored effectively, but its convenience has contributed to its growing use. Further, by generating electrical energy in very large power plants, an *economy of scale* in such production can be achieved, that is, the unit cost of electrical energy goes down with increasing plant size. In general, the use of electrical energy may include various kinds of conversion equipment in addition to transmission lines and control devices.

The structure of an electrical power or energy† system is very large and complicated. However, it can be represented basically *by five* main components. The *energy source* may be coal, natural gas, or oil burned in a furnace to heat water and generate steam in a boiler; it may be water in a dam; it may be oil or gas burned in a combustion turbine; or it may be fissionable material, which in a nuclear reactor will heat water to produce steam. The *generation system* converts the energy source into electrical energy. The *transmission system* transports this bulk electrical energy from the generation system to principal load centers where it is distributed through (usually extra) high-voltage lines. The *distribution system* distributes such energy to consumers by using lower-voltage networks. Finally, the last component, that is, *load*, utilizes the energy by converting it to a required form for lights, motors, heaters, or other equipment, alone or in combination.

Figure 1.1 shows a detailed view of an *electrical power system* that delivers energy from the source to the load connected to it. Note that the first transformer in the system (the one next to the power plant) is called *a step-up transformer*, and the second transformer (the one at the end of the transmission line) is called *a step-down transformer*.

According to the *energy conservation principle* of thermodynamics, energy is never used up; it is simply converted to different forms. Presently available energy conversion methods can be categorized into four different groups. The first group includes the conventional methods that generate more than 99% of today's electrical energy. They convert thermal energy from fossil fuels or from

---

* As of 1991, the number of electrical motors in the 1–120 hp range was more than 125 million, according to a recent study by the U.S. Department of Energy. This study also stated that 53%–58% of the electrical energy generated is consumed by electric motor–driven systems.

† The term *energy* is being increasingly used in the electrical power industry to replace the conventional term *power*, depending on the context. Here, the terms are used interchangeably.

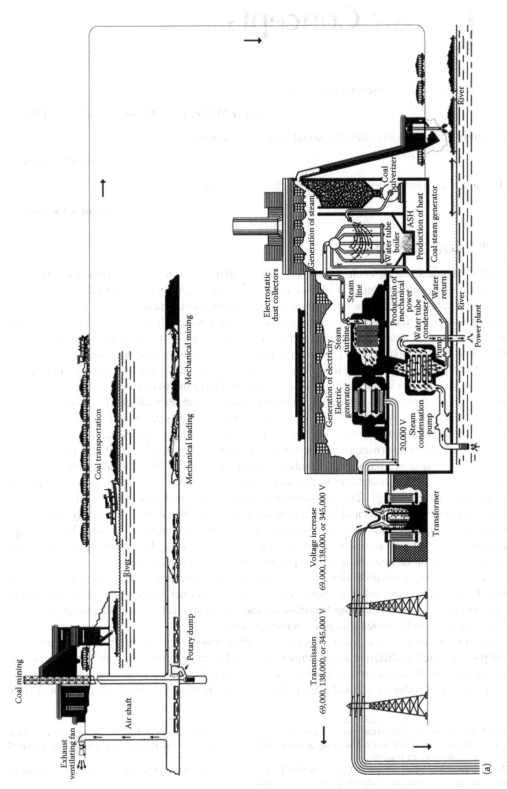

FIGURE 1.1 A detailed view of an electric power system: (a) general fuel supply system and power plant and (b) transmission and distribution systems.

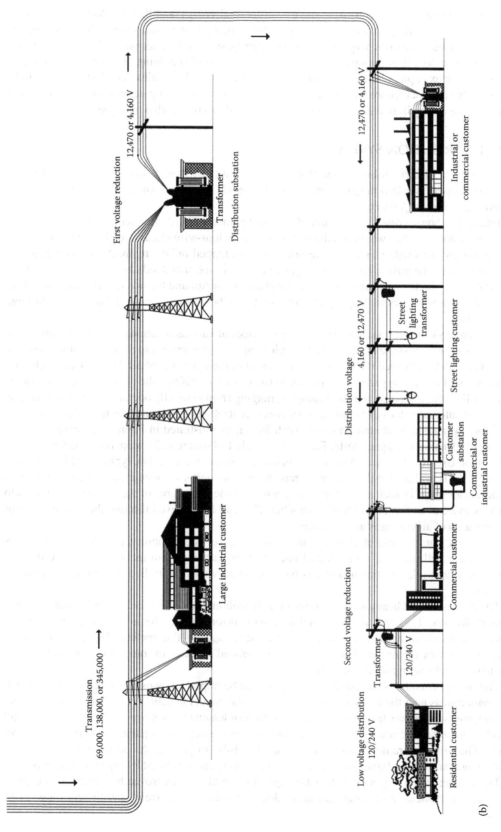

FIGURE 1.1 (continued)

nuclear fission energy to mechanical energy via thermal energy and then to electrical energy; or they convert hydro energy to electrical energy. The second group contains methods that are technically possible but have low-energy conversion efficiency, such as the internal combustion engine and the gas turbine. The third group covers the methods capable of supplying only very small amounts of energy, for example, photovoltaic solar cells, fuel cells, and batteries. The last group includes methods that are not technologically feasible but appear to have great potential, for example, fusion reactors, magnetohydrodynamic (MHD) generators, and electrogas-dynamic generators.

1.2 DISTRIBUTION SYSTEM

The part of the electric utility system that is between the distribution substation and the distribution transformers is called the *primary* system. It is made of circuits known as *primary feeders* or *primary distribution feeders*.

Figure 1.2 shows a *one-line diagram* of a typical primary distribution feeder. A feeder includes a *"main"* or main feeder, which usually is a three-phase, four-wire circuit, and branches or laterals, which usually are single-phase or three-phase circuits tapped off the main. Also sublaterals may be tapped off the laterals as necessary. In general, laterals and sublaterals located in residential and rural areas are single phase, and consist of one phase conductor and the neutral. The majority of the distribution transformers are single phase and connected between the phase and the neutral through fuse cutouts.

A given feeder is sectionalized by reclosing devices at various locations in such a manner as to remove as little as possible of the faulted circuit so as to hinder service to as few consumers as possible. This can be achieved through the coordination of the operation of all the fuses and reclosers.

It appears that, due to growing emphasis on the service reliability, the protection schemes in the future will be more sophisticated and complex, ranging from manually operated devices to remotely controlled automatic devices based on supervisory-controlled or computer-controlled systems.

Typically, a residential area served by such feeder, as illustrated in Figure 1.2, serves approximately 1000 homes per square mile. The feeder area is 1–4 square miles, depending on load density of the area. Usually, there are 15 to single-phasor laterals per feeder. Also, typically 150–500 short-circuit MVA is available at the substation bus. A given feeder is sectionalized by reclosing devices at various locations in such a manner as to remove as little as possible of the faulted circuit so as to hinder service to as few consumers as possible. This can be achieved through the coordination of the operation of all the fuses and reclosers.

It appears that, because of growing emphasis on the service reliability, the protection schemes in the future will be more sophisticated and complex, ranging from manually operated devices to remotely controlled automatic devices based on supervisory-controlled or computer-controlled systems.

The congested and heavy-load locations in metropolitan areas are served by using underground primary feeders. They are usually radial three-conductor cables. The improved appearance and less-frequent trouble expectancy are among the advantages of this method. However, it is more expensive, and the repair time is longer than the overhead systems. In some cases, the cable can be less than that of underground installation.

The voltage conditions on distribution systems can be improved by using shunt capacitors which are connected as near the loads as possible to derive the greater benefit. The use of shunt capacitors also improves the power factor involved, which in turn lessens the voltage drops and currents, and therefore losses, in the portions of a distribution system between the capacitors and the bulk power buses. The capacitor ratings should be selected carefully to prevent the occurrence of excessive overvoltages at times of light loads because of the voltage rise produced by capacitor currents.

The voltage overvoltages on distribution systems can also be improved by using series capacitors. But the application of series capacitors does not reduce the currents and therefore losses, in the system.

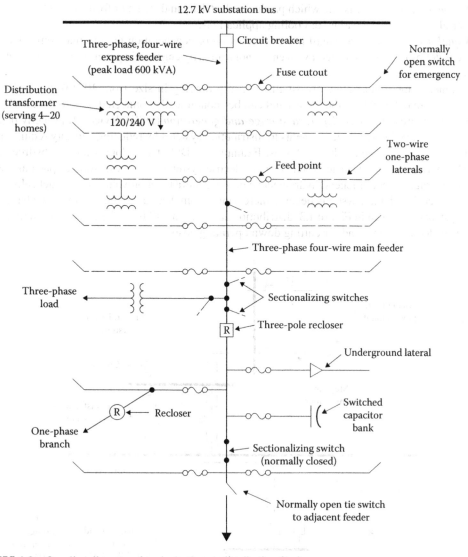

FIGURE 1.2   One-line diagram of typical primary distribution feeders.

1.3   IMPACT OF DISPERSED STORAGE AND GENERATION

Following the oil embargo and the rising prices of oil, the efforts toward the development of alternative energy sources (preferably renewable resources) for generating electric energy have been increased. Furthermore, opportunities for small power producers and cogenerators have been enhanced by recent legislative initiatives, for example, the *Public Utility Regulatory Policies Act* (PURPA) of 1978, and by the subsequent interpretations by the *Federal Energy Regulatory Commission* (FERC) in 1980.

The following definitions of the criteria affecting facilities under PURPA are given in Section 201 of PURPA:

A small power production facility is one which produces electric energy solely by the use of primary fuels of biomass, waste, renewable resources, or any combination thereof. Furthermore, the capacity of such production sources together with other facilities located at the same site must not exceed 80 MW.

A cogeneration facility is one which produces electricity and steam or forms of useful energy for industrial, commercial, heating, or cooling applications.

A qualified facility is any small power production or cogeneration facility that conforms to the previous definitions and is owned by an entity not primarily engaged in generation or sale of electric power.

In general, these generators are small (typically ranging in size from 100 kW to 10 MW and connectable to either side of the meter) and can be economically connected only to the distribution system. They are defined as *dispersed-storage-and-generation* (DSG) devices. If properly planned and operated, DSG may provide benefits to distribution systems by reducing capacity requirements, improving reliability, and reducing losses. Examples of DSG technologies include hydroelectric, diesel generators, wind electric systems, solar electric systems, batteries, storage space and water heaters, storage air conditioners, hydroelectric pumped storage, photovoltaics, and fuel cells.

As power-distribution systems become increasingly complex due to the fact that they have more DSG systems, as shown in Figure 1.3, distribution automation will be indispensable for maintaining a reliable electric supply and for cutting down operating costs.

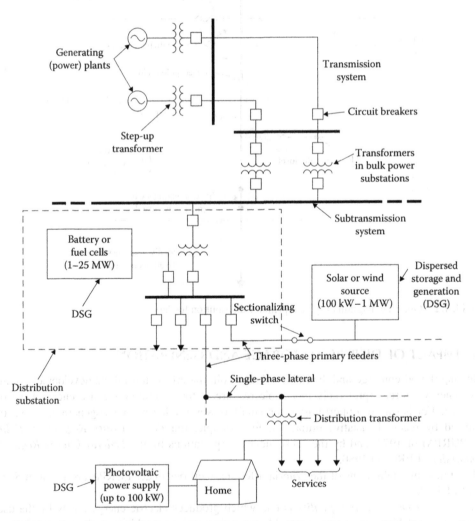

FIGURE 1.3 In the future, small, dispersed-energy-storage-and-generation units attached to a customer's home, a power-distribution feeder, or a substation would require an increasing amount of automation and control.

In distribution systems with DSG, the feeder or feeders will no longer be radial. Consequently, a more complex set of operating conditions will prevail for both steady state and fault conditions. If the dispersed generator capacity is large relative to the feeder supply capacity, then it might be considered as backup for normal supply. If so, this could improve service security in instances of loss of supply. In a given fault, a more complex distribution of higher magnitude fault currents will occur due to multiple supply sources. Such systems require more sophisticated detection and isolation techniques than those adequate for radial feeders. Therefore, distribution automation, with its multiple point monitoring and control capability, is well suited to the complexities of a distribution system with DSG.

1.4 BRIEF OVERVIEW OF BASIC ELECTRICAL MACHINES

In general, an electrical machine can be defined as an apparatus that can be used either to convert electrical energy into mechanical energy or to convert mechanical energy into electrical energy. If such a machine is used to convert electrical energy into mechanical energy, it is called a *motor*; however, if it is used to convert mechanical energy into electrical energy, it is called *a generator*.

Any given machine can convert energy in either direction and can therefore be used either as a motor or as a generator. Such conversion is facilitated through the action of a magnetic field. A generator has a rotary motion provided by a prime mover which supplies mechanical energy input. The relative motion between the conductors and the magnetic field of a generator produces an electrical energy output.

On the other hand, a motor has electrical energy supplied to its windings and a magnetic field that generates an electromagnetic interaction to produce mechanical energy or torque. Figure 1.4 shows an installed 1300 MW cross-compound turbine generator.

In general, each machine has a nonmoving (i.e., stationary) part and a moving (i.e., nonstationary) part. Depending on whether such a machine functions as a generator or motor, the moving part that is attached to a mechanical system receives mechanical input or provides mechanical output. The motion of such a moving part can be linear (e.g., as in *linear motors*), vibrating or reciprocating (e.g., as in various electrical razors), or rotating.

FIGURE 1.4 An installed 1300 MW cross-compound turbine generator.

In this book, only the rotating electrical machines are reviewed. They include (1) polyphase synchronous machines, (2) polyphase induction (also called *asynchronous*) machines, and (3) dc machines. However, there are other rotating and linear machines that are not included here. They operate basically on the same principles. Examples include

- *Reluctance machines,* which are synchronous machines without the dc excitation and are used in timers, electrical clocks, and recording applications.
- *Hysteresis machines,* which are similar to reluctance machines with a solid cylinder rotor made up of a permanent magnet material that needs only one electrical input. They are used in phonograph turntables and in other constant speed applications, such as electrical clocks.
- *Rotating rectifiers,* which have the same performance as regular synchronous machines except that field excitation is provided by an ac auxiliary generator and by rectifiers that are stationed on the rotor.
- *Permanent magnet machines,* which are ordinary synchronous machines with the field excitation provided by a permanent magnet. They have a very high efficiency since there are no field losses.
- *Becky Robinson* and *Nadyne–Rice machines,* which are brushless synchronous machines that operate based on rotor magnetic structure with a changing reluctance. They are mainly used in aerospace applications.
- *Lundell machines,* which are also brushless synchronous machines (but need slip rings to supply a dc field) that operate based on rotor magnetic structures with a changing reluctance. They are mainly used in automotive alternators.
- *Inductor and flux-switch machines,* which are inductor flux-switch configurations based on a changeable-reluctance principle similar to the reluctance machines, and a function of rotor position accomplished by the rotor design. They can be used as brushless synchronous motors and generators in aerospace and traction applications.

In addition to basic rotating electrical machines, transformers are also discussed. Even though a transformer involves the interchange of ac electrical energy from one voltage level to another, some of its principles of operation constitute the foundation for the study of electromechanical energy conversion. Thus, many of the relevant equations and conclusions of transformer theory are equally applicable to electromechanical energy conversion theory.

Rotating electrical machines have an outside (i.e., *stationary*) part that is called the stator and an inner (i.e., *rotating*) part that is called the rotor. As shown in Figure 2.1a, the rotor is centered within the stator, and the space that is located between the outside of the rotor and the inside of the stator is called the air gap. The figure shows that the rotor is supported by a steel rod that is called a *shaft*.

In turn, the shaft is supported by bearings so that the rotor can turn freely. Both the rotor and the stator of a rotating machine, as well as a transformer, have windings. The terminology that is commonly used to describe the windings of basic electrical machines and transformers is given in Table 1.1.

It is important to point out that in the study of any electromechanical apparatus, there is a need to model its electric circuit and one should be very familiar with the ac circuit analysis applicable to power circuits. Each electric circuit concept is analogous to a corresponding magnetic circuit concept.* Thus, to understand electrical machines, one needs a good knowledge of the concepts of both magnetic circuits and electrical power circuits. Therefore, a brief review of phasor representation is included in Appendix A. Also, the concepts of real and reactive powers in single-phase ac circuits are briefly reviewed in the following section. In Chapters 2 and 3, the concepts associated

---

* In Table 1.1, "ac current" is grammatically a redundant statement. Nevertheless, ac and dc, originally used as abbreviations, are now commonly used as adjectives in engineering vocabulary.

TABLE 1.1

Terminology Used to Describe the Windings of Basic Electrical Machines and Transformers

Apparatus	Name of Winding	Location of Winding	Function of Winding	Type of Current in Winding
Synchronous	Armature	Stator	Input/output	ac
Synchronous	Field	Rotor	Magnetizing	dc
Induction machine	Stator	Stator	Input	ac
dc machine	Armature	Rotor	Input/output	ac in winding'
	Field	Stator	Magnetizing	dc at brushes
Transformer	Primary	—	Input	ac
	Secondary	—	Output	ac

with three-phase circuits and magnetic circuits are reviewed. It is hoped that such reviews are sufficient to provide a common base, in terms of notation and references, in order to be able to follow the subsequent chapters.

1.5   REAL AND REACTIVE POWERS IN SINGLE-PHASE AC CIRCUITS

If the sinusoidal voltage across the terminals of a *single-phase* ac circuit is used *as a reference* and designated by the phasor $V = |V| \angle 0°$ and the phasor of the alternating current in the circuit is $|I| \angle \theta$, then the *real power* (i.e., *average* or *active power*) can be expressed as

$$P = |V||I| \cos \theta \qquad (1.1)$$

or

$$P = V_{rms} I_{rms} \cos \theta \qquad (1.2)$$

and

$$Q = |V||I| \sin \theta \qquad (1.3)$$

or

$$Q = V_{rms} I_{rms} \sin \theta \qquad (1.4)$$

Since $\sin \theta$ is dimensionless, Q has the dimension of volt-ampere. However, to help distinguish between real and reactive powers, Q is measured in var, which stands for volt-ampere reactive. The relation of a voltage with respect to a lagging current can be observed in the phasor diagram shown in Figure 1.5a. The term *power factor* is used for the factor $\cos \theta$, and sometimes the term *reactive factor* is used for the factor $\sin \theta$.

Apparent power S is the product of the phasor voltage magnitude and current magnitude. Therefore, it can be expressed as

$$S = |V||I| \qquad (1.5a)$$

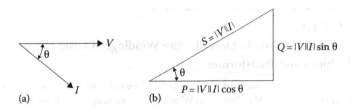

FIGURE 1.5 For a lagging current: (a) current and voltage phasor diagram and (b) power triangle.

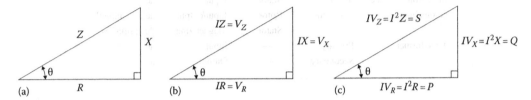

FIGURE 1.6 Evolution of the power triangle: (a) impedance triangle, (b) voltage triangle, and (c) power triangle.

or

$$S = VI \tag{1.5b}$$

The relationship between the real, reactive, and apparent powers can be represented by a triangle, known as the *power triangle*, as shown in Figure 1.5b. Figure 1.6 illustrates the evolution of the power triangle. Notice that the θ angle is greater when Q is greater and it is smaller when Q is smaller. Of course, when θ is greater the power factor, that is, $PF = \cos \theta$ is smaller; or when θ is smaller, $PF = \cos \theta$ is greater. The *power factor* is defined as

$$PF = \cos\theta = \frac{R}{Z} = \frac{P}{S} \tag{1.6}$$

or the *power factor angle* θ is a function of the power *factor PF* in a circuit so that

$$\theta = \arccos(PF) = \cos^{-1}(PF) \tag{1.7}$$

If the phasor voltage and phasor current are of a *purely* resistive network, the real power (i.e., average or active power) can be expressed as

$$P = VI \cos\theta = \frac{V^2}{R} = I^2 R \tag{1.8}$$

since $\cos \theta = 1$.

Similarly, the reactive power for a *purely capacitive* network can be expressed as

$$Q = VI \sin\theta = \frac{V^2}{X} = I^2 X \tag{1.9}$$

since $\sin \theta = 1$.

TABLE 1.2

Relationship between Lagging and Leading Currents and Loads in Single Phase

	Lagging	Leading
Current	$I = I\angle-\theta = I(\cos\theta - j\sin\theta)$	$I = I\angle\theta = I(\cos\theta + j\sin\theta)$
	where $I = \dfrac{V\angle0°}{Z\angle\theta} = \dfrac{V}{Z}\angle-\theta = I\angle-\theta$	where $I = \dfrac{V\angle0°}{Z\angle-\theta} = \dfrac{V}{Z}\angle+\theta = I\angle\theta$
Load	$S_L = S_L(\cos\theta + j\sin\theta) = P_L + jQ_L$	$S_L = S_L(\cos\theta - j\sin\theta) = P_L - jQ_L$
	since $S = P + jQ = VI^*$	since $S = P - jQ = VI^*$
	where $V = V\angle0°$	where $V = V\angle0°$
	and $I = I\angle-\theta$ or $I^* = I\angle\theta$	and $I = I\angle\theta$ or $I^* = I\angle-\theta$

In power system computations, it has been customary to define *a complex power for single phase** as

$$S_\phi = V_\phi I_\phi^* = P_\phi + jQ_\phi \tag{1.10}$$

where

S_ϕ is the complex (or phasor) power for single phase

I_ϕ^* is the conjugate of current phasor I_ϕ per phase

When the resultant S_ϕ *single-phase complex* (or *phasor*) *power* is in rectangular form, then the related real and reactive power can be expressed as

$$P_\phi = \mathcal{Re}(VI_\phi^*) = V_\phi I_\phi \cos\theta \tag{1.11}$$

$$Q_\phi = \mathcal{Im}(VI_\phi^*) = V_\phi I_\phi \cos\theta \tag{1.12}$$

Assume that a single-phase load is connected to a bus and being fed by current and power by the bus and that bus voltage is $V = V\angle0°$ and the load current is $I = I\angle-\theta$ or, in other words, the load has a lagging power factor and as a result of it, it can also be said that it has a lagging current. Also assume that in the lagging case, the current lags its voltage by the power factor angle of θ and in the leading case, the current leads its voltage by the power factor angle of θ, as given in Table 1.2.

Notice that the angle θ depends on the impedance, or its components, or specifically its inductance or capacitance values.

Three-phase complex power $S_{3\phi}$ can be found from

$$S_{3\phi} = \sqrt{3}V_L I_L^* = P_{3\phi} + jQ_{3\phi} \tag{1.13}$$

---

* The algebraic sign of reactive power has been a subject of debate for 25 years. Finally, the convention defining S as VI^* was adapted by the American Institute of Electrical Engineers, approved by the American Standards Association, and published in *Electrical Engineering*. Therefore, to obtain the proper sign for Q, it is necessary to calculate S as VI^*, which would reverse the sign for Q.

where

V_L is the line-to-line (or line) voltage

I_L^* is the line current

Note that $I_L = \sqrt{3}I_\phi$ or $I_\phi = I_L/\sqrt{3}$ for the balanced systems.

The three-phase apparent power is

$$S_{3\phi} = \sqrt{3}V_L I_L \tag{1.14}$$

When the resultant $S_{3\phi}$ *three-phase complex* (or *phasor*) *power* is in rectangular form, then the related *three-phase* real and *three-phase* reactive power for a *balanced* system can be expressed as

$$P_{3\phi} = \mathcal{Re}(\sqrt{3}V_L I_L^*)$$

$$= \sqrt{3}V_L I_L \cos\theta$$

$$= 3V_\phi I_\phi \cos\theta$$

$$= 3S_\phi \cos\theta$$

$$= S_{3\phi}\cos\theta \tag{1.15}$$

$$Q_{3\phi} = \mathcal{Im}(\sqrt{3}V_L I_L^*)$$

$$= \sqrt{3}V_L I_L \sin\theta$$

$$= 3V_\phi I_\phi \sin\theta$$

$$= 3S_\phi \sin\theta$$

$$= S_{3\phi}\sin\theta \tag{1.16}$$

Assume that a *balanced* three-phase load is connected to a set of three-phase buses and being fed by currents and powers by each bus and that bus voltages are given as $V_L = V_L \angle 0°$ and the load currents are given as $I_L = I_L \angle -\theta$ or, in other words, the *balanced* three-phase loads have lagging power factors as a result of also having lagging currents. Also assume that in the lagging case, each current lags its voltage by the power factor angle of θ on each phase and in the leading case, each current leads its voltage by the power factor angle of θ, as given in Table 1.3.

Consider the circuit shown in Figure 1.7a which is made of an ideal single-phase source connected to a single-phase load over a power line. Assume that there are two additional voltage sources that have equal voltages that are 120° out of phase with respect to one another, as shown in Figure 1.7b. Notice that there are now three neutral wires in this setup. Since the neutral wire is going to carry only the residual return current of phase currents, it can be reduced down to only one wire, as shown in Figure 1.3c. This approach obviously, in turn, saves money. However, if the system is made up of balanced loads, the residual current on the neutral wire becomes zero, thus the neutral wire, theoretically, can totally be eliminated, as shown in Figure 1.8a. This will save even more money in turn. But, such balanced situation rarely ever exists in real life.

If, the wye-connected load impedances are also be replaced by their equivalent delta-connected impedances, as shown in Figure 1.8b, so that each of the delta-connected impedances are now equal to three times of each of the wye-connected impedances. Hence, the system in Figure 1.8b will still

TABLE 1.3

Relationship between Lagging and Leading Currents and Loads in Three Phase

	Lagging	Leading
Current	$I_L = I_L\angle -\theta = I_L(\cos\theta - j\sin\theta)$	$I_L = I_L\angle\theta = I_L(\cos\theta + j\sin\theta)$
Load	$S_{3\phi} = S_{3\phi}(\cos\theta + j\sin\theta) = P_L + jQ_L$	$S_{3\phi} = S_{3\phi}(\cos\theta - j\sin\theta) = P_L - jQ_L$
	since $S_{3\phi} = P_{3\phi} + jQ_{3\phi} = \sqrt{3}V_L I_L^*$	since $S_{3\phi} = P_{3\phi} - jQ_{3\phi} = \sqrt{3}V_L I_L^*$
	where $V_L = V_L\angle 0°$	where $V_L = V_L\angle 0°$
	and $I_L = I_L\angle -\theta$ or $I_L^* = I_L\angle\theta$	and $I_L = I_L\angle\theta$ or $I_L^* = I_L\angle -\theta$

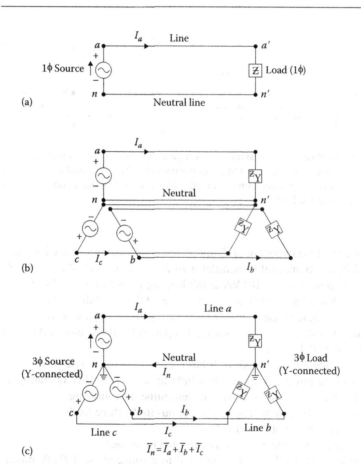

FIGURE 1.7 Evolution of three-phase system: (a) a single-phase system, (b) three single-phase system, and (c) a three-phase system.

act just like the system in Figure 1.8a. It is also possible to keep the load in wye connection as before, but instead connecting the ideal three single-phase voltage sources in delta instead of wye connection, as shown in Figure 1.8c. Furthermore, it is also possible to connect the loads in delta as well, as shown in Figure 1.8d. Such delta connection is only done in rare applications such as in the electrical power systems of ships. Also, the three-phase generators are in general connected in wye so that impedance can be inserted between their neutral point and the ground to reduce any future fault currents.

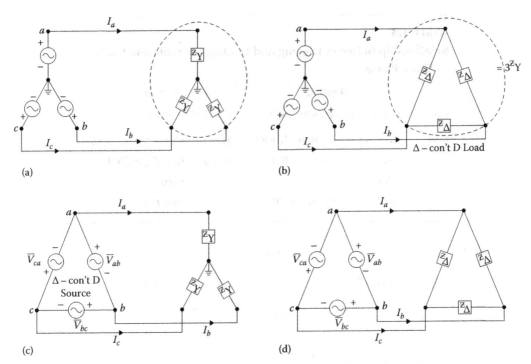

FIGURE 1.8 Various three-phase systems: (a) a wye-connected source connected to a wye-connected load, (b) a wye-connected source connected to an equivalent delta-connected load, (c) an equivalent delta-connected source connected to a wye-connected load, and (d) equivalent delta-connected source connected to an equivalent delta-connected load.

PROBLEMS

1.1 Assume that two load impedances Z_1 and Z_2 are connected in series with respect to each other but Z_1 and Z_2 are connected in parallel with Z_3 and that loads Z_1, Z_2, and Z_3 require 5 kVA at 0.8 lagging power factor, 10 kVA at 0.9 lagging power factor, 5 kW at unity power factor, respectively. Determine the kW required by the total load, if the frequency is 60 Hz.

1.2 A 10 kW 220 V single-phase ac motor is operating at 0.7 lagging power factor. Find the value of the capacitor that needs to be connected in parallel with the motor, if the power factor is to be improved to 0.95 lagging.

1.3 Assume that a single-phase $2400\angle 0°$ V bus is connected to a single-phase 100 kW motor operating at a lagging power factor of 0.9, a lighting load of 100 kW operating at a unity power factor, and a static capacitor of 100 kvar. Determine the following:

(a) The total complex power supplied by the bus to the three loads

(b) The total current supplied to the bus

(c) The power factor of the total load connected to the bus

1.4 A single-phase $4800\angle 0°$ V bus is connected to a single-phase 100 kW motor operating at a lagging power factor of 0.8, a lighting load of 200 kW operating at a unity power factor, and static capacitors of 150 kvar. Determine the following:

(a) The total complex power supplied by the bus to the loads

(b) The total current supplied to the bus

(c) The power factor of the total load connected to the bus

1.5 Consider Problem 1.3 and use the bus voltage of $2400\angle 0°$ V as the reference phasor and determine the following:

(a) The phasor current of the lighting load.

(b) The phasor current of the motor.

(c) The phasor current of the capacitor.

(d) The total phasor load current.

(e) Draw the phasor diagram of the voltage, all three load currents, and show how the three load currents combine to become the total load current in terms of phasor addition.

1.6 Consider Problem 1.4 and use the bus voltage of $4800\angle0°$ V as the reference phasor. Determine the following:

(a) The phasor current of the lighting load.

(b) The phasor current of the motor.

(c) The phasor current of the capacitor.

(d) The total phasor load current.

(e) Draw the phasor diagram of the voltage, all three load currents, and show how the three load currents combine to become the total load current in terms of phasor addition.

2 Three-Phase Circuits

Imagination is more important than knowledge.

Albert Einstein

The great end of learning is nothing else but to seek for the lost mind.

Mencius, Works, 299 BC

Earn your ignorance!

Learn something about everything

Before you know nothing about anything!

Turan Gönen

2.1 INTRODUCTION

In a single-phase ac circuit, instantaneous power to a load is of a *pulsating* nature. Even at unity power factor (i.e., when the voltage and the current are in phase with respect to each other), the instantaneous power is less than unity (i.e., when the voltage and the current are not in phase). The instantaneous power is not only zero four times in each cycle but it is also negative twice in each cycle. Therefore, because of economy and performance, almost all electrical power is produced by *polyphase* sources (i.e., by those generating voltages with more than one phase*).

A polyphase generator has two or more single phases connected so that they provide loads with voltages of equal magnitudes and equal phase differences.[†] For example, in a balanced *n-phase* system, there are *n* voltage sources connected together. Each phase voltage (or source) alternates sinusoidally, has the same magnitudes, and has a phase difference of 360/*n*° (where *n* is the number of phases) from its adjacent voltage phasors, except in the case of two-phase systems.[‡] Generators of 6, 12, or even 24 phases are sometimes used with *polyphase rectifiers* to supply power with low levels of ripples in voltage on the do side in the range of kilowatts. Today, virtually all the power produced in the world is three-phase power with a frequency of 50 or 60 Hz. In the United States, 60 Hz is the standard frequency. Recently, six-phase power transmission lines have been proposed because of their ability to increase power transfer over existing lines and reduce electrical environmental impact.[§]

2.2 THREE-PHASE SYSTEMS

As previously stated, even though other polyphase systems are feasible, the power utility industry has adopted the use of *three-phase* systems as the standard. Consequently, most of the generation,

---

* *A phase* is one of three branches making up a three-phase circuit. For example, in a wye connection, a phase is made up of those circuit elements connected between one line and neutral; however, in a delta connection, a phase consists of those circuit elements connected between two lines.

[†] Therefore, a polyphase generator is somewhat analogous to a multicylinder automobile engine in that the power delivered is steadier. Consequently, there is less vibration in the rotating machinery, which, in turn, performs more efficiently.

[‡] In a *two-phase generator*, the two equal voltages differ in phase by 90°, but in the *three-phase generator*, the three equal voltages have a phase-angle difference of 120°. However, the use of two-phase systems is very uncommon.

[§] In six-phase transmission lines, the conductor potential gradients are lower, which reduces both audible noise and electrostatic effects without requiring additional insulation.

transmission, distribution, and heavy-power utilization of electrical energy are done using three-phase systems. A three-phase system is supplied by a three-phase generator (i.e., *alternator*), which consists essentially of three single-phase systems displaced in time phase from each other by one-third of a period, or 120 electrical degrees. The advantages of three-phase systems over single-phase systems are as follows:

- Less conductor material is required in the three-phase transmission of power and therefore it is more economical.
- Constant rotor torque and therefore steady machine output can be achieved.
- Three-phase machines (generators or motors) have higher efficiencies.
- Three-phase generators may be connected in parallel to supply greater power more easily than single-phase generators.

Figure 2.1a shows the structure of an elementary three-phase and two-pole ac generator (also called an *alternator*). Its structure has basically two parts: the stationary outside part which is called the *stator* and the rotating inside part which is called the *rotor*. The field winding is located on the rotor and is excited by a direct current source through slip rings located on the common shaft. Thus, an alternator has a rotating electromagnetic field; however, its stator windings are *stationary*. The elementary generator shown in Figure 2.1a has three identical stator coils (*aa'*, *bb'*, and *cc'*), of one or more turns, displaced by 120° in space with respect to each other. If the rotor is driven counterclockwise at a constant speed, voltages will be generated in the three phases according to Faraday's law, as shown in Figure 2.1b. Notice that the stator windings constitute the *armature of* the generator (unlike *dc* machines where the armature is the rotor). Thus, the field rotates inside the armature. Each *of* the three stator coils makes up one phase in this single generator. Figure 2.1b shows the generated voltage waveforms in time domain, while Figure 2.1c shows the corresponding phasors *of* the three voltages. The stator phase windings can be connected in either wye or delta, as shown in Figure 2.2a and b, respectively. In wye configuration, *if* a neutral conductor is brought out, the system is defined as a *four-wire three-phase system*; otherwise, it is a *three-wire, three-phase system*. In a delta connection, no neutral exists and therefore it is a *three-wire three-phase system*.

2.2.1 IDEAL THREE-PHASE POWER SOURCES

An ideal and balanced three-phase voltage source has three equal voltages that are 120° out *of* phase with respect to one another, as shown in Figure 2.3. Therefore, the balanced three-phase voltages given in the *abc phase sequence** (or phase order) can be expressed as

$$V_a = V_\phi \angle 0° \tag{2.1}$$

$$V_b = V_\phi \angle 240° = V_\phi \angle -120° \tag{2.2}$$

$$V_c = V_\phi \angle 120° \tag{2.3}$$

where V_ϕ is the magnitude of the phase voltage given in rms value. Similarly, the balanced three-phase voltages given in the *acb phase sequence* can be expressed as

$$V_a = V_\phi \angle 0° \tag{2.4}$$

---

* All phasors of a phasor diagram are assumed to rotate counterclockwise. A simple way of defining the phase sequence is to locate a point on any phasor in the system, for example, V_a, and then move clockwise until the next two phasors are met, that is, V_b and V_c. The phase sequence is then *abc*. In the United States, almost all utility systems have the *abc* phase sequence.

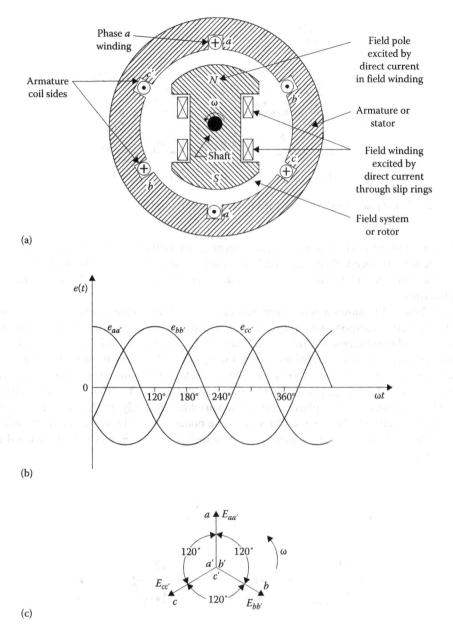

(a)

(b)

(c)

FIGURE 2.1 (a) Basic structure of an elementary three-phase, two-pole ac generator; (b) generated voltage waveforms in time domain; and (c) corresponding voltage phasors.

$$\boldsymbol{V}_b = V_\phi \angle 120° \tag{2.5}$$

$$\boldsymbol{V}_c = V_\phi \angle 240° = V_\phi \angle -120° \tag{2.6}$$

Furthermore, in the balanced three-phase systems, each phase has equal impedance so that the resulting phase currents are equal and phase-displaced from each other by 120°. The term *balanced* describes three-phase voltages or currents, which are equal in magnitude and are 120°

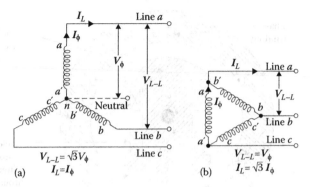

FIGURE 2.2 Stator phase windings connected in (a) wye and (b) delta.

out of phase with respect to each other, and form a symmetrical three-phase set. If the three-phase system is balanced, then equal real and reactive power flow in each phase. On the other hand, if the three-phase system is not balanced, it may lack some or all of the aforementioned characteristics.

Figure 2.4a and b shows a wye-connected and a delta-connected[*] ideal three-phase source, respectively. The corresponding voltage and current phasor diagrams are shown in Figure 2.5. The use of double-subscript notation greatly simplifies the phasor analysis. In the case of voltages, the subscripts indicate the points between which the voltage exists. Here, the first subscript is defined as positive with respect to the second. Therefore, the order of subscripts indicates the direction in which the voltage rise is defined. For example, $V_{an} = -V_{na}$. Hence, switching the order of the subscript causes a 180° phase shift in the variable. Similarly, in the case of currents, the subscript order indicates the *from–to* direction. The nodes *a*, *b*, and *c* are called the *terminals* or *lines*, and the point *n* is called the *neutral*. The branches *a–n*, *b–n*, and *c–n* are defined as the *phases* of the source.

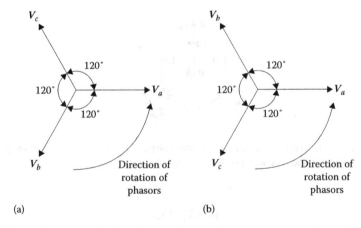

FIGURE 2.3 Phasor diagrams for balanced three-phase voltages, arranged in (a) positive (or *abc*) phase sequence and (b) negative (or *acb*) phase sequence.

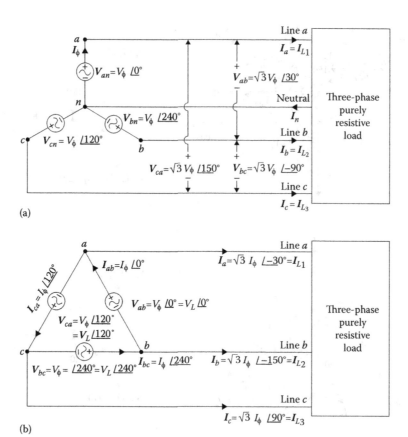

FIGURE 2.4   Ideal three-phase source connected in (a) wye and (b) delta.

The voltages V_{an}, V_{bn}, and V_{cn} are defined as the *line-to-neutral voltages* or *line-to-ground voltages* or simply *phase voltages*. The voltages V_{ab}, V_{bc}, and V_{ca} are defined as the *line-to-line voltages* or *phase-to-phase voltages* or simply *line voltages*. In general, whenever a three-phase voltage level is given, it is understood that it is *a line voltage* unless otherwise specified.

2.2.1.1   Wye-Connected Ideal Three-Phase Source

Figure 2.5a illustrates how to determine the line voltages graphically from the given balanced phase voltages, if the source is connected in wye (or star). The line voltages can be determined mathematically as

$$V_{ab} = V_{an} + V_{nb}$$

$$= V_{an} - V_{bn}$$

$$= V_\phi \angle 0° - V_\phi \angle -120°$$

$$= V_\phi(1 + j0) - V_\phi\left(-\frac{1}{2} - j\frac{\sqrt{3}}{2}\right)$$

$$= \sqrt{3}V_\phi \angle 30°$$

$$= V_L \angle 30° \qquad (2.7)$$

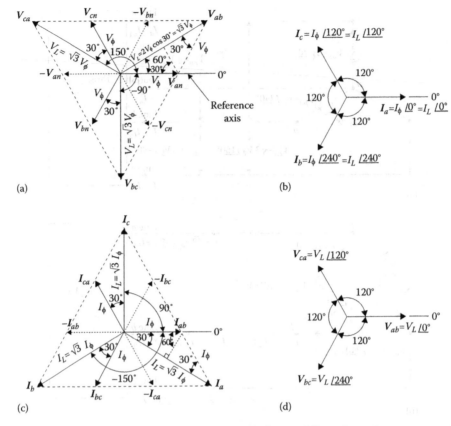

FIGURE 2.5 Phasor diagrams for three-phase sources: (a) phase and line voltages in a wye connection, (b) line currents in a wye connection, (c) phase and line currents in a delta connection, and (d) line voltages in a delta connection.

Similarly,

$$V_{bc} = V_{bn} + V_{nc}$$

$$= V_{bn} - V_{cn}$$

$$= \sqrt{3}V_\phi \angle -90°$$

$$= V_L \angle -90° \tag{2.8}$$

and

$$V_{ca} = V_{cn} + V_{na}$$

$$= V_{cn} - V_{an}$$

$$= \sqrt{3}V_\phi \angle 150°$$

$$= V_L \angle 150° \tag{2.9}$$

where

    V_ϕ is the the magnitude of the phase voltage

    V_L is the magnitude of the line voltage

$$V_L = \sqrt{3}V_\phi \qquad (2.10)$$

so that

$$V_\phi = |\mathbf{V}_{an}| = |\mathbf{V}_{bn}| = |\mathbf{V}_{cn}| \qquad (2.11)$$

and

$$V_L = |\mathbf{V}_{ab}| = |\mathbf{V}_{bc}| = |\mathbf{V}_{ca}| = \sqrt{3}V_\phi \qquad (2.12)$$

The line voltages are also 120° out of phase with respect to each other and form a symmetrical three-phase set. Figure 2.5b shows that each current lags its phase voltage by an equal phase angle. However, the phase and line currents are the same in a wye connection. Figure 2.6a and b shows alternative ways of drawing the phasors given in Figure 2.5a and b, respectively. As can be seen in those figures, the sum of the line voltages is zero for a balanced system, that is,

$$\mathbf{V}_a + \mathbf{V}_b + \mathbf{V}_c = 0 \qquad (2.13)$$

and similarly,

$$\mathbf{I}_a + \mathbf{I}_b + \mathbf{I}_c = 0 \qquad (2.14)$$

Therefore, the neutral conductor does not carry any current (i.e., $\mathbf{I}_n = 0$) if the source and load are both balanced. Otherwise,*

$$\mathbf{I}_n = - (\mathbf{I}_a + \mathbf{I}_b + \mathbf{I}_d) \qquad (2.15)$$

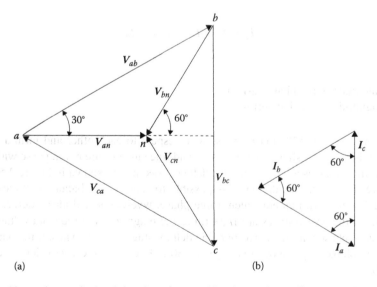

(a)   (b)

FIGURE 2.6 Alternative methods of drawing phasors: (a) voltage phasor diagram and (b) current phasor diagram.

---

* Most commercial generators are wye connected, with their neutral grounded through a resistor. Such a grounding resistor (typically, 700 Ω) limits ground fault (i.e., short circuit) currents, and therefore substantially reduces the amount of possible damage to apparatus in the event of a ground fault.

2.2.1.2 Delta-Connected Ideal Three-Phase Source

Figure 2.5c illustrates how to determine the line currents graphically from the given balanced phase currents, if the source is connected in delta (or mesh). The line currents can also be determined mathematically. For example, if the balanced phase currents are given in *abc phase sequence* as

$$I_{ab} = I\angle 0° \tag{2.16}$$

$$I_{bc} = I_\phi\angle 240° = I_\phi\angle -120° \tag{2.17}$$

$$I_{ca} = I_\phi\angle 120° \tag{2.18}$$

as shown in Figure 2.5c. The corresponding line currents can be found from the KCL as

$$I_a = I_{ab} - I_{ca}$$
$$= I_\phi\angle 0° - I_\phi\angle 120° \tag{2.19}$$

$$I_b = I_{bc} - I_{ab} \tag{2.20}$$

$$I_c = I_{ca} - I_{bc} \tag{2.21}$$

Since

$$I_L = \sqrt{3}I_\phi \tag{2.22}$$

then

$$|I_a| = |I_b| = |I_c| = I_L = \sqrt{3}I_\phi \tag{2.23}$$

where
  I_ϕ is the magnitude of the phase current
  I_L is the magnitude of the line current

The line currents are also 120° out of phase with respect to each other and form a symmetrical three-phase set. Figure 2.5c shows that the phase and line currents are not in phase with each other. The phase and line voltages are the same in a delta connection, as shown in Figure 2.5d.

Furthermore, for easier calculation, it is possible to replace any balanced, three-phase delta-connected ideal source with an equivalent three-phase, wye-connected ideal source. In this case, the magnitudes of the wye voltages are $1/\sqrt{3}$ times the magnitudes of the delta voltages. The wye voltages are out of phase with the corresponding delta voltages by 30°. Thus, if the phase sequence is *abc*, the wye voltages lag the delta voltages by 30°. Otherwise, the wye voltages lead the delta voltages by 30°.

2.2.2 Balanced Three-Phase Loads

Three-phase loads may be connected either in wye connections or delta connections, as shown in Figure 2.7a and b, respectively. In a wye connection, the line voltages are 30° ahead of the corresponding phase voltages. However, the line currents and the corresponding phase currents are the

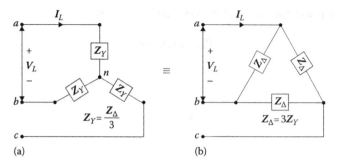

FIGURE 2.7 Balanced, three-phase loads: (a) wye-connected load and (b) delta-connected load.

same, as shown in Figure 2.5b. The magnitudes of line voltages are $\sqrt{3}$ times those for phase voltages. In a delta connection, the line currents are 30° behind the corresponding phase currents, as shown in Figure 2.5c. Here, the magnitudes of line currents lag the line-to-neutral voltages. The line currents also lag the line-to-neutral voltages by the phase-impedance angle, regardless of whether the circuit is wye or delta.

When the impedances in all three phases are identical, the load is said *to be balanced*. If a balanced three-phase source is connected to a balanced load over an inherently balanced transmission or distribution lines, then the total system is balanced. If the balanced three-phase load is wye connected,

$$Z_{an} = Z_{bn} = Z_{cn} = Z_Y \tag{2.24}$$

then

$$I_{an} = \frac{V_{an}}{Z_{an}} = \frac{V}{|Z_Y|} \angle -\theta \tag{2.25}$$

$$I_{bn} = \frac{V_{bn}}{Z_{bn}} = \frac{V}{|Z_Y|} \angle 240° -\theta \tag{2.26}$$

$$I_{cn} = \frac{V_{cn}}{Z_{cn}} = \frac{V}{|Z_Y|} \angle 120° -\theta \tag{2.27}$$

where

$$Z_Y = Z_Y \angle \theta \tag{2.28}$$

By applying the KCL at the point *n*,

$$I_n = I_{an} + I_{bn} + I_{cn} = 0 \tag{2.29}$$

and therefore the neutral conductor does not exist (from a theoretical point of view) and has no effect on the system. Since

$$V_\phi = V_{an} = V_{bn} = V_{cn} \tag{2.30}$$

and

$$I_\phi = I_{an} = I_{bn} = I_{cn} \tag{2.31}$$

then the total three-phase real power of the load can be expressed as

$$P_{3\phi} = 3V_\phi I_\phi \cos\theta \tag{2.32}$$

where

$$V_\phi = \frac{V_L}{\sqrt{3}} \tag{2.33}$$

and

$$I_\phi = I_L \tag{2.34}$$

Therefore,

$$P_{3\phi} = \sqrt{3}V_L I_L \cos\theta \tag{2.35}$$

Similarly, the total three-phase power of the load can be expressed as

$$Q_{3\phi} = 3V_\phi I_\phi \sin\theta \tag{2.36}$$

or

$$Q_{3\phi} = \sqrt{3}V_L I_L \sin\theta \tag{2.37}$$

Therefore, the total apparent power of the load can be found as

$$S_{3\phi} = \left(P_{3\phi}^2 + Q_{3\phi}^2\right)^{1/2} \tag{2.38}$$

$$S_{3\phi} = 3S_{1\phi} \tag{2.39}$$

or

$$S_{3\phi} = 3V_\phi I_\phi \tag{2.40}$$

Substituting Equations 2.33 and 2.34 into Equation 2.40,

$$S_{3\phi} = \sqrt{3}V_L I_L \angle\theta \tag{2.41}$$

where
  V_L is the magnitude of the line voltage
  I_L is the magnitude of the line current
  θ is the power factor angle by which the line current lags or leads the line voltage (or the angle of the impedance in each phase)

The power factor of the three-phase load is still $\cos\theta$. If the balanced three-phase load is delta connected,

$$Z_{ab} = Z_{bc} = Z_{ca} = Z_\Delta = Z_\Delta \angle \theta \qquad (2.42)$$

then

$$I_{ab} = \frac{V_{ab}}{Z_{ab}} = \frac{V_L}{Z_\Delta} \angle - \theta \qquad (2.43)$$

$$I_{bc} = \frac{V_{bc}}{Z_{bc}} = \frac{V_L}{Z_\Delta} \angle 240° - \theta \qquad (2.44)$$

$$I_{ca} = \frac{V_{ca}}{Z_{ca}} = \frac{V_L}{Z_\Delta} \angle 120° - \theta \qquad (2.45)$$

Therefore, the line currents can be found from

$$I_a = I_{ab} - I_{ca} \qquad (2.46)$$

$$I_b = I_{bc} - I_{ab} \qquad (2.47)$$

$$I_c = I_{ca} - I_{bc} \qquad (2.48)$$

Since

$$I_L = I_a = I_b = I_c = \sqrt{3} I_\phi \qquad (2.49)$$

$$V_L = V_\phi \qquad (2.50)$$

then the (*total*) three-phase *real* power of the load can be expressed as

$$P_{3\phi} = 3 V_\phi I_\phi \cos \theta \qquad (2.51)$$

or

$$P_{3\phi} = \sqrt{3} V_L I_L \cos \theta \qquad (2.52)$$

which is identical to Equation 2.35. Similarly, the (*total*) three-phase reactive power of the load connected in delta can be found from Equations 2.36 or 2.37. The total apparent power of the load can be found from Equations 2.40 or 2.41. The power expressions are independent of the type (i.e., wye or delta) of load connection, as long as

$$Z_\Delta = 3 Z_Y \qquad (2.53)$$

and since

$$Z = Z \angle \theta$$

$$\theta = \arg\left|Z_\Delta\right| = \arg\left|Z_Y\right| \tag{2.54}$$

The complex power can be found directly from the real and reactive powers per phase as

$$S_{1\phi} = P_{1\phi} + jQ_{1\phi} = V_\phi I_\phi^* \tag{2.55}$$

$$S_{1\phi} = P_{1\phi} + jQ_{1\phi} = Z_\phi^*\left|I_\phi\right|^2 \tag{2.56}$$

or

$$S_{1\phi} = P_{1\phi} + jQ_{1\phi} = Y_\phi^*\left|V_\phi\right|^2 \tag{2.57}$$

Thus, the three-phase complex, real, and reactive powers can be found from

$$S_{3\phi} = 3S_{1\phi} = 3P_{1\phi} + j3Q_{1\phi} \tag{2.58}$$

or

$$S_{3\phi} = P_{3\phi} + jQ_{3\phi} = \sqrt{3}V_L I_L^* \tag{2.59}$$

Table 2.1 provides a summary comparison of the basic variables of delta- and wye-connected, balanced, three-phase loads. Notice that the connection type does not affect the power calculations.

The virtue of working with balanced systems is that they can be analyzed on a single-phase basis. The current in any phase is always the phase-to-neutral voltage divided by the per-phase load impedance. Therefore, it is not necessary to calculate the currents in the remaining phase separately. Calculating the current, voltage, and power in only one phase is sufficient in an analysis because of the complete symmetry that exists between the three phases. The knowledge of these variables in one phase, which is referred to as the *reference phase*, directly provides information about all phases. This type of analysis is called *per-phase analysis*. This characteristic of balanced

TABLE 2.1

Comparison of Balanced, Three-Phase Loads Connected in Delta and Wye

Three-Phase Load	Δ-Connected Load	Y-Connected Load
Load impedance	$Z_\Delta = 3Z_Y$	$Z_Y = Z_\Delta/3$
Line current	$I_L = \sqrt{3}I_\phi$	$I_L = I_\phi$
Line-to-line voltage	$V_L = V_\phi$	$V_L = \sqrt{3}V_\phi$
Three-phase real power	$P_{3\phi} = 3V_\phi I_\phi \cos\theta$	$P_{3\phi} = \sqrt{3}V_L I_L \sin\theta$
Three-phase reactive power	$Q_{3\phi} = 3V_\phi I_\phi \sin\theta$	$Q_{3\phi} = \sqrt{3}V_L I_L \sin\theta$
Three-phase apparent power	$S_{3\phi} = 3V_\phi I_\phi$	$S_{3\phi} = \sqrt{3}V_L I_L$

three-phase systems is the basis for the use of *one-line diagrams*. In these diagrams, a circuit composed of three or more conductors is pictorially represented by a single line with standard symbols for transformers, switchgear, and other system components.

Example 2.1

Assume that the phase voltages of a wye-connected source (at its terminals) are given as $V_{an} = 277.13\angle0°$ V, $V_{bn} = 277.13\angle240°$ V, and $V_{cn} = 277.13\angle120°$ V. Determine the following:

(a) The line voltages of V_{ab}, V_{bc}, and V_{ca}.
(b) If a balanced, wye-connected, three-phase load of $Z_{an} = Z_{bn} = Z_{cn} = 10\angle30°$ Ω is connected to the source, find all the phase and line currents.

Solution

(a) The line voltages are found as

$$V_{ab} = V_{an} - V_{bn} = 277.13\angle0° - 277.13\angle240° = 480\angle30° \text{ V}$$

$$V_{bc} = V_{bn} - V_{cn} = 277.13\angle240° - 277.13\angle120° = 480\angle-90° \text{ V}$$

$$V_{ca} = V_{cn} - V_{an} = 277.13\angle120° - 277.13\angle0° = 480\angle150° \text{ V}$$

(b) Since in a wye-connected, three-phase load the phase and line currents are the same,

$$I_a = I_{an} = \frac{V_{an}}{Z_{an}} = \frac{277.13\angle0° \text{ V}}{10\angle30° \text{ Ω}} = 27.713\angle-30° \text{ A}$$

$$I_b = I_{bn} = \frac{V_{bn}}{Z_{bn}} = \frac{277.13\angle240° \text{ V}}{10\angle30° \text{ Ω}} = 27.713\angle210° \text{ A}$$

$$I_c = I_{cn} = \frac{V_{cn}}{Z_{cn}} = \frac{277.13\angle120° \text{ V}}{10\angle30° \text{ Ω}} = 27.713\angle90° \text{ A}$$

Example 2.2

A balanced delta-connected, three-phase load withdraws 200 A per phase with a leading power factor of 0.85 from a 12.47 kV line. Determine the following:

(a) The line current of the load
(b) The phase voltage of the load
(c) The three-phase apparent power
(d) The three-phase real power
(e) The three-phase reactive power
(f) The three-phase complex power

Solution

(a) Since the load is connected in delta, the line current of the load is

$$I_L = \sqrt{3}I_\phi = \sqrt{3} \times 200 = 346.41 \text{ A}$$

(b) The phase voltage of the load is

$$V_\phi = V_L = 12,470 \text{ V}$$

(c) The three-phase (or total) apparent power is

$$S_{3\phi} = \sqrt{3}V_L I_L = \sqrt{3}(12{,}470 \text{ V})(346.41 \text{ A}) = 7{,}482 \text{ kVA}$$

or

$$S_{3\phi} = 3V_\phi I_\phi = 3(12{,}470 \text{ V})(200 \text{ A}) = 7{,}482 \text{ kVA}$$

(d) The three-phase (or total) real power is

$$P_{3\phi} = \sqrt{3}V_L I_L \cos\theta = \sqrt{3}(12{,}470 \text{ V})(346.41 \text{ A})0.85 = 6{,}359.7 \text{ kW}$$

or

$$P_{3\phi} = S_{3\phi} \cos\theta = (7{,}482 \text{ kVA})(0.85) = 6{,}359.7 \text{ kW}$$

(e) The three-phase (or total) reactive power is

$$Q_{3\phi} = \sqrt{3}V_L I_L \sin\theta = \sqrt{3}(12{,}470 \text{ V})(346.41 \text{ A})0.5268 = 3{,}941.52 \text{ kvar}$$

or

$$Q_{3\phi} = S_{3\phi} \sin\theta = (7{,}482 \text{ kVA})0.5268 = 3{,}941.52 \text{ kvar}$$

(f) The three-phase (or *total*) complex power is

$$S_{3\phi} = \sqrt{3}V_L I_L^* = P_{3\phi} - jQ_{3\phi} = (6{,}359.7 \text{ kW}) - j(3{,}941.52 \text{ kvar}) = 7{,}482.01\angle -31.79° \text{ kVA}$$

Example 2.3

A balanced three-phase load of 8000 kW with a lagging power factor of 0.90 is supplied by a three-phase 34.5 kV* line. If the line resistance and inductive reactance are given as 5 and 7 Ω per phase (i.e., line to neutral), determine the following:

(a) The line current of the load
(b) The power factor angle of the load
(c) The line-to-neutral voltage of the line at the receiving end (i.e., *the load side*)
(d) The voltage drop due to line impedance
(e) The line-to-line voltage of the line at the sending end
(f) The power loss due to line impedance

Solution

(a) From Equation 2.52, the line current of the load is

$$I_L = \frac{P_{3\phi}}{\sqrt{3}V_L \cos\theta} = \frac{8{,}000 \text{ kW}}{\sqrt{3}(34.5 \text{ kV})0.90} = 148.75 \text{ A}$$

(b) The power factor of the load is

$$\theta = \cos^{-1} 0.90 = 25.8°$$

---

* Unless otherwise specified, it is customary to assume a phase-to-phase voltage or line-to-line (i.e., line) voltage.

(c) The line-to-neutral voltage of the line at the receiving end is

$$V_{R(L-N)} = \frac{V_L}{\sqrt{3}} = \frac{34,500 \text{ V}}{\sqrt{3}} = 19,918.6 \text{ V}$$

(d) The voltage drop in the line due to line impedance is

$$I_L Z_L = \left[148.75(\cos 25.8° - j\sin 25.8°)\right](5 + j7)$$

$$= 1,279.3\angle 28.6° \text{ V}$$

or

$$I_L Z_L = (148.75\angle -25.8°)(8.6023\angle 54.46°)$$

$$= 1,279.3\angle 28.6° \text{ V}$$

(e) The line-to-neutral voltage at the sending end is

$$V_{S(L-N)} = V_{R(L-N)} + I_L Z$$

$$= 19,918.6\angle 0° + 1,279.3\angle 28.6$$

$$= 21,206.7\angle 1.7° \text{ V}$$

Therefore, the line-to-line voltage is

$$V_{S(L)} = \sqrt{3}V_{S(L-N)}$$

$$= \sqrt{3}(21,206.7 \text{ V}) = 36,731.1 \text{ V}$$

(f) The power loss due to line resistance is

$$P_{3\phi} = 3I_L^2 R = 3(148.75)^2(5) = 331.9 \text{ kW}$$

Example 2.4

A balanced, three-phase, delta-connected load is supplied by a balanced, wye-connected source over a balanced three-phase line. The source voltage data are given in *abc* phase sequence in which V_{an} is $7.62\angle 0°$ kV and the line impedance is $1 + j7\,\Omega$. If the balanced load consists of three equal impedances of $15 + j10\,\Omega$, determine the following:

(a) The line currents I_a, I_b, and I_c
(b) The phase voltages V_{ab}, V_{bc}, and V_{ca} of the delta-connected load
(c) The phase currents I_{ab}, I_{bc}, and I_{ca} of the delta-connected load
(d) The phasor diagram of the phasors found in parts (a), (b), and (c)

Solution

(a) Converting the given delta-connected load to its equivalent wye-connected form,

$$Z_Y = \frac{Z_\Delta}{3} = \frac{15 + j10}{3} = 5 + j3.33 \,\Omega \text{ per phase}$$

Therefore,

$$I_a = \frac{7620\angle 0°}{6 + j10.33} = 637.7\angle -59.9° \text{ A}$$

$$I_b = \frac{7620\angle -120°}{6 + j10.33} = 637.7\angle -179.9° \text{ A}$$

$$I_c = \frac{7620\angle 120°}{6 + j10.33} = 637.7\angle 60.1° \text{ A}$$

(b) *The* line-to-neutral voltages at the wye-connected load *can* be found as

$$V_{an} = I_a Z_a = (637.7\angle -59.9°)(6\angle 33.7°) = 3830.9\angle -26.2° \text{ V}$$

$$V_{bn} = I_b Z_b = (637.7\angle -179.9°)(6\angle 33.7°) = 3830.9\angle -146.2° \text{ V}$$

$$V_{cn} = I_c Z_c = (637.7\angle 60.1°)(6\angle 33.7°) = 3830.9\angle 93.8° \text{ V}$$

Therefore,

$$V_{bc} = V_{bn} - V_{cn} = V_{ab}e^{-j120°}$$
$$= 6635.3\angle -116.2° \text{ V}$$

$$V_{ca} = V_{cn} - V_{an} = V_{ab}e^{j120°}$$
$$= 6635.3\angle 123.8° \text{ V}$$

Alternatively,

$$V_{ab} = \sqrt{3}V_{an}\angle\theta_{an} + 30°$$
$$= \sqrt{3}(3830.9\angle -26.2° + 30°) \text{ V} = 6635.3\angle 3.8° \text{ V}$$

$$V_{bc} = \sqrt{3}V_{bn}\angle\theta_{bn} + 30°$$
$$= \sqrt{3}(3830.9\angle -146.2° + 30°) \text{ V} = 6635.3\angle -116.2° \text{ V}$$

$$V_{ca} = \sqrt{3}V_{cn}\angle\theta_{cn} + 30°$$
$$= \sqrt{3}(3830.9\angle 93.8° + 30°) \text{ V} = 6635.3\angle 123.8° \text{ V}$$

(c) Thus,

$$I_{ab} = \frac{V_{ab}}{Z_\Delta} = \frac{6635.3\angle 3.8° \text{ V}}{18.03\angle 33.7° \text{ }\Omega} = 368\angle -29.9° \text{ A}$$

$$I_{bc} = \frac{V_{bc}}{Z_\Delta} = \frac{6635.3\angle -116.2° \text{ V}}{18.03\angle 33.7° \text{ }\Omega} = 368\angle -149.9° \text{ A}$$

$$I_{ca} = \frac{V_{ca}}{Z_\Delta} = \frac{6635.3\angle 123.8° \text{ V}}{18.03\angle 33.7° \text{ }\Omega} = 368\angle 90.1° \text{ A}$$

(d) The phasor diagram is shown in Figure 2.8.

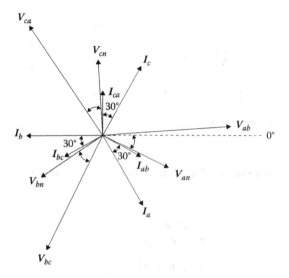

FIGURE 2.8 Phasor diagram for Example 2.4.

2.3 UNBALANCED THREE-PHASE LOADS

If an unbalanced three-phase load connected in delta or wye is present in an otherwise balanced three-phase system, the method of *symmetrical components** is normally used to analyze the system. However, in simple situations, the direct application of conventional circuit theory can be used without much difficulty, as in the following example.

Example 2.5

An unbalanced, three-phase, delta-connected load is supplied by a balanced three-phase source through a power line. The load impedances Z_{ab}, Z_{bc}, and Z_{ca} are given as $5 + j5\,\Omega$, $5 - j5\,\Omega$, $5 + j0\,\Omega$, respectively. The power line impedance is given as $2 + j2\,\Omega$ per phase. In the event that the line-to-line voltages $V_{a'b'}$, $V_{b'c'}$, and $V_{c'a'}$ are $110\angle 0°$, $110\angle 240°$, and $110\angle 120°$ V, respectively, determine the following:

(a) The line currents $I_{a'a}$, $I_{b'b}$, and $I_{c'c}$
(b) The line-to-line voltages V_{ab}, V_{bc}, and V_{ca}

Solution

(a) First, convert the delta-connected load shown in Figure 2.9a to its equivalent wye-connected form, as shown in Figure 2.9b,

$$Z_a = \frac{Z_{ab}Z_{ca}}{Z_{ab} + Z_{bc} + Z_{ca}}$$

$$= \frac{(5 + j5)(5 + j0)}{(5 + j5) + (5 - j5) + (5 + j0)} = \frac{5(5 + j5)}{15} = 1.67 + j1.67\,\Omega$$

$$Z_b = \frac{Z_{ab}Z_{bc}}{Z_{ab} + Z_{bc} + Z_{ca}}$$

$$= \frac{(5 + j5)(5 - j5)}{15} = 3.33 + j0\,\Omega$$

---

* For the theory and applications of symmetrical components, see *Electric Power Transmission System Engineering* and *Modern Power System Analysis* of Gönen (1988, 2000).

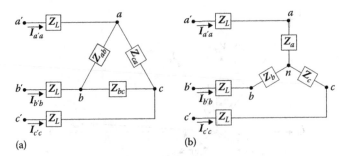

FIGURE 2.9 Illustration for Example 2.5: (a) delta-connected load and (b) equivalent wye-connected load.

$$Z_c = \frac{Z_{bc}Z_{ca}}{Z_{ab} + Z_{bc} + Z_{ca}}$$

$$= \frac{(5 - j5)(5 + j0)}{15} = 1.67 + j1.67 \ \Omega$$

By using KVL,

$$V_{a'b'} = V_{d'a} + V_{an} + V_{nb} + V_{bb'} = 110\angle 0°$$

$$V_{b'c'} = V_{b'b} + V_{bn} + V_{nc} + V_{cc'} = 110\angle 240°$$

where

$$V_{d'a} + V_{an} = I_{d'a}(Z_L + Z_a)$$

$$V_{d'a} + V_{an} = I_{d'a}(Z_L + Z_a)$$

$$V_{d'a} + V_{an} = I_{d'a}(Z_L + Z_a)$$

but

$$I_{c'c} = -(I_{d'a} + I_{b'b})$$

Therefore, $I_{c'c}$ can be eliminated so that

$$\begin{bmatrix} Z_L + Z_a & -Z_L - Z_b \\ Z_L + Z_c & 2Z_L + Z_b + Z_c \end{bmatrix} \begin{bmatrix} I \\ I \end{bmatrix} = \begin{bmatrix} 110\angle 0° \\ 110\angle 240° \end{bmatrix}$$

$$\begin{bmatrix} 3.67 + j3.67 & -5.33 - j2 \\ 3.67 + j0.33 & 9 + j2.33 \end{bmatrix} \begin{bmatrix} I_{d'a} \\ I_{b'b} \end{bmatrix} = \begin{bmatrix} 110\angle 0° \\ 110\angle 240° \end{bmatrix}$$

By using determinants and *Cramer's rule,*

$$Q_{old} = P \times \tan\theta_{old}$$

$$= (346.84 \text{ kW}) \times \tan(\cos^{-1}0.8)$$

$$= (346.84 \text{ kW}) \times 0.75$$

$$= 260.13 \text{ kvar}$$

$$I_{a'a} = \frac{\Delta_1}{\Delta} = \frac{\begin{vmatrix} 110\angle 0° & -5.33 - j2 \\ 110\angle 240° & 9 + j2.33 \\ 3.67 + j3.67 & -5.33 - j2 \\ 3.67 + j0.33 & 9 + j2.33 \end{vmatrix}}{}$$

$$= \frac{110\angle 0°(9.2967\angle 14.51°) - (5.6929\angle 200.57°)110\angle 240°}{5.1902\angle 45°(9.2967\angle 14.51°) - (5.6929\angle 200.57°)(3.6848\angle 5.14°)}$$

$$= 13.85\angle -6.87° \text{ A}$$

Similarly,

$$I_{b'b} = \frac{\Delta_2}{\Delta} = \frac{\begin{vmatrix} 3.67 + j3.67 & 110\angle 0° \\ 3.67 + j0.33 & 110\angle 240° \end{vmatrix}}{69.178\angle 29.04°}$$

$$= 9.61\angle -162.9° \text{ A}$$

Therefore,

$$I_{c'c} = -(13.85\angle -6.87° + 9.27\angle 217.48°) = 17.01\angle 74.18° \text{ A}$$

As a check,

$$I_{a'a} + I_{b'b} + I_{c'c} = 13.85\angle -6.87° + 9.61\angle -162.9° + 17.01\angle 74.18° = 0$$

(b) The line-to-neutral voltages can be found as

$$V_{an} = I_{a'a}Z_a = (13.85\angle -6.87° \text{ A})(2.36\angle 45°) = 33\angle -26° \text{ V}$$

$$V_{bn} = I_{b'b}Z_b = (9.61\angle -162.9° \text{ A})(3.33\angle 0°) = 32\angle -162° \text{ V}$$

$$V_{cn} = I_{c'c}Z_c = (17.01\angle 74.18° \text{ A})(2.36\angle -45°) = 40\angle 29° \text{ V}$$

Therefore, the line-to-line voltages can be found as

$$V_{ab} = V_{an} - V_{bn} = 33\angle -26° - 32\angle -162° = 61\angle -5.4° \text{ V}$$

$$V_{bc} = V_{bn} - V_{cn} = 32\angle -162° - 40\angle 29° = 71.9\angle -15.6° \text{ V}$$

$$V_{ca} = V_{cn} - V_{an} = 40\angle 29° - 33\angle 26° = 35.17\angle 81.98° \text{ V}$$

Notice that the unbalanced loads destroy the symmetry between the phasors, and cause the resulting currents and voltages not to have the simplicity that is characteristic of a balanced three-phase system.

Example 2.6

An unbalanced, three-phase, four-wire, wye-connected load, as shown in Figure 2.10, is connected to a balanced three-phase, four-wire source. The load impedances Z_a, Z_b, and Z_c are given as $100\angle50°$, $150\angle-140°$, and $50\angle-100°$ Ω per phase, respectively. If the line voltage is 13.8 kV, determine the following:

(a) The line and neutral currents
(b) The total power delivered to the loads

Solution

(a) Taking the line-to-line neutral voltages of phase-a voltage as the reference,

$$V_\phi = \frac{V_L}{\sqrt{3}} = \frac{13,800\ \text{V}}{\sqrt{3}} = 7,967.4\ \text{V}$$

so that

$$V_{an} = 7,967.4\angle0°\ \text{V}$$

$$V_{bn} = 7,967.4\angle-120°\ \text{V}$$

$$V_{cn} = 7,967.4\angle120°\ \text{V}$$

Therefore, the line currents can be found as

$$I_a = \frac{V_{an}}{Z_a} = \frac{7,967.4\angle0°\ \text{V}}{100\angle50°\ \Omega} = -79.7\angle-50°\ \text{A}$$

$$I_b = \frac{V_{bn}}{Z_b} = \frac{7,967.4\angle-120°\ \text{V}}{150\angle-140°\ \Omega} = 53.1\angle20°\ \text{A}$$

$$I_c = \frac{V_{cn}}{Z_c} = \frac{7,967.4\angle120°\ \text{V}}{50\angle-100°\ \Omega} = 159.3\angle220°\ \text{A}$$

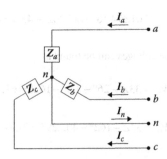

FIGURE 2.10 An unbalanced, three-phase, four-wire, wye-connected load.

Thus,

$$I_n = -(I_a + I_b + I_c)$$

$$= (79.7\angle - 50° + 53.1\angle 20° + 159.3\angle 220°) = 146.8\angle 81.8° \text{ A}$$

(b) Therefore, the power delivered by each phase is

$$P_a = V_{an}I_a \cos\theta_a = (7,967.4 \text{ V})(79.7 \text{ A})\cos 50° = 408,171.3 \text{ W}$$

$$P_b = V_{bn}I_b \cos\theta_b = (7,967.4 \text{ V})(53.1 \text{ A})\cos 140° = 324,089.6 \text{ W}$$

$$P_c = V_{cn} I_c \cos\theta_c = (7,967.4 \text{ V})(159.3 \text{ A})\cos 100° = 220,395.5 \text{ W}$$

Thus, the total power delivered is

$$P_{3\phi} = P_a + P_b + P_c = 952,656.3 \text{ W} \cong 952.6 \text{ kW}$$

2.4 MEASUREMENT OF AVERAGE POWER IN THREE-PHASE CIRCUITS

A *wattmeter* is a device that has a potential coil and a current coil, which are designed and connected in such a way that its pointer's deflection is proportional to $VI\cos\theta$. Here, V is the rms value of the voltage applied across the potential coil, I is the rms value of the current passing through the current coil, and θ is the angle between the voltage and the current phasors involved. The direction in which the pointer deflects depends on the instantaneous polarity of the current-coil current and the potential-coil voltage. Thus, each coil has one terminal with a polarity mark ±, as shown in Figure 2.11. The wattmeter deflects in the right direction when the polarity-marked terminal of the potential coil is connected to the phase in which the current coil has been inserted.

If a separate wattmeter is used to measure the average (real) power in each phase, the total real power in a three-phase system can be found by adding the three wattmeter readings. If the load is delta connected, each wattmeter has its current coil connected on one side of the delta and its potential coil connected line to line. If the load is wye connected and the neutral wire does exist, the potential coil of each wattmeter is connected between each phase and the neutral wire. However, in actual practice, it may not be possible to have access to either the neutral of the wye connection or the individual phases of the delta connection in order to connect a wattmeter in each of the phases.

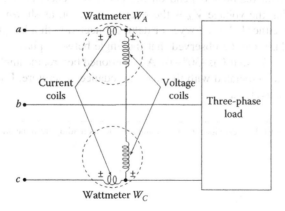

FIGURE 2.11 Connection diagram for the two-wattmeter method of measuring three-phase power.

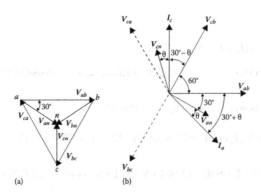

FIGURE 2.12   Phasor diagrams for the two-wattmeter method: (a) voltage-phasor diagram and (b) voltage and current phasor diagram.

In these cases, three wattmeters can be connected as shown in Figure P2.7. The common point 0 is a floating potential point. The total real power used by the load, whether it is delta or wye connected, balanced or unbalanced, is the sum of the three wattmeter readings. Therefore,

$$P_{3\phi} = W_a + W_b + W_c \tag{2.60}$$

It is also possible in delta- and wye-connected loads to use only two wattmeters to measure the total real power, as shown in Figure 2.11. This method is known as the *two-wattmeter method* of measuring three-phase power. The algebraic sum of the readings of the two wattmeters* will give the total average power consumed by the three-phase load. Thus,

$$P_{3\phi} = W_a + W_c \tag{2.61}$$

where the wattmeter connected on phase a provides the reading of

$$W_a = V_{ab} I_a \cos\theta_a \tag{2.62}$$

where θ_a is the angle between phasors V_{ab} and I_a. Similarly,

$$W_c = V_{cb} I_c \cos\theta_c \tag{2.63}$$

where θ_c is the angle between phasors V_{ab} and I_c. Notice that the reading of the wattmeter W_c is determined by V_{ab} and I_c. Even though the sum of the two readings depends only on the total power of the load, the individual readings depend on the phase sequence. Now assume that the phase sequence is *abc* and that the voltage V_{ab} is the reference phasor, as shown in Figure 2.12b. Also assume that the load is either balanced wye- or delta connected with a lagging power factor angle of θ. From Figure 2.12a, it can be observed that the angle between phasors V_{ab} and I_a is $(30° + \theta)$ and that between phasors V_{cb} and I_c is $(30° - \theta)$. As mentioned before, the angle θ is the *load power-factor angle* or the angle associated with the load impedance. Therefore, Equations 2.62 and 2.63 can be expressed, respectively,[†] as

---

* If one wattmeter reads backward, reverse its current coil and subtract its reading from the other wattmeter.
[†] If the phase sequence is *acb*,
  $W_a = V_L I_L \cos(30° - \theta)$
  and
  $W_c = V_L I_L \cos(30° + \theta)$
  so that
  $P_{3\phi} = W_a + W_c$
  and that the total reactive power is $Q_{3\phi} = \sqrt{3}\,(W_a - W_c)$

$$W_a = V_L I_L \cos (30° + \theta) \tag{2.64}$$

$$W_c = V_L I_L \cos (30° - \theta) \tag{2.65}$$

where

$$\theta_a = 30° + \theta \tag{2.66}$$

$$\theta_c = 30° - \theta \tag{2.67}$$

and where V_L and I_L are magnitudes of the line-to-line voltage and line current, respectively. Thus, the total average power can be determined as

$$P_{3\phi} = (W_a + W_c) = V_L I_L \left[\cos(30° + \theta) + \cos (30° - \theta) \right] = \sqrt{3} V_L I_L \cos\theta \tag{2.68}$$

and the total reactive power can be determined as

$$Q_{3\phi} = (W_a - W_c) = V_L I_L \left[\cos(30° - \theta) - \cos(30° + \theta) \right] = \sqrt{3} V_L I_L \sin\theta \tag{2.69}$$

By observing Equations 2.64 and 2.65, the following conclusions can be reached for the two-watt-meter method of measuring three-phase power in a balanced circuit:

- If the load power-factor equals 0.5, one of the wattmeters shows zero.
- If the load power-factor is less than 0.5, one of the wattmeters shows a negative value.
- If the load power-factor is greater than 0.5, both wattmeters show a positive value.
- Reversing the phase sequence will interchange the readings on the wattmeters.

Example 2.7

An unbalanced, three-phase, delta-connected load, as shown in Figure 2.13, is supplied by a balanced three-phase system given in abc phase sequence in which V_{ab} is $220\angle 0°$ V. The load impedances Z_{ab}, Z_{bc}, and Z_{ca} are given as $10\angle 0°$, $5\angle 60°$, and $15\angle -20°$ Ω, respectively. Determine the following:

(a) The phase currents I_{ab}, I_{bc}, and I_{ca}
(b) The line currents I_a, I_b, and I_c

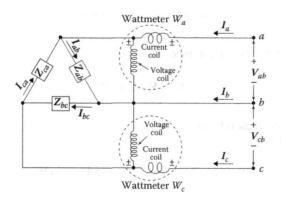

FIGURE 2.13 Circuit for Example 2.7.

(c) The powers absorbed by the individual impedances of the load
(d) The total power absorbed by the load
(e) The power recorded on each wattmeter

Solution

(a) The phase currents can be found as

$$I_{ab} = \frac{V_{ab}}{Z_{ab}} = \frac{220\angle 0° \text{ V}}{10\angle 0° \text{ }\Omega} = 22\angle 0° \text{ A}$$

$$I_{bc} = \frac{V_{bc}}{Z_{bc}} = \frac{220\angle -120° \text{ V}}{5\angle 60° \text{ }\Omega} = 44\angle -180° \text{ A}$$

$$I_{ca} = \frac{V_{ca}}{Z_{ca}} = \frac{220\angle 120° \text{ V}}{15\angle -20° \text{ }\Omega} = 14.67\angle 140° \text{ A}$$

(b) Therefore, the line currents are

$$I_a = I_{ab} - I_{ca} = 22\angle 0° - 14.67\angle 140° = 34.5\angle -15.8° \text{ A}$$

$$I_b = I_{bc} - I_{ab} = 44\angle -180° - 22\angle 0° = 66\angle 180° \text{ A}$$

$$I_c = I_{ca} - I_{bc} = 14.67\angle 140° - 44\angle -180° = 34.1\angle 16° \text{ A}$$

(c) The powers absorbed by the individual impedances of the load can be found as

$$P_{ab} = Re(V_{ab}I_{ab}^*) = Re\left[(220\angle 0°)(22\angle 0°)\right] = 4.84 \text{ kW}$$

$$P_{bc} = Re(V_{bc}I_{bc}^*) = Re\left[(220\angle -120°)(22\angle 180°)\right] = 4.84 \text{ kW}$$

$$P_{ca} = Re(V_{ca}I_{ca}^*) = Re\left[(220\angle 120°)(22\angle -140°)\right] = 3.02 \text{ kW}$$

(d) The total power absorbed by the load is

$$P = P_{ab} + P_{bc} + P_{ca} = 12.7 \text{ kW}$$

(e) The power recorded by the wattmeter *a* is

$$W_a = |V_{ab}||I_{ab}|\cos\theta_a$$

where θ_a is the angle between V_{ab} and I_a.
Therefore,

$$W_a = |220\angle 0°||34.5\angle 15.8°|\cos 15.8° = 7.3 \text{ kW}$$

or alternatively,

$$W_a = \text{Re}\left(V_{ab}I_a^*\right) = \text{Re}\left[(220\angle 0°)(34.5\angle 15.8°)\right] = 7.3 \text{ kW}$$

Similarly,

$$W_C = Re\left(V_{bc}I_c^*\right) = Re\left[(220\angle -120°)(34.1\angle -16°)\right] = 5.4 \text{ kW}$$

Therefore, the total power read by the wattmeters is 12.7 kW.

2.5 POWER FACTOR CORRECTION

In general, loads on electric utility systems have two components: *active power* (measured in kilowatts) and *reactive power* (measured in kilovars). Active power has to be generated at power plants, whereas reactive power can be supplied by either power plants or capacitors. If reactive power is supplied only by power plants, each system component, including generators, transformers, and transmission and distribution lines, has to be increased in size accordingly. However, by using capacitors, the reactive power demand as well as line currents are reduced from the capacitor locations all the way back to power plants.* As a result, losses and loadings are reduced in distribution lines, substation transformers, and transmission lines. The power-factor correction generates savings in capital expenditures and fuel expenses through a release of power capacity and a decrease in power losses in all the equipment between the point of installation of the capacitors and the source power plants.

The *economic power factor* is the power factor at which the economic benefits of using capacitors equals the cost of capacitors. However, the correction of power factor to unity becomes more expensive with respect to the marginal cost of the capacitors installed. It has been found in practice that the economic power factor is about 0.95. In distribution systems, including industrial applications, *shunt* capacitors are used and are connected in delta or wye. However, in transmission systems, the capacitors are connected in *series* with the line involved.

Example 2.8

Assume that a 2.4 kV, single-phase circuit supplies a load of 294 kW at lagging power factor and that the load current is 175 A. To improve the power factor, determine the following:

(a) The uncorrected power factor and reactive load
(b) The new corrected power factor after installing a shunt capacitor unit with a rating of 200 kvar

Solution

(a) Before the power factor correction,

$$S_{old} = VI = (2.4 \text{ kV})(175 \text{ A}) = 420 \text{ kVA}$$

Therefore, the *uncorrected* power factor can be found as

$$PF_{old} = \cos\theta = \frac{P}{S_{old}}$$

$$= \frac{294 \text{ kW}}{420} = 0.7$$

---

* For further information, see Chapter 8 of *Electric Power Distribution System Engineering* of Gönen (2008).

and the reactive load is

$$Q_{old} = S_{old} \times \sin(\cos^{-1} PF_{old})$$

$$= (420 \text{ kVA})(0.7141) = 300 \text{ kvar}$$

(b) After the installation of the 200 kvar capacitors,

$$Q_{new} = Q_{old} - Q_{cap}$$

$$= (300 \text{ kvar}) - (200 \text{ kvar}) = 100 \text{ kvar}$$

and therefore the *new* (or *corrected*) power factor is

$$PF_{new} = \cos\theta_{new} = \frac{P}{\left(P2 + Q_{new}^2\right)^{1/2}}$$

$$= \frac{294 \text{ kW}}{\left[(294 \text{ kW})^2 + (100 \text{ kvar})^2\right]^{1/2}} = 0.95\% \text{ or } 95\%$$

Example 2.9

A three-phase, 400 hp, 60 Hz, 4.16 kV wye-connected induction motor has a full-load efficiency of 86%, and a lagging power factor of 0.8. If it is necessary to correct the power factor of the load to a lagging power factor of 0.95 by connecting three capacitors at the load, find the following:

(a) The rating of such a capacitor bank in kvar
(b) The capacitance of each single-phase unit, if the capacitors are connected in delta in µF
(c) The capacitance of each single-phase unit, if the capacitors are connected in wye in µF

Solution

(a) The input power of the induction motor is

$$P = \frac{(400 \text{ hp})(0.7457 \text{ kW/hp})}{0.86} = 346.84 \text{ kW}$$

The reactive power of the motor at the *uncorrected* power factor is

$$Q_{old} = P \times \tan\theta$$

$$= (346.83 \text{ kW}) \times \tan(\cos^{-1} 0.8)$$

$$= (346.83) \times 0.75$$

$$= 260.13 \text{ kvar}$$

The reactive power of the motor at the *corrected* power factor is

$$Q_{new} = P \times \tan\theta_{new}$$

$$= (346.84 \text{ kW}) \times \tan(\cos^{-1} 0.95)$$

$$= (346.84 \text{ kW}) \times 0.3287$$

$$= 114 \text{ kvar}$$

Thus, the reactive power provided by the capacitor bank is

$$Q_{cap} = Q_{old} - Q_{new}$$

$$= (260.13 \text{ kvar}) - (114 \text{ kvar})$$

$$= 146.13 \text{ kvar}$$

Therefore, the rating of the capacitor bank is 146.13 kvar.

(b) If the capacitors are connected in delta, the line current is

$$I_L = \frac{Q_{cap}}{\sqrt{3}V_L}$$

$$= \frac{146.13 \text{ kvar}}{\sqrt{3}(4.16 \text{ kV})} = 20.28 \text{ A}$$

and thus, the current in each capacitance of the delta-connected capacitor bank is

$$I_{cap} = \frac{I_L}{\sqrt{3}}$$

$$= \frac{20.28 \text{ A}}{\sqrt{3}} = 11.71 \text{ A}$$

Therefore, the reactance of each capacitor is

$$X_{cap} = \frac{V_L}{I_{cap}}$$

$$= \frac{4160 \text{ V}}{11.71 \text{ A}} = 355.25 \ \Omega$$

and the capacitance of each unit is

$$C = \frac{10^6}{\omega X_{cap}} = \frac{10^6}{(2\pi f)X_{cap}}$$

$$= \frac{10^6}{2\pi(60 \text{ Hz})(355.55 \ \Omega)}$$

$$= 7.47 \ \mu F$$

Note that

$$C = \frac{1}{\omega X_{cap}} \text{ in F}$$

or

$$C = \frac{10^6}{\omega X_{cap}} \text{ in } \mu F$$

Note that the aforementioned equation gives the capacitance in μF. This equation is modified from the previous equation, which gives the answer in F, by dividing both sides by 10^6, as it can be seen easily.

(c) If the capacitors are connected in wye in the capacitor bank,

$$I_{cap} = I_L = 20.28\,\text{A}$$

and therefore,

$$X_{cap} = \frac{V_{L-N}}{I_{cap}} = \frac{4160\,\text{V}}{\sqrt{3}(20.28\ \text{A})} = 118.43\,\Omega$$

Therefore, the capacitance of each unit is

$$C = \frac{10^6}{(2\pi f)X_{cap}}$$

$$= \frac{10^6}{2\pi(60\,\text{Hz})(118.43\ \Omega)}$$

$$= 22.4\,\mu\text{F}$$

PROBLEMS

2.1 A three-phase, wye-connected induction motor is supplied by a three-phase and four-wire system with line-to-line voltage of 220 V and the impedance of the motor is $6.3 + j3.05\,\Omega$ per phase. Determine the following:
  (a) The magnitude of the line current
  (b) The power factor of the motor
  (c) The three-phase average power consumed by the motor
  (d) The current in the neutral wire.

2.2 A balanced three-phase load of 15 MVA with a lagging load factor of 0.85 is supplied by a 115 kV subtransmission line. If the line impedance is $50 + j100\,\Omega$ per phase, determine the following:
  (a) The line current of the load
  (b) The power loss due to the line impedance
  (c) The power factor angle of the load
  (d) The line-to-neutral voltage of the line at the receiving end
  (e) The voltage drop due to the line impedance
  (f) The line-to-line voltage of the line at the sending end

2.3 A balanced, three-phase, delta-connected load is supplied by a balanced, three-phase, wye-connected source over a balanced three-phase line. The source voltages are in abc phase sequence in which V_{an} is $19.94\angle 0°$ kV and the line impedance is $10 + j80\,\Omega$ per phase. If the balanced load consists of three equal impedances of $60 + j30\,\Omega$, determine the following:
  (a) The line currents I_a, I_b, and I_c
  (b) The phase voltages V_{ab}, V_{bc}, and V_{ca} of the load
  (c) The phase currents I_{ab}, I_{bc}, and I_{ca} of the load

2.4 Assume that the impedance of a power line connecting buses 1 and 2 is $50\angle 70°\,\Omega$, and that the bus voltages are $7560\angle -10°$ and $7200\angle 0°$ V per phase, respectively. Determine the following:
  (a) The real power per phase that is being transmitted from bus 1 to bus 2
  (b) The reactive power per phase that is being transmitted from bus 1 to bus 2

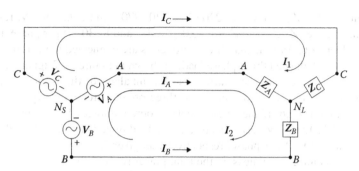

FIGURE P2.6 Circuit of Problem 2.6.

(c) The complex power per phase that is being transmitted

2.5 Solve Problem 2.4 assuming that the line impedance is $50\angle 26°$ per phase.

2.6 An unbalanced three-phase, three-wire, wye-connected load is connected to a balanced, three-phase, three-wire, wye-connected source, as shown in Figure P2.6. If the line-to-neutral source voltages V_a, V_b, and V_c are $220\angle 30°$, $220\angle 270°$, and $220\angle 150°$ V, respectively, and the load impedances Z_a, Z_b, and Z_c are $4\angle 0°$, $5\angle 90°$, and $8\angle 30°$ Ω per phase, respectively. Determine the following:

(a) The mesh currents I_1 and I_2 using determinants and *Cramer's rule*

(b) The line currents I_a, I_b, and I_c.

(c) The potential difference between the source neutral N_S and the common node of the load, that is, N_L

(d) Whether or not a neutral wire connecting the neutral point N_S and N_L is required

2.7 An unbalanced, three-phase, three-wire, wye-connected load is connected to a balanced, three-phase, three-wire, wye-connected source, as shown in Figure P2.6. If the line-to-neutral source voltages V_a, V_b, and V_c are $220\angle 30°$, $220\angle 270°$, and $220\angle 150°$ V, respectively, and the load impedances Z_a, Z_b, and Z_c are $4\angle 0°$, $5\angle 90°$, and $8\angle 30°$ Ω per phase, respectively, as given in Problem 2.6. Assume that three wattmeters are connected to measure the total power received by the unbalanced three-phase load, as shown in Figure P2.7. Ignore the small impedance of the current coils in the wattmeters and determine the following:

(a) The power recorded on each wattmeter

(b) The total power absorbed by the load

2.8 An unbalanced, three-phase, three-wire, wye-connected load is connected to a balanced, three-phase, three-wire, wye-connected source, as shown in Figure P2.6. If the line-to-neutral

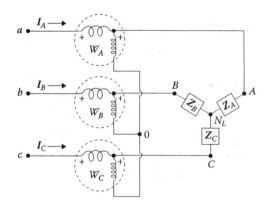

FIGURE P2.7 Circuit of Problem 2.7.

source voltages V_a, V_b, and V_c are 220∠30°, 220∠270°, and 220∠150° V, respectively, and
the load impedances Z_a, Z_b, and Z_c are 4∠0°, 5∠90°, and 8∠30° Ω per phase, respectively, as
given in Problem 2.6. Assume that three wattmeters are connected to measure the total power
received by the unbalanced three-phase load, as shown in Figure P2.7. Ignore the small impedance of the current coils in the wattmeters. An unbalanced, three-phase, three-wire, wye-
connected load is connected to a balanced three-phase, three-wire, wye-connected source,
as shown in Figure P2.6. If the line-to-neutral source voltages V_a, V_b, and V_c are 220∠30°,
220∠270°, and 220∠150° V, respectively, and the load impedances Z_a, Z_b, and Z_c are 4∠0° Ω,
5∠90° Ω, and 8∠30° Ω per phase, respectively, as given in Problem 2.6. Assume that three
wattmeters are connected to measure the total power received by the unbalanced three-phase
load, as shown in Figure P2.7. Ignore the small impedance of the current coils in the wattmeters and assume that only two wattmeters are used and connected, as shown in Figure P2.8.
Determine the following:
  (a) The power recorded on each wattmeter
  (b) The total power absorbed by the load
2.9 If the impedance of a transmission line connecting buses 1 and 2 is 50∠80° Ω, and the bus
voltages are 70∠215° and 68∠0° kV per phase, respectively, determine the following:
  (a) The complex power per phase that is being transmitted from bus 1 to bus 2
  (b) The active power per phase that is being transmitted
  (c) The reactive power per phase that is being transmitted
2.10 A three-phase motor is connected to a three-phase line that has an *abc* phase sequence and is
supplied by 15 A current at a 0.85 lagging power factor. If a single-phase motor withdrawing
5 A current at a 0.707 lagging power factor is connected across lines *a* and *b* of the three-phase
power line, determine the total current in each line.
2.11 Three loads are connected to a 208 V, three-phase power source that has an *abc* phase
sequence. The first load is a wye-connected, three-phase motor withdrawing a line current of
20 A at a 0.8 lagging power factor. The second load is a single-phase load between lines *a* and
b and withdraws a 10 A current at a 0.8 leading power factor. The third load is also a single-
phase motor connected between lines *b* and *c* and withdraws a 7 A current at a 0.707 lagging
power factor. Use the voltage V_{ab} as the reference phasor and determine the following:
  (a) All line and phase voltages
  (b) All line currents
  (c) The total input power in watts
2.12 A three-phase, 60 Hz, wye-connected synchronous generator is providing power to two
balanced three-phase loads. The first load is delta-connected and made up of three 12∠45°

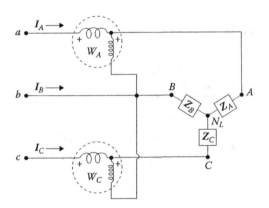

FIGURE P2.8   Circuit of Problem 2.8.

impedances, while the second load is wye connected and made up of three $5\angle60°$ Ω impedances. Determine the following:

(a) Total (i.e., equivalent) load impedance per phase (i.e., line to neutral).

(b) The line current I_a at the generator terminal. Use $V_a = \left(208/\sqrt{3}\right)\angle0° \cong 120\angle0°$ V.

(c) The total complex power provided by the generator.

2.13 A three-phase, 60 Hz, wye-connected synchronous generator has balanced line-to-line voltages of 480 V at its terminals. The generator is supplying power to two balanced and delta-
connected, three-phase loads. The first load is made up of three $15\angle-30°$ Ω impedances, while
the second load is made up of three $18\angle50°$ Ω impedances. Determine the following:

(a) Total (i.e., equivalent) load impedance per phase (i.e., line to neutral).

(b) The line current I_a at the generator terminal. Use a phase voltage of $V_a=277.1281\angle0°$ V
(since $480/\sqrt{3}V = 277.1281$ V).

(c) The total complex power provided by the generator.

2.14 If a balanced, three-phase, 15 MW total load is fed by a 138 kV power line at a 0.85 lagging
power factor, determine the following:

(a) The line current

(b) The value of the capacitor in μF per phase, if a wye-connected capacitor bank is used to
correct the power factor to a 0.95 lagging power factor

2.15 If a balanced, three-phase, 20 MW total load is fed by a 138 kV (line-to-line) power source at
a 0.8 lagging power factor, determine the following:

(a) The line current

(b) The value of the capacitor in μF per phase, if a wye-connected capacitor bank is used to
correct the power factor to a 0.9 lagging power factor

(c) If a delta-connected capacitor bank is used in part (b), the value of each capacitor in such
a bank. [Hint: Use the results of part (b).]

2.16 Consider the balanced and delta-connected three-phase load that is shown in Figure P2.16.
Assume that each impedance is $9+j9$ Ω and that the line-to-line voltage, with a magnitude of
4160 V, is supplied by a three-phase source where $V_{bc}=V_{bc}\angle0°$. In other words, assume that the
voltage source phasors are aligned in the same fashion as the circuit symbols between nodes
a, b, and c. Determine the following:

(a) The line currents I_a, I_b, and I_c

(b) The total (i.e., three-phase) average power in watts

2.17 Consider the balanced and delta-connected three-phase load that is shown in Figure P2.16.
Assume that each impedance is $3+j6$ Ω and that the load is supplied by a balanced three-
phase source where $V_{ab}=12,470\angle30°$ V so that $V_a=V_{an}\cong7,200\angle0°$ V. (In other words, line
voltages lead phase voltages by 30°.) Determine the following:

(a) The line currents I_a, I_b, and I_c

(b) The total (i.e., three-phase) average power in watts

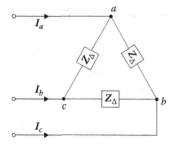

FIGURE P2.16 Circuit of Problem 2.16.

2.18 Consider the delta-connected, three-phase load that is shown in Figure P2.18. Assume that the delta-connected unbalanced load is supplied by the same three-phase source that is given in Problem 2.16.

Determine the following:

(a) The line currents I_a, I_b, and I_c

(b) The voltage V_{de}.

(c) The total (i.e., three-phase) average power in watts

2.19 Assume that the delta-connected load is an unbalanced three-phase load, and is supplied by the same balanced, three-phase power source that is given in Problem 2.17. If $Z_{ab} = 5 + j5\,\Omega$, $Z_{bc} = 5 - j5\,\Omega$, and $V_{ab} = 12{,}470\angle 30°$ V, determine the following:

(a) The line currents I_a, I_b, and I_c

(b) The total (i.e., three-phase) average power in watts

2.20 A three-phase, 60 Hz, 5 MVA synchronous generator has balanced line-to-line voltages of 8320 V at its terminals. If the generator is supplying power to a wye-connected, balanced load of 5 MWA at a 0.8 lagging power factor, determine the following:

(a) The total complex power absorbed by the wye-connected load

(b) The load impedance per phase

2.21 A three-phase, 60 Hz, 5 MVA synchronous generator has balanced line-to-line voltages of 8320 V at its terminals. If the generator is supplying power to a delta-connected, balanced load of 5 MVA at a 0.8 lagging power factor, determine the following:

(a) The total complex power absorbed by the wye-connected load

(b) The load impedance per phase

2.22 A three-phase, 60 Hz, 3000 kVA wye-connected synchronous generator has balanced line-to-line voltages of 4160 V at its terminals. The internal impedance of the generator is $Z_s = j3.5\,\Omega$ per phase. A delta-connected, balanced, three-phase load of $15 + j30\,\Omega$ per phase is connected to the generator over an S switch. The line to-line voltage at the switch before it is closed is 4160 V. Use phase-a voltage as the *reference phasor* and determine the following:

(a) The percent voltage drop of no-load voltage of the terminal voltage at the switch when the load is connected

(b) The total complex power delivered to the load

2.23 Assume that a balanced, three-phase, wye-connected source is connected to a balanced, three-phase, wye-connected load and that a neutral conductor connects the neutral points of the source and the load. Assume *abc* phase sequence and $V_a = 230\angle 0°$ V at the load terminal. If the load impedances are $Z_a = Z_b = Z_c = 2.3\angle 30°\,\Omega$, determine the following:

(a) The line currents I_a, I_b, and I_c

(b) The current in the neutral conductor, that is, I_n

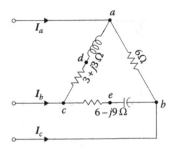

FIGURE P2.18   Circuit of Problem 2.18.

2.24 A balanced, three-phase, delta-connected load is connected to a balanced three-phase source of abc sequence. If $V_{ab}=460\angle30°$ V and the load impedances $Z_{ab}=Z_{bc}=Z_{ca}=4.6\angle45°\,\Omega$, determine the following:

(a) The line voltages of V_{bc} and V_{ca}

(b) The currents I_{ab}, I_{bc}, and I_{ca} of the load

(c) The line currents I_a, I_b, and I_c of the load

2.25 A three-phase, 60 Hz, 5 MVA, wye-connected synchronous generator has balanced line-to-line voltages of 4160 V at its terminals. The internal impedance of the generator is $Z_s=j2\,\Omega$ per phase. A delta-connected, balanced three-phase load of $3+j5\,\Omega$ per phase is connected to the generator over an S switch. The line to-line voltage at the switch before it is closed is 4160 V. Use the phase-a voltage of $2400\angle0°$ V. Use phase-a voltage as the *reference phasor* and determine the following:

(a) The percent voltage drop of no-load voltage of the terminal voltage at the switch when the load is connected

(b) The total complex power delivered to the load

2.26 A three-phase, 60 Hz, wye-connected synchronous generator has balanced line-to-line voltages of 480 V at its terminals. The generator is supplying power to two balanced and delta-connected, three-phase loads. The first load is made up of three $12\angle40°\,\Omega$ impedances, while the second load is made up of three $18\angle80°\,\Omega$ impedances. Use a phase voltage of $V_a=277.1281\angle0°$ V (since $V_a = 480/\sqrt{3}$V $= 277.1281$ V). Determine the following:

(a) Total (i.e., equivalent) load impedance per phase (i.e., line to neutral)

(b) The line current I_a at the generator terminal

(c) The total complex power provided by the generator

2.27 A three-phase, 60 Hz, wye-connected synchronous generator is providing power to two balanced three-phase loads. The first load is wye connected and made up of three $6\angle45°\,\Omega$ impedances, while the second load is delta connected and made up of three $9\angle75°\,\Omega$ impedances. Use a phase voltage of $V_a=277.1281\angle0°$ V (since $480 / \sqrt{3}$V $= 277.1281$ V).

Determine the following:

(a) Total (i.e., equivalent) load impedance per phase (i.e., line to neutral)

(b) The line current I_a at the generator terminal

(c) The total complex power provided by the generator

2.28 Two three-phase generators are supplying the same load bus, as shown in Figure P2.28. Both generators produce balanced voltages of abc phase sequence. Use $V_a=120\angle0°$ V and $V_a' =115\angle0°$ V as the reference voltages for the left and right generators in the figure,

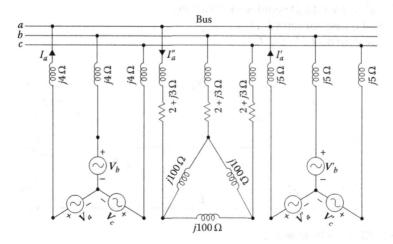

FIGURE P2.28 Circuit of Problem 2.28.

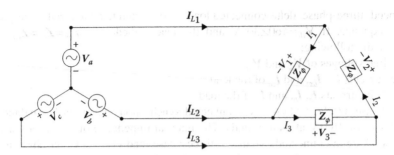

FIGURE P2.30 Circuit of Problem 2.30.

respectively. If a balanced three-phase load is connected in the middle of the bus, as shown in the figure, determine the following:

(a) The phasor currents I_a, I'_a, and I''_a (Hint: First convert the delta-connected load into its equivalent wye-connected load and then apply the *nodal analysis method* using the *line-to-neutral approach* since both the generators and the load are balanced.)

(b) The total complex power supplied to the load

(c) The line-to-line voltage at the terminals of the load

(d) The complex power supplied by the left generator to the bus

(e) The complex power supplied by the right generator to the bus

2.29 Solve Example 2.5, but use the following data. Let $Z_{ab}=6-j6\,\Omega$, $Z_{bc}=0+j6\,\Omega$, $Z_{ca}=6+j6\,\Omega$, and use a line impedance $Z_L=3+j3\,\Omega$ per phase. Use $V_{a'b'}=120\angle0°$ V, $V_{b'c'}=120\angle240°$ V and $V_{c'a'}=120\angle120°$ V as the balanced source voltages.

2.30 Consider Figure P2.30 and assume that the balanced delta-connected load is made up of three impedances of $Z_\phi=27.71\angle-40°\,\Omega$ per phase and that source voltages V_a, V_b, and V_c are $277.1\angle0°$, $277.1\angle-120°$, and $277.1\angle120°$ V, respectively. Determine the following:

(a) The load voltages V_1, V_2, and V_3

(b) The load (phase) currents I_1, I_2, and I_3

(c) The line currents I_{L1}, I_{L2}, and I_{L3}

2.31 Solve Example 2.7, but use the following data. Assume that $V_{ab}=480\angle0°$ V, and that the load impedances Z_{ab}, Z_{bc}, and Z_{ca} are given as $20\angle30°$, $10\angle-60°$, and $15\angle45°\,\Omega$, respectively.

2.32 A balanced and delta-connected, three-phase voltage source is supplying power to a balanced and delta-connected, three-phase load, as shown in Figure P2.32. The source voltages V_{ab}, V_{bc}, V_{ca} are $208\angle30°$, $208\angle270°$, and $208\angle150°$ V, respectively. The load impedances are $Z_{ab}=Z_{bc}=Z_{ca}=2+j3\,\Omega$. Determine the following:

(a) The load (phase) currents I_{ab}, I_{bc}, and I_{ca}

(b) The line currents I_a, I_b, and I_c

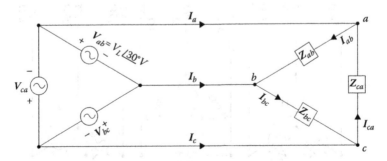

FIGURE P2.32 Circuit of Problem 2.32.

(c) The total real and reactive power supplied to the load

(d) The total complex power supplied to the load

2.33 An unbalanced, three-phase, three-wire, wye-connected load is connected to a balanced, three-phase, wye-connected source, and only two wattmeters are used and connected, as shown in Figure P2.8. The line-to-neutral source voltages V_a, V_b, and V_c are $220\angle30°$, $220\angle270°$, and $220\angle150°$ V, respectively, and the line currents I_a, I_b, and I_c are $71.62\angle-11°$, $61.28\angle16.2°$, and $13.26\angle133.6°$ A, respectively. Determine the following:

(a) The power recorded on each wattmeter

(b) The total power absorbed by the load

2.34 Consider Figure P2.32 and assume that a delta-connected source is supplying power to a delta-connected balanced load. If the generator has $V_{ab}=480\angle30°$ V, $V_{bc}=480\angle270°$ V, $V_{ca}=480\angle150°$ V, and load impedances of $Z_{ab}=Z_{bc}=Z_{ca}=5+j5\,\Omega$, determine the following:

(a) The load (phase) currents I_{ab}, I_{bc}, and I_{ca}.

(b) The line currents I_a, I_b, and I_c.

(c) The total real and reactive power supplied to the load.

(d) The total complex power supplied to the load.

(e) Draw the current and voltage diagrams.

2.35 Consider Figure P2.32 and assume that a delta-connected source is supplying power to a delta-connected balanced load. If the generator has $V_{ab}=380\angle30°$ V, $V_{bc}=380\angle270°$ V, $V_{ca}=380\angle150°$ V and load impedances of $Z_{ab}=Z_{bc}=Z_{ca}=2+j3\,\Omega$, determine the following:

(a) The load (phase) currents I_{ab}, I_{bc}, and I_{ca}.

(b) The line currents I_a, I_b, and I_c.

(c) The total real and reactive power supplied to the load.

(d) The total complex power supplied to the load.

(e) Draw the current and voltage diagrams.

2.36 Assume that a three-phase, 480 V power line is supplying a delta-connected load and that $V_{ab}=480\angle30°$ V, $V_{bc}=480\angle270°$ V, $V_{ca}=480\angle150°$ V. Also assume that the current in each phase of the delta-connected load is 10 A at a 0.85 lagging power factor. If V_{an} is used as a *reference phasor* and the phase sequence is *abc*, determine the following:

(a) Draw the circuit, as given.

(b) The line-to-line voltages of V_{ab}, V_{bc}, and V_{ca}.

(c) The line-to-neutral voltages of V_{an}, V_{bn}, and V_{cn}.

(d) All phase and line currents.

(e) Draw the phasor diagram for the phase and line voltages.

(f) Draw the phasor diagram for the phase and line currents.

2.37 Assume that a three-phase, 380 V power line is supplying a delta-connected load and that $V_{ab}=380\angle30°$ V, $V_{bc}=380\angle270°$ V, and $V_{ca}=380\angle150°$ V. Also assume that the current in each phase of the delta-connected load is 5 A at a 0.75 lagging power factor. If V_{an} is used as a *reference phasor* and the phase sequence is *abc*, determine the following:

(a) Draw the circuit, as given.

(b) The line-to-line voltages of V_{ab}, V_{bc}, and V_{ca}.

(c) The line-to-neutral voltages of V_{an}, V_{bn}, and V_{cn}.

(d) All phase and line currents.

(e) Draw the phasor diagram for the phase and line voltages.

(f) Draw the phasor diagram for the phase and line currents.

2.38 A 2.4 kV, single-phase circuit supplies a load of 250 kW at a lagging power factor, and the load current is 160 A. If it is necessary to improve the power factor, determine the following:

(a) The uncorrected power factor and reactive load

(b) The new corrected power factor after installing a shunt capacitor unit with a rating of 250 kvar

2.39 A three-phase, 50 hp, 60 Hz, 480 V, wye-connected induction motor has a full-load efficiency of 0.85%, and a lagging power factor of 0.75. If it is required to correct the power factor of the load to a lagging power factor of 0.95 by connecting three capacitors, find the following:
   (a) The rating of such a capacitor bank, in kVA
   (b) The capacitance of each single-phase unit, if the capacitors are connected in delta, in μF
   (c) The capacitance of each single-phase unit, if the capacitors are connected in wye, in μF

2.40 Redo Example 2.4 (except part c) by using MATLAB®. Use the other given values and determine the following:
   (a) Write the MATLAB program script.
   (b) Give the MATLAB program output.

2.41 Redo Example 2.6 by using MATLAB. Use the other given values and determine the following:
   (a) Write the MATLAB program script.
   (b) Give the MATLAB program output.

3 Magnetic Circuits

All is flux, nothing stays still.

Heraclitus, 500 BC

Time crumbles things; everything grows old under the power of Time and is forgotten through the lapse of Time.

Aristotle, 352 BC

Lord, why is it that wisdom comes so late and life is so short?

Turan Gönen

3.1 INTRODUCTION

Today, the phenomenon of *magnetism** is used in the operation of a great number of electrical apparatus including generators, motors, transformers, measuring instruments, televisions, radios, telephones, tape recorders, computer memories, computer magnetic tapes, car ignition tapes, refrigerators, air conditioners, heating equipment, and power tools. A material that has the ability to attract iron and steel is called a *magnet*. Magnets can be categorized as being *permanent* or *temporary*, based on their ability to retain magnetism. Figure 3.1a shows a permanent (bar) magnet and its magnetic field. Notice that the magnetic *flux* (Φ) lines (i.e., the magnetic *lines of force*[†]) are continuous, and come from the north pole and go toward the south pole. The direction of this field can be established using a compass (which is simply a freely suspended magnetized steel needle) since the marked end[‡] of a compass needle always points to the earth's magnetic north pole. As shown in Figure 3.1b, when a permanent magnet is placed near a metal, the magnetic lines go through the metal and magnetize it. If two permanent magnets are located close together as shown in Figure 3.1c, the magnets are attracted toward each other since the direction of the magnetic lines of force of both magnets is the same. However, if the two magnets are located in the opposite direction as shown in Figure 3.1d, the two magnets are repelled and forced apart since the magnetic lines of force go from north to south and are opposing.

3.2 MAGNETIC FIELD OF CURRENT-CARRYING CONDUCTORS

As illustrated in Figure 3.2a, when a conductor carries an electric current I, a magnetic field is created around the conductor.[§] The direction of magnetic lines of force (or field) is determined using

---

* The phenomenon of magnetism has been recognized since 600 BC (by the ancient Greeks). However, the first experimental work was performed in the sixteenth century by the English physician, Gilbert, who discovered the existence of a magnetic field around earth. Also, Oersted recognized that current-carrying conductors could have magnetic effects. Further studies, done by Ampère, on the magnetic field around current-carrying loops, led to the theory of magnetism to a great extent. This and other experiments that were performed by Henry and Faraday established the foundation for the development of modern electrical machinery.

† It is interesting to note that the lines of force in reality do not exist, but the concept is sometimes beneficial in describing the properties of magnetic fields.

‡ Since, according to the rule of magnetic attraction and repulsion; unlike magnetic poles attract and like poles repel, the marked end of the compass needle is really a south pole.

§ Oersted discovered a definite relationship that exists between electricity and magnetism in 1819.

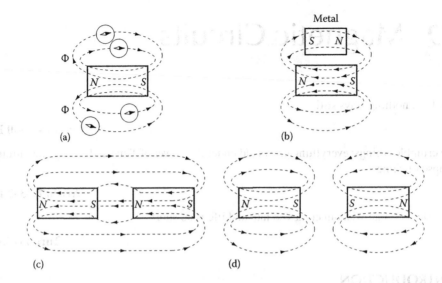

FIGURE 3.1 Magnetic field of (a) single permanent magnet, (b) metal in the vicinity of a permanent magnet, (c) two permanent magnets with unlike poles, and (d) two permanent magnets with like poles.

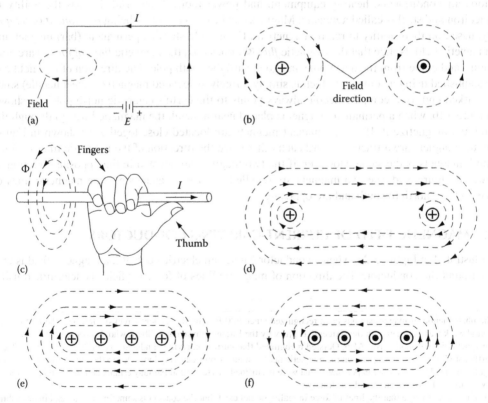

FIGURE 3.2 Magnetic lines of field (a) around a current-carrying conductor, (b) around a current-carrying conductor toward the reader and away from the reader, (c) determined by using Ampère's right-hand rule, (d) around two parallel conductors, (e) around four conductors all carrying current away from the reader, and (f) around four conductors all carrying current toward the reader.

Ampère's *right-hand rule*, which is illustrated in Figure 3.2c. It shows that *if the conductor is held in the right hand with the thumb pointing in the direction of the current flow, the fingers then point in the direction of the magnetic field around the conductor.* Thus, the conversion of energy between mechanical and electrical forms is achieved through magnetic fields.

Figure 3.2b shows the magnetic fields around a conductor carrying current toward the reader and away from the reader, respectively. Note that, in the figure, the symbol of "dot in a circle" denotes a cross-sectional view of a conductor carrying current toward the reader, while the symbol "+ in a circle" denotes the current flowing away from the reader. Figure 3.2d illustrates the magnetic field around a coil made up of two parallel conductors. Similarly, Figure 3.2e and f shows the magnetic fields around a coil made up of four conductors all carrying current away from the reader and toward the reader, respectively.

Figure 3.3 shows a current-carrying coil that is formed by wrapping a conductor (or wire) on a hollow cardboard or fiber cylinder. The magnetic lines of force (i.e., flux) are concentrated within the cylinder. Each turn of the wire develops a magnetic field in the same direction. Because the direction of current flow is the same in all turns of the wire, the resultant magnetic field generated inside the coil is all in the same direction. The polarity of such a coil can also be found by using the right-hand rule, as illustrated in Figure 3.3. *If the coil is held in the right hand with the fingers pointing in the direction of the current in the coil, the thumb then points toward the north pole of the coil.* In Figure 3.3, the end of the coil where the flux comes out is the *north pole* of the coil. Figure 3.4 shows the magnetic circuit of a two-pole dc generator. A required strong magnetic field is produced by the two field coils wound around the iron pole cores. As the armature, which is located on the rotor, is rotating through the magnetic field, an *electromotive force* (emf) is generated in the armature conductors. The measure of a coil's ability to produce flux is called *magnetomotive force* (mmf). The mmf of a magnetic circuit corresponds to the emf in an electric circuit. The mmf of a coil depends on the amount of the current flowing in the coil and the number of turns in the coil. The product of the current in amperes and the number of turns is called the *ampère-turns* of the coil.

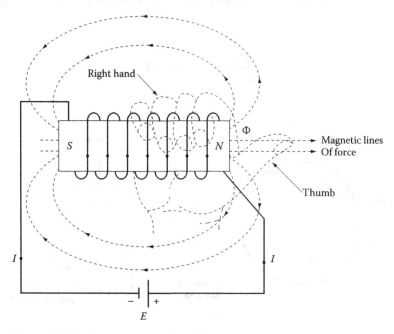

FIGURE 3.3   Magnetic field of a current-carrying coil.

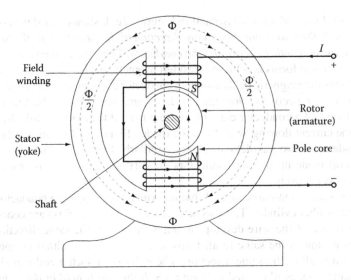

FIGURE 3.4 The magnetic circuit of a two-pole generator.

3.3 AMPÈRE'S MAGNETIC CIRCUITAL LAW

Figure 3.5a shows a ring-shaped coil of N turns supplied by a current I. The dotted circular line represents an arbitrary closed path that has the same *magnetic field intensity* value, H, over an elementary length, $d\ell$, at any location on the path. In contrast, Figure 3.5b shows a magnetic circuit with a ring-shaped magnetic core called a *toroid* with a coil wound around the entire core. If a current I flows through a coil of N turns, a field is created in the toroid.

According to *Ampère's circuital law*, the magnetic potential drop around a closed path is balanced by the mmf, giving rise to the field (i.e., the mmf encircled by the closed path). For the average path at a mean radius r, the magnetic field intensity H is related to its source NI by Ampère's circuital law,

$$\oint H \cdot d\ell = \mathcal{F} = N \times I \tag{3.1}$$

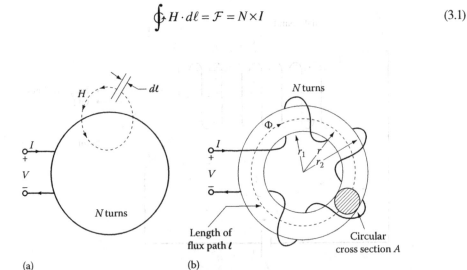

FIGURE 3.5 Illustration of a magnetic field of (a) a current-carrying coil of N turns and (b) toroidal coil.

The line integral can be solved easily if the closed path of integration is selected along the direction of H, and as previously noted, H has a constant value along the mean path ℓ. Thus,

$$H \times \ell = N \times I \tag{3.2}$$

$$H 2\pi r = N \times I \tag{3.3}$$

from which

$$H = \frac{N \times I}{2\pi r} \tag{3.4}$$

The quantity NI is called the *magnetomotive force*, and its unit is *ampere-turn* (A·turn) or *ampere* (A) since N has no dimensions.

$$\mathcal{F} = H \times \ell = N \times I \tag{3.5}$$

from which

$$H = \frac{\mathcal{F}}{\ell} \; \text{A·turns/m} \tag{3.6}$$

$$H = \frac{NI}{\ell} \; \text{A·turns/m} \tag{3.7}$$

The magnetic field intensity H describes the field produced by the mmf, and its unit is ampere-turn per meter. Also, the magnetic field intensity H produces a *magnetic flux density B* everywhere it exists. They are related to each other by

$$B = \mu H \; \text{Wb/m}^2 \; \text{or T} \tag{3.8}$$

The value of B depends not only on H (and thus the current) but also on the medium in which H is located. The SI unit* of B is weber/m^2 (Wb/m^2) or tesla (T). The effect of the medium is presented by its permeability[†] μ in henrys/m (H/m). Here, the μ represents the relative ease of establishing a magnetic field in a given material. The permeability of free space is called μ_0 and it has a value of

$$\mu_0 = 4\pi \times 10^{-7} \; \text{H/m} \tag{3.9}$$

which is approximately the same for air. The permeability of any other material with respect to the permeability of free space is called its *relative permeability* μ_r. Therefore,

$$\mu_r = \frac{\mu}{\mu_0} \tag{3.10}$$

---

* Older units of magnetic flux density (i.e., the flux per unit area) that are still in use include lines/in.^{2}, kilolines/in.^{2}, and gausses (G). Note that $1\,G = 1\,\text{Mx/cm}^2$ and $1\,T = 10\text{kG} = 10^4\,G$. Therefore, if a flux density is given in lines/in.^{2}, it must be multiplied by 1.55×10^{-5} to convert it to Wb/m^2 or T.

[†] Permeability, based on Equation 3.8, can be defined as the ratio of change in magnetic flux density to the corresponding change in magnetic field intensity. Therefore, in a sense, permeability is not a constant parameter but depends on the flux density or on the applied mmf that is used to energize the magnetic circuit.

The relative permeability μ_r is dimensionless* and equal to 1.0 for free space. The permeability of any material can be expressed as

$$\mu = \mu_r \times \mu_0 = \frac{B}{H} \tag{3.11}$$

For materials used in electrical machines, the value of μ_r can be as high as several thousands. The larger the value of μ_r, the smaller the current that is needed to produce a given flux density B in the machine. By substituting Equation 3.7 into 3.8, the magnitude of the flux density can be expressed as

$$B = \mu \times H = \frac{\mu NI}{\ell} \tag{3.12}$$

The total flux crossing of a given cross-sectional area A can be found from

$$\Phi = \int_A B \cdot dA \tag{3.13}$$

here dA is the differential unit of area. Therefore,

$$\Phi = B \times A \text{ Wb} \tag{3.14}$$

This equation is correct if the flux density vector is perpendicular to the place of area A, and if the flux density is constant at each location in the given area. For the toroid shown in Figure 3.5b, the average flux density may correspond to the path at the mean radius of the toroid. Thus, the total flux[†] in the core is

$$\Phi = B \times A = \frac{\mu NIA}{\ell} \text{ Wb} \tag{3.15}$$

$$\Phi = Bnr^2 = \frac{\mu NI\pi r^2}{\ell} \text{ Wb} \tag{3.16}$$

The product of the winding turns N and the flux Φ that links them is called the *flux linkage*. Flux linkage is usually denoted by the Greek letter λ (lambda) and expressed as

$$\lambda = N \times \Phi \text{ Wb} \tag{3.17}$$

3.4 MAGNETIC CIRCUITS

Consider the simple magnetic core shown in Figure 3.6a, by substituting Equation 3.5 into 3.15,

---

* From the relative permeability point of view, all materials can be classified into four distinct groups: (1) *diametric* ($\mu_r = 1.0^-$), (2) *nonmagnetic* ($\mu_r = 1.0$), (3) *paramagnetic* ($\mu_r = 1.0^+$), and (4) *ferromagnetic* ($1.0^+ < \mu_r < \infty$). Furthermore, a special type of diamagnetism has started to make the headlines recently. It is a case of *perfect diamagnetism* (known as the *Meissner effect*) that takes place in particular types of materials (which are called *superconductors*) at temperatures near absolute zero. They are increasingly used in certain types of electromagnetic devices, including various rotating machines and switching devices. In such superconductive materials, the flux density approaches zero and the relative permeability is basically zero. Therefore, a magnetic field cannot be developed in the superconducting material.
[†] The SI unit for magnetic flux is *webers* (Wb). The older unit of flux was the *line* or *maxwell*. Thus, 1 Wb $= 10^8$ Mx $= 10^8$ lines $= 10^5$ kilolines.

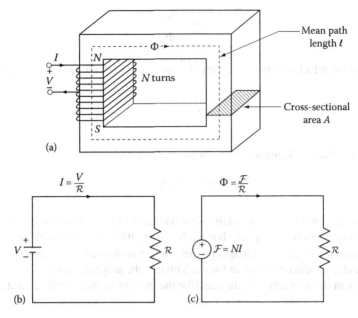

(a)

$$I = \frac{V}{R}$$

$$\Phi = \frac{\mathcal{F}}{\mathcal{R}}$$

V

R

$\mathcal{F} = NI$

$\mathcal{R}$

(b)   (c)

FIGURE 3.6   (a) A simple magnetic core, (b) its electric circuit, and (c) its magnetic circuit analogue.

$$\Phi = \frac{\mu NIA}{\ell}$$

$$= \frac{NI}{\ell/\mu A}$$

$$= \frac{N \times I}{\mathcal{R}} \tag{3.18}$$

$$= \frac{\mathcal{F}}{\mathcal{R}} \tag{3.19}$$

from which

$$\mathcal{F} = \Phi \times \mathcal{R} \tag{3.20}$$

where $\mathcal{R}$ is the reluctance of the magnetic path, and therefore

$$\mathcal{R} = \frac{\ell}{\mu A} \tag{3.21}$$

for uniform permeability μ, cross-sectional area A, and mean path length ℓ of the magnetic circuit. The reluctance* can also be expressed as

* In the SI system, no specific name is given to the dimension of reluctance except to refer to it as so many units of reluctance. One can observe from Equation 3.22 that its real dimensions are *ampere-turns/weber*. In some old literature, the word *rels* has been used as the unit of reluctance.

$$\mathcal{R} = \frac{\mathcal{F}}{\Phi} \tag{3.22}$$

The reciprocal of the reluctance is called the *permeance* of the magnetic circuit and is expressed as

$$\mathcal{P} = \frac{1}{\mathcal{R}} \tag{3.23}$$

Therefore, the flux given by Equation 3.19 can be expressed as

$$\Phi = \mathcal{F} \times \mathcal{R} \tag{3.24}$$

Figure 3.7 shows a cutaway view of an oil distribution transformer that reveals its iron core magnetic circuit and coil construction. Figure 3.8 shows a 630 kVA, 10/0.4 kV solid dielectric transformer.

In many aspects, the electric and magnetic circuits are analogous. For example, notice the analogy between the electric circuit shown in Figure 3.6b and the magnetic circuit shown in Figure 3.6c. Note that the flux in the magnetic circuit acts like the current in the electric circuit, the reluctance

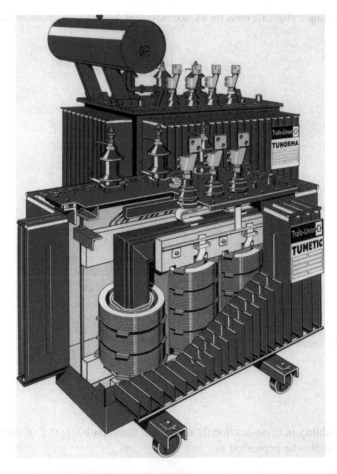

FIGURE 3.7 (See color insert.) Oil distribution transformers: Cutaway of a TUMETIC transformer with an oil expansion tank shown in the foreground, and a TUNORMA with an oil expansion tank shown in the background. (Courtesy of Siemens AG, Munich, Germany.)

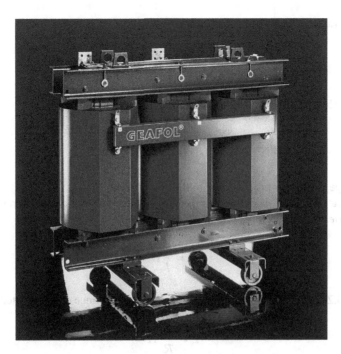

FIGURE 3.8 (See color insert.)   A 630 kVA, 10/0.4 kV GEAFOL solid dielectric transformer. (Courtesy of Siemens AG, Munich, Germany.)

in the magnetic circuit can be treated like the resistance* in the electric circuit, and the mmf in the magnetic circuit can be treated like the emf in the electric circuit. Equation 3.20 is often referred to as *Ohm's law of the magnetic circuit*. However, electric and magnetic circuits are not analogous in all respects. For instance, *energy* must be continuously provided when a direct current is flowing in an electric circuit, whereas in the case of a magnetic circuit, once the flux is established, then it remains constant. Similarly, there are no *magnetic insulators, only electric insulators*.

It is interesting that reluctances and permeances connected in series and parallel are treated in the same manner as resistances and conductances connected in series and parallel, respectively. For example, the equivalent reluctance of a number of reluctances connected in series with respect to each other is

$$\mathcal{R}_{eq} = \mathcal{R}_1 + \mathcal{R}_2 + \mathcal{R}_3 + \cdots \tag{3.25}$$

Similarly, the equivalent reluctance of a number of reluctances connected in parallel with respect to each other is

$$\frac{1}{\mathcal{R}_{eq}} = \frac{1}{\mathcal{R}_1} + \frac{1}{\mathcal{R}_2} + \frac{1}{\mathcal{R}_3} + \cdots \tag{3.26}$$

Alternatively, first the equivalent permeance is found by using

$$\mathcal{P}_{eq} = \mathcal{P}_1 + \mathcal{P}_2 + \mathcal{P}_3 + \cdots \tag{3.27}$$

---

* The resistance of a wire of length *e* and cross-sectional area *A* is given by $\mathcal{R} = \ell / \rho A$ where ρ is the conductivity of the material in S/m.

and then the equivalent reluctance is determined as

$$\mathcal{R}_{eq} = \frac{1}{\mathcal{P}_{eq}} \tag{3.28}$$

Also, Equation 3.15 is substituted into Equation 3.17

$$\lambda = \frac{\mu N^2 IA}{\ell} \, \text{Wb} \tag{3.29}$$

so that flux linkage is directly proportional to the coil current.

The *inductance L* of a coil is defined as the flux linkage per ampere of current in the coil and measured in *henries* (H). Therefore,

$$L = \frac{\lambda}{I} \, \text{H} \tag{3.30}$$

From Equations 3.21, 3.29, and 3.30, it can be shown that inductance can be related to reluctance as

$$L = \frac{N^2}{\mathcal{R}} \, \text{H} \tag{3.31}$$

Alternatively, inductance can be expressed in terms of permeance as

$$L = N^2 \times \mathcal{P} \, \text{H} \tag{3.32}$$

Example 3.1

Consider the toroid that is shown in Figure 3.5b with outside and inside radiuses of 5 and 4 cm, respectively. Assume that 500 turns are wound around the toroid of ferromagnetic material to produce a total flux of 16.85856×10^{-5} Wb in the core. If the magnetic field intensity in the core is 1000 A·turns/m, determine the following:

(a) The length of the average flux path in the toroid and the cross-sectional area perpendicular to the flux
(b) The flux density in the core
(c) The required mmf
(d) The amount of current that must flow in the turns of the toroid

Solution

(a) Since the mean radius r is

$$r = \frac{r_1 + r_2}{2}$$

$$= \frac{4+5}{2} = 4.5 \text{ cm or } 4.5 \times 10^{-2} \text{ m}$$

the length of the average flux path is

$$\ell = 2\pi r$$

$$= 2\pi(4.5 \times 10^{-2}) = 0.2827 \text{ m}$$

Thus, the cross-sectional area is

$$A = \pi \left(\frac{d}{2}\right)^2 = \pi \left(\frac{r_2 - r_1}{2}\right)^2 = \frac{\pi}{4}(r_2 - r_1)^2$$

$$= \frac{\pi}{4}(0.05 - 0.04)^2$$

$$= 7.854 \times 10^{-5} \text{ m}^2$$

(b) The flux density in the core is

$$B = \frac{\Phi}{A}$$

$$= \frac{16.85856 \times 10^{-5} \text{ Wb}}{7.854 \times 10^{-5} \text{ m}^2}$$

$$= 21.465 \text{ Wb/m}^2 \text{ or } 21.465 \text{ T}$$

(c) Since the magnetic field intensity is given as 1000 A·turns/m, the required mmf is

$$\mathcal{F} = H \times \ell$$

$$= (1000 \text{ A} \times \text{turns/m})(0.2827 \text{ m}) = 282.7 \text{ A} \cdot \text{turns}$$

(d) Since

$$\mathcal{F} = N \times I$$

therefore,

$$I = \frac{\mathcal{F}}{N}$$

$$= \frac{282.7 \text{ A} \cdot \text{turns}}{500 \text{ turns}} = 0.5654 \text{ A}$$

Example 3.2

Resolve Example 3.1 but assume that the toroid is made of plastic. The permeability of plastic is the same as that for free air. Assume that the total flux amount is the same as before but that the magnetic field intensity is unknown.

Solution

(a) As before,

$$r = 4.5 \times 10^{-2} \text{ m}, \ \ell = 0.2527 \text{ m, and } A = 7.854 \times 10^{-5} \text{ m}^2$$

(b) The flux density in the core is still

$$B = 21.4657 \text{ Wb / m}^2 \text{ or } 21.4657 \text{ T}$$

(c) Since

$$B = \mu_0 \times H$$

the magnetic field intensity in the plastic core is

$$H = \frac{B}{\mu_0} = \frac{21.4657 \text{ Wb/m}^2}{4\pi \times 10^{-7} \text{ H/m}} = 1.7081 \times 10^7 \text{ A} \cdot \text{turns/m}$$

and since

$$\mathcal{F} = H \times \ell = N \times I$$

or

$$\mathcal{F} = H \times \ell = (1.7081 \times 10^7 \text{ A} \cdot \text{turns/m})(0.2827 \text{ m}) = 4,828,870.11 \text{ A} \cdot \text{turns}$$

(d) Therefore,

$$I = \frac{\mathcal{F}}{N}$$

$$= \frac{4,828,870.11 \text{ A} \cdot \text{turns}}{500 \text{ turns}}$$

$$= 9,657.7 \text{ A}$$

Note that the current required to produce the same amount of flux has increased by almost 210.9 times from that which was required in Example 3.1, when the core was made of soft-steel casting.

Example 3.3

Consider the coil wound on the plastic toroidal of Example 3.2 and assume that the plastic ring has a rectangular cross section. The thickness of the core is 1 cm. Its outside diameter is 40 cm and the inside diameter is 30 cm. The coil has 200 turns of round copper wire which has a 3 mm diameter. Determine the following:

(a) For a current of 50 A, find the magnetic flux density at the mean diameter of the coil.
(b) The inductance of the coil, if the flux density within it is uniform and equal to the amount at the average diameter.
(c) Assume that the practical equivalent circuit of the coil that is made of R and L of the coil that is connected in series with each other. If the volume resistivity of copper is 17.2×10^{-9} Ω m, find the values of the R and L in the equivalent circuit.

Solution

(a) At the average diameter of the toroid (where the average flux length is located), the magnetic field intensity in the plastic core is

$$H = \frac{NI}{2\pi r}$$

$$= \frac{(200 \text{ turns}) \times (50 \text{ A})}{0.35r}$$

$$= 9095 \text{ A/m}$$

$$B = \mu_0 H$$

$$= (4\pi \times 10^{-7})(9095 \text{ A/m})$$

$$= 11.43 \times 10^{-3} \text{ Wb/m}^2$$

(b) Assuming $B_{avg} = 11.43 \times 10^{-3}$ Wb/m², the average flux is

$$\phi = BA$$

$$= (11.43 \times 10^{-3} \text{ Wb/m}^2)[(0.5\text{m})(0.01\text{m})]$$

$$= 57.15 \times 10^{-6} \text{ Wb}$$

The flux linkage is

$$\lambda = N\phi$$

$$= (200 \text{ turns}) (57.5 \times 10^{-6} \text{ Wb})$$

$$= 11.43 \times 10^{-3} \text{ Wb}$$

Since

$$L = \frac{\lambda}{I}$$

$$= \frac{11.43 \times 10^{-3} \text{ Wb}}{50}$$

$$= 0.2286 \times 10^{-3} \text{ H}$$

Alternatively,

$$\mathcal{R} = \frac{\ell}{\mu_0 A}$$

$$= \frac{2\pi r}{\mu_0 A}$$

$$= \frac{\pi(0.35 \text{ m})}{(4\pi \times 10^{-7})(0.1 \times 0.05)}$$

$$= 175 \times 10^6 \text{ A/Wb}$$

$$L = \frac{N^2}{\mathcal{R}}$$

$$= \frac{200^2 \text{ turns}}{175 \times 10^6 \text{ A/Wb}}$$

$$= 0.2286 \times 10^{-3} \text{ H}$$

(c) At radius r, where $0.15\,\text{m} < r < 0.20\,\text{m}$, flux density is

$$B = \frac{\mu_0 NI}{2\pi r}$$

$$= \frac{(4\pi \times 10^{-7})(200 \text{ turns})I}{2\pi r} \text{ T}$$

The flux is

$$\phi = \int_{0.15}^{0.20} B \times 0.1 dr \text{ Wb}$$

$$\lambda = N\phi$$

$$= \frac{0.1\mu_0 N^2 I}{2\pi} \int_{0.15}^{0.20} \frac{dr}{dt} \text{ Wb}$$

and

$$L = \frac{\lambda}{I}$$

$$= \frac{0.1(4\pi \times 10^{-7})(200^2)}{2\pi} ln\left(\frac{0.2}{0.15}\right)$$

$$= 0.2301 \times 10^{-3} \text{ H}$$

$$\text{Error} = \frac{0.2301 - 0.2286}{0.2301} \times 100 = 0.651\%$$

(d)

$$R = \frac{\rho l_{wire}}{A_{wire}}$$

$$= \frac{(17.2 \times 10^{-9} \ \Omega\text{m})200 \times 0.3}{(\pi \times 3^2 \times 10^{-6/4})/4}$$

$$= 0.1460 \ \Omega$$

Therefore, the equivalent parameters of an approximate equivalent circuit are $R = 0.1360 \ \Omega$ and $L = 0.2286 \text{ mH}$

3.5   MAGNETIC CIRCUIT WITH AIR GAP

Air gaps are fundamental in many magnetic circuits currently in use. As shown in Figure 3.4, every electromechanical energy converter is made up of two parts, namely, (1) the stator and (2) the rotor embedded in the air gap of the stator.

As shown in Figure 3.9a, essentially the same flux is present in the magnetic core and the air gap. To sustain the same flux density, the air gap must have much more mmf than the magnetic core. If the flux density is high, the magnetic core section of the magnetic circuit may show the saturation effect. However, the air-gap section of the magnetic circuit will remain unsaturated due to the fact that the $B-H$ curve for the air medium is linear, with a constant permeability.* In Figure 3.9a, ℓ_c is the length of the magnetic core, while ℓ_g is the length of the air gap. Since there is more than one material involved, such a magnetic circuit is said to be made of a *composite*

---

* The flux density in the air gap can be easily measured by the use of an instrument known as a *gauss meter*. The principle of the design of such an instrument is known as the *Hall effect*.

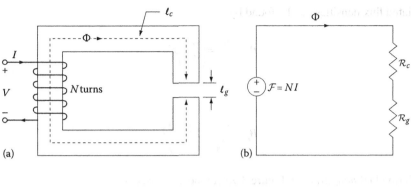

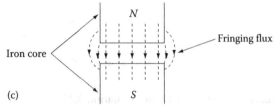

FIGURE 3.9 (a) A simple magnetic circuit with an air gap, (b) its magnetic circuit analogue, and (c) the fringing effect of magnetic flux at an air gap.

structure. Figure 3.9b shows the equivalent magnetic circuit that has the reluctance of the air gap $\mathcal{R}_g$ in series with the reluctance of the core $\mathcal{R}_c$. Applying Ampère's law, the required mmf can be found from

$$\mathcal{F} = H_c \times \ell_c + H_g \times \ell_g \tag{3.33}$$

$$N \times I = H_c \times \ell_c + H_g \times \ell_g \tag{3.34}$$

The resulting flux can be found from

$$\Phi = \frac{N \times I}{\mathcal{R}_c + \mathcal{R}_g} \tag{3.35}$$

where the reluctances for the core and air gap are

$$\mathcal{R}_c = \frac{\ell_c}{\mu_0 A_c} = \frac{\ell_c}{\mu_0 \mu_r A_c} \tag{3.36}$$

$$\mathcal{R}_g = \frac{\ell_g}{\mu_0 A_g} \tag{3.37}$$

where
   ℓ_c is the mean length of the core
   ℓ_g is the length of the air gap
   A_c is the cross-sectional area of the core
   A_g is the cross-sectional area of the air gap

The associated flux densities can be found by

$$B_c = \frac{\Phi_c}{A_c} = \left(\frac{\mu_c}{\ell_c}\right)\mathcal{F} \tag{3.38}$$

and

$$B_g = \frac{\Phi_g}{A_g} = \left(\frac{\mu_0}{\ell_g}\right)\mathcal{F} \tag{3.39}$$

Since the individual *mmf drops* in Figure 3.9b can be expressed as

$$H_c \times \ell_c = \Phi \times \mathcal{R}_c \tag{3.40}$$

$$H_g \times \ell_g = \Phi \times \mathcal{R}_g \tag{3.41}$$

by substituting Equations 3.40 and 3.41 into Equation 3.33

$$\mathcal{F} = N \times I = \Phi(\mathcal{R}_c + \mathcal{R}_g) \tag{3.42}$$

If there is an air gap in a magnetic circuit, there is a tendency for the flux to bulge outward (or spread out) along the edges of the air gap, as shown in Figure 3.9c, rather than to flow straight through the air gap parallel to the edges of the core. This phenomenon is called *fringing* and is taken into account by assuming a larger (effective) air-gap cross-sectional area. The common practice is to use an (effective) air-gap area by adding the air-gap length to each of the two dimensions that make up the cross-sectional area. Thus, the new (effective) air-gap area becomes

$$A_g = (a + \ell_g)(b + \ell_g) \tag{3.43}$$

where a and b are the actual core dimensions of a given rectangular-shaped core. The corrected gap area slightly reduces the gap reluctance. The relative effect of fringing increases with the length of the air gap.

3.6 BRIEF REVIEW OF FERROMAGNETISM

Magnetic materials that include certain forms of iron and its alloys in combination with cobalt, nickel, aluminum, and tungsten are called *ferromagnetic materials.** They are relatively easy to magnetize since they have a high value of relative permeability μr. These ferromagnetic materials are classified as *hard* or *soft* materials. Soft ferromagnetic materials include most of the soft steels, iron, nickel, cobalt, and one rare-earth element, as well as many alloys of the four elements. Hard ferromagnetic materials comprise the permanent magnetic materials such as alnico (which is iron alloyed with aluminum, nickel, and cobalt), the alloys of cobalt with a rare-earth element such as samarium, the copper–nickel alloys, the chromium steels, and other metal alloys.

---

* There are also other magnetic materials that have been in use in recent years. They include (1) ferromagnetic materials, (2) superparamagnetic materials, and (3) ferrofluidic materials. The *ferromagnetic materials* are ferrites, and therefore made up of iron oxides. They include permanent magnetic ferrites (e.g., strontium or barium ferrites), manganese–zinc ferrites, and nickel–zinc ferrites. The *superparamagnetic materials* are made up of powdered iron (or other magnetic material) particles that are mixed in a nonferrous epoxy or other plastic material. Permalloy is an example of such a material and is made up of molybdenum–nickel–iron powder. Finally, the *ferrofluidic materials* are magnetic fluids that are made up of three components: a carrier fluid, iron oxide particles suspended in the fluid, and a stabilizer.

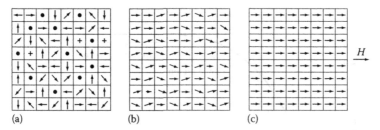

FIGURE 3.10 (a) Magnetic domains oriented randomly, (b) magnetic domains becoming magnetized, and (c) magnetic domains fully magnetized (lined up) by the magnetic field H.

The atoms of a ferromagnetic material tend to have their magnetic fields closely aligned. Within the crystals of such materials, there are many tiny (usually of microscopic) regions called *domains*. In any such given domain, all the atoms are aligned with their magnetic fields pointing in the same direction. Each domain behaves like a small permanent magnet. However, if the material is not magnetized, it will not have any flux within it, since each tiny domain is oriented randomly. This is illustrated in Figure 3.10a where the arrows represent the magnetic-moment direction within each domain. Notice that the domain alignments may be randomly distributed in three dimensions. The size of the domains is such that a single crystal may have many domains, each aligned with an axis of the crystal. If an electric field is applied to such a metal piece, the number of domains aligned with the magnetic field will increase since the individual atoms within each domain will physically switch orientation to align themselves with the magnetic field. This, in turn, increases the flux in the iron as well as the strength of the magnetic field, causing more atoms to switch orientation, as shown in Figure 3.10b. Figure 3.10c shows that all domains are aligned with the magnetic field, and any increase in the strength of the field will not cause any change in orientation. Thus, the material is referred to as *saturated*. If the material becomes saturated as the magnetizing field intensity is increased, the flux density changes very little and eventually not at all.* Figure 3.11 shows a typical dc *magnetization curve*[†] of a ferromagnetic material. It shows the behavior of the three regions of domain: the linear region, the knee region, and the saturation region. Figure 3.12 shows the magnetization curves of two typical ferromagnetic materials used in the manufacture of power apparatus.

Consider the magnetic circuit shown in Figure 3.6 and assume that its magnetic core is initially unmagnetized. Assume that, instead of applying a *dc* current to the coil, an ac current is applied. Since the core is initially unmagnetized, the flux in the core is zero. As shown in Figure 3.13a, if the current in the coil is increased, the flux in the core will increase. As a result, its magnetic field intensity H will also increase and follow the initial magnetization curve (along Oa) until saturation is reached. At the saturation point a, the flux density has reached its maximum value, B_{sat} and the magnetic material is fully saturated. The corresponding value of the magnetic field intensity is H_{sat}. If the current is now decreased in the coil, thereby decreasing the magnetizing force (i.e., magnetic field intensity) H, the initial curve will not be retraced. A different path (along the ab curve) will be followed indicating that there is a lag or delay in the reversal of domains. (Note that B does not

---

* If the external magnetic field is removed, the orientation of individual domains will not become totally randomized again since shifting back the orientation of atoms will need additional energy that may not be available. Therefore, the metal piece will remain magnetized. Such energy requirement may be fulfilled by applying (1) mmf in the opposite direction, (2) a large enough mechanical shock, or (3) heat. For example, a permanent magnet can easily be demagnetized if it is hit by a hammer, dropped, or heated. Such additional energy can cause domains to lose their alignment. Also note that at a very high temperature known as the *Curie point*, magnetic moments cease to exist, and therefore, the magnetic material involved loses its magnetic properties. For example, the Curie point for iron is about 775°C.
[†] The terms *magnetization curve* and *saturation curve* are used interchangeably in practice.

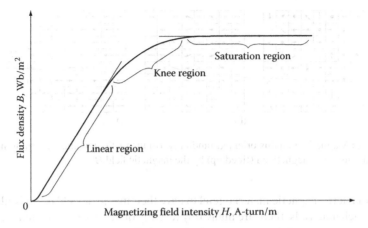

FIGURE 3.11 Typical magnetization curve showing the behavior of the three regions of domain behavior for a ferromagnetic core.

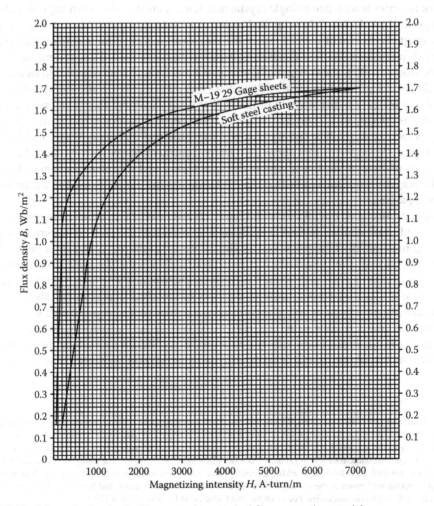

FIGURE 3.12 Magnetization (or B–H) curves of two typical ferromagnetic materials.

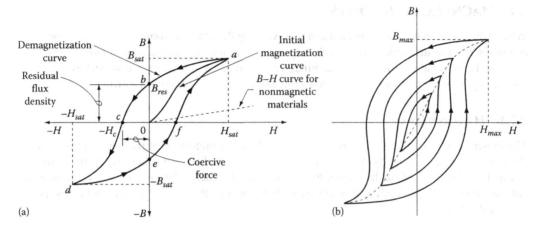

FIGURE 3.13 (a) Hysteresis loop for a typical ferromagnetic core and (b) family of hysteresis loops corresponding to different peak values of excitation.

decrease as quickly as it increased.) This irreversibility is called *hysteresis*,[*] which simply means that the flux density B *lags behind* the field intensity H.[†] When the current in the coil is zero, that is, H is equal to zero, there is still a residual value of magnetic flux density B_{res} in the core, the magnitude of which depends on the material. This is called *residual flux density* or *remanence* and in effect creates a permanent magnet. To decrease the flux density B to zero requires a *coercive field intensity* (also called *coercive force*) H_c. Any further increase in H in the reverse direction causes the magnetic core to be magnetized with the opposite polarity. This is achieved with a reversed current flow in the coil. Increasing the current in the negative direction further will result in a saturation level at which point (i.e., point d) it can be assumed that all domains in the magnetic core are aligned in the opposite direction. As the current or the field intensity H is brought to zero, the magnetic flux density B in the magnetic core will again be equal to its residual magnetism (at point e). Again, the direction of the current in the coil has to be reversed to make the magnetic flux density in the core equal to zero. Therefore, if the process continues in this manner, the *hysteresis loop* shown in Figure 3.13a will be traced out.

Figure 3.13 shows that for each maximum value of the ac magnetic field intensity cycle, there is a separate steady-state loop. Therefore, the complete magnetization characteristic is made up of a set (or family) of loops for different peak values of excitation. In Figure 3.13b, the dashed curve that connects the tips of the loops is the dc magnetization curve of the magnetic material.

The shape of the hysteresis loop is a function of the type of magnetic material. As stated previously, magnetically soft materials have very low residual flux density and low coercive field intensity. Therefore, these materials (such as silicon iron with 3% or 4% S_i content) are used in the manufacture of electric machines and transformers. In such materials, the magnetization can be changed quickly without much friction. Thus, their hysteresis loops usually have a tall, narrow-shaped small area. At 50 or 60 Hz, such hysteresis loops have a narrower shape than the ones for higher frequencies. In other words, the higher the frequency is, the broader the associated hysteresis loop will be. Therefore, in dc current, the hysteresis loop almost turns into a curve known as the *magnetization curve*.

---

[*] The term originates from the Greek word *hysterein* to be behind or to lag. It represents the failure to retrace the initial magnetization curve.

[†] Since any change of B lags behind the change of magnetizing field intensity H which produces it, there will be an angular displacement between the rotating mmf wave of the stator winding and the alternating field induced in the rotor iron. As a result of this, there will be a hysteresis torque whenever the iron is moving relative to the inducing mmf wave. This is the principle on which small *hysteresis motor* operation is based.

3.7   MAGNETIC CORE LOSSES

When a magnetic material is subjected to a time-varying flux, there is some energy loss in the material in the form of magnetic losses. Such magnetic losses are also called *iron* or *core losses*.* The cores of the armatures of dc and ac machines, transformers, and reactors are subject to these core losses. In general, core losses are defined as the sum of hysteresis and eddy-current losses.

3.7.1   Hysteresis Loss

The hysteresis loss is caused by continuous reversals in the alignment of the magnetic domains in the magnetic core. Succinctly put, the energy that is required to cause these reversals is the hysteresis loss. The area of the hysteresis loop represents the energy loss during one cycle in a unit cube of the core material. According to Charles P. Steinmetz, the hysteresis loss can be determined empirically from

$$P_h = v \times f \times k_h \times B_m^n \, \text{W} \tag{3.44}$$

where
  v is the volume of ferromagnetic material, m^3
  f is the frequency, Hz
  k_h is the proportionality constant depending upon the core material (typically, soft iron, silicon sheet steel, and permalloy are 0.025, 0.001, and 0.0001, respectively)
  B_m is the maximum flux density
  n is the Steinmetz exponent, which varies from 1.5 to 2.5 depending upon the core material, varies from 1.5 to 2.5 (typically, a value of 2.0 is used for estimating purposes)

3.7.2   Eddy-Current Loss

Because iron is a conductor, time-varying magnetic fluxes induce opposing voltages and currents called *eddy currents* that circulate within the iron core, as shown in Figure 3.14b.[†] In the solid iron core, these undesirable circulating currents flow around the flux and are relatively large because they encounter very little resistance. Therefore, they produce power losses with associated heating effects and cause demagnetization. As a result of this demagnetization, the flux distribution in the core becomes *nonuniform*, since most of the flux is pushed toward the outer surface of the magnetic material. As shown in Figure 3.14a, the eddy currents always tend to flow perpendicular to the flux and in a direction that opposes any change in the magnetic field due to Lenz's law. In other words, the induced eddy currents tend to establish a flux that opposes the original change imposed by the source.

To significantly increase the resistance encountered by these eddy currents so that the associated power losses can be minimized,[‡] the magnetic core is usually built up from stackings of thin steel sheet *laminations*, as shown in Figure 3.14c. The surfaces of such sheet laminations are coated with an oxide or a very thin layer of electrical insulation (usually an insulating varnish or sometimes paper). As a general rule, the thinner the laminations are, the lower the losses are, and since the eddy current losses are proportional to the square of the lamination thickness. In addition, as previously stated, the resistivity of steel laminations is substantially increased by the addition of a small amount of silicon.

---

* No core losses take place in iron cores carrying flux that does not vary with time.
[†] At very high frequencies, the interior of the magnetic core is practically unused because of the large (and circulating) eddy currents induced and their inhibiting effect. This phenomenon that takes place in magnetic circuits is known as the *magnetic skin effect*.
[‡] However, there are devices that are built based on the use of such eddy currents, such as eddy-current brakes, automobile speedometers (i.e., drag cup tachometers), and others.

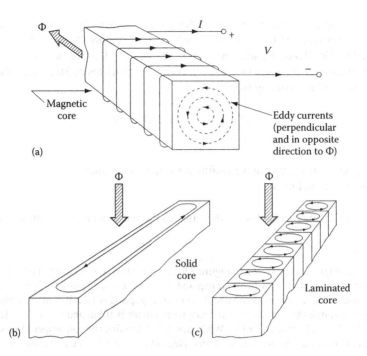

FIGURE 3.14 Development of eddy currents in magnetic cores.

Figure 3.15 shows the various shapes of steel laminations that are in use. The use of laminated cores makes the *actual* (or effective) cross-sectional area of the magnetic core being less than the *gross* (or *apparent*) cross-sectional area of the core represented by the stack of laminations. In actual calculations, this is taken into account by using the following stacking factor:

$$\text{Stacking factor} = \frac{\text{Actual magnetic cross} - \text{Sectional area}}{\text{Gross magnetic cross} - \text{Sectional area}} \tag{3.45}$$

FIGURE 3.15 Various shapes of steel laminations.

As the lamination thickness increases, the stacking factor approaches unity. For example, lamination thickness ranges from 0.0127 to 0.36 mm with corresponding stacking factors (at 60 Hz) that range from 0.50 to 0.95. Thus, the stacking factor approaches 1.0 as the lamination and the lamination surface insulation thicknesses increase. According to Charles P. Steinmetz, the eddy-current loss can be determined empirically from

$$P_e = k_e \times v(f \times t_\ell \times B_m)^2 \text{ W} \qquad (3.46)$$

where
    k_e is the proportionality constant depending upon the core material
    t_ℓ is the lamination thickness

The definitions of the other variables are the same as the ones given for Equation 3.44.

Example 3.4

Consider the magnetic core shown in Figure 3.9a. Assume that it is made up of a square-shaped, uniform cross-sectional area with an air gap and a core of soft-steel casting. The square-shaped cross-sectional area of the core is equal to that of the air gap and is 1.4×10^{-3} m^2. The mean length of flux path through the steel core of the magnetic circuit is 0.4 m and the air-gap length ℓ_g is 2 mm (or 0.002 m). If a flux of 2×10^{-3} Wb is needed, determine the coil ampere-turns (i.e., coil mmf) that are necessary to produce such flux. Neglect the flux fringing at the air gap.

Solution

Since the fringing of the flux across the air gap is neglected, the cross-sectional areas of the core and the gap are the same. Thus,

$$A_c = A_g = 1.4 \times 10^{-3} \text{ m}^2$$

The reluctance of the air gap is

$$\mathcal{R} = \frac{\ell_g}{\mu_0 A_g}$$

$$= \frac{2 \times 10^{-3} \text{ m}}{(4\pi \times 10^{-7})(1.4 \times 10^{-3} \text{ m}^2)}$$

$$= 1{,}136{,}821 \text{ A} \cdot \text{turns/Wb}$$

Thus, the needed mmf for the air gap is

$$\mathcal{F}_g = \Phi \times \mathcal{R}_g$$

$$= (2 \times 10^{-3} \text{ Wb})(1{,}136{,}821 \text{ A} \cdot \text{turns/Wb})$$

$$= 2{,}273.64 \text{ A} \cdot \text{turns}$$

The flux density in the steel core can be found from

$$B_c = \frac{\Phi}{A_c}$$

$$= \frac{2 \times 10^{-3} \text{ Wb}}{1.4 \times 10^{-3} \text{ m}^2} = 1.4286 \text{ Wb/m}^2$$

From the magnetization curve* for soft-steel casting shown in Figure 3.12, the corresponding magnetizing intensity is found as

$$H_c = 2{,}200 \text{ A} \cdot \text{turns/m}$$

Therefore, the required mmf to overcome the reluctance of the core can be found as

$$\mathcal{F}_c = H_c \times \ell_c$$

$$= (2{,}200 \text{ A} \cdot \text{turns/m})(0.4 \text{ m})$$

$$= 880 \text{ A} \cdot \text{turns}$$

Hence, the total required mmf from the coil is

$$\mathcal{F}_{coil} = \mathcal{F}_g + \mathcal{F}_c$$

$$= 2{,}273.64 + 880$$

$$= 3{,}153.64 \text{ A} \cdot \text{turns}$$

Example 3.5

Repeat Example 3.4 assuming that the core is made up of M-19 29 Gage sheets. Use a stacking factor of 0.90 for the laminations to determine the actual (i.e., the effective) cross-sectional area of the core. In the air gap, the cross-sectional area of flux is larger than in the iron core. To correct this *fringing* in the air gap, add the gap length to each of the two dimensions that make up its area.

Solution

From Equation 3.45, the actual area is found as

$$A_{c,actual} = (A_{c,gross})(f_{stacking})$$

$$= (1.4 \times 10^{-3} \text{ m}^2)(0.90)$$

$$= 1.26 \times 10^{-3} \text{ m}^2$$

Thus, the flux density in the core is

$$B_c = \frac{\Phi}{A_{c,actual}}$$

$$= \frac{2 \times 10^{-3} \text{ Wb}}{1.26 \times 10^{-3} \text{ m}^2}$$

$$= 1.5873 \text{ Wb/m}^2$$

From the magnetism curve for M-19 Gage sheets shown in Figure 3.12, the corresponding magnetizing intensity is found as

$$H_c = 2700 \text{ A} \cdot \text{turns/m}$$

---

* In practice, it is also referred to as the *B–H curve.*

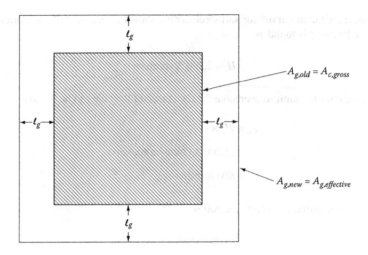

FIGURE 3.16   Illustration for Example 3.6.

Thus, the required mmf to overcome the reluctance of the core is

$$\mathcal{F}_c = H_c \times \ell_c$$

$$= (2700 \text{ A} \cdot \text{turns/m})(0.4\,\text{m})$$

$$= 1080 \text{ A} \cdot \text{turns}$$

At the air gap, the cross-sectional area increases due to the flux fringing. Therefore, the *new* (i.e., *effective*) area in the air gap can be found by adding the air-gap length to each of the two dimensions which make up the square-shaped cross-sectional area, as shown in Figure 3.16. Thus, the new air-gap area is found from

$$A_{g,new} = (a + \ell_g)(b + \ell_g) \tag{3.43}$$

but $a = b$, since the area is a square. Therefore,

$$A_{g,new} = (a + \ell_g)^2 \tag{3.47}$$

where

$$a = (A_{g,old})^{1/2} = (A_{c,gross})^{1/2} \tag{3.48}$$

Hence,

$$A_{g,new} = \left[(A_{g,old})^{1/2} + \ell_g \right]^2 \tag{3.49}$$

Thus,

$$A_{g,new} = \left[(1.4 \times 10^{-3})^{1/2} + (0.002) \right]^2 = 1.7153 \times 10^{-3} \text{ m}^2$$

Therefore, the resulting reluctance of the air gap is

$$\mathcal{R}_g = \frac{\ell_g}{\mu_0 \times A_{g,new}}$$

$$= \frac{2 \times 10^{-3} \text{ m}}{(4\pi \times 10^{-7})(1.7153 \times 10^3 \text{ m}^2)}$$

$$= 927{,}854.9 \text{ A} \cdot \text{turns/Wb}$$

Hence, the required mmf to overcome the reluctance of the air gap is

$$\mathcal{F}_g = \Phi \times \mathcal{R}_g$$

$$= (2 \times 10^{-3} \text{ Wb})(927{,}854.9 \text{ A} \cdot \text{turns/Wb})$$

$$= 1{,}855.7 \text{ A} \cdot \text{turns}$$

Thus, the total mmf required from the coil is

$$\mathcal{F}_{coil} = \mathcal{F}_g + \mathcal{F}_c$$

$$= 1{,}855.7 + 1{,}080$$

$$= 2{,}935.7 \text{ A} \cdot \text{turns}$$

Alternatively, the permeability of the core is calculated as

$$\mu_c = \frac{B_c}{H_c}$$

$$= \frac{1.5873 \text{ Wb/m}^2}{2{,}700 \text{ A} \cdot \text{turns/m}}$$

$$= 58.7889 \times 10^{-5} \text{ Wb/A} \cdot \text{turns m} \ (= \text{H/m})$$

Thus, the reluctance of the core is

$$\mathcal{R}_c = \frac{\ell_c}{\mu_c \times A_{c,actual}}$$

$$= \frac{0.4 \text{ m}}{(58.7889 \times 10^{-5})(1.26 \times 10^{-3} \text{ m}^2)}$$

$$\cong 540{,}000.4 \text{ A} \cdot \text{turns/Wb}$$

Therefore, the total reluctance is

$$\mathcal{R}_{tot} = \mathcal{R}_c + \mathcal{R}_g$$

$$= 540{,}000.4 + 927{,}854.9$$

$$= 1{,}467{,}855.3 \text{ A} \cdot \text{turns/Wb}$$

Hence, the total required mmf from the coil is

$$\mathcal{F}_{coil} = \mathcal{R}_{tot} \times \Phi$$

$$= (1,467,855.3)(2 \times 10^{-3})$$

$$= 2,935.71 \text{ A} \cdot \text{turns}$$

Example 3.6

Consider the solid ferromagnetic core shown in Figure 3.17a. The depth of the solid core is 5 cm. Assume that the relative permeability of the core is 1000 and remains constant and that the current value in the coil is 4 A. Ignore the fringing effects at the air gap and determine the following:

(a) The equivalent circuit of the given core shown in Figure 3.17a
(b) The individual values of all reluctances
(c) The value of the total reluctance of the core
(d) The value of the total flux that exists in the core
(e) The individual value of each flux that exists in each leg of the core
(f) The value of flux density in each leg

Solution

(a) The equivalent circuit is shown in Figure 3.17b.
(b) The reluctance values are

$$\mathcal{R}_1 = \frac{\ell_1}{\mu \times A_1} = \frac{\ell_1}{\mu_r \times \mu_0 \times A_1}$$

$$= \frac{2\left[(0.05/2) + 0.15 + 0.05\right] + 0.20}{1,000(4\pi \times 10^{-7})(0.05 \times 0.05)}$$

$$= 206,901.4 \text{ A} \cdot \text{turns/Wb}$$

$$\mathcal{R}_2 = \frac{\ell_2}{\mu \times A_2} = \frac{\ell_2}{\mu_r \times \mu_0 \times A_2}$$

$$= \frac{0.20 - 0.0002}{1,000(4\pi \times 10^{-7})(0.10 \times 0.05)}$$

$$= 31,799.16 \text{ A} \cdot \text{turns/Wb}$$

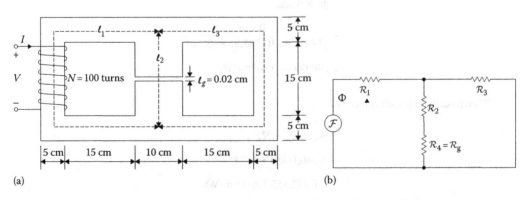

(a)   (b)

FIGURE 3.17   (a) Magnetic core for Example 3.6 and (b) its equivalent circuit (analog).

$$\mathcal{R}_3 = \mathcal{R}_1 = 206{,}901.4 \text{ A} \cdot \text{turns/Wb}$$

$$\mathcal{R}_4 = \frac{\ell_4}{\mu_r \times \mu_0 \times A_4}$$

$$= \frac{0.0002}{1\left(4\pi \times 10^{-7}\right)\left(0.10 \times 0.05\right)}$$

$$= 31{,}830.99 \text{ A} \cdot \text{turns/Wb}$$

(c) The total reluctance of the core is

$$\mathcal{R}_{tot} = \frac{\mathcal{R}_3(\mathcal{R}_2 + \mathcal{R}_4)}{\mathcal{R}_2 + \mathcal{R}_3 + \mathcal{R}_4}$$

$$= 48{,}645.4 \text{ A} \cdot \text{turns/Wb}$$

(d)

$$\Phi_{tot} = \frac{N \times I}{\mathcal{R}_{tot}}$$

$$= \frac{100 \times 4}{48{,}645.4}$$

$$= 8.2228 \times 10^{-3} \text{ Wb}$$

(e)

$$\Phi_1 = \Phi_{tot} = 8.2228 \times 10^{-3} \text{ Wb}$$

$$\Phi_2 = \Phi_{tot}\left(\frac{\mathcal{R}_3}{\mathcal{R}_2 + \mathcal{R}_3 + \mathcal{R}_4}\right)$$

$$= 6.29 \times 10^{-3} \text{ Wb}$$

$$\Phi_3 = \Phi_{tot}\left(\frac{\mathcal{R}_2}{\mathcal{R}_2 + \mathcal{R}_3 + \mathcal{R}_4}\right)$$

$$= 9.67 \times 10^{-4} \text{ Wb}$$

(f)

$$B_1 = \frac{\Phi_1}{A_1}$$

$$= \frac{0.0082228}{0.0025 \text{ m}^2}$$

$$= 3.289 \text{ Wb/m}^2$$

$$B_2 = \frac{\Phi_2}{A_2}$$

$$= \frac{0.00629}{0.005 \text{ m}^2}$$

$$= 1.258 \text{ Wb/m}^2$$

$$B_3 = \frac{\Phi_3}{A_3}$$

$$= \frac{0.000967}{0.0025 \text{ m}^2}$$

$$= 0.387 \text{ Wb/m}^2$$

Example 3.7

Figure 3.18a shows a cross section of the magnetic structure of a four-pole dc machine. On each of the four stator poles, there is a coil with equal turns. Since the four coils are connected in series, all carry the same current. The stator poles and rotor are made up of laminations of silicon steel sheets, while the stator yoke is made up of cast steel. Based on the given information, do the following:

(a) Draw an equivalent magnetic circuit.
(b) Derive an equation to determine the mmf produced by each winding.

Solution

(a) The equivalent magnetic circuit of the dc machine is shown in Figure 3.18b. In the figure, the subscripts r, s, p, and g denote rotor, stator yoke, stator pole, and air gap, respectively. Since the magnetic structure is symmetric, an analysis of one-quarter is sufficient. Therefore, the mmf produced by each winding is $\mathcal{F} = NI$ and provides the required flux on a per-pole basis. If the flux in the air-gap region is known, the flux densities in all sections of the machine can be found.

(b) As can be observed in Figure 3.18a, the flux supplied by each pole is the same in the pole, the pole face, and the air-gap area. Because the mmf drop in both halves of the yoke or rotor must be the same, the flux is divided equally when it flows through the stator yoke or the rotor. The equivalent magnetic circuit of the dc machine can be represented in terms of the reluctances, as shown in Figure 3.18b. The required mmf per pole can be determined by using Ampère's law for any one of the flux paths shown in Figure 3.18b. Therefore,

$$2\mathcal{F} = \Phi(2\mathcal{R}_p + 2\mathcal{R}_g) + \frac{\Phi}{2}(\mathcal{R}_r + \mathcal{R}_s) \tag{3.50}$$

or

$$\mathcal{F} = \Phi\left[(\mathcal{R}_p + \mathcal{R}_g) + 0.25(\mathcal{R}_r + \mathcal{R}_s)\right] \tag{3.51}$$

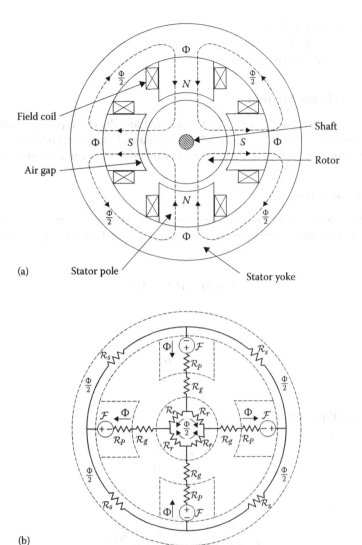

FIGURE 3.18   (a) Magnetic structure of a four-pole dc machine and (b) its equivalent circuit.

where
$\mathcal{R}_p$ is the reluctance of the stator pole
$\mathcal{R}_g$ is the reluctance of the air gap
$\mathcal{R}_r$ is the reluctance of the rotor
$\mathcal{R}_s$ is the reluctance of the stator yoke

Thus, the flux Φ is known and the reluctance of the magnetic circuit can be calculated from its physical dimensions and known permeability.

3.8   HOW TO DETERMINE FLUX FOR A GIVEN MMF

In the previous examples, the problem was as follows: given a magnetic circuit, find the mmf required to produce a given flux. The nonlinearity of iron presents the following more difficult problem: given an applied mmf, find the flux in a magnetic circuit. This problem can be solved by the following methods:

- The trial-and-error method
- The graphical method
- The magnetization curve method

3.8.1 Trial-and-Error Method

In the *trial-and-error method*, a value for Φ is selected and the corresponding mmf is computed. It is compared with NI, then a new value of Φ is selected and the corresponding new mmf value is computed. This procedure is repeated until the determined mmf is equal (or almost equal) to NT.

3.8.2 Graphical Method

This procedure is also called the *load line method*. Consider the magnetic circuit shown in Figure 3.9a. For a magnetic circuit with a core length ℓ_c and an air-gap length ℓ_g, for a given value of mmf,

$$\mathcal{F} = NI = H_c \ell_c + H_g \ell_g \tag{3.52}$$

since

$$H_g = \frac{B_g}{\mu_0} \tag{3.53}$$

Substituting it into Equation 3.52,

$$NI = H_c \ell_c + \frac{N_g}{\mu_0} \ell_g \tag{3.54}$$

and rearranging Equation 3.54,

$$B_g = B_c = -\mu_0 \left(\frac{\ell_c}{\ell_g} \right) H_c + \frac{NI\mu_0}{\ell_g} \tag{3.55}$$

Equation 3.52 represents a straight line since it is in the form of

$$y = mx + c \tag{3.56}$$

The resulting straight line is called the *load line* and can be plotted on the magnetization (i.e., the B–H) curve of the core. The slope of such a line can be expressed as

$$m = -\mu_0 \left(\frac{l_c}{\ell_g} \right) \tag{3.57}$$

The intersection of this line on the B ordinate is

$$c = \frac{NI\mu_0}{\ell_g} \tag{3.58}$$

Also, its intersection on the H axis is

$$H_c = \frac{NI}{\ell_c} \tag{3.59}$$

The intersection of the load line with the magnetization curve provides the value of B_c. Therefore, the value of the flux is found from

$$\Phi = B_c A \tag{3.60}$$

An alternative method of developing the load line is based on two steps:

Step 1. Assume that all the mmf is in the air gap, that is, $H_c = 0$. Therefore, the air-gap flux density can be expressed as

$$B_g = \left(\frac{NI}{\ell_g}\right)\mu_0 \tag{3.61}$$

The resulting value of B_g is the intersection of the load line on the B ordinate.

Step 2. Assume that all the mmf is in the core, that is, $B_g = 0$. Thus, the magnetizing intensity of the core can be expressed as

$$H_c = \frac{NI}{\ell_c} \tag{3.62}$$

The resulting value of H_c is the intersection of the load line on the H axis.

3.8.3   MAGNETIZATION CURVE METHOD

In this method, various values of flux Φ are chosen and the corresponding values of mmf are determined. The values of Φ versus mmf are plotted. The resulting curve is called the *magnetization curve* of the apparatus. Finally, by using the magnetization curve and the given value of current I, the value of flux Φ corresponding to $\mathcal{F} = NI$ is determined.

Example 3.8

Consider the magnetic core shown in Figure 3.9a. Assume that it is made up of a square-shaped, uniform cross-sectional area with an air gap and that a core of soft-steel casting. The square-shaped cross-sectional area of the core is equal to that of the air gap and is 1.5×10^{-3} m^2. The mean length of the flux path through the steel core of the magnetic circuit is 0.5 m, and the air-gap length is 3 mm (or 0.003 m). If the coil has 500 turns and the coil current is 7 A, find the flux density in the air gap. Neglect the flux fringing at the air gap and use the *trial-and-error method*.

Solution

The following steps can be used:

Step 1. Assume a flux density (since $B = \Phi/A$).
Step 2. Find He (from the B–H curve) and H_g (from $H_g = B_g/\mu_0$).
Step 3. Find $\mathcal{F}_c$ (from $\mathcal{F}_c = H_c\ell_c$), $\mathcal{F}_g$ (from $\mathcal{F}_g = H_g\ell_g$), and $\mathcal{F}$ (from $\mathcal{F} = \mathcal{F}_c + \mathcal{F}_g$).
Step 4. Find I from $I = \mathcal{F}/N$.
Step 5. If the found I is different from the given current, select a new appropriate value for the flux density. Continue this process until the calculated value of current is close to the given current value of 7 A.

TABLE 3.1

Table for Example 3.9

B	H_c	H_g	$\mathcal{F}_c$	$\mathcal{F}_g$	$\mathcal{F}$	I
1.3	1630	1.0345×10^{-6}	815	3103.5	3918.5	7.84
1.2	1280	0.9549×10^{-6}	640	2864.8	3504.8	7.0

Note that if all the mmf were only in the air gap, the resulting flux density would be

$$B = \frac{NI\mu_0}{\ell_g} = 1.4661 \text{ Wb/m}^2$$

However, since this is not the case, the actual flux density will be less than this value. This calculation process is illustrated in Table 3.1. The value of the flux is found from

$$\Phi = B_g A_g = B_c A_c = (1.2 \text{ Wb/m}^2)(1.5\times10^{-3} \text{ m}^2) = 0.0018 \text{ Wb}$$

Example 3.9

Solve Example 3.8 using the *graphical method*.

Solution

The intersection of the load line on the B ordinate is found using Equation 3.58 as

$$c_c = \frac{NI\mu_0}{\ell_c}$$

$$= \frac{(500 \text{ turns})(7 \text{ A})(4\pi\times10^{-7})}{0.003 \text{ m}} = 1.4661 \text{ A}\cdot\text{turns/m}$$

The intersection of this line on the H axis is

$$H_c = \frac{NI}{\ell_c}$$

$$= \frac{(500 \text{ turns})(7 \text{ A})}{0.5 \text{ m}} = 7000 \text{ A}\cdot\text{turns/m} \qquad (3.62)$$

Its slope is

$$m = -\mu_0\left(\frac{\ell_c}{\ell_g}\right) \qquad (3.57)$$

or

$$m = (4\pi\times10^{-7})\left(\frac{0.5 \text{ m}}{0.003 \text{ m}}\right) = 2094\times10^{-4}$$

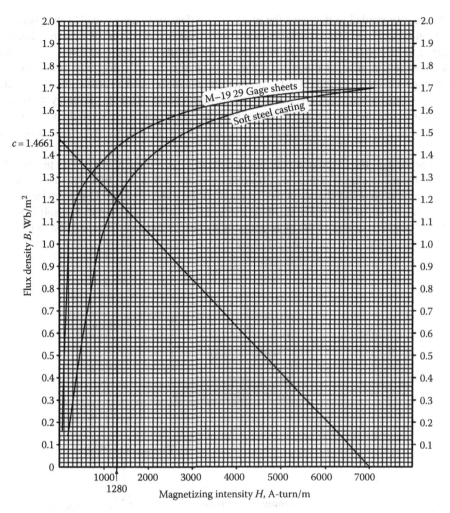

FIGURE 3.19 *B–H* curves for two materials.

As shown in Figure 3.19, the intersection of the load line with the magnetization curve gives the value of flux density in the air gap as

$$B_g = B_c = 1.2 \text{ Wb/m}^2$$

Thus, the value of flux is found from

$$\Phi = B_c A_c = B_g A_g$$

$$= (1.2 \text{ Wb/m}^2)(1.5 \times 10^{-3} \text{ m}^2)$$

$$= 0.0018 \text{ Wb}$$

3.9 PERMANENT MAGNETS

In general, after the removal of the excitation current, all ferromagnetic cores retain some flux density called the *residual flux density* B_r. To return the magnetic core to its original state, it has to be demagnetized by applying the magnetizing field intensity in the opposite direction. The value of the field intensity needed to decrease the residual flux density to zero is called the *coercive force*. Materials suitable for permanent magnets are known as *magnetically hard materials* because they are difficult to magnetize, but have high residual flux density and coercive force. The various categories of permanent magnet materials include (1) ductile metallic magnets, such as *Cunife*; (2) ceramic magnets, such as *Indox*; (3) brittle metallic magnets, such as *Alnico* (cast and sintered); and (4) rare-earth cobalt magnets, such as *samarium–cobalt*. Among the many applications of permanent magnets are loudspeakers, small generators, magnetic clutches and couplings, measuring instruments, magnetrons, television-focusing units, video recording, and information storage in computers.

In magnet design, the shape of a permanent magnet affects the amount of flux produced. Equal volumes of magnetic core will provide different amounts of flux as a function of their shape. For example, the most common shapes of *Alnico magnets* are rods, bars, and U shapes. However, since these materials are difficult to machine, their shapes are usually made simple, and soft-iron parts are added to the magnetic circuit in the more complex shapes, as shown in Figure 3.20a. It is interesting

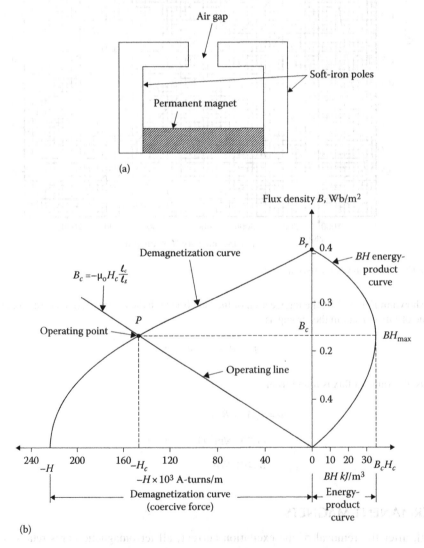

FIGURE 3.20 (a) Typical configuration made up of a permanent magnet, soft-iron pole pieces and air gap; (b) typical demagnetization and associated energy-product curves.

to note that such a magnet would be excited by first placing a magnetically soft-iron part in the air gap. This part, known as the *keeper*, is removed during use and replaced afterward.

Figure 3.20b shows the demagnetization curve and corresponding energy-product curve (or *B–H* curve) of the permanent magnet, for the magnetic circuit shown in Figure 3.20a. Ignoring the reluctance of the soft-iron parts and applying Ampère's law

$$H_c \ell_c + H_g \ell_g = 0 \tag{3.63}$$

where H_c is the magnetizing force within the core. If fringing at the air gap is ignored, the flux density inside the core is equal to that in the air gap. Therefore,

$$B_c = B_g = \mu_0 H_g = -\mu_0 H_0 \left(\frac{\ell_c}{\ell_g} \right) \tag{3.64}$$

This is the equation of a straight line. Its intersection with the magnetization characteristic provides the optimum operating point* P, as shown in Figure 3.20b. It determines the values of B_c and H_c for the permanent magnet. Here, having an air gap has the same effect as inserting a negative field inserted into the magnetic circuit. From Equation 3.61, the magnetizing force of the magnet can be expressed as

$$H_c = -H_g \left(\frac{\ell_g}{\ell_c} \right) \tag{3.65}$$

which can also be expressed in terms of the flux density B_c of the magnet by

$$H_c = -\frac{B_c A_c \ell_g}{\mu_0 A_g \ell_c} \tag{3.66}$$

Because magnetic leakage is negligibly small, the flux has to be the same in all parts of the magnetic circuit. Therefore,

$$\Phi = B_c A_c = B_g A_g \tag{3.67}$$

Furthermore, based on the assumption that the cross-sectional area of the magnet is uniform, the volume is found from

$$A_c \ell_c = A_c \left(-\ell_g \frac{H_g}{H_c} \right) = -A_g \ell_g \frac{H_g}{H_c} \tag{3.68}$$

if fringing and leakage are ignored. Also note that the product of B and H represents the energy density within the core. Thus, at the maximum energy density, the volume of the core is

$$Vol_c = \left(\frac{B_g^2}{\mu_0 B_c H_c} \right) (Vol_g) \tag{3.69}$$

where Vol_g is the volume of the air gap.

---

* To establish the maximum energy in the air gap, the point of operation must correspond to the maximum energy product of the magnet.

PROBLEMS

3.1 Consider the toroid shown in Figure 3.5b with inside and outside radii at 6 and 7 cm, respectively. Assume that 250 turns are wound around the toroid of soft-steel casting to produce a total flux of 8.5609×10^{-5} Wb in the core. If the magnetic field intensity in the core is 1000 A·turns/m, determine the following:
   (a) The length of the average flux path in the toroid and the cross-sectional area perpendicular to the flux
   (b) The flux density in the core
   (c) The required mmf
   (d) The amount of current that must flow in the turns of the toroid

3.2 Consider the results of Problem 3.1 and determine the following:
   (a) The flux linkage in the core
   (b) The inductance of the coil by using Equation 3.30
   (c) The inductance of the coil by using Equation 3.31
   (d) The permeance of the coil

3.3 Consider the magnetic core shown in Figure 3.9a. Assume that it is made up of a square-shaped, uniform cross-sectional area with an air gap and a core of soft-steel casting. The square-shaped cross-sectional area of the core is equal to that of the air gap and is 2×10^{-3} m^2. The mean length of the flux path through the steel core of the magnetic circuit is 0.6 m and the air-gap length ℓ_g is 3 mm (or 0.003 m). If a flux of 3×10^{-3} Wb is needed, determine the coil ampere-turns (i.e., coil mmf) necessary to produce such flux. Neglect the flux fringing at the air gap.

3.4 Solve Problem 3.3 but assume that the core is made up of M-19 29 Gage sheets. Therefore, use a stacking factor of 0.95 for the laminations to determine the actual (i.e., the effective) cross-sectional area of the core. The cross-sectional area of the flux in the air gap is larger than the one in the iron core. To correct this *fringing* in the air gap, add the gap length to each of the two dimensions which make up its area.

3.5 Consider Example 3.4 but ignore the iron reluctance. Find the amount of flux that flows in the magnetic circuit if the mmf of the coil is
   (a) 2000 A·turns
   (b) 1000 A·turns

3.6 Assume that the magnetic core shown in Figure 3.6a is made up of soft-steel casting with a cross-sectional area of 200 cm^2 and an average flux length of 100 cm. If the coil has 500 turns, determine the following:
   (a) The amount of current required to produce 0.02 Wb of flux in the core
   (b) The relative permeability of the core at the current level found in Part (a)
   (c) The reluctance of the core

3.7 Solve Problem 3.6 but assume that the amount of flux in the core is 0.03 Wb.

3.8 Solve Problem 3.6 but assume that the cross-sectional area of the core is 50 cm^2 and that the average flux length is 50 cm. Also assume that the amount of flux needed is 55×10^{-4} Wb and that the coil has 200 turns.

3.9 Solve Problem 3.6 but assume that the magnetic core is made up of M-19 29 Gage sheets with an actual cross-sectional core area of 25 cm^2 and an average flux length of 60 cm. The amount of flux that is needed in the core is 39×10^{-4} Wb and that the coil has 350 turns.

3.10 Solve Problem 3.6 but assume that the cross-sectional area of the core is 100 cm^2 and that the average flux length is 70 cm. Also assume that the amount of flux needed is 125×10^{-4} Wb and that the coil has 800 turns.

3.11 Solve Problem 3.10 but assume that the magnetic core is made up of M-19 29 Gage sheets rather than soft-steel casting. Use an actual cross-sectional core area of 100 cm^2.

3.12 Solve Problem 3.10 but assume that the coil has 200 turns rather than 800 turns.

3.13 Solve Problem 3.6 but assume that the cross-sectional area of the core is $150 \, cm^2$ and that the average flux length is $60 \, cm$. Also assume that the amount of flux needed is 25×10^{-3} Wb and that the coil has 500 turns.

3.14 Solve Problem 3.13 but assume that the magnetic core is made up of M-19 29 Gage sheets rather than soft-steel casting. Use an actual cross-sectional area of $150 \, cm^2$.

3.15 Solve Problem 3.13 but assume that the coil has 250 turns rather than 500 turns.

3.16 Consider the magnetic core shown in Figure 3.9a. Assume that it is made up of M-19 29 Gage steel laminations with a stacking factor of 0.9. Let the gross cross-sectional area of the core of $0.02 \, m^2$ be equal to the cross-sectional area at the air gap, ignoring the fringing of fluxes around the gap. Also let the lengths of the gap and the average flux path in the iron be 0.001 in. and 0.4 m, respectively. Assume that the coil has 2000 turns and determine the following:
  (a) The current needed to produce a flux of 0.03 Wb in the air gap, ignoring iron reluctance
  (b) Repeat Part (a) without ignoring iron reluctance

3.17 Assume that a magnetic circuit made up of soft-steel casting has a uniform cross-sectional area containing an air gap, as shown in Figure 3.9a. Let the cross-sectional areas of iron core and air gap be $1.5 \times 10^{-3} \, m^2$, ignoring the fringing of fluxes across the air gap. If the lengths of the average flux path in the iron and the gap are 0.4 and 0.002 m, respectively, determine the coil mmf to produce a flux of 2.5×10^{-3} Wb.

3.18 Assume that a magnetic circuit of uniform cross-sectional area of $1.75 \times 10^{-3} \, m^2$ has an air gap, and that the cross-sectional areas of the core and the air gap are the same. Ignore any effects of fringing around the gap. The magnetic core is made up of M-19 29 Gage sheets, the average length of flux path through the steel part of the magnetic circuit is 0.6 in., and the air-gap length is 2×10^{-3} m. The permeability of air is $4\pi \times 10^{-7}$ H/m. If a flux of 3×10^{-3} Wb is needed, determine the following:
  (a) Air-gap reluctance
  (b) Air-gap mmf
  (c) Flux density in the air gap
  (d) Flux density in the iron
  (e) Field intensity in the iron
  (f) mmf in the iron
  (g) mmf in the coil
  (h) The required current in the coil, if the coil has 3000 turns

3.19 Consider Problem 3.18 and assume that the uniform cross-sectional area is $1.5 \times 10^{-3} \, m^2$, the average length of flux path through the core is 0.4 m, and the air-gap length is 2.5×10^{-3} m. Assume that the core is made up of soft-steel casting and that the permeability of air is $4\pi \times 10^{-7}$ H/m and that the coil has 3000 turns. If a flux of 2×10^{-3} Wb is needed, answer the questions in Problem 3.18.

3.20 Consider the magnetic core shown in Figure P3.20 and notice that three sides of the core are of uniform width, whereas the fourth side is somewhat thinner. The depth of the core into the page is 10 cm. The coil has 300 turns, and the relative permeability of the core is 2000. Find the amount of flux that will be produced in the core by a 5 A input current.

3.21 Consider the magnetic core given in Problem 3.20 and assume that there is a small gap of 0.06 cm at point A (i.e., at midpoint of ℓ_1 distance) shown in Figure P3.20. Assume that due to fringing, the effective cross-sectional area of the air gap has increased by 6%. Use the given information here and in Problem 3.20 and determine:
  (a) The total reluctance of the flux path (i.e., including the iron core and the air gap)
  (b) The current necessary to produce a flux density of $1 \, Wb/m^2$ in the air gap

3.22 Consider the magnetic core shown in Figure P3.22 and notice that three sides of the core are of uniform width, whereas the fourth side is somewhat thinner. The depth of the core into the page is 15 cm. The coil has 250 turns and the relative permeability of the core is 2500. Find the amount of flux that will be produced in the core by a 10 A input current.

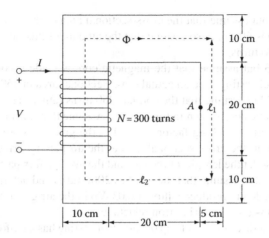

FIGURE P3.20 Magnetic core for Problem 3.20.

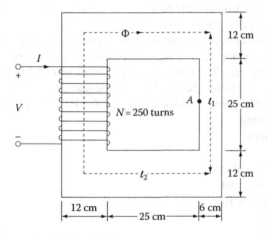

FIGURE P3.22 Magnetic core for Problem 3.22.

3.23 Consider the magnetic core given in Problem 3.22 and assume that there is a small gap of 0.04 cm at point A (i.e., at the midpoint of distance ℓ_1) shown in Figure P3.22. Assume that due to fringing, the effective cross-sectional area of the air gap has increased by 4%. The core is made up of soft-steel casting. Use the information given here and in Problem 3.22 and determine the following:

  (a) The total reluctance of the flux path (i.e., including the iron core and the air gap)
  (b) The current necessary to produce a flux density of 2 Wb/m² in the air gap

3.24 Consider the magnetic core and its equivalent circuit shown in Figure P3.24a and b, respectively, and assume that it represents an elemental stator and rotor setup of a dc motor. The stator has a square-shaped cross-sectional area (A_s) of 25 cm² and average path length (ℓ_s) of 100 cm. The rotor has a cross-sectional area (A_r) of 25 cm² and an average path length (i.e., the diameter of the cylindrical rotor (ℓ_r)) of 5 cm. Each air gap (on each side of the rotor) is 0.03 cm wide. The cross-sectional area of the air gap with fringing is 27.5625 cm². The relative permeability of the iron core used for both the stator and the rotor is 3000. The coil is located on the stator has 400 turns. If the current in the coil is 2 A, determine the resulting flux density in the air gaps.

3.25 Solve Problem 3.24 but assume that the length of each air gap (ℓ_g) is 0.01 cm rather than 0.03 cm.

3.26 Solve Problem 3.24 but assume that the length of each air gap (ℓ_g) is 0.06 cm rather than 0.03 cm.

3.27 Solve Problem 3.24 but assume that the number of turns of the coil is 200 turns rather than 400.

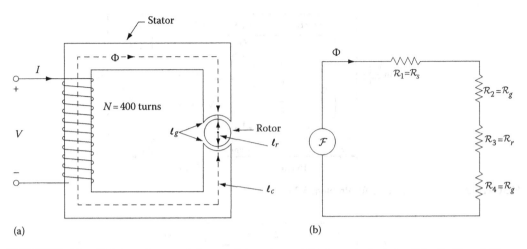

(a)

(b)

FIGURE P3.24 (a) Magnetic core of elemental dc motor and (b) its equivalent circuit for Problem 3.24.

3.28 Consider the elemental stator and rotor setup of the dc motor given in Problem 3.24. Assume that the stator has a square-shaped cross-sectional area (A_s) of 150 cm² and an average path length (ℓ_s) of 150 cm. The rotor has a cross-sectional area (A_r) of 150 cm² and an average path length (i.e., the diameter of cylindrical rotor) (ℓ_r) of 15 cm. Each air gap (on each side of the rotor) is 0.02 cm wide. The cross-sectional area of the air gap with fringing is 158 cm². The relative permeability of the iron core, used for both the stator and rotor, is 4000. The coil located on the stator has 600 turns. If the current in the coil is 4 A, determine the resulting flux density in the air gaps.

3.29 Solve Problem 3.28 but assume that the length of each air gap (fig) is 0.01 cm rather than 0.02 cm.

3.30 Solve Problem 3.23 but assume that the length of each air gap (Qg) is 0.04 cm rather than 0.02 cm.

3.31 Solve Problem 3.28 but assume that the number of turns of the coil is 300 turns rather than 600.

3.32 Consider the magnetic core shown in Figure P3.32. Assume that the depth of the core is 10 cm and that the relative permeability of the solid core is 2000. Determine the following:
  (a) The value of the current that will produce a flux of 0.01 Wb in the core
  (b) The flux density at the bottom of the core if the current amount found in Part (a) is used
  (c) The flux density at the right side of the core, if the current amount found in Part (a) is used

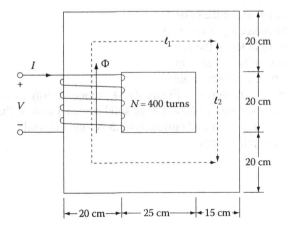

FIGURE P3.32 Magnetic core for Problem 3.32.

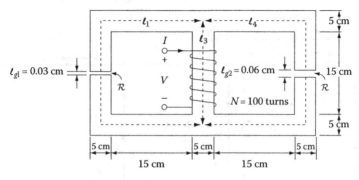

FIGURE P3.33   Magnetic core for Problem 3.33.

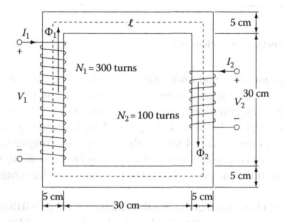

FIGURE P3.34   Magnetic core for Problem 3.34.

3.33 Consider the magnetic core shown in Figure P3.33. Assume that the depth of the solid core is 5 cm and that the relative permeability of the core is 1000. Also assume that the air gaps on the left and right legs are 0.03 and 0.06 cm, respectively. Take the fringing effects at each gap into account by calculating the effective area of each air gap as 5% greater than its actual physical size. If the coil has 100 turns and 2 A current in it, determine the following:

    (a) The total reluctance of the core
    (b) The total flux in the core
    (c) The flux in the left leg of the core
    (d) The flux in the center leg of the core
    (e) The flux in the right leg of the core
    (f) The flux density in the left air gap
    (g) The flux density in the right air gap

3.34 Consider the magnetic core shown in Figure P3.34. Notice the way each coil is wound on each leg. Use a relative permeability of 2000 for the solid core and assume that it remains constant. If the value of current $I_1 = 1.0$ A and current $I_2 = 1.5$ A, find the total flux produced by them in the core. Use a core depth of 5 cm.

4 Transformers

Education is the best provision for old age.

Aristotle, 365 BC

A teacher affects eternity.

Author Unknown

Education is…hanging around until you've caught on.

Will Rogers

4.1 INTRODUCTION

In general, a transformer is *a static** electromagnetic machine (i.e., it has no moving parts). Transformers are commonly used for changing the voltage and current levels in a given electrical system, establishing electrical isolation, impedance matching, and measuring instruments.

Power and distribution transformers are used extensively in electrical power systems to generate the electrical power at the most economical generator-voltage level, to transmit and distribute electrical power at the most economical voltage level, and to utilize power at the most economical, suitable, and safe voltage level.

Isolating transformers are used to electrically isolate electric circuits from each other or to block dc signals while maintaining ac continuity between the circuits, and to eliminate electromagnetic noise in many types of circuits.

Transformers are widely used in communication systems that vary in frequency from audio to radio to video levels. They perform various tasks, such as impedance matching for improved power transfer, and are used as input transformers, output transformers, and insulation apparatus between electric circuits, and interstage transformers. Transformers are used in the whole frequency spectrum in electrical circuits, from near dc to hundreds of megahertz, including both continuous sinusoidal and phase waveforms. For example, they can be found in use at power-line frequencies (between 60 and 400 Hz), audio frequencies (20–20,000 Hz), ultrasonic frequencies (20,000–100,000 Hz), and radio frequencies[†] (over 300 kHz).

Transformers are also used in measuring instruments. Instrument transformers are used to measure high voltages and large currents with standard small-range voltmeters (120 V), ammeters (5 A), and wattmeters, and to transform voltages and currents to activate relays for control and protection.

Voltage transformers[‡] (VTs) (also known as PTs, i.e., *potential transformers*) are single-phase transformers that are used to step down the voltage to be measured to a safe value.

---

* However, there are some special transformers in which some motion takes place in components of the electromagnetic structure. Examples include the variable autotransformer, which has a tap that moves between primary and secondary, as well as some types of voltage regulators that are employed in distribution systems.

† In general, the size of a transformer can be significantly reduced by using it with higher frequencies. (Because of this fact, aircraft generators are designed to produce power at 400 Hz rather than at 50 or 60 Hz.) Also, a transformer designed for use at 50 or 60 Hz can always be used at higher frequencies, whereas a transformer designed for use at 400 Hz would not operate properly at lower frequencies, because its core would saturate and the secondary voltage would not be similar to nor proportional to the primary voltage.

‡ They are also called *pot transformers*.

Current transformers (CTs) are used to step down currents to measurable levels. The secondaries of both voltage and CTs are normally grounded.

A transformer consists of a primary winding and a secondary winding linked by a mutual magnetic field. Transformers may have an air core, an iron core, or a variable core, depending on their operating frequency and application.

Transformers are also quite different in size and shape depending on the application. In power system applications, the single- or three-phase transformers with ratings up to 500 kVA are defined as *distribution transformers*, whereas those transformers with ratings over 500 kVA at voltage levels of 69 kV and above are defined as *power transformers*.* Figure 4.1 shows a cutaway view of a single-phase, overhead pole-mounted distribution transformer. Notice that it has two high-voltage bushings since it is built to operate under line-to-line voltage rather than line-to-neutral voltage. Figure 4.2 shows a three-phase, 345/161 kV autotransformer used as a power transformer. Its power ratings are 214/285/357 MVA for its OA/FA/FOA operations. Note that the OA/FA/FOA means *oil-immersed, self-cooled/forced-air-cooled/forced-oil-cooled.*[†]

A transformer is basically made up of two or more windings coupled by a mutual magnetic field. Ferromagnetic cores are employed to develop tight magnetic coupling and high flux densities. When such a coupling exists, the transformer is called an *iron-core transformer*.

Most distribution and power transformers are immersed in a tank of oil for better insulation[‡] and cooling purposes. The leads of the windings are brought to the outside of the tank through insulating bushings which are attached to the tank, as shown in Figure 4.1.

Such transformers are used in high-power applications.[§]

When there is no ferromagnetic material but only air present, such a transformer is called an *air-core transformer*.[¶] These transformers have poor magnetic coupling and are usually used in lower-power applications such as in electronic circuits. In this chapter, the focus is set exclusively on *iron-core transformers*.

However, the American National Standards Institute (ANSI) ratings were revised in the year 2000 to make them more consistent with IEC designations. This system has four-letter code that indicates the cooling (IEEE C57.12.00–2000):

First letter—Internal cooling medium in contact with the windings:

   O: Mineral oil or synthetic insulating liquid with fire point = 300°C

   K: Insulating liquid with fire point >300°C

   L: Insulating liquid with no measurable fire point

---

* For further information, see Gönen (2008).

[†] Today, various methods are in use to get the heat out of the tank more effectively. Historically, as the transformer sizes increased, the losses outgrew any means of self-cooling that was available at the time, thus a water-cooling method was put into practice. This was done by placing a coil of metal tubing in the top oil, around the inside of the tank. Water was pumped through this cooling coil to get rid of the heat from the losses. Another method was circulating the hot oil through an external oil-to-water heat exchanger. This method is called forced-oil-to-water cooling (FOW). Today, the most common of these forced-oil-cooled transformers employs an external bank of oil-to-air heat exchangers through which the oil is continuously pumped. It is known as type *FOA*. In present practice, fans are automatically used for the first stage and pumps for the second, in triple-rated transformers which are designated as type *OA/FA/FOA*. These transformers carry up to about 60% of maximum nameplate rating (i.e., *FOA* rating) by natural circulation of the oil (OA) and 80% of maximum nameplate rating by forced cooling which consists of fans on the radiators (FA). Finally, at maximum nameplate rating (*FOA*), not only is oil forced to circulate through external radiators, but fans are also kept on to blow air onto the radiators as well as into the tank itself.

[‡] The National Electric Manufacturers Association (NEMA) has grouped various types of insulation into classes and assigned a maximum permissible hottest spot temperature to each class. Since the hottest spot temperature is usually at some inaccessible spot within a coil, the maximum permissible average temperature (determined by measuring the resistance of the coil) is somewhat lower. The difference between the ambient temperature and the average temperature of the coil is its temperature rise. The sum of the ambient temperature plus the maximum temperature rise plus the hot-spot allowance equals the maximum temperature rating of the insulation.

[§] Today, it is technically possible to manufacture large power transformers having sheet-wound coils that are insulated by compressed gas (e.g., SF_6, i.e., *sulfur hexafluoride*) and cooled by forced circulation of liquid.

[¶] They are also known as *dry-type transformers*.

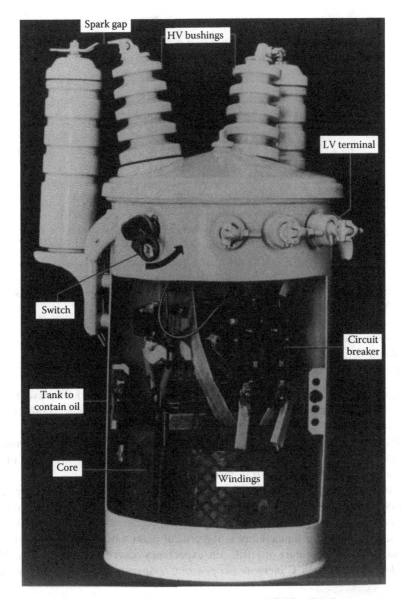

FIGURE 4.1   Cutaway view of a single-phase distribution transformer. (Courtesy of ABB Corporation, Zurich, Switzerland.)

Second letter—Circulation mechanism for internal cooling medium:

  N: Natural convection flow through cooling equipment and in windings

  F: Forced circulation through cooling equipment (i.e., *coolant pumps*); natural convection flow in windings (also called *nondirected flow*)

  D: Forced circulation through cooling equipment, directed from the cooling equipment into at least the main windings

Third letter—External cooling medium:

  A: Air

  W: Water

FIGURE 4.2   Installed view of a three-phase, 345/161 kV autotransformer with power ratings of 214/285/357 MVA for OA/FA/FOA. (Courtesy of North American Transformer, Milpitas, CA.)

Fourth letter—Circulation mechanism for external cooling medium:
   N: Natural convection
   F: Forced circulation—fans (*air cooling*), pumps (*water cooling*)

Therefore, *OA/FA/FOA* is equivalent to *ONAA/ONAF/OFAF*. Each cooling level typically provides an extra one-third capability: 21/28/35 MVA. Table 4.1 shows equivalent cooling classes in old and new naming schemes.

Utilities do not overload substation transformers as much as distribution transformers, but they do not run them hot at times. As with distribution transformers, the trade-off is loss of life versus the immediate replacement cost of the transformer.

Ambient conditions also affect loading. Summer peaks are much worse than winter peaks. IEEE Std. C57.91–1995 provides detailed loading guidelines and also suggests an approximate adjustment of 1% of the maximum nameplate rating for every degree Celsius above or below 30°C.

The hottest-spot-conductor temperature is the critical point where insulation degrades. Above the hot-spot-conductor temperature of 110°C, life expectancy decreases exponentially. *The life of a transformer halves for every 8°C increase in operating temperature.* Most of the time, the hottest

TABLE 4.1
Equivalent Cooling Classes

Year 2000 Designations	Designation prior to Year 2000
ONAN	OA
ONAF	FA
ONAN/ONAF/ONAF	OA/FA/FA
ONAN/ONAF/OFAF	OA/FA/FOA
OFAF	FOA
OFWF	FOW

Source: IEEE Std. C57.12.00-2000, IEEE Standard General Requirements for Liquid-Immersed Distribution, Power, and Regulating Transformers.

temperatures are nowhere near this. The impedance of substation transformers is normally about 7%–10%. This is the impedance on the base rating, the self-cooled rating (OA or ONAN).

4.2 TRANSFORMER CONSTRUCTION

The magnetic cores of transformers used in power systems are built either in core type or shell type, as shown in Figure 4.3. In either case, the magnetic cores are made up of stacks of laminations cut from silicon-steel sheets. Silicon-steel sheets usually contain about 3% silicon and 97% steel.

The silicon content decreases the magnetizing losses, especially the ones due to hysteresis loss. The laminations are coated with a nonconducting and insulating varnish on one side. Such a laminated core substantially reduces the core loss due to eddy currents.

Most laminated materials are cold-rolled and often specially annealed to orient the grain or iron crystals. This causes a very high permeability and low hysteresis to flux in the direction of rolling. Thus, in turn, it requires a lower exciting current. The laminations for the core-type transformer, shown in Figure 4.3a, may be made up of L-shaped, or U- and I-shaped laminations.

The core for the shell-type transformer, shown in Figure 4.3b, is usually made up of E- and I-shaped laminations. It is necessary to clamp laminations and impregnate the coils because of the cyclic magnetic forces and other forces that exist between parallel conductors carrying current.

The lack of clamping, or improper clamping, may cause an objectionable audible noise* that can be characterized as a humming sound. To minimize the use of copper and decrease copper loss, the magnetic cores of large transformers are built in stepped cores, as shown in Figure 4.4.

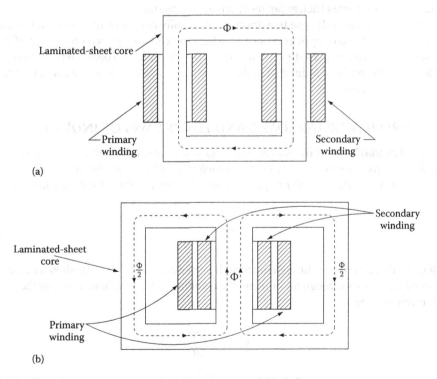

FIGURE 4.3 Transformer core construction: (a) core type and (b) shell type.

---

* The source of the audible sound is mechanical vibration of the core, produced by a steel characteristic known as magnetostriction. Because the magnetostrictive motion grows with increased flux density, the audible sound can be minimized by decreasing flux density.

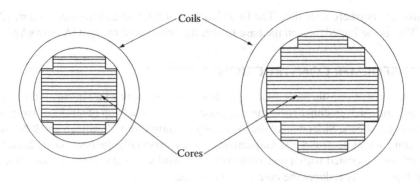

FIGURE 4.4   Stepped transformer cores.

Figure 4.3a shows *a core-type construction* that has the total number of primary winding turns located on one leg of the core and the total number of secondary winding turns placed on the other leg. This design causes large leakage flux and therefore results in a smaller mutual flux for a given primary voltage.

To keep the leakage flux within a few percent of the mutual flux, each winding may be divided into two coils; the two half coils are then mounted on two sides of the rectangle.* A larger reduction in leakage flux can be obtained by further subdividing and sandwiching the primary and secondary turns, however, at considerable cost. Leakage flux can be greatly decreased by using the *shell-type construction* shown in Figure 4.3b. However, the steel-to-copper weight ratio is greater in the shell-type transformer. It is more efficient but more costly in material.

The coils employed in shell-type transformers are usually of a "*pancake*" form unlike the cylindrical forms used in the core-type transformer, where the coils are placed one on top of the other, the low-voltage winding is placed closer to the core with the high-voltage winding on top. This design simplifies the problem of insulating the high-voltage winding from the core and reduces the leakage flux considerably.

4.3   BRIEF REVIEW OF FARADAY'S AND LENZ'S LAWS OF INDUCTION

According to **Faraday's law of induction**, *whenever a flux passes through a turn of a coil, a voltage (i.e., an electromotive force [em]) is induced, in each turn of that coil, that is directly proportional to the rate of change in the flux with respect to time.* Therefore, induced voltage can be found from

$$e_{ind} = \frac{d\Phi}{dt} \tag{4.1}$$

where Φ is the flux that passes through the turn. If such a coil has N turns, as shown in Figure 4.5a, and the same flux passes through all of them, the resulting induced voltage between the two terminals of the coil becomes

$$e_{ind} = N \frac{d\Phi}{dt} \tag{4.2}$$

However, according to **Lenz's law of induction**, if the coil ends were connected together, the voltage built-up would produce a current that would create *a new flux opposing the original flux change.*

---

* Such design is especially beneficial for laboratory use because each pair of coils can be connected in series or in parallel, and therefore four different primary and secondary potential differences can be provided.

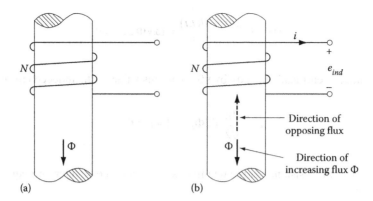

FIGURE 4.5 Illustration of Lenz's law: (a) a flux Φ passing through a coil and (b) corresponding voltage buildup in the coil.

Therefore, such a voltage buildup in the coil has to be in the proper direction to facilitate this, as shown in Figure 4.5b. Thus, Equations 4.1 and 4.2 can be reexpressed as

$$e_{ind} = -\frac{d\Phi}{dt} \tag{4.3}$$

and

$$e_{ind} = -N\frac{d\Phi}{dt} \tag{4.4}$$

where the negative sign in the equations signifies that the polarity of the induced voltage opposes the change that produced it.*

Alternatively, the magnitude of the induced voltage can be determined using the *flux linkage A* of a given coil. Thus,

$$e_{ind} = \frac{d\lambda}{dt} \tag{4.5}$$

where

$$\lambda = \sum_{i=1}^{N} \Phi_i = Li \tag{4.6}$$

Furthermore, because the induced voltage equals the rate of change of flux linkages, an applied sinusoidal voltage has to produce a sinusoidally changing flux, provided that the resistive voltage drop is negligible. Thus, if the flux as a function of time is given as

$$\Phi = \Phi_m \sin \omega t \tag{4.7}$$

where
  Φ_m is the maximum value of the flux
  ω *is 2nf*
  f is the frequency in hertz, then the induced voltage is given as

---

* Therefore, Lenz's law is very useful in determining the polarity of the voltage induced in the secondary winding of a transformer.

$$e(t) = N\frac{d\Phi}{dt} = \frac{d(Li)}{dt} = \omega N\Phi_m \cos\omega t \qquad (4.8)$$

Notice that the induced emf leads the flux by 90°. The rms value of the induced emf is given as

$$E = \frac{2\pi}{\sqrt{2}} fN\Phi_m = 4.44 fN\Phi_m \qquad (4.9)$$

If voltage drop due to the resistance of the winding is neglected, the counter-emf equals the applied voltage. Therefore,

$$V = E = 4.44 fN\Phi_m \qquad (4.10)$$

$$\Phi_m = \frac{V}{4.44 fN} \qquad (4.11)$$

where V is the rms value of the applied voltage. (Note that Equation 4.10 is known as the *emf equation** of a transformer.)

The flux is determined solely by the applied voltage, the frequency of the applied voltage, and the number of turns in the winding. The excitation (or exciting) current adjusts itself to produce the maximum flux required.

Therefore, if the maximum flux density takes place in a saturated core, the current has to increase disproportionately during each half period to provide this flux density. For this reason, inductors with ferromagnetic cores end up having nonsinusoidal excitation currents.

Therefore, if the maximum flux density takes place in a saturated core, the current has to increase disproportionately during each half period to provide this flux density. For this reason, inductors with ferromagnetic cores end up having nonsinusoidal excitation currents.

If the core is unsaturated and the resistance of the coil is negligible, the maximum value of the magnetizing current[†] can be found from

$$I_m = \frac{N\Phi_m}{L} = \left(\frac{N}{L}\right)\left(\frac{V}{4.44 fN}\right) = \frac{\sqrt{2}V}{\omega L} \qquad (4.12)$$

In the phasor form, the magnetizing current that produces the mutual flux is

$$I_m = \frac{V}{j\omega L} = \frac{V}{jX_m} \qquad (4.13)$$

where X_m is the magnetizing reactance of the coil.

---

* It is also known as the *general transformer equation.*

† It may be important to keep in mind that for a voltage to be induced across the secondary winding, there must be a changing current in the primary winding. In the event that a do source were connected to the primary winding, the current would become so large that the transformer would burn out. This is because on do (i.e., when $f = 0$ Hz), the primary winding acts like a low resistance due to the fact that once the current reaches a steady-state value, the inductive reactance is equal to zero ohms.

4.4 IDEAL TRANSFORMER

Consider a transformer with two windings, a primary winding of N_1 turns and a secondary winding of N_2 turns, as shown in Figure 4.6a. The core is made up of a ferromagnetic material. Assume that the transformer is an **ideal transformer*** with the following properties:

- The winding resistances are negligible.
- All magnetic flux is confined to the ferromagnetic core and links both windings, that is, leakage fluxes do not exist.
- The core losses are negligible.
- The permeability of the core material is almost infinite so that negligible. The net mmf is required to establish the flux in the core. In other words, the excitation current required to establish flux in the core is negligible.
- The magnetic core material does not saturate.

If the primary winding is connected to an energy source with a time-varying voltage v_1, a time-varying flux Φ and a flux linkage λ_1 of winding N_1 is established in the core. If v_1 varies over time, then i_1, Φ, and λ_1 will vary over time, and an emf e_2 will be induced in winding N_1. Therefore,

$$v_1 = e_1 = \frac{d\lambda_1}{dt} = N_1 \frac{d\Phi}{dt} \tag{4.14}$$

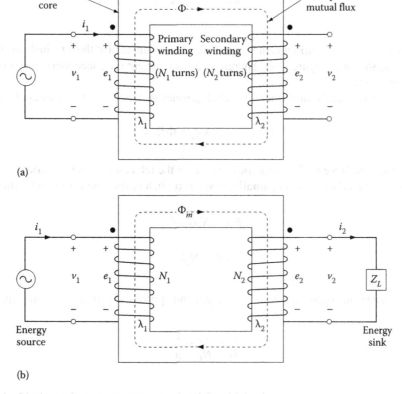

(a)

(b)

FIGURE 4.6   Ideal transformer: (a) with no load and (b) with load.

---

* The ideal transformer, although it is fictitious, is a very useful device in power and communication systems analysis.

Because there is no leakage flux, the flux Φ must link all N_2 turns of the secondary winding. Since the resistance of the secondary winding is assumed to be zero in an ideal transformer, it induces a voltage e_2 which is the same as the terminal voltage v_2. Thus,

$$v_2 = e_2 = \frac{d\lambda_2}{dt} = N_2 \frac{d\Phi}{dt} \tag{4.15}$$

From Equations 4.14 and 4.15,

$$\frac{v_1}{v_2} = \frac{e_1}{e_2} = \frac{N_1}{N_2} = a \tag{4.16}$$

which may also be written in terms of rms values as

$$\frac{V_1}{V_2} = \frac{E_1}{E_2} = \frac{N_1}{N_2} = a \tag{4.17}$$

where a *is* known as the **turns ratio**.* Note that the potential ratio is equal to the turns ratio. (Here, lowercase letters are used for instantaneous values and uppercase letters are used for rms values.) From Equation 4.17,

$$V_1 = \left(\frac{N_1}{N_2}\right) V_2 = aV_2 \tag{4.18}$$

Assume that a load (energy sink) with an impedance Z_L is connected at the terminals of the secondary winding, as shown in Figure 4.6b. Therefore, a load current (i.e., secondary current) will flow in the secondary winding.

Since the core of an ideal transformer is infinitely permeable, the net mmf will always be zero. Thus,

$$\mathcal{F}_{net} = N_1 i_1 - N_2 i_2 = \Phi \mathcal{R} = 0 \tag{4.19}$$

where $\mathcal{R}$ is the reluctance of the magnetic core. Since the reluctance of a magnetic core of a well-designed modern transformer is very small (almost zero) before the core is saturated, then

$$N_1 i_1 - N_2 i_2 = 0 \tag{4.20}$$

$$N_1 i_1 = N_2 i_2 \tag{4.21}$$

That is, the primary and secondary mmfs are equal and opposite in direction.† From Equation 4.21,

$$\frac{i_1}{i_2} = \frac{N_2}{N_1} = \frac{1}{a} \tag{4.22}$$

---

* It is also known as the *ratio of transformation*.
† The net mmf acting on the core is thus zero.

which may be written in terms of rms values as

$$\frac{I_1}{I_2} = \frac{N_2}{N_1} = \frac{1}{a} \tag{4.23}$$

Hence, the currents in the windings are inversely proportional to the turns of the windings. From Equation 4.23,

$$I_1 = \left(\frac{N_2}{N_1}\right)I_2 = \frac{I_2}{a} \tag{4.24}$$

From Equations 4.16 and 4.22,

$$v_1 i_1 = v_2 i_2 \tag{4.25}$$

or in terms of rms values

$$V_1 I_1 = \left(\frac{N_1}{N_2} V_2\right)\left(\frac{N_2}{N_1} I_2\right) = V_2 I_2 \tag{4.26}$$

That is, in an ideal transformer, the input power (VA) is equal to the output power (VA). In other words, the value of the apparent power remains the same.* This is the *power invariance principle* which means that the volt-amperes are conserved.

Furthermore, the complex power supplied to the primary is equal to the complex power delivered by the secondary to the load. Thus,

$$V_1 I_1^* = V_2 I_2^* \tag{4.27}$$

In the event that the primary and secondary turns are equal, these transformers are usually known as the **isolating transformers**, as previously stated.

In power systems, if the number of turns of the secondary winding is greater than the number of turns of the primary winding, the transformer is known as a **step-up transformer**.

On the other hand, if the number of turns of the primary winding is greater than those of the secondary winding, the transformer is known as a **step-down transformer**.

Example 4.1

Determine the number of turns of the primary and the secondary windings of a 60 Hz, 240/120 V ideal transformer, if the flux in its magnetic core is no more than 5 mWb.

Solution

From Equation 4.10, the number of turns that the primary winding must have is

$$N_1 = \frac{V_1}{4.44 f \Phi_m}$$

$$= \frac{240\ \text{V}}{4.44(60\ \text{Hz})(5 \times 10^{-3}\ \text{Wb})} = 180\ \text{turns}$$

---

* This transfer of power happens without any direct electrical connection between the primary and secondary windings. Such *electrical isolation* is, in certain applications, mandated for safety reasons. Examples include some medical apparatus and instrumentation designs for systems operating at a very high voltage.

and the number of turns that the secondary winding must have is

$$N_2 = \frac{V_2}{4.44 f \Phi_m}$$

$$= \frac{120 \text{ V}}{4.44 (60 \text{ Hz})(5 \times 10^{-3} \text{ Wb})} = 90 \text{ turns}$$

or simply,

$$N_2 = \frac{N_1}{a}$$

$$= \frac{180}{2} = 90 \text{ turns}$$

4.4.1 Dot Convention in Transformers

Notice that the primary and secondary voltages that are shown in Figure 4.6a have the same polarities. The dots near the upper end of each winding are known as the *polarity marks.*

Such dots point out that the upper or marked terminals have the same polarities, at a given instant of time when current enters the primary terminal and leaves the secondary terminal.

In other words, the **dot convention** implies that (1) *currents entering at the dotted terminals will result in mmfs that will produce fluxes in the same direction,* and (2) *voltages from the dotted to undotted terminals have the same sign.**

Therefore, in Figure 4.6a, since the current i_1 flows into the dotted end of the primary winding and the current i_2 flows out of the dotted end of the secondary winding, the mmfs will be subtracted from each other. Thus, it can be said that the transformer has a **subtractive polarity**. Here, current i_2 is flowing in the direction of the induced current, according to Lenz's law.

As shown in Figure 4.7a, for single-phase transformer windings, the terminals on the high-voltage side are labeled H_1 and H_2, while those on the low-voltage side are identified as X_1 and X_2. The terminal with subscript 1 in this convention (known as the *standard method of marking transformer terminals*) is equivalent to the dotted terminal in the dot-polarity notation. In a transformer where H_1 and X_1 terminals are adjacent, as shown in Figure 4.7a, the transformer is said to have *subtractive polarity*. If terminals H_1 and X_1 are diagonally opposite, the transformer is said to

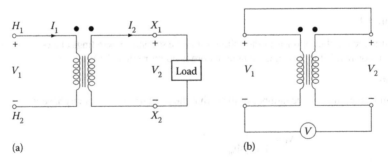

(a)   (b)

FIGURE 4.7 Polarity determination: (a) polarity markings of a single-phase two-winding transformer and (b) polarity test.

---

* Once a dot is assigned arbitrarily to a terminal of a given coil, the dotted terminals of all other coils coupled to it are found by Lenz's law, and therefore cannot be chosen at random. Sometimes, the identical terminals are marked by the ± sign instead of the dot sign.

have **additive polarity**.* Note that having the polarity markings in both dot convention as well as standard marking is really an unnecessary duplication.

Transformer polarities can be found by performing a simple test in which two adjacent terminals of high- and low-voltage windings are connected together and a small voltage is applied to the high-voltage winding, as shown in Figure 4.7b. Then the voltage between the high- and low-voltage winding terminals that are not connected together are measured. *The polarity is **subtractive** if the voltage V reading is less than the voltage V_1 which is applied to the high voltage winding. The polarity is **additive**† if the voltage V reading is greater than the applied voltage V_1.*

4.4.2 Impedance Transfer through a Transformer

Consider Figure 4.8a which shows an ideal transformer with a load impedance Z_L (of an apparatus or a circuit element) connected to its secondary terminals. Assume that all variables involved are given in phasors. Therefore, the impedance Z_L is defined as the ratio of the phasor voltage across it to the phasor current flowing through it. Hence,

$$Z_L = \frac{V_2}{I_2}$$

(4.28)

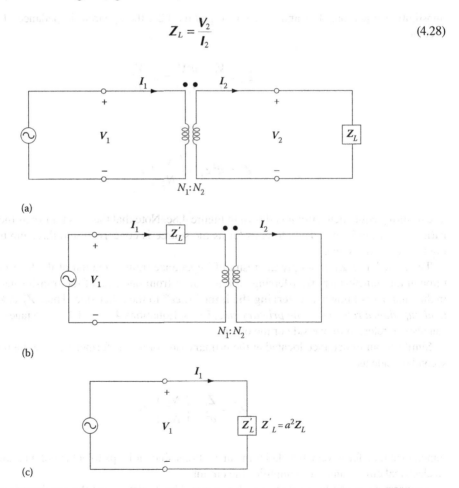

(a)

(b)

(c)

FIGURE 4.8 Illustration of impedance transfer across an ideal transformer: (a) an ideal transformer with a load impedance, (b) after the transfer of the impedance to the source side, and (c) the resultant equivalent circuit.

---

* Polarities result from the relative directions in which the two windings are wound on the core.

† According to the ANSI, additive polarities are required in large (greater than 200 kVA) high-voltage (higher than 8660 V) power transformers. To reduce voltage stress between adjacent leads, small transformers have subtractive polarities. For further information, see Gönen (2008).

Here, not only the voltages V_1 and V_2 are in phase, but also the currents I_1 and I_2. As shown in Figure 4.8b, the apparent impedance of the primary circuit of the transformer is

$$Z'_L = \frac{V_1}{I_1} \tag{4.29}$$

where the primary voltage and current, respectively, are

$$V_1 = aV_2 \tag{4.18}$$

$$I_1 = \frac{I_2}{a} \tag{4.24}$$

Substituting Equations 4.18 and 4.24 into Equation 4.29, the apparent impedance of the primary becomes

$$Z'_L = \frac{V_1}{I_1} = \frac{a^2 V_2}{I_2/a} = a^2 \frac{V_2}{I_2} \tag{4.30}$$

or

$$Z'_L = a^2 Z_L = \left(\frac{N_1}{N_2}\right)^2 Z_L \tag{4.31}$$

The resulting equivalent circuit is shown in Figure 4.8c. Note that the a^2 is known as the **impedance ratio** of the transformer. Therefore, as far as the source is concerned, the three circuits shown in Figure 4.8 are the same.

The impedance Z'_L is simply the result of impedance transformation of the load impedance Z_L through the transformer. Transferring an impedance from one side of the transformer to the other in this manner is known as **referring the impedance*** to the other side. Thus, Z'_L is known as the *load impedance referred to the primary side*. Using Equations 4.18 and 4.24, voltages and currents can also be referred to one side or the other.

Similarly, an impedance located at the primary side of a transformer can also be referred to the secondary side as

$$Z'_L = \frac{Z_1}{a^2} = \left(\frac{N_2}{N_1}\right)^2 Z_1 \tag{4.32}$$

Impedance transfer is very beneficial in calculations since it helps to get rid of a coupled circuit in an electrical circuit and thus simplifies the circuit.

Furthermore, it can be used in *impedance matching* to determine the *maximum power transfer* from a source with an internal impedance Z_s to a load impedance Z_L. Here, it is necessary to select the turns ratio so that

---

* It s also known as *reflecting, transferring, or scaling* the impedance.

$$\mathbf{Z}'_L = \left(\frac{N_1}{N_2}\right)^2 Z_L = a^2 Z_L = Z_s \tag{4.33}$$

For maximum power transfer when Z_L may be complex,

$$\mathbf{Z}'_L = \left(\frac{N_1}{N_2}\right)^2 Z_L = a^2 Z_L = Z_s \tag{4.34}$$

where $\mathbf{Z}^*_s$ is the conjugate of Z_s.

4.4.3 RELATIONSHIP BETWEEN INPUT AND OUTPUT POWERS OF AN IDEAL TRANSFORMER

The input power provided to a transformer by its primary circuit is

$$P_{in} = V_1 I_1 \cos\theta_1 \tag{4.35}$$

where θ_1 is the angle between the primary voltage and the primary current. The power output of a transformer through its secondary circuit to its load is

$$P_{out} = V_2 I_2 \cos\theta_2 \tag{4.36}$$

where θ_2 is the angle between the secondary voltage and the secondary current. In an ideal transformer,

$$\theta_1 = \theta_2 = 0$$

Therefore, the same power factor is seen by both the primary and secondary windings. Also, since $V_2 = V_1/a$ and $I_2 = aI_1$, substituting them into Equation 4.36,

$$P_{out} = V_2 I_2 \cos\theta = \left(\frac{V_1}{a}\right)(aI_1)\cos\theta$$

or

$$P_{out} = V_1 I_1 \cos\theta = P_{in} \tag{4.37}$$

Therefore, in an ideal transformer, the output power is equal to its input power. This makes sense because, by definition, an ideal transformer has no internal power losses. One can extend the same argument to reactive and apparent powers. Therefore,

$$Q_{in} = V_1 I_1 \sin\theta = V_2 I_2 \sin\theta = Q_{out} \tag{4.38}$$

$$S_{in} = V_1 I_1 = V_2 I_2 = S_{out} \tag{4.39}$$

Example 4.2

Assume that a 60 Hz, 250 kVA, 2400/240 V distribution transformer is an ideal transformer and determine the following:

(a) Its turns ratio.
(b) The value of load current (i.e., I_2), if a load impedance connected to its secondary (i.e., low-voltage side) terminals makes the transformer fully loaded.
(c) The value of the primary-side (i.e., high-voltage side) current.
(d) The value of the load impedance referred to the primary side of the transformer.

Solution

(a) The turns ratio of the transformer is

$$a = \frac{N_1}{N_2} = \frac{V_1}{V_2} = \frac{2,400 \text{ V}}{240 \text{ V}} = 10$$

(b) Since the transformer is an ideal transformer, it has no losses.

$$S = V_1 I_1 = V_2 I_2$$

from which

$$I_2 = \frac{S}{V_2} = \frac{250,000 \text{ VA}}{240 \text{ V}} = 1,041.67 \text{ A}$$

(c) The corresponding primary current

$$I_1 = \frac{I_2}{a} = \frac{1,041.67 \text{ A}}{10} = 104.167 \text{ A}$$

or

$$I_1 = \frac{S}{V_1} = \frac{250,000 \text{ VA}}{2,400 \text{ V}} = 104.167 \text{ A}$$

(d) The value of the load impedance at the secondary side is

$$Z_L = \frac{V_2}{I_2} = \frac{240 \text{ V}}{1,041.67 \text{ A}} = 0.2304 \ \Omega$$

Thus, its value referred to the primary side is

$$Z_L' = \left(\frac{N_1}{N_2}\right)^2 Z_L = a^2 Z_L = 10^2 \times 0.2304 = 23.04 \ \Omega$$

Example 4.3

A single-phase, 60 Hz transformer is supplying power to a load of $3 + j5\,\Omega$ through a short power line with an impedance of $0.2 + j0.6\,\Omega$, as shown in Figure 4.9a. The voltage at the generator bus (i.e., bus 1) is $277\angle 0°$ V. Determine the following:

(a) If the current at bus 1 is equal to the current at bus 2 (i.e., $I_{line} = I_{load}$), find the voltage at the load bus and the power losses that take place in the power line.
(b) If two ideal transformers T_1 and T_2 are inserted at the beginning and end of the line, as shown in Figure 4.9b, find the voltage at the load bus and the power losses in the line.

Solution

(a) Figure 4.10 shows the line-to-neutral diagram of the one-line diagram of the given system. Using the generator terminal voltage as the reference phasor, the line current can be found as

$$I_{line} = \frac{V_G}{Z_{line} + Z_{load}} = \frac{277\angle 0°\ \text{V}}{3.2 + j5.6\ \Omega} = 42.9471\angle -60.26°\ \text{A}$$

Since $I_{line} = I_{load}$, the voltage at the load bus (i.e., bus 2) is

$$V_{load} = I_{load} \times Z_{load} = (42.9471\angle -60.26°\ \text{A})(3 + j5\ \Omega) = 250.4245\angle -1.22°\ \text{V}$$

and the line losses (i.e., the copper losses) are

$$P_{line\ loss} = I_{line}^2 R_{line} = (42.9471\ \text{A})^2(0.2\ \Omega) = 368.8901\ \text{W}$$

(b) Figure 4.9b shows the one-line diagram of the system with the step-up transformer T_1 (with a turns ratio* of $a_1 = 1/10$) and the step-down transformer T_2 (with a turns ratio of $a_2 = 10/1$). The load impedance referred to the power-line side of transformer T_2 is

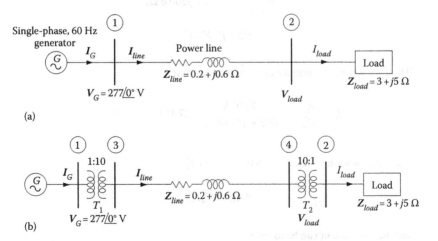

(a)

(b)

FIGURE 4.9 One-line diagram of the power system given in Example 4.3: (a) without the step-up and step-down transformers and (b) with the transformers.

---

* The given numbers in 10:1 and 1:10 simply represent the turns ratios, respectively, rather than representing the actual number of turns in each winding.

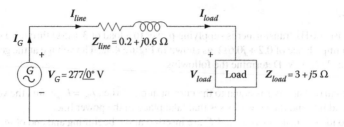

FIGURE 4.10   The line-to-neutral diagram of the system given in Figure 4.9a.

$$\mathbf{Z}'_{load} = a_2^2 \mathbf{Z}_{load} = \left(\frac{N_1}{N_2}\right)^2 \mathbf{Z}_{load} = \left(\frac{10}{1}\right)^2 (3 + j5\ \Omega) = 300 + j500\ \Omega$$

The resulting equivalent impedance is

$$\mathbf{Z}_{eq} = \mathbf{Z}_{line} + \mathbf{Z}_{load}$$

$$= (0.2 + j0.6\ \Omega) + (300 + j500\ \Omega)$$

$$= 300.2 + j500.6\ \Omega$$

Referring this $\mathbf{Z}_{eq}$ to the generator side of transformer T_1, the new equivalent impedance is found as

$$\mathbf{Z}'_{eq} = a_2^2 \mathbf{Z}_{eq}$$

$$= \left(\frac{N_1}{N_2}\right)^2 \mathbf{Z}_{eq}$$

$$= \left(\frac{1}{10}\right)^2 (300.2 + j500.6\ \Omega)$$

$$= 3.002 + j5.006\ \Omega$$

Thus, the generator current can be calculated from

$$\mathbf{I}_G = \frac{\mathbf{V}_G}{\mathbf{Z}'_{eq}} = \frac{277\angle 0°\ \text{V}}{3.002 + j5.006\ \Omega} = 47.4549\angle -59.05\ \text{A}$$

Therefore, working back through transformer T_1,

$$N_1 \mathbf{I}_G = N_2 \mathbf{I}_{line}$$

and the line current can be found as

$$\mathbf{I}_{line} = \frac{N_1}{N_2}\mathbf{I}_G = \frac{1}{10}(47.4549\angle -59.05°\ \text{A}) = 4.7455\angle -59.05°\ \text{A}$$

Similarly, working back through transformer T_2,

$$N_{line}I_{line} = N_2 I_{load}$$

and the load current can be found as

$$I_{load} = \frac{N_2}{N_1} I_{load} = \frac{10}{1}(4.7455\angle - 59.05° \text{ A}) = 47.455\angle - 59.05° \text{ A}$$

Hence, the voltage at the load bus is

$$V_{load} = I_{load}Z_{load} = (47.455\angle - 59.05° \text{ A})(3 + j5 \text{ } \Omega) = 276.7093\angle - 0.01° \text{ V}$$

The line losses are

$$P_{line\ loss} = I_{line}^2 R_{line} = (4.7455 \text{ A})^2(0.2 \text{ } \Omega) = 4.5039 \text{ W}$$

Notice that the percent reduction in the line losses, after adding the step-up and the step-down transformers is

$$\text{Reduction in } P_{line\ loss} = \frac{368.8901 \text{ W} - 4.5039 \text{ W}}{368.8901 \text{ W}} \times 100 = 98.78\%$$

Example 4.4

Assume that the impedances of the transformers T_1 and T_2, given in Part (b) of Example 4.3, are not small enough to ignore, and that they are $Z_{T_1} = 0 + j0.15 \text{ } \Omega$ and $Z_{T_2} = 0 + j0.15 \text{ } \Omega$, respectively. Also assume that they are referred to the high-voltage sides of each transformer, respectively. Solve Part (b) of Example 4.3, accordingly.

Solution

Figure 4.11a shows the one-line diagram of the given system that includes T_1 the step-up and T_2 the step-down transformer. Figure 4.11b shows the line-to-neutral diagram of the same system. Figure 4.11c shows the load impedance referred to power-line side of transformer T_2 as well as the impedance of the transformer (which is given as already referred to its high-voltage side). Therefore, the resulting equivalent impedance can be found as

$$Z_{eq} = Z_{T_1} + Z_{line} + Z_{T_2} + Z'_{load}$$
$$= Z_{T_1} + Z_{line} + Z_{T_2} + a_2^2 Z_{load}$$
$$= j0.15 + (0.2 + j0.6) + j0.15 + 10^2(3 + j5)$$
$$= 300.2 + j500.9 \text{ } \Omega$$

Referring this Z_{eq} to the generator side of transformer T_1, as shown in Figure 4.11d, the new equivalent impedance is found as

$$Z'_{eq} = a_1^2 Z_{eq} = \left(\frac{1}{10}\right)^2 (300.2 + j500.9) = 3.002 + j5.009 \text{ } \Omega$$

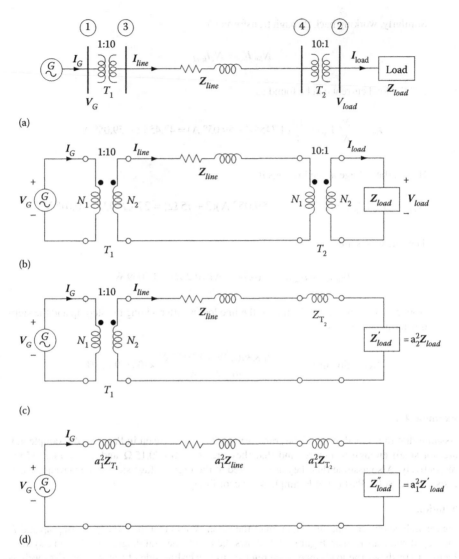

FIGURE 4.11   One-line diagram of the power system given in Example 4.4.

Therefore, the generator current can be found from

$$I_G = \frac{V_G}{Z'_{eq}} = \frac{277\angle 0° \text{ V}}{3.002 + j5.009 \text{ } \Omega} = 47.4339\angle -59.06° \text{ A}$$

Thus, working back through transformer T_1,

$$I_{line} = \frac{N_1}{N_2} I_G = \frac{1}{10}(47.4339\angle -59.06°) = 4.74339\angle -59.06° \text{ A}$$

Similarly, working back through transformer T_2,

$$I_{load} = \frac{N_1}{N_2} I_{line} = \left(\frac{10}{1}\right)(4.74339\angle -59.06°) = 47.4339\angle -59.06° \text{ A}$$

Hence, the voltage at the load bus is

$$V_{load} = I_{load}Z_{load} = (47.4339\angle -59.06°)(3+j5) = 276.5873\angle -0.02°$$

The line losses are

$$P_{line\ loss} = I_{line}^2 R_{line} = (4.74339)^2(0.2) = 4.4999\ \text{W}$$

The percent reduction in line losses, after adding the step-up and the step-down transformer, is

$$\text{Reduction in } P_{line\ loss} = \frac{368.8901\ \text{W} - 4.4999\ \text{W}}{368.8901\ \text{W}} \times 100 = 98.78\%$$

4.5 REAL TRANSFORMER

A real transformer differs from an ideal transformer in many respects. For example, as illustrated in Figure 4.12a, (1) the primary and secondary winding resistances R_1 and R_2 are not negligible, (2) the leakage fluxes and Φ_{ℓ_1} and Φ_{ℓ_2} exist, (3) the core losses are not negligible, (4) the permeability of the core material is not infinite and therefore a considerable mmf is required to establish mutual flux Φ_m in the core, and (5) the core material saturates. The resulting representation of this transformer is shown in Figure 4.12b. Here, X_{ℓ_1} and X_{ℓ_2} are the leakage fluxes, respectively. Therefore,

$$X_{\ell_1} = \omega L_{\ell_1}$$
$$= \omega N_1^2 \mathcal{P}_{\ell_1} \tag{4.40}$$

$$X_{\ell_2} = \omega L_{\ell_2}$$
$$= \omega N_2^2 \mathcal{P}_{\ell_2} \tag{4.41}$$

where
$\omega = 2\pi f$
L_{ℓ_2} is the leakage inductance of the primary winding

$$L_{\ell_1} = N_1^2 \mathcal{P}_{\ell_1}$$
$$= \frac{N_1 \Phi_{\ell_1}}{I_1} \tag{4.42}$$

L_{ℓ_1} is the leakage inductance of the secondary winding

$$L_{\ell_2} = N_2^2 \mathcal{P}_{\ell_2}$$
$$= \frac{N_2 \Phi_{\ell_2}}{I_2} \tag{4.43}$$

$\mathcal{P}_{\ell_1}$ is the permeance of the leakage flux path of the primary winding
$\mathcal{P}_{\ell_2}$ is the permeance of the leakage flux path of the secondary winding

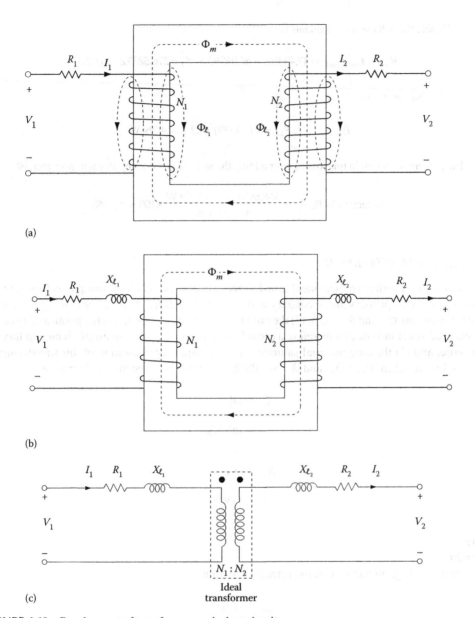

(a)

(b)

(c)

FIGURE 4.12 Development of transformer-equivalent circuits.

In such a representation, the transformer windings are tightly coupled by a mutual flux, and represented as shown in Figure 4.12c.

As illustrated in Figure 4.13a, the primary current I_1 must be large enough to compensate the demagnetizing effect of the load current (i.e., the secondary current), but also provide for adequate mmf to develop the resultant mutual flux. Note that I_2' is the load component in the primary and can be expressed as

$$I_2' = \frac{N_2}{N_1} I_2 = \frac{I_2}{a} \tag{4.44}$$

In other words, I_2' is the secondary current referred to the primary, as it is in the ideal transformer. Therefore, the primary current can be expressed in terms of phasor summation as

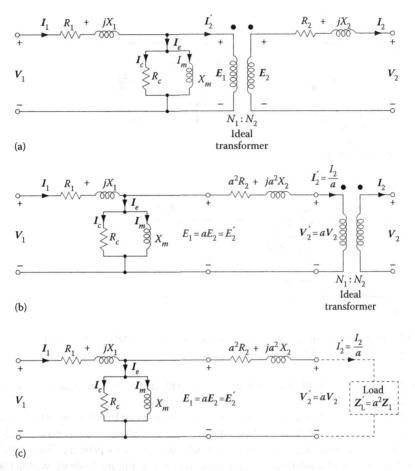

FIGURE 4.13 Transformer-equivalent circuits: (a) the real transformer, (b) referred to the primary, and (c) referred to the primary (without showing the ideal transformer).

$$I_1 = I_2' + I_e = \frac{I_2}{a} + I_e \tag{4.45}$$

where I_e is the excitation current (i.e., the additional primary current) needed to develop the resultant mutual flux. Such excitation current
I_e is nonsinusoidal and can be expressed as

$$I_e = I_c + I_m \tag{4.46}$$

where

    I_c is the core-loss component of the excitation current supplying the hysteresis and eddy-current losses in the core
    I_m is the magnetizing component of the excitation current needed to magnetize the core

Here, I_c is in phase with the counter-emf E_1 and I_m lags E_1 by 90°. Therefore, the core-loss component and the magnetizing component are modeled by a resistance R_c and an inductance X_m, respectively, that are connected across the primary voltage source.

Even though *Faraday's law* dictates the core flux to be very nearly sinusoidal for a sinusoidal terminal voltage, saturation causes the excitation current to be very nonsinusoidal. Thus, in reality,

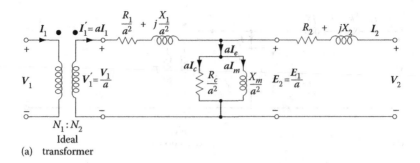

(a)

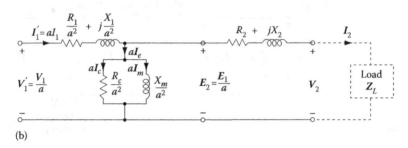

(b)

FIGURE 4.14 Transformer-equivalent circuits: (a) referred to the secondary and (b) referred to the secondary (without showing the ideal transformer).

both I_c and I_m are actually nonlinear, and hence the resistance R_c and the reactance X_m are at best approximations.

The ideal transformer* shown in Figure 4.13a can be eliminated by referring all secondary quantities to the primary. Figure 4.13b and c shows the first and second steps of this process. Figure 4.13c shows the equivalent circuit of a transformer (model) referred to its primary.[†] Figure 4.14a shows the equivalent circuit of a transformer referred to its secondary. Figure 4.14b shows the transformer equivalent circuit referred to its secondary without showing the ideal transformer.

4.6 APPROXIMATE EQUIVALENT CIRCUIT OF A REAL TRANSFORMER

The transformer equivalent circuits developed in the previous section (and shown in Figures 4.13c and 4.14b) are often more accurate than is necessary in practice. This is especially true in power system applications.

The excitation branch has a very small current in comparison to the load current of the transformer. Of course, such a small excitation current, I_e, causes a negligibly small voltage drop in the primary winding impedance $(R_1 + jX_1)$. Therefore, by moving the excitation admittance (i.e., the shunt branch) from the middle of the T-circuit to either the left (as shown in Figure 4.15a) or the right, the primary and secondary impedances are left in series with each other so that they can be added together as shown in Figure 4.15b.

---

[*] Such a representation is known as the *Steinmetz circuit model of* a transformer. Steinmetz had the brilliant idea *of* separating the linear phenomenon by which leakage-flux voltage is induced, from the nonlinear phenomenon by which mutual-flux voltage is induced in an iron-core transformer. His approach, based on linear circuit theory, provided an easy solution for developing a circuit model for an iron-core transformer.

[†] The physical meaning of referring secondary quantities to the primary implies that the real and reactive powers in an impedance Z_L through which the secondary current I_2 flows is the same when the primary current I_1 flows through an equivalent impedance Z'_L. Therefore, there cannot be any difference in the performance of a transformer determined from an equivalent circuit referred to the primary or the secondary.

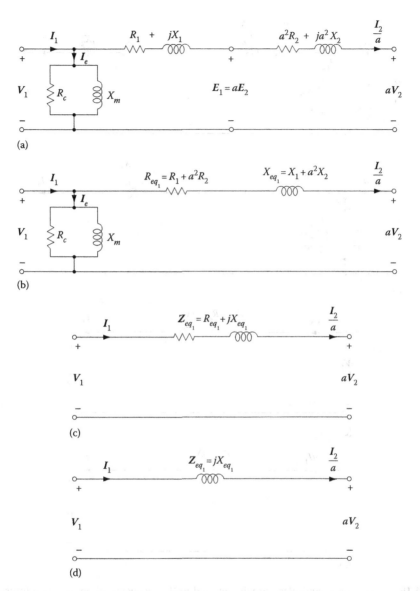

FIGURE 4.15 Approximate equivalent circuits referred to the primary of an iron-core transformer: (a) referred to the primary, (b) referred to the primary (collecting R's and X's together), (c) after the elimination of the shunt branches, and (d) after neglecting the resistance involved.

The equivalent impedance of such an approximate equivalent circuit of the transformer depends on whether its equivalent circuit is referred to the primary or secondary. As shown in Figure 4.15b, if the equivalent impedance is referred to the primary,

$$\boldsymbol{Z}_{eq_1} = R_{eq_1} + jX_{eq_1} \tag{4.47}$$

where

$$R_{eq} = R_1 + a^2 R_2 \tag{4.48}$$

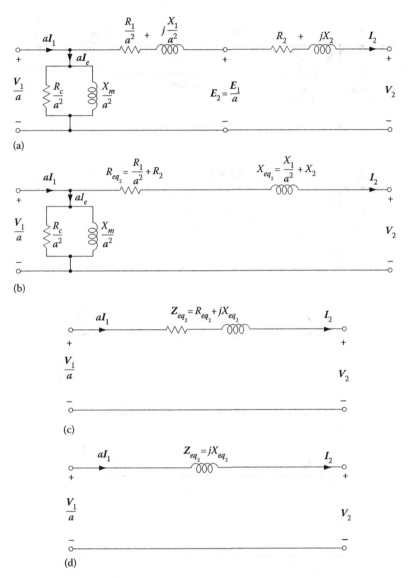

FIGURE 4.16 Approximate equivalent circuits referred to the secondary of an iron-core transformer: (a) referred to the secondary, (b) referred to the secondary (collecting R's and X's together), (c) after the elimination of the shunt branches, and (d) after neglecting the resistance involved.

$$X_{eq_1} = X_1 + a^2 X_2 \tag{4.49}$$

Here, the R_{eq_1} and X_{eq_1} are the equivalent resistance and reactance referred to the primary, respectively. As shown in Figure 4.16b, if the equivalent impedance is referred to the secondary,

$$Z_{eq_2} = R_{eq_2} + jX_{eq_2} \tag{4.50}$$

where

$$R_{eq_2} = \frac{R_1}{a^2} + R_2 \tag{4.51}$$

$$X_{eq_2} = \frac{X_1}{a^2} + X_2 \tag{4.52}$$

Here, the terms R_{eq_2} and X_{eq_2} represent the equivalent resistance and reactance values referred to the secondary, respectively. It is interesting to notice that

$$\frac{Z_{eq_1}}{Z_{eq_2}} = \frac{R_{eq_1}}{R_{eq_2}} = \frac{X_{eq_1}}{X_{eq_2}} = a^2 \tag{4.53}$$

A further approximation of the equivalent circuit can be made by removing the excitation branch, as shown in Figures 4.15c and 4.16c. The resultant error is very small since the excitation current I_e is very small in comparison to the rated current of the transformer.

Furthermore, in power transformers, the equivalent resistance R_{eq} *is* small in comparison to the equivalent reactance X_{eq}. Therefore, the transformer can only be represented by its equivalent reactance X_{eq}, as shown in Figures 4.15d and 4.16d. Thus, the corresponding equivalent impedances can be expressed as

$$Z_{eq_1} = jX_{eq_1} \tag{4.54}$$

$$Z_{eq_2} = jX_{eq_2} \tag{4.55}$$

Example 4.5

Consider a 75 kVA, 2400/240 V, 60 Hz distribution transformer with $Z_1 = 0.612 + j1.2\,\Omega$ and $Z_2 = 0.0061 + j0.0115\,\Omega$ for its high-voltage and low-voltage windings, respectively. Its excitation admittance referred to the 240 V side is $Y_{e_2} = 0.0191 - j0.0852\,\Omega$. The transformer delivers its rated I_L at 0.9 lagging PF.

(a) Draw the equivalent circuit with the excitation admittance referred to the primary side.
(b) Find the emfs of E_1 and E_2 induced by the equivalent mutual flux, I_e, I_1 at 0.9 lagging PF, and the applied V_1 when the transformer delivers a rated load at rated V_2 voltage.

Solution

(a) Figure 4.17 shows the equivalent circuit of the transformer, with the excitation admittance referred to the primary side. Therefore,

$$Y_{e_1} = \frac{Y_{e_2}}{a^2} = \frac{0.0191 - j0.0852\text{ S}}{10^2} = 1.91 \times 10^{-4} - j8.52 \times 10^{-4}\text{ S}$$

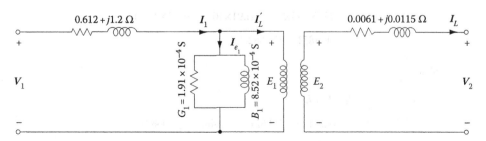

FIGURE 4.17 Circuit for Example 4.16.

(b) Here,

$$I_L = I_2 = \frac{S}{V_2} = \frac{75,000 \text{ VA}}{240 \text{ V}} = 312.5 \text{ A}$$

Let V_2 be the reference phasor so that

$$V_2 = 240\angle 0° \text{ V}$$

Thus,

$$I_L = I_2 = 312.5(0.9 - j0.4359) = 281.25 - j136.2156 \text{ A}$$

Hence,

$$E_2 = V_2 + I_L(R_2 + jX_2) = 240\angle 0° + (281.25 - j136.2156)(0.0061 + j0.0115) = 243.294\angle 0.57° \text{V}$$

so that,

$$E_1 = aE_2 = 10\left[240\angle 0° + (281.25 - j136.2156)(0.0061 + j0.0115) \right]$$

$$= (10)[243.294\angle 0.57° \text{ V}]$$

$$= 2432.94\angle 0.57° \text{ V}$$

Therefore, the load current referred to the primary side is

$$I'_L = \frac{I_L}{a}$$

$$= \frac{281.25 - j136.2156 \text{ A}}{10}$$

$$= 28.125 - j13.6216 \text{ A}$$

Notice that

$$I_1 = I'_L + I_{e_1}$$

where I_{e_1} produces the mutual flux in the core. Since

$$I_{e_1} = E_1 Y_{e_1}$$

$$= (2432.94\angle 0.57°)(1.91 \times 10^{-4} - j8.52 \times 10^{-4})$$

$$= 0.4851 - j2.0682 \text{ A}$$

also

$$I_1 = I'_L + I_{e_1}$$

$$= (28.125 - j13.6216) + (0.4851 - j2.0682)$$

$$= 32.6299\angle -28.74° \text{ A}$$

Therefore,

$$V_1 = E_1 + I_1(R_1 + jX_1)$$

$$= 2432.94\angle 0.57° + (32.6299\angle - 28.74°)(0.612 + j1.2)$$

$$= 2469.6396\angle 1.13° \text{ V}$$

4.7 DETERMINATION OF EQUIVALENT-CIRCUIT PARAMETERS

The equivalent circuits of a given transformer can be used to predict and evaluate its performance. If the complete design data of a transformer are available (such data are usually available only to its designer), the necessary parameters can be computed from the dimensions and properties of the materials used.

However, once the transformer is manufactured, it may be desirable to verify the accuracy of the performance predictions. This can be achieved by means of two tests designed to determine the parameters of the equivalent circuit. These two tests* are known as the *open-circuit test* and the *short-circuit test*.

4.7.1 OPEN-CIRCUIT TEST

The purpose of the open-circuit test[†] is to determine the excitation admittance of the transformer-equivalent circuit, the no-load loss, the no-load excitation current, and the no-load power factor. Such an open-circuit test is performed by applying rated voltage to one of the windings, with the other winding (or windings) open-circuited.

The input power, current, and voltage are measured, as shown in Figure 4.18a. (*However, for reasons of safety and convenience, usually the high-voltage winding is open-circuited and the test is conducted by placing the instruments on the low-voltage side of the transformer.*)

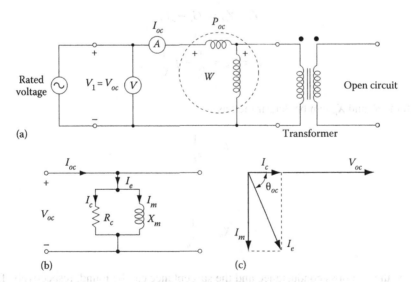

FIGURE 4.18 Open-circuit test: (a) wiring diagram for the open-circuit test, (b) equivalent circuit, and (c) no-load phasor diagram.

---

* Each test can be done by exciting either winding. However, in large transformers with high levels of both voltage and current, it may be a good idea to excite the low-voltage winding for the open-circuit test and to excite the high-voltage winding for the short-circuit test.

[†] It is also known as the *core-loss test*, the *iron-loss test*, the *no-load test*, the *excitation test*, or the *magnetization test*.

Once such information is collected, one can determine the magnitude and the angle of excitation impedance after finding the open-circuit (i.e., no-load) power factor. Here, the voltage drop in the leakage impedance of the winding (which is excited in the open-circuit test) caused by the normally small excitation current is usually ignored.

This results in an approximate equivalent circuit, as shown in Figure 4.18b. Also ignored is the (primary) power loss due to the excitation current. Therefore, the excitation admittance can be expressed as

$$Y_e = Y_{oc} = \frac{I_{oc}}{V_{oc}} \angle -\theta_{oc} \tag{4.56}$$

where θ_{oc} is the angle of the admittance found from the open-circuit power factor PF_{oc} as

$$PF_{oc} = \cos\theta_{oc} = \frac{P_{oc}}{V_{oc}I_{oc}} \tag{4.57}$$

so that

$$\theta_{oc} = \cos^{-1}\frac{P_{oc}}{V_{oc}I_{oc}} \tag{4.58}$$

or

$$\theta_{oc} = \cos^{-1} PF_{oc} \tag{4.59}$$

For a given transformer, the P_{oc} is always lagging. For this reason, there is a negative sign in front of θ_{oc} in Equation 4.56. If the excitation admittance is expressed in rectangular coordinates,

$$Y_e = Y_{oc} = G_c - jB_m \tag{4.60}$$

$$Y_e = Y_{oc} = \frac{1}{R_c} - j\frac{1}{X_m} \tag{4.61}$$

from which the R_c and X_m can be determined as

$$R_c = \frac{1}{G_c} \tag{4.62}$$

and

$$X_m = \frac{1}{B_m} \tag{4.63}$$

Alternatively, the core-loss conductance and the susceptance can be found, respectively, from

$$G_c \cong G_{oc} = \frac{P_{oc}}{V_{oc}^2} \tag{4.64}$$

$$B_m \cong B_{oc} = \left(Y_{oc}^2 - G_{oc}^2\right)^{1/2} \tag{4.65}$$

The values of R_c and X_m can be determined from Equations 4.62 and 4.63, respectively, as before. The no-load phasor diagram can be drawn as shown in Figure 4.18c.

4.7.2 SHORT-CIRCUIT TEST

The purpose of the short-circuit test* is to determine the equivalent resistance and reactance of the transformer under rated conditions. This test is performed by short-circuiting one winding (usually the low-voltage winding) and applying a *reduced* voltage to the other winding, as shown in Figure 4.19a.

The reduced input voltage is adjusted until the current in the shorted winding is equal to its rated value. The input voltage, current, and power are measured as before. The applied voltage V_{sc} is only a small percentage of the rated voltage and is sufficient to circulate rated current in the windings of the transformer. Usually, this voltage is about 2%–12% of the rated voltage.

Therefore, the excitation current is small enough to be ignored. If it is neglected, then one can assume that all the voltage drop will take place in the transformer and is due to the series elements in the circuit, as shown in Figure 4.19b. The shunt branch representing excitation admittance does not appear in this equivalent circuit. The series impedance Z_{sc} can be found from

$$Z_{sc} = Z_{eq_1} = \frac{V_{sc}}{I_{sc}} = \frac{V_{sc}\angle 0°}{I_{sc}\angle -\theta_{sc}}$$

(4.66)

The short-circuit power factor is lagging and determined from

$$PF_{sc} = \cos\theta_{sc} = \frac{P_{sc}}{V_{sc}I_{sc}}$$

(4.67)

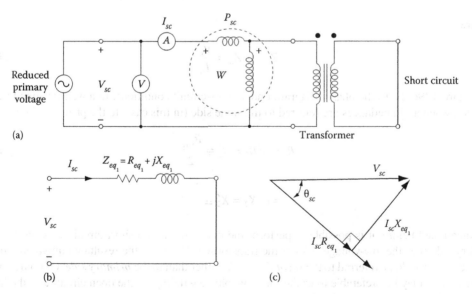

FIGURE 4.19 Short-circuit test: (a) wiring diagram for the short-circuit test, (b) equivalent circuit, and (c) phasor diagram.

---

* It is also known as the *impedance* test or the *copper-loss test*. Since the voltage applied under short-circuit conditions is small, the core losses are ignored and the wattmeter reading represents the copper losses in the windings. Therefore, for all practical purposes $P_{cu} = P_{sc}$.

so that

$$\theta_{sc} = \cos^{-1} \frac{P_{sc}}{V_{sc}I_{sc}} \tag{4.68}$$

$$\theta_{sc} = \cos^{-1} PF_{sc} \tag{4.69}$$

For a given transformer, the PF_{sc} is *always* lagging. For this reason, there is a negative sign in front of θ_{sc} in Equation 4.66. Equation 4.66 can be expressed as

$$Z_{eq_1} = \frac{V_{sc}}{I_{sc}} \angle \theta_{sc} \tag{4.70}$$

$$= R_{eq_1} + jX_{eq_1}$$

where

$$R_{eq_1} = R_1 + a^2 R_2 \tag{4.71}$$

$$X_{eq_1} = X_1 + a^2 X_2 \tag{4.72}$$

Alternatively, the equivalent circuit resistance and reactance can be found, respectively, from

$$R_{eq_1} \cong R_{sc} + \frac{P_{sc}}{I_{sc}^2} \tag{4.73}$$

$$X_{eq_1} = X_{sc} = \left(Z_{eq}^2 - R_{eq}^2 \right)^{1/2} \tag{4.74}$$

where

$$Z_{eq_1} \cong Z_{sc} = \frac{V_{sc}}{I_{sc}} \tag{4.75}$$

Figure 4.19c shows the phasor diagram under short-circuit conditions. In a well-designed transformer, when all impedances are referred to the same side (in this case, to the primary side),

$$R_1 = a^2 R_2 = R_2' \cong \frac{R_{eq_1}}{2} \tag{4.76}$$

$$X_1 = a^2 X_2 = X_2' \cong \frac{X_{eq_1}}{2} \tag{4.77}$$

As mentioned before, it is possible to perform the open-circuit and short-circuit tests on the secondary side (i.e., the low-voltage side) of the transformer. However, the resultant equivalent circuit impedances would be referred to the *secondary side* rather than to the *primary side*. Of course, with large units it may be preferable to excite the low-voltage winding on the open circuit and the high-voltage winding on the short circuit.

4.8 TRANSFORMER NAMEPLATE RATING

Among the information provided by the nameplate of a transformer are its apparent power (in terms of the kVA rating or the MVA rating), voltage ratings, and impedance.

For example, a typical transformer may have 25 WA, 2400/120 V. Here, the voltage ratings point out that the transformer has two windings: one rated for 2400 V and the other for 120 V. Since the voltage ratio also represents the turns ratio, the turns ratio of the transformer is

$$a = \frac{N_1}{N_2} = \frac{V_1}{V_2} = \frac{2400 \text{ V}}{120 \text{ V}} = 20$$

Also, the given 25 kVA rating indicates that each winding is designed to carry 25 kVA. Thus, the current rating for the high-voltage winding is 25,000 VA/2,400 V =10.42 A, but for the low-voltage winding is 25,000 VA/120 V = 208.33A.

When a current of 208.33 A flows through the secondary winding, there will be a current of 10.42 A in the primary winding, ignoring the additional small excitation current that flows through the primary winding.

The kVA rating always refers to the *output kVA* measured at the secondary (load) terminals. The input kVA will be slightly more due to the losses involved.

Transformer impedance is always provided on the nameplate in percentage. For example, 5% means 0.05 per unit based on its nameplate ratings. In terms of percentage or per unit, the given figure could be referred to the primary winding or secondary winding. Nevertheless, in either case, it would still be 5%.

Example 4.6

Consider a 15 kVA, 7500/480 V, 60 Hz distribution transformer. Assume that the open-circuit and short-circuit tests were performed on the primary side of the transformer and that the following data were obtained:

	Open-Circuit Test (on Primary)	Short-Circuit Test (on Primary)
Voltmeter	$V_{oc} = 7500$ V	$V_{sc} = 366$ V
Ammeter	$I_{oc} = 0.2006$ A	$I_{sc} = 2$ A
Wattmeter	$P_{oc} = 180$ W	$P_{sc} = 300$ W

Determine the impedance of the approximate equivalent circuit referred to the primary side, and draw the corresponding simplified equivalent circuit.

Solution

The power factor during the open-circuit test is

$$PF_{oc} = \cos\theta_{oc} = \frac{P_{oc}}{V_{oc}I_{oc}} = \frac{180 \text{ W}}{(7500 \text{ V})(0.2006 \text{ A})} = 0.1196 \text{ lagging}$$

The excitation admittance *is*

$$\mathbf{Y}_e = \mathbf{Y}_{oc} = \frac{I_{oc}}{V_{oc}} \angle - \cos^{-1} PF_{oc}$$

$$= \frac{0.2006 \text{ A}}{7500 \text{ V}} \angle - \cos^{-1}(0.1196)$$

$$= 0.0000267 \angle -83.129° \text{ S}$$

$$= 0.0000032 - j0000265 \text{ S}$$

$$= \frac{1}{R_c} - j\frac{1}{X_m}$$

Therefore,

$$R_c = \frac{1}{0.0000032}$$

$$= 312,500 \, \Omega$$

$$\cong 312.5 \, k\Omega$$

$$X_m = \frac{1}{0.0000265}$$

$$= 37,658.32 \, \Omega$$

$$\cong 37.3 \, k\Omega$$

The power factor during *the short-circuit test is*

$$PF_{sc} = \cos\theta_{sc} = \frac{P_{sc}}{V_{sc}I_{sc}}$$

$$= \frac{300 \, W}{(366 \, V)(2 \, A)}$$

$$= 0.41 \, \text{lagging}$$

The series (i.e., the equivalent) impedance *is*

$$Z_{eq_1} = Z_{sc} = \frac{V_{sc}}{I_{sc}} \angle \cos^{-1} PF_{sc}$$

$$= \frac{366 \, V}{2 \, A} \angle \cos^{-1} 0.41$$

$$= 183 \angle 65.81° \, \Omega$$

$$= 75 + j166.93 \, \Omega$$

Therefore, the equivalent resistance and reactance are

$$R_{eq_1} = 75 \, \Omega \quad \text{and} \quad X_{eq_1} = 166.9352$$

The corresponding simplified equivalent circuit is shown in Figure 4.20a.

However, if an equivalent T-circuit is needed, the values of individual primary and secondary resistances and leakage reactances, referred to the same side, are usually assumed to be equal. Therefore,

$$R_1 = a^2 R_2 = R_2'$$

$$\cong \frac{R_{eq_1}}{2}$$

$$= \frac{75 \, \Omega}{2}$$

$$= 37.5 \, \Omega$$

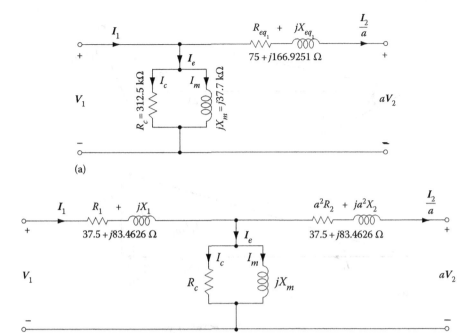

FIGURE 4.20   Simplified equivalent circuit: (a) referred to the primary side and (b) its equivalent t-circuit representation.

and

$$X_1 = a^2 X_2 = X'_2$$

$$\cong \frac{X_{eq_1}}{2}$$

$$= \frac{166.93\ \Omega}{2}$$

$$= 83.46\ \Omega$$

where

$$a = \frac{V_1}{V_2}$$

$$= \frac{7500\ \text{V}}{480\ \text{V}}$$

$$= 15.625$$

Therefore, as shown in Figure 4.21b,

$$R_2 = \frac{R'_2}{a^2}$$

$$= \frac{37.5\ \Omega}{15.625^2} \cong 0.154\ \Omega$$

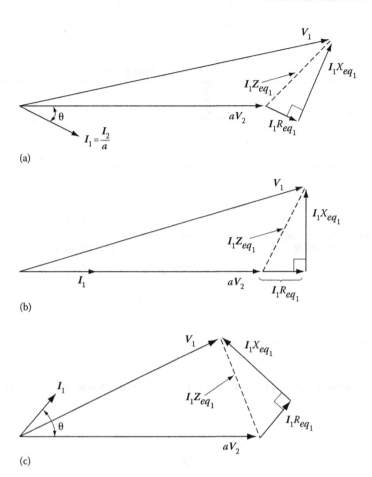

FIGURE 4.21 Phasor diagram of a transformer operating at (a) lagging power factor, (b) unity power factor, and (c) leading power. All quantities are referred to the primary side of the transformer.

and

$$X_2 = \frac{X_2'}{a^2}$$

$$= \frac{83.46\ \Omega}{15.625^2} \cong 0.342\ \Omega$$

Therefore, as shown in Figure 4.21b,

$$R_2 = \frac{R_2'}{a^2}$$

$$= \frac{37.5\ \Omega}{15.625^2}$$

$$= 0.154\ \Omega$$

and

$$X_2 = \frac{X_2'}{a^2}$$

$$= \frac{83.46\,\Omega}{15.63^2}$$

$$= 0.342\,\Omega$$

4.9 PERFORMANCE CHARACTERISTICS OF A TRANSFORMER

The main use of the equivalent circuit of a given transformer is to determine its *performance characteristics*, which are basically its voltage regulation and its efficiency.

4.9.1 VOLTAGE REGULATION OF A TRANSFORMER

The **voltage regulation** of a transformer is the change in the magnitude of secondary terminal voltage from no load to full load when the primary voltage is constant. It is usually expressed as a percentage of the full-load value* as

$$\%Voltage\ Regulation = \frac{V_{2(no\ load)} - V_{2(full\ load)}}{V_{2(full\ load)}} \times 100 \qquad (4.78)$$

Here, the full load is the rated load of the secondary.

At no load, the secondary terminal voltage may change from the rated voltage[†] value due to the effect of the impedance of the transformer. Also, since at no load,

$$V_{2\ (no\ load)} = \frac{V_1}{a}$$

then

$$\%Voltage\ regulation = \frac{\dfrac{V_1}{a} - V_{2(full\ load)}}{V_{2(full\ load)}} \times 100 \qquad (4.79)$$

The voltage regulation is affected by the magnitude and power factor of the load as well as by the internal impedance (i.e., the *leakage impedance*) of the transformer.

Even though in electric power engineering applications it is usually considered good practice to have a small voltage regulation, under certain circumstances transformers with high impedance and high-voltage regulation are used to decrease the fault currents in a circuit.

As shown in the phasor diagrams of Figure 4.21, depending on the power factor of the load, the voltage regulation can be positive, zero, or negative. Here, all circuit parameters are *referred to the primary side* of the transformer. At a lagging power factor, the voltage regulation is positive, as shown in Figure 4.21a. With certain exceptions, as mentioned before, it is usually good practice to minimize the voltage regulation. As shown in Figure 4.21a, the voltage regulation is *positive* at *lagging* power factors, whereas it is *negative* at *leading* power factors, as shown in Figure 4.21c.

---

* For further information, see Chapter 9 of Gönen (2008).
[†] It is also called the *nameplate voltage, nominal transformer voltage, or full-load voltage.*

This means that the secondary terminal voltage is greater under full load than under no load. Such a situation takes place when the power-factor-correction capacitor banks remain in the circuit while the load is low. (This causes a partial resonance between the capacitance of the load and the leakage inductance of the transformer.)

The solution is to adjust the capacitor sizes and/or to use some of them as *switchable* capacitors.* As can be seen from Figure 4.21a, the primary voltage can be expressed as

$$V_1 = aV_2 + I_1 Z_{eq_1} \tag{4.80}$$

$$V_1 = aV_2 + I_1 R_{eq_1} + jI_1 X_{eq_1} \tag{4.81}$$

If all circuit parameters are *referred to the secondary side* of the transformer, then the phasor diagrams corresponding to lagging, unity, and leading power factors are as shown in Figure 4.22. For example, as can be seen from Figure 4.22a, the primary voltage can be expressed as

$$\frac{V_1}{a} = V_2 + I_2 Z_{eq_2} \tag{4.82}$$

$$\frac{V_1}{a} = V_2 + I_2 R_{eq_2} + jI_2 X_{eq_2} \tag{4.83}$$

It is possible to use an approximate value for the primary voltage by taking into account only the horizontal components in the phasor diagram, as shown in Figure 4.23. Therefore, when all quantities are referred to the secondary side of the transformer, the approximate value of the primary voltage is

$$\frac{V_1}{a} = V_2 + I_2 R_{eq_2} \cos\theta + I_2 X_{eq_2} \sin\theta \tag{4.84}$$

Transformers used in power system applications are usually designed with taps on one winding in order to change its turns ratio over a small range. Such *tap changing* is frequently achieved automatically in large power transformers to maintain a reasonably constant secondary-side voltage[†] as the magnitude and power factor of the load connected to the secondary side terminals change.

Tap changing is also used to compensate for the deviations in primary-side voltage as a result of feeder impedance. In distribution transformers, however, the tap changing is normally done manually.

Example 4.7

A 75 kVA, 2400/240 V, 60 Hz distribution transformer has equivalent resistance and reactance of 0.009318 Ω and 0.058462 Ω, respectively, which are both referred to its secondary side. Use the exact equation for V_1 and determine the full-load voltage regulation:

---

[*] For further information, see Chapter 8 of Gönen (2008).

[†] Certain types of loads such as incandescent lamps and motors require rated voltage and frequency for their optimum operation. Otherwise, at voltages above the rated voltage, the lives of the incandescent lamps are shortened. Similarly, when providing a rated load at subnormal voltage, motors draw overcurrents, which in turn cause the motors to overheat. Therefore, such loads must be served by transformers that have a *small* voltage regulation. However, arc welding transformers require a *large* voltage regulation so that they can operate at almost constant current.

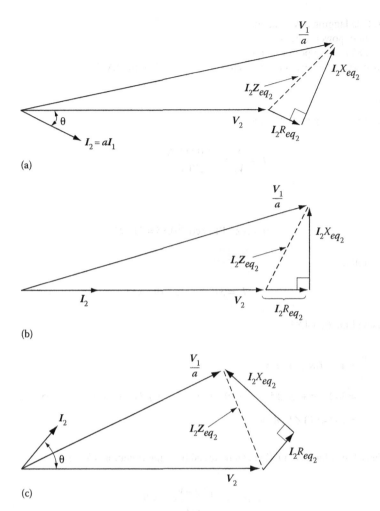

(a)

(b)

(c)

FIGURE 4.22   Phasor diagram of a transformer operating at (a) lagging power factor, (b) unity power factor, and (c) leading power factor. All quantities are referred to the secondary side of the transformer.

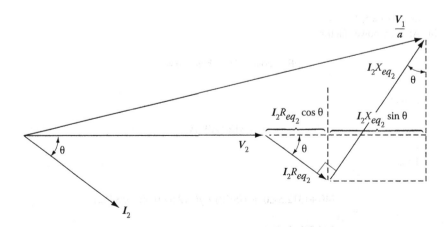

FIGURE 4.23   Phasor diagram showing the derivation of the approximate equation for V_1/a.

(a) At 0.85 lagging power factor.
(b) At unity power factor.
(c) At 0.85 leading power factor.
(d) Also write the necessary codes to solve the problem in MATLAB®.

Solution

(a) At 0.85 lagging power factor,

$$I_2 = \frac{S}{V_2} = \frac{75,000 \text{ VA}}{240 \text{ V}} = 312.5 \text{ A}$$

and

$$\theta = \cos^{-1} \text{PF} = \cos^{-1} 0.85 = 31.79°$$

Therefore,

$$I_2 = 312.5 \angle -31.79° \text{ A}$$

Using Equation 4.83,

$$\frac{V_1}{a} = V_2 + I_2 R_{eq_2} + jI_2 X_{eq_2}$$

$$= 240 \angle 0° + (312.5 \angle -31.79°)(0.009318) + j(312.5 \angle -31.79°)(0.058462)$$

$$= 252.4873 \angle 3.18° \text{ V}$$

Thus, the voltage regulation can be found by using Equation 4.79 as

$$\%V \ Reg = \frac{V_1 / a - V_{2,FL}}{V_{2,FL}} \times 100$$

$$= \frac{252.4873 - 240}{240} \times 100$$

$$= 5.2$$

or $V \ Reg = 5.2\%$

(b) At unity power factor,

$$\text{PF} = \cos\theta = 1.0 \quad \text{thus} \quad \theta = 0°$$

so that

$$I_2 = 312.5 \angle 0° \text{ A}$$

Thus,

$$\frac{V_1}{a} = 240 + (312.5 \angle 0°)(0.0918) + j(312.5 \angle 0°)(0.058462)$$

$$= 243.598 \angle 4.3° \text{ V}$$

Therefore, the voltage regulation is

$$\%V\ Reg = \frac{243.598 - 240}{240} \times 100 = 1.5$$

or $V\ Reg = 1.5\%$

(c) At 0.85 leading power factor,

$$I_2 = 312.5\angle 31.79° \text{ A}$$

Hence,

$$\frac{V_1}{a} = 240 + (321.5\angle 31.79°)(0.009318) + j(312.5\angle 31.79°)(0.058462)$$

$$= 233.4755\angle 4.19° \text{ V}$$

Thus, the voltage regulation is

$$\%V\ Reg = \frac{233.4755 - 240}{240} \times 100 = -2.72$$

or

$$\%\ V\ Reg = -2.72\%$$

(d) Here is the MATLAB script:

```
clc
clear
%System Parameters
S = 75000; % in VA
V1 = 2400; % in Volts
V2 = 240; % in Volts
n = V1/V2;
Req2 = 0.009318; % in Ohms
Xeq2 = 0.058462; % in Ohms
Zeq = Req2 + j*Xeq2; % in Ohms
% Solution for part a
PF = 0.85; % lagging
I2_mag = S/V2;
theta = acosd(PF)
I2 = I2_mag*(PF - j*sind(theta))
% NOTE: V1_a represents V1/a
V1_a = V2 + I2*Req2 + j*I2*Xeq2
Vreg = (abs(V1_a) - V2)/V2 * 100
% Solution for part b
PF = 1;
I2_mag = S/V2;
theta = acosd(PF)
```

```
I2 = I2_mag*(PF + j*sind(theta))
% NOTE: V1_a represents V1/a
V1_a = V2 + I2*Req2 + j*I2*Xeq2;
Vreg = (abs(V1_a) - V2)/V2 * 100
% Solution for part c
PF = 0.85% leading
I2_mag = S/V2;
theta = acosd(PF)
I2 = I2_mag*(PF + j*sind(theta))
% NOTE: V1_a represents V1/a
V1_a = V2 + I2*Req2 + j*I2*Xeq2
Vreg = (abs(V1_a) - V2)/V2 * 100
```

Here is the MATLAB Output for Example 4.7

```
theta=
31.7883
I2=
2.6563e+002-1.6462e+002i
V1_a=
2.5210e+002+1.3995e+001i
Vreg=
5.2030
theta=
0
I2=
312.5000
Vreg=
1.4991
PF=
0.8500
theta=
31.7883
I2=
2.6563e+002+1.6462e+002i
V1_a=
2.3285e+002+1.7063e+001i
Vreg=
-2.7186
>>
```

4.9.2 TRANSFORMER EFFICIENCY

The efficiency of any equipment can be defined as the ratio of output power to input power. Therefore, the efficiency (η) is

$$\eta = \frac{P_{out}}{P_{in}} = \frac{P_{out}}{P_{out} + P_{loss}} \tag{4.85a}$$

or

$$\eta = \frac{P_{in} - P_{loss}}{P_{in}} = 1 - \frac{P_{loss}}{P_{in}} \tag{4.85b}$$

The power losses in a given transformer are the *core losses,* which can be considered constant for a given voltage and frequency, and the *copper losses,** caused by the resistance of the windings.

The *core losses* are the sum of hysteresis and eddy-current losses. Therefore, the input power can be expressed as

$$P_{in} = P_{out} + P_{core} + P_{cu} \tag{4.86}$$

where

$$P_{out} = V_2 I_2 \cos\theta = S_{out} \cos\theta \tag{4.87}$$

Here, the $\cos\theta$ is the load power factor. Therefore, the percent efficiency of the transformer is

$$\eta = \frac{V_2 I_2 \cos\theta}{V_2 I_2 \cos\theta + P_{core} + P_{cu}} \times 100 \tag{4.88}$$

where

$$P_{cu} = I_1^2 R_1 + I_2^2 R_2 \tag{4.89}$$

$$= I_1^2 R_{eq_1}$$

$$= I_2^2 R_{eq_2} \cong P_{sc}$$

$$P_{core} = P_{cu} \tag{4.90}$$

The current, voltage, and equivalent circuit parameters must be referred to the same side of the transformer. The *maximum efficiency is* achieved when the core loss is equal to the copper loss, that is,

$$P_{core} = P_{cu} \tag{4.91}$$

In general, the efficiency of transformers at a rated load is very high and increases with their ratings. For example, transformers as small as 1 kVA may have an efficiency of 90%. Power transformer efficiencies vary from 95% to 99%. In a well-designed transformer, both core losses and copper losses are extremely small, so that efficiency is very high.[†] For example, efficiency for very large transformers is about 99%.

In contrast to power transformers, distribution transformers operate well below the rated power output most of the time. Therefore, their efficiency performance is approximately evaluated based on *all-day* (or *energy*) *efficiency,* which is defined as

---

* The term *copper loss* is still used for the losses caused by the resistances of the windings, regardless of whether they are copper or aluminum.

[†] The maximum allowable temperature that the transformer may be permitted to reach is imposed by the temperature rating of the insulation used for the coils. Because of this, the losses in the transformer must not be permitted to remain at excessively high temperatures for too long. The copper losses dictate a maximum allowable continuous current value and the iron losses set a maximum voltage value. (Because of the saturation of the iron core, operating a transformer above the rated voltage causes the no-load current to increase drastically. This also dictates a maximum allowable operating voltage.) The two limitations are independent of each other and the load power factor. Because of this, transformers are rated in kVA rather than kW.

$$\eta_{AD} = \frac{\text{Energy output over 24 h}}{\text{Energy input over 24 h}} \times 100 \tag{4.92}$$

or

$$\eta_{AD} = \frac{\text{Energy output over 24 h}}{\text{Energy output over 24 h} + \text{losses over 24 h}} \times 100 \tag{4.93}$$

Hence, if the load cycle of the transformer is known, the all-day efficiency can easily be found. Here, the load cycle is segmented into periods where the load is approximately constant, and the energy and losses for each period are calculated.

Example 4.8

Consider a 100 kVA, 7200/240 V, 60 Hz transformer. Assume that the open-circuit and short-circuit tests were performed on the transformer and that the following data were obtained:

	Open-Circuit Test (on Primary)	Short-Circuit Test (on Primary)
Voltmeter	$V_{oc} = 7200\,\text{V}$	$V_{sc} = 250\,\text{V}$
Ammeter	$I_{oc} = 0.65\,\text{A}$	$I_{sc} = 13.889\,\text{A}$
Wattmeter	$P_{oc} = 425\,\text{W}$	$P_{sc} = 1420\,\text{W}$

Also assume that the transformer operates at full load with a 0.90 lagging power factor. If the given power factor belongs to the load, not to the transformer, determine the following:

(a) The equivalent impedance, resistance, and reactance of the transformer all referred to the primary side
(b) Total losses, including the copper and core losses, at full load
(c) The efficiency of the transformer
(d) Percent voltage regulation of the transformer
(e) The phasor diagram of the transformer

Solution

(a) Referred to the primary side,

$$Z_{sc} = Z_{eq_1} = \frac{V_{sc}}{I_{sc}} = \frac{250\,\text{V}}{13.889\,\text{A}} = 18\,\Omega$$

and

$$R_{eq_1} = \frac{P_{sc}}{I_{sc}^2} = \frac{1{,}420\,\text{W}}{13.889^2\,\text{A}} = 7.36\,\Omega$$

Therefore,

$$X_{eq_1} = \left(Z_{eq_1}^2 - R_{eq_1}^2\right)^{1/2} = (18^2 - 7.36^2)^{1/2} = 16.43\,\Omega$$

(b) The full load current is

$$I_1 = \frac{S}{V_1} = \frac{100{,}000\,\text{VA}}{7{,}200\,\text{V}} = 13.89\,\text{A}$$

thus,

$$P_{cu} = I_1^2 R_{eq_1} = (13.89 \text{ A})^2 (7.36 \ \Omega) = 1,419.98 \text{ W}$$

and

$$P_{core} = P_{oc} = 425 \text{ W}$$

Hence, the total loss at full load is

$$P_{loss} = P_{cu} + P_{core} = 1,419.98 + 425 = 1,844.98 \text{ W}$$

(c) To find the input power, let us find the output power first, which is

$$P_{out} = S\cos\theta$$
$$= (100,000 \text{ VA}) \times 0.90$$
$$= 90,000 \text{ W}$$

then the input power can be found as

$$P_{in} = P_{out} + P_{loss} = 90,000 + 1,844.98 = 91,844.98 \text{ W}$$

Hence, the efficiency of the transformer is

$$\eta = 1 - \frac{P_{loss}}{P_{in}} = 1 - \frac{1,844.98}{91,844.98} = 0.9799\% \text{ or } 97.99\%$$

(d) By using Equation 4.81,

$$V_1 = aV_2 + I_1\left(R_{eq_1} + jX_{eq_1}\right) = 7,200\angle 0° + (13.89\angle -25.84°)(7.36 + j16.43) = 7,393.19\angle 1.25° \text{ V}$$

Note that since $PF = \cos\theta = 0.90$ lagging, then

$$\theta = \arccos(PF) = \cos^{-1}(PF) = \cos^{-1}(0.90) = -25.84\angle 0°$$

Therefore, the percent voltage regulation is

$$\% V \ Reg = \frac{V_1 - aV_2}{aV_2} \times 100 = \frac{7,393.19 - 7,200}{7,200} \times 100 = 2.68$$

or

$$V \ Reg = 2.68\%$$

(e) The phasor diagram of the transformer is as shown in Figure 4.21a. Note that $\theta = 25.84°$,

$$I_1 = 13.89\angle -25.84° \text{ A}, aV_2 = 7,200\angle 0° \text{V}, \text{ and } V_1 = 7,293.19\angle 1.25° V.$$

4.10 THREE-PHASE TRANSFORMERS

Today, for reasons of efficiency and economy, most electrical energy is generated, transmitted, and distributed using a three-phase system rather than a single-phase system. Three-phase power may be transformed either by the use of a single three-phase transformer or three single-phase transformers, which are properly connected with each other for a three-phase operation. A three-phase transformer, in comparison to a bank of three single-phase transformers, weighs less, costs less, needs less floor space, and has a slightly higher efficiency. In the event of failure, however, the entire three-phase transformer must be replaced.

On the other hand, if three separate single-phase units (i.e., *a three-phase transformer bank*) are used, only one of them needs to be replaced.* Also, a standby three-phase transformer is more expensive than a single-phase spare transformer. Figure 4.24 shows the two versions of three-phase core construction that are normally used: core type and shell type.

In the *core-type* design, both the primary and secondary windings of each phase are placed only on one leg of each transformer, as shown in Figure 4.24a. For balanced, three-phase sinusoidal voltages, the sum of the three-core fluxes at any given time must be zero. This is a requirement that does not have to be met in the *shell-type* construction. In the *core-type construction,* the magnetic reluctance of the flux path of the center phase is less than that of the outer two phases.

Therefore, there is some imbalance in the magnetic circuits of the three phases of the transformer. This in turn results in unequal magnetizing currents that affect their harmonic composition. In essence, such a design prevents the existence of the third-harmonic flux, and thus pretty much avoids third-harmonic voltages.

For example, third-harmonic flux components (which are in time phase in this design) of three-phase core type-transformers are reduced to at least one-tenth of what shell-type or single-phase transformer cores have. In the case of wye–wye connected windings with isolated neutrals, no third-harmonic excitation current components are present.

A *shell-type transformer* is quite different in character from a core-type transformer. As shown in Figure 4.24b, the flux in the outside paths of the core is reduced by 42% since in *a shell-type*

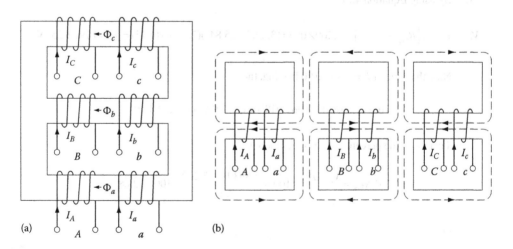

(a)   (b)

FIGURE 4.24 Three-phase, two-winding transformer core construction: (a) core type and (b) shell type.

---

* However, it is not possible to use transformers to convert a single-phase system to a three-phase system for a large amount of power. Relatively very small amounts of power can be developed from a single-phase system using R–C phase shift networks (or an induction phase converter) to produce two-phase power which in turn can be transformed into three-phase power.

FIGURE 4.25 (See color insert.) A 40MVA, 110kV ± 16%/21 kV, three-phase, core-type transformer, 5.2 m high, 9.4 m long, 3 m wide, weighing 80 tons. (Courtesy of Siemens AG, Munich, Germany.)

construction the center phase windings are wound in the opposite direction of the other two phases.

Since all yoke cross sections are equal, not only is the amount of core requirement reduced, but also the manufacturing process involved is simplified. Furthermore, in a shell-type transformer, the no-load losses are less than those in a core-type transformer.

Figure 4.25 shows a 40MVA, 110kV ± 16%/21 kV, three-phase core-type transformer. Notice that its primary-side voltage can be adjusted by ±16%. Figure 4.26 shows an 850/950/1100MVA, 415 kV ± 11%/27 kV, three-phase shell-type transformer. Figure 4.27 shows a 10MVA and a 40WA core-type three-phase transformer with *GEAFOL solid dielectric core.* Finally, Figure 4.28 shows a typical core and coil assembly of a three-phase core-type power transformer. Notice that its core

FIGURE 4.26 (See color insert.) A 850/950/1100 XWA, 415 kV ± 11%/27 kV, three-phase, shell-type transformer, 11.3 in high, 14 in long, 5.7 in wide, weighing (without cooling oil) 552 tons. (Courtesy of Siemens AG, Munich, Germany.)

FIGURE 4.27 (See color insert.)   10 MVA and 50 kVA, core-type, three-phase transformers with GEAFOL solid dielectric cores. (Courtesy of Siemens AG, Munich, Germany.)

and coil assembly are rigidly supported and clamped by heavy, fabricated clamping structures. The windings are concentrically placed on the core legs and laterally braced by inserting kiln-dried, maple wood dowels between the windings and the core. The complete assembly is rigidly braced to withstand the mechanical forces experienced under fault conditions and to resist vibration and shock forces encountered during shipment and installation. All high-voltage leads are brought to tap changers, terminal blocks, or bushings.

4.11   THREE-PHASE TRANSFORMER CONNECTIONS

As previously stated, in a three-phase power system, it is often necessary to step up or step down the voltage levels at various locations in the system. Such transformations can be achieved by means of transformer banks that have three identical single-phase transformer units, one for each phase, or by the use of three-phase transformer units. In either case, each phase has one primary and secondary winding associated with it. These primary and secondary windings may be connected independently in either *delta* (Δ) or *wye* (Y) configurations.* There are four possible connections for a three-phase transformer, namely, **wye–wye** (Y–Y), **delta–delta** (Δ–Δ), **wye–delta** (Y–Δ), and **delta–wye** (Δ–Y), as shown in Figure 4.29.

---

* The delta and wye configurations are also known as the *mesh* and *star* configurations, respectively.

FIGURE 4.28 (See color insert.) Atypical core and coil assembly of a three-phase, core-type, power transformer. (Courtesy of North American Transformer, Milpitas, CA.)

The primary benefit of using a wye-connected winding in a transformer is that it *provides a neutral point* so that phase voltages are also available. In a *wye–wye connection*, there is no phase displacement between the primary and secondary line-to-line voltages, even though it is possible to shift the secondary voltages 180° by reversing all three secondary windings.

The use of a wye–wye connection creates no problem as long as it has *solidly grounded neutrals* (especially the neutral for the primary side*). Here, the addition of a primary neutral connection makes each transformer independent of the other transformer units, one for each phase, or by the use of three-phase transformer units. In either case, each phase has one primary and secondary winding associated with it. These primary and secondary windings may be connected independently in either delta (Δ) (A) or wye (Y) configurations.[†]

There are four possible connections for a three-phase transformer, namely, wye–wye (Y–Y), *delta–delta* (Δ–Δ), *wye–delta* (Y–Δ), and *delta–wye* (Δ–Y), as shown in Figure 4.29.

The primary benefit of using a wye-connected winding in a transformer is that it provides a neutral point so that phase voltages are also available. In a *wye–wye connection*, there is no phase

---

* In the event that the neutral point of the primary windings is connected to the neutral point of the power source, there will be no difference between the behavior of a wye–wye connected, core-type or shell-type, three-phase transformer and a three-phase transformer bank.

[†] The delta and wye configurations are also known as the *mesh* and *star* configurations, respectively.

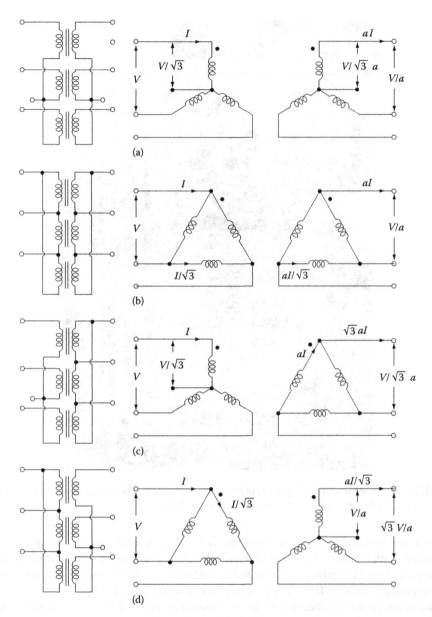

FIGURE 4.29 Basic three-phase transformer connections: (a) wye–wye, (b) delta–delta, (c) wye–delta, and (d) delta–wye.

displacement between the primary and secondary line-to-line voltages, even though it is possible to shift the secondary voltages 180° by reversing all three secondary windings. The use of a wye–wye connection creates no problem as long as it has *solidly grounded neutrals* (especially the neutral for the primary side*). Here, the addition of a primary neutral connection makes each transformer independent of the other two. Also, dissimilar transformers will not cause voltage unbalance under no-load conditions. Due to the neutrals, the additive third-harmonic components cause a current flow in the neutral rather than building up large voltages. If there is a delta-connected *tertiary* wind-

---

* In the event that the neutral point of the primary windings is connected to the neutral point of the power source, there will be no difference between the behavior of a wye–wye connected core-type or shell-type three-phase transformer and a three-phase transformer bank.

ing, in addition to the primary and secondary windings, the third-harmonic voltages are suppressed by trapping third-harmonic (circulating) currents within the delta tertiary winding.

However, *if such a neutral is not provided,* the phase voltages become drastically unbalanced when the load is unbalanced. This causes neutral instability that makes unbalanced loading impractical, even though the line-to-line voltages remain normal. There are also problems with third harmonics. In summary, any attempt to operate a wye–wye connection of transformers without the presence of a primary neutral connection will lead to difficulty and potential failure.

In fact, trouble occurs even under no-load conditions. Therefore, such a wye–wye connection is seldom used in practice.*

In the *delta–delta connection,* under balanced conditions, the line currents are u times the currents in the windings when the third harmonics in the excitation current are ignored. There is no phase shift and no problem with unbalanced loads or harmonics. If a center tap is available on one transformer secondary, the bank may be used to supply a *three-phase, four-wire delta system.*† Also, if one transformer fails in service, the remaining two transformers in the bank can be operated as an *open-delta* (or V–V) connection at about 58% of the original capacity of the bank. However, in a complete delta–delta bank, transformers tend to share the load inversely to their internal impedances, and, therefore, identical transformers have to be used.

In the *wye–delta connection,* there is no problem with third-harmonic components in its voltages, since they are absorbed in a circulating current on the delta side. This connection can be used with unbalanced loads. In high-voltage transmission systems, the high-voltage side is connected in delta and the low-voltage side is connected in wye. Due to the delta connection, the secondary voltage is shifted 30° with respect to the primary voltage.

In the United States, it is standard practice to make the secondary voltage (i.e., the lower voltage) lag the primary voltage (i.e., the higher voltage) by 30°. This connection is basically used to step down a high voltage to a lower voltage.

In the *delta–wye connection,* there is also no problem with third-harmonic components in its voltages. It has the same advantages and the same phase shift as the wye–delta connection. *The secondary voltage lags the primary voltage by* 30°, as is the case for the wye–delta connection. This connection is basically used to step up a low voltage to a high voltage. In general, when a wye–delta or delta–wye connection is used, the wye is preferably on the high-voltage side, and the neutral is grounded. Thus, the transformer insulation can be manufactured to withstand 11 V times the line voltage, instead of the total line voltage.

Example 4.9

Consider a three-phase, 15 MVA, 138/13.8 kV distribution substation transformer that is being used as a step-down transformer. Determine the ratings and turn ratios of the transformer, if it is connected in

(a) Wye–delta
(b) Delta–wye

---

* In general, all single-phase transformers when excited at rated voltage produce a third harmonic. This is due to the fact that their cores saturate fast and, because of this, their magnetization currents become distorted. Therefore, when a perfectly sinusoidal voltage (e.g., at 60 Hz) is applied to the primary of a transformer, it produces a magnetization current that has the fundamental component. Luckily, in single-phase transformers, the magnetization current is small in comparison to the load current. Therefore, the resulting distortion in the current waveform is negligible, whereas in three-phase transformers, the three fundamental magnetization currents are displaced by 120°. The third harmonic currents, however, are *in phase with respect to each other* (as are the 6th, 9th, 12th, etc. harmonics). Therefore, in an ungrounded wye–wye connection, such a tripled third-harmonic component induces a secondary voltage waveform in each winding that has a large third-harmonic voltage. Therefore, the output voltage waveforms are distorted. To prevent this, a neutral line to ground at either primary or secondary (or both) must be provided. However, if the connection is wye–delta, delta–delta, or delta–wye, the third harmonics circulate within the delta, and thus the harmonic voltage is suppressed and no secondary voltage distortion is taking place.
† For further information, see Gönen (2008).

(c) Delta–delta
(d) Wye–wye

Solution

The rated primary line current is

$$I_{L_1} = \frac{S_{3\phi}}{\sqrt{3}V_{L_1}}$$

$$= \frac{15 \times 10^6 \text{ VA}}{\sqrt{3}(138{,}000 \text{ V})}$$

$$= 62.7555 \text{ A}$$

The rated secondary line current is

$$I_{L_2} = \frac{S_{3\phi}}{\sqrt{3}V_{L_2}}$$

$$= \frac{15 \times 10^6 \text{ VA}}{\sqrt{3}(13{,}800 \text{ V})}$$

$$= 627.555 \text{ A}$$

(a) If the transformer is connected in wye–delta:

$$\textit{Rated total kVA} = S_{3\phi} = \frac{15 \times 10^6 \text{ VA}}{1{,}000} = 15{,}000 \text{ kVA}$$

$$\text{Rated kVA per phase} = \frac{S_{3\phi}}{3} = \frac{15{,}000 \text{ kVA}}{3} = 5{,}000 \text{ kVA}$$

$$\textit{Rated } I_1 = I_{L_1} = 62.7775 \text{ A}$$

$$\textit{Rated } I_2 = \frac{I_{L_2}}{\sqrt{3}} = 362.3188 \text{ A}$$

$$\textit{Rated } V_{L_1} = 138 \text{ kV}$$

$$\textit{Rated } V_{L_2} = 13.8 \text{ kV}$$

$$\textit{Rated } V_1 = \frac{V_{L_1}}{\sqrt{3}} = \frac{13{,}800\text{V}}{\sqrt{3}} = 79{,}674.3 \text{ V}$$

$$\textit{Rated } V_2 = V_{L_2} = 13{,}800 \text{ V}$$

$$\textit{Turns ratio} = a = \frac{V_1}{V_2} = \frac{79{,}674.3 \text{ V}}{13{,}800 \text{ V}} = 5.7735$$

(b) If the transformer is connected in delta–wye:

$$Rated\ total\ kVA = \frac{S_{3\phi}}{1,000} = 15,000\ kVA$$

$$Rated\ per\ phase\ = 5,000\ kVA$$

$$Rated\ I_1 = \frac{I_{L_1}}{\sqrt{3}} = \frac{62.7555\ A}{\sqrt{3}} = 36.2319\ A$$

$$Rated\ I_2 = I_{L_2} = 627.555\ A$$

$$Rated\ V_{L_1} = 138\ kV\ \text{and Rated}\ V_{L_2} = 13.8\ kV$$

$$Rated\ V_1 = V_{L_1} = 138\ kV$$

$$Rated\ V_2 = \frac{V_{L_2}}{\sqrt{3}} = \frac{13,800\ V}{\sqrt{3}} = 7,967.4337\ V$$

$$Rated\ V_2 = \frac{V_{L_2}}{\sqrt{3}} = \frac{13,800\ V}{\sqrt{3}} = 7,967.4337\ V$$

$$Turns\ ratio = a = \frac{V_1}{V_2} = \frac{138,000\ V}{7,967.4337\ V} = 17.3205$$

(c) If the transformer is connected in delta–delta:

$$Rated\ total\ kVA = \frac{S_{3\phi}}{1,000} = 15,000\ kVA$$

$$Rated\ kVA\ per\ phase = 5,000\ kVA$$

$$Rated\ I_1 = \frac{I_{L_1}}{\sqrt{3}} = \frac{62.7555\ A}{\sqrt{3}} = 36.2319\ A$$

$$Rated\ I_2 = \frac{I_{L_2}}{\sqrt{3}} = \frac{627.555\ A}{\sqrt{3}} = 362.319\ A$$

$$Rated\ V_{L_1} = 138\ kV\ \text{and}\ Rated\ V_{L_2} = 13.8\ kV \quad Rated\ V_1 = V_{L_1} = 138\ kV$$

$$Rated\ V_2 = V_{L_2} = 13.8\ kV$$

$$Turns\ ratio = a = \frac{V_1}{V_2} = \frac{138\ kV}{13.8\ kV} = 10$$

(d) If the transformer is connected in wye–wye:

$$Rated\ total\ kVA = \frac{S_{3\phi}}{1,000} = 15,000\ kVA$$

$$Rated\ kVA\ per\ phase = 5,000\ kVA$$

$$Rated\ I_1 = I_{L_1} = 62.7555\ A$$

$$Rated\ I_2 = I_{L_2} = 627.555\ A$$

$$Rated\ V_{L_1} = 138\ kV\ and\ Rated\ V_{L_2} = 13.8\ kV$$

$$Rated\ V_1 = \frac{138,000\ V}{\sqrt{3}} = 79,674.3\ V$$

$$Rated\ V_2 = \frac{V_{L_2}}{\sqrt{3}} = \frac{13,800\ V}{\sqrt{3}} = 7,967.4337\ V \quad a = \frac{V_1}{V_2} = \frac{79.67\ kV}{7.967\ kV} = 10$$

or

$$a = \frac{V_{L_1}}{V_{L_2}} = \frac{138\ kV}{13.8\ kV} = 10$$

4.12 AUTOTRANSFORMERS

The two windings in the usual two-winding transformer are not connected to each other; in other words, they are electrically isolated from each other. Therefore, power is transferred inductively from one side to the other. However, an autotransformer has a single winding, part of which is common to both the primary and the secondary simultaneously.

Thus, *in an autotransformer, there is no electrical isolation between the input side and the output side.* As a result, the power is transferred from the primary to the secondary through both *induction* and *conduction*. As shown in Figure 4.30, an autotransformer can be used as a step-down or step-up transformer. Consider the step-down connection shown in Figure 4.30a.

The *common winding is* the winding between the low-voltage terminals, while the remainder of the winding belonging exclusively to the high-voltage circuit is called the *series winding*. This combined with the common winding forms *the series-common winding* between the high-voltage terminals.

In a sense, an autotransformer is just a normal two-winding transformer connected in a special way. The only structural difference is that the *series winding must have extra insulation in order to be just as strong as the one on the common winding.* In a *variable autotransformer,* the tap is movable.

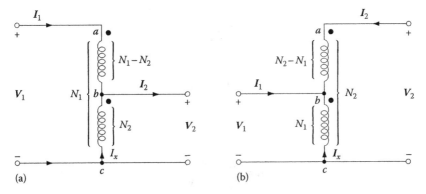

FIGURE 4.30   Autotransformers used as (a) step-down or (b) step-up transformers.

Autotransformers are increasingly used to interconnect two high-voltage transmission lines operating at different voltages, as shown in Figure 4.2. They can be used as *step-down* or *step-up* transformers.

Consider the equivalent circuit of an ideal transformer (neglecting losses) shown in Figure 4.30a. The output voltage V_2 is related to the input voltage as it is in a two-winding transformer. Therefore,

$$\frac{V_1}{V_2} = \frac{N_1}{N_2} = a \tag{4.94}$$

where $a > 1$ for *a step-down transformer,* since $N_1 > N_2$. Also, since an ideal transformer is assumed,

$$V_1 I_1 = V_2 I_2 \tag{4.95}$$

$$\frac{V_1}{V_2} = \frac{I_2}{I_1} = a \tag{4.96}$$

Since the excitation current is neglected, then I_1 and I_2 are in phase, and the current in the common section of the winding is

$$I_x = I_2 - I_1 \tag{4.97}$$

Also, the mmfs of the two windings are equal. Thus, according to the mmf balance equation

$$N_2 I_x = (N_1 - N_2) I_1 \tag{4.98}$$

or

$$I_x = \frac{N_1 - N_2}{N_2} = (a-1)I_1 = I_2 - I_1 \tag{4.99}$$

Hence,

$$I_1 = \frac{I_2}{a} \tag{4.100a}$$

$$= \frac{N_2}{N_1} I_2 \qquad (4.100b)$$

or

$$\frac{I_2}{I_1} = a \qquad (4.101)$$

Since

$$a = \frac{N_1}{N_2} = \frac{N_c + N_s}{N_c} \qquad (4.102)$$

then

$$\frac{I_2}{I_1} = \frac{N_c + N_s}{N_c} \qquad (4.103)$$

where
  N_c is the number of turns in common winding $= N_2$
  N_s is the number of turns in series winding $= N_1 - N_2$

Similarly, it can be shown that

$$\frac{V_2}{V_1} = \frac{N_c}{N_c + N_s} \qquad (4.104)$$

The apparent power delivered to the load is S_{out} and can be expressed as

$$S_{out} = V_2 I_2 \qquad (4.105)$$

From Equation 4.97,

$$I_2 = I_1 + I_x = I_1 + (I_2 - I_1) \qquad (4.106)$$

Substituting Equation 4.106 into Equation 4.105,

$$S_{out} = V_2 I_1 + V_2 (I_2 - I_1) \qquad (4.107a)$$

$$= S_{cond} + S_{ind} \qquad (4.107b)$$

where
  S_{cond} is the conductively transferred power to the load through N_2 winding

$$= V_2 I_1 \qquad (4.108)$$

S_{ind} is the inductively transferred power to the load through $N_1 - N_2$ winding

$$= V_2 (I_2 - I_1) \qquad (4.109)$$

The S_{cond} and S_{ind} are related to S_{out} by

$$\frac{S_{ind}}{S_{out}} = \frac{I_2 - I_1}{I_2} = \frac{a-1}{a} \tag{4.110a}$$

$$= \frac{N_s}{N_c + N_s} \tag{4.110b}$$

$$= \frac{N_1 - N_2}{N_1} \tag{4.110c}$$

and

$$\frac{S_{cond}}{S_{out}} = \frac{I_1}{I_2} = \frac{1}{a} \tag{4.111a}$$

$$= \frac{N_2}{N_1} \tag{4.111b}$$

where $a > 1$ for *a step-down transformer.*

Similarly, for *a step-up transformer,* as shown in Figure 4.30b, the current in the common section of the winding is

$$I_x = I_2 - I_1 \tag{4.97}$$

so that

$$I_1 = I_2 - I_x = I_2 - (I_2 - I_1) = I_2 + (I_1 - I_2) \tag{4.112}$$

Substituting Equation 4.112 into Equation 4.95,

$$S_{out} = V_2 I_2 = V_1 I_1 = S_{ind} \tag{4.113a}$$

$$= V_1 I_2 + V_1 (I_1 - I_2) \tag{4.113b}$$

$$= S_{cond} + S_{ind} \tag{4.113c}$$

where

S_{cond} is the conductively transferred power to the load through N_2 winding

$$= V_1 I_2 \tag{4.114}$$

S_{ind} is the inductively transferred power to the load through $N_2 - N_1$ winding

$$= V_1 (I_1 - I_2) \tag{4.115}$$

The S_{cond} and S_{ind} are related to S_{out} by

$$\frac{S_{ind}}{S_{out}} = \frac{I_1 - I_2}{I_1} = 1 - a \tag{4.116}$$

and

$$\frac{S_{cond}}{S_{out}} = \frac{I_2}{I_1} = a \tag{4.117}$$

where

$$a = \frac{N_c}{N_c + N_s} \tag{4.118}$$

and $a < 1$ for *a step-up transformer.*

The advantages of autotransformers include lower leakage reactances, lower losses, and smaller excitation current requirements. Most of all, an autotransformer is cheaper than the equivalent two-winding transformer (especially when the voltage ratio does not vary too greatly from 1 to 1).

The disadvantages of autotransformers are that there is no electrical isolation between the primary and secondary and that there is a greater short-circuit current than for the two-winding transformer.

Three-phase autotransformer banks generally have wye-connected main windings with the neutral normally connected solidly to ground. In addition, it is common practice to include a third winding connected in delta, called the *tertiary winding.*

Example 4.10

Assume that a single-phase, 100 kVA, 2400/240 V two-winding transformer is connected as an autotransformer to step down the voltage from 2640 to 2400 V. The transformer connection is as shown in Figure 4.30a, with 240 and 2400 V windings for sections *ab* and *bc*, respectively. Compare the kVA rating of the autotransformer with that of the original two-winding transformer, and determine all three currents as well as S_{out}, S_{ind}, and S_{cond}.

Solution

The rated current in the 240 V winding (or in the *ab* section) is

$$I_1 = \frac{100,000 \text{ VA}}{240 \text{ V}} = 416.6667 \text{ A}$$

Similarly, the rated current in the 2,400 V winding (or in the *be* section) is

$$I_x = I_2 - I_1$$

$$= \frac{100,000 \text{ VA}}{2,400 \text{ V}} = 41.6667 \text{ A}$$

Therefore, the load current is

$$I_2 = I_1 + I_x$$

$$= 416.6667 + 41.6667 = 458.3334 \text{ A}$$

Alternatively, by first calculating the turns ratio as

$$a = \frac{2,640 \text{ V}}{2,400 \text{ V}} = 1.10$$

then

$$I_2 = aI_1$$

$$= \frac{2,640 \text{ V}}{2,400 \text{ V}}(416.6667 \text{ A})$$

$$= 458.3334 \text{ A}$$

as before. The kVA (or output) rating of the autotransformer is

$$S_{auto} = V_1 I_1 = V_2 I_2$$

$$= \frac{2,640 \times 416.6667 \text{ A}}{1,000}$$

$$= 1,100 \text{ kVA}$$

Notice that the two-winding transformer rating was 100 kVA. Therefore, the ratio of the auto-transformer capacity to the two-winding transformer capacity is

$$\frac{S_{auto}}{S_{two\,wing}} = \frac{1,100 \text{ kVA}}{1,000 \text{ kVA}} = 11$$

In other words, the kVA capacity of the transformer increased 12.1 times when it was connected as an autotransformer. Here, S_{auto} and S_{out} are at the same rating.

Also, the *inductively supplied power* to the load is

$$S_{ind} = V_2(I_2 - I_1)$$

$$= 2,400(458.3334 - 416.66667)$$

$$= 100 \text{ kVA}$$

or

$$S_{ind} = \frac{a-1}{a} S_{auto}$$

$$= \frac{1.10-1}{1.10} \times 1,100$$

$$= 100 \text{ kVA}$$

The *conductively supplied power* to the load is

$$S_{cond} = \frac{S_{auto}}{a}$$

$$= \frac{1,100 \text{ kVA}}{1.10}$$

$$= 1,000 \text{ kVA}$$

4.13 THREE-WINDING TRANSFORMERS

Figure 4.31a shows a single-phase, three-winding transformer. Three-winding transformers are usually used in bulk power (transmission) substations to lower the transmission voltage to the subtransmission voltage level. They are also frequently used at distribution substations.

If excitation impedance is neglected, the equivalent circuit of a three-winding transformer can be represented by a wye of impedances, as shown in Figure 4.31b, where the primary, secondary, and tertiary windings are denoted by 1, 2, and 3, respectively. Note that the common point 0 is fictitious and is not related to the neutral of the system.

While the primaries and secondaries are usually connected in wye–wye, the tertiary windings of a three-phase and three-winding transformer bank are connected in delta.

The tertiaries are used for (1) providing a path for the third harmonics and their multiples in the excitation and the zero-sequence currents (the zero-sequence currents are trapped and circulate in the delta connection), (2) in-plant power distribution, and (3) the application of power factor correcting capacitors or reactors.

If the three-winding transformer can be considered an ideal transformer, then

$$\frac{V_2}{V_1} = \frac{N_2}{N_1} \tag{4.119}$$

$$\frac{V_3}{V_1} = \frac{N_3}{N_1} \tag{4.120}$$

$$N_1 I_1 = N_2 I_2 + N_3 I_3 \tag{4.121}$$

where V_1, V_2, and V_3 are the primary, secondary, and tertiary terminal voltages, respectively, and N_1, N_2, and N_3 are the turns in the respective windings. Also, I_1, I_2, and I_3 are the currents in the three windings.

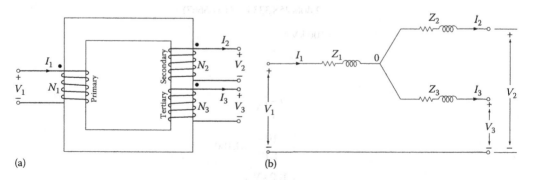

(a)   (b)

FIGURE 4.31 A single-phase, three-winding transformer: (a) winding diagram and (b) equivalent circuit.

The impedance of any of the branches shown in Figure 4.31b can be determined by considering the short-circuit impedance between pairs of windings with the third winding open. Therefore,

$$Z_{12} = Z_1 + Z_2 \tag{4.122}$$

$$Z_{13} = Z_1 + Z_3 \tag{4.123}$$

$$Z_{23} = Z_2 + Z_3 \tag{4.124}$$

If the leakage impedances Z_1, Z_2, and Z_3 are referred to the primary, they are then expressed as

$$Z_1 = \frac{1}{2}(Z_{12} + Z_{13} - Z_{23}) \tag{4.125}$$

$$Z_2 = \frac{1}{2}(Z_{23} + Z_{12} - Z_{13}) \tag{4.126}$$

$$Z_3 = \frac{1}{2}(Z_{13} + Z_{23} - Z_{12}) \tag{4.127}$$

where
Z_{12} is the leakage impedance measured in primary with secondary short-circuited and tertiary open
Z_{13} is the leakage impedance measured in primary with tertiary short-circuited and secondary open
Z_{23} is the leakage impedance measured in secondary with tertiary short-circuited and primary open
Z_1 is the leakage impedance of primary winding
Z_2 is the leakage impedance of secondary winding
Z_3 is the leakage impedance of tertiary winding

In most large transformers, the value of Z_2 is very small and can be negative.

In contrast to the situation with a two-winding transformer, the kVA ratings of the three windings of a three-winding transformer bank are not usually equal. Therefore, *all impedances* as defined earlier *should be expressed based on the same kVA base*. Figure 4.32 shows a completely assembled 910 MVA, 20.5/500 kV, three-phase step-up transformer.

4.14 INSTRUMENT TRANSFORMERS

In general, instrument transformers are of two types: *current transformers* and *voltage transformers*.* They are used in ac power circuits to provide safety for the operator and equipment from high voltage; they permit proper insulation levels and current-carrying capacity in relays, meters, and other instruments. In the United States, the standard instruments and relays are rated at 5 A and/or 120 V, 60 Hz.

Regardless of the type of instrument transformer in use, the external load applied to its secondary is referred to as its *burden*. The burden usually describes the impedance connected to the transformer's secondary winding, but may specify the volt-amperes supplied to the load.[†] For example, a transformer supplying 5 A to a resistive burden of 0.5 Ω may also be said to have a burden of 12.5 VA at 5 A.

---

* VTs were formerly called *potential transformers* (PTs).
[†] For further information, see Gönen (2008).

FIGURE 4.32 A completely assembled 910 MVA, 20.5/500 kV, three-phase, step-up transformer, 39 ft high, 36 ft long, 20 ft wide, weighing 562 tons. (Courtesy of ABB Power T&D Company, Raleigh, NC.)

CTs are connected in series with the line, as shown in Figure 4.33a. They are used to step down the current at a rated value of 5 A for ammeters, wattmeters, and relays. As shown in the figure, frequently the primary is not an integral part of the transformer itself, but is part of the line in which current is being measured.

It is very important to note that CTs can be very dangerous. During the operation of a CT, *its secondary terminals must never be open-circuited!* Unlike other types of transformers, the number of primary ampere-turns is constant for any given primary current.

When the secondary is open-circuited, the primary mmf is not balanced by a corresponding secondary mmf. (In other words, there will be no secondary mmf to oppose the primary mmf.) Therefore, all of the primary current becomes excitation current. *Consequently, a very high flux density is produced in the core, causing a very high voltage to be induced in the secondary.* In addition to endangering the user, it may damage the transformer insulation and also cause overheating due to excessive core losses.

Furthermore, if such high magnetizing forces are suddenly removed from the core, they may leave behind substantial amounts of residual magnetism, causing the turns ratio to be different from the one that existed before. As shown in Figure 4.33a, if the ammeter needs to be removed, the proper procedure is to *close the shorting switch first.*

The VT primary is connected across the potential difference to be measured, as shown in Figure 4.33b. The secondary is connected to the voltmeter, wattmeter, or relay potential winding. The VTs are specially designed to be very accurate step-down transformers.

The rated output of a VT seldom exceeds a few hundred volt-amperes. As shown in Figure 4.33b, for safety reasons, the secondary side of a VT is always grounded and well insulated from the high-voltage side.

4.15 INRUSH CURRENT

Occasionally, upon energizing a power transformer, a *transient phenomenon* (due to magnetizing current characteristics) takes place even if there is no load connected to its secondary. As a result, *its magnetizing current peak may be several times* (about 8–10 times) the rated transformer current, or it may be practically unnoticeable.

Because of losses in the excited winding and magnetic circuit, this current ultimately decreases to the normal value of the excitation current (i.e., to about 5% or less of the rated transformer current). Such a transient event is known as the **inrush current** phenomenon.

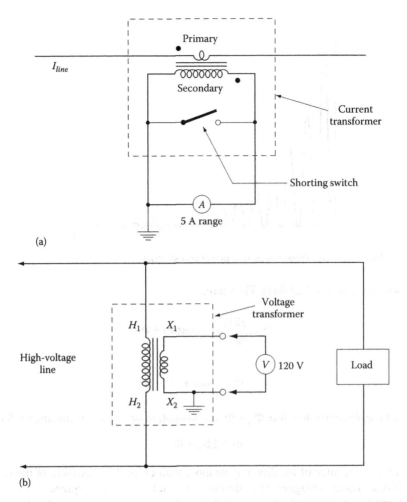

FIGURE 4.33 Instrument transformer connections: (a) current transformer connection and (b) voltage transformer connection.

It may cause (1) a momentary dip in the voltage if the impedance of the excitation source is significant, (2) undue stress in the transformer windings, or (3) improper operation of protective devices (e.g., tripping overload or common differential relays*).

The magnitude of such an inrush current depends on the magnitude, polarity, and rate of change in applied voltage at the time of switching. For example, assume that the applied voltage, at $t=0$, happens to be

$$v(t) = \sqrt{2}V_1 \sin \omega t = \frac{d\lambda_1}{dt} = N_1 \frac{d\phi}{dt} \tag{4.128}$$

The resultant flux is

$$W\Phi = \frac{\sqrt{2}V_1}{N_1} \int_0^t \sin \omega t dt + \Phi(0) \tag{4.129}$$

---

* For further information, see Blume et al. (1951).

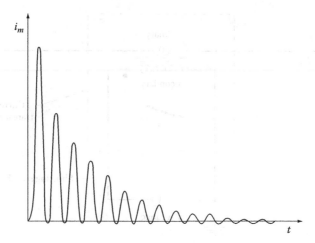

FIGURE 4.34 Inrush current phenomenon in a power transformer.

where $\Phi(0)=\Phi_r$ (i.e., the residual flux). Therefore,

$$\Phi = \frac{\sqrt{2}V_1}{\omega N_1}(1-\cos\omega t)+\Phi_r \qquad (4.130)$$

or

$$\Phi = -\Phi_m \cos\omega t + \Phi_m + \Phi_r \qquad (4.131)$$

Assuming that the *dc* component flux $\Phi_m + \Phi_r$ is constant, at $\omega t = \pi$, the instantaneous flux is

$$\Phi = 2\Phi_m + \Phi_r \qquad (4.132)$$

That is, the maximum value of the flux may be more than twice the maximum of the normal flux, since there is often residual magnetism in the core when it is initially energized.

Obviously, such doubling of the maximum flux in the core causes a tremendously large magnetization current. As shown in Figure 4.34, as time progresses (i.e., in about a few cycles), there will be a fast decay in the inrush current. Luckily, the probability of the occurrence of this theoretical maximum inrush current is relatively small.

PROBLEMS

4.1 Assume that an ideal transformer is used to step down 13.8–2.4 kV and that it is fully loaded when it delivers 100 kVA. Determine the following:
(a) Its turns ratio.
(b) The rated currents for each winding.
(c) The load impedance, referred to the high-voltage side, corresponding to full load.
(d) The load impedance referred to the low-voltage side, corresponding to full load.

4.2 A single-phase, 2500/250 V, two-winding ideal transformer has a load of $10\angle 40°\,\Omega$ connected to its secondary. If the primary of the transformer is connected to a 2400 V line, determine the following:
(a) The secondary current
(b) The primary current
(c) The input impedance as seen from the line
(d) The output power of the transformer in kVA and in kW
(e) The input power of the transformer in kVA and in kW

4.3 Assume that a three-phase, two-winding transformer is rated 50 MVA, 345/138 kV, and has 10% impedance. When the transformer has a current of 0.8 per unit in its high-voltage winding, determine the following:

(a) The corresponding primary and secondary currents in per unit and in amperes

(b) The internal impedance of the transformer, in ohms, referred to the high- and the low-voltage windings, respectively

(c) If the low-voltage terminals of the transformer are short-circuited and a 0.3 pu voltage is applied to the high-voltage winding, the resulting high-voltage side and low-voltage side currents in amperes and in per units

(d) If a current of 1.0 pu flows in the high-voltage winding, the resultant (internal) IZ voltage drop in the transformer in volts and in per units

4.4 A 60 Hz, 50 kVA, 2400/240 V, single-phase ideal transformer has 50 turns on its secondary winding. Determine the following:

(a) The values of its primary and secondary currents

(b) The number of turns on its primary windings

(c) The maximum flux Φ_m in the core

4.5 A 60 Hz, 120/24 V, single-phase transformer has 300 turns in its primary winding, determine the following:

(a) The number of turns in its secondary winding

(b) Its turns ratio

(c) The value of the mutual flux in its core

4.6 A 60 Hz, 75 kVA, 2400/240 V, single-phase transformer is used to step down the voltage of a distribution system. If the low voltage is to be kept constant at 240 V, determine the following:

(a) The value of the load impedance connected to the low-voltage side that will cause the transformer to be fully loaded

(b) The value of such a load impedance referred to the high-voltage side

(c) The values of the load current referred to the low-voltage and high-voltage sides

4.7 An audio frequency transformer is employed to couple a 100 Ω resistive load to an electronic source that can be represented by a constant voltage of 6 V in series with an internal resistance of 4000 Ω. Assume that the transformer is an ideal transformer and determine the following:

(a) The turns ratio needed to provide maximum power transfer by matching the load and source impedances

(b) The values of the current, voltage, and power at the load under such conditions

4.8 Repeat Example 4.3 but assume that the impedance of the power line is $0.5 + j1.6\,\Omega$ and that the impedance of the load is $4 + j7\,\Omega$.

4.9 Repeat Example 4.3 but assume that the impedances of the power line and the load are $4 + j9$ and $3 + j10\,\Omega$, respectively. Also assume that the generator bus voltage V_G is $220\angle0°$ V.

4.10 Use the data given in Problem 4.8 and repeat Example 4.4. The impedances of the transformers T_1 and T_2 are and $Z_{T_1} = j0.10\,\Omega$ and $Z_{T_2} = j0.10\,\Omega$, respectively. Both *of them* are referred to the high-voltage side of each transformer, respectively.

4.11 Use the data given in Problem 4.9 and repeat Example 4.4. The impedances of the transformers T_1 and T_2 are $Z_{T_1} = j0.10\,\Omega$ and, $Z_{T_2} = j0.10\,\Omega$, respectively. Both of them are referred to the high-voltage side of each transformer, respectively. The turns ratios of transformers 1 and 2 are 1/5 and 5/1, respectively.

4.12 Repeat Example 4.3 but assume that the generator has an internal impedance as shown in Figure P4.12 and that $Z_G = jX_d = j9.5\,\Omega$.

4.13 Repeat Example 4.4 but assume that the generator has an internal impedance of $Z_G = jXd = j9.5\,\Omega$.

4.14 Repeat Example 4.3 but assume that the generator has an internal impedance of $Z_G = jXd = j6.4\,\Omega$. Also assume that the impedance of the power line is $0.5 + j1.6\,\Omega$ and that the impedance of the load is $4 + j7\,\Omega$.

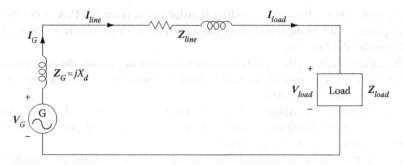

FIGURE P4.12   Figure for Problem 4.12.

4.15 Repeat Example 4.4 but assume that the generator has an internal impedance of $Z_G = jXd = j6.4\,\Omega$. The impedance of the power line is $0.5 + j1.6\,\Omega$ and that the impedance of the load is $4 + j7\,\Omega$. The impedances of the transformers T_1 and T_2 are $Z_{T_1} = j0.10\,\Omega$ and $Z_{T_2} = j0.10\,\Omega$, respectively. Both of them are referred to the high-voltage side of each transformer, respectively.

4.16 Consider a 50 kVA, 2400/240 V, 60 Hz distribution transformer. Assume that the open-circuit and short-circuit tests were performed on the primary side of the transformer and that the following data were obtained:

	Open-Circuit Test (on Primary)	Short-Circuit Test (on Primary)
Voltmeter	$V_{oc} = 2400\,\text{V}$	$V_{sc} = 52\,\text{V}$
Ammeter	$I_{oc} = 0.2083\,\text{A}$	$I_{sc} = 20.8333\,\text{A}$
Wattmeter	$P_{oc} = 185\,\text{W}$	$P_{sc} = 615\,\text{W}$

Determine the impedances of the approximate equivalent circuit referred to the primary side.

4.17 Consider a 100 kVA, 7200/240 V, 60 Hz distribution transformer. Assume that the open-circuit and short-circuit tests were performed on the primary side of the transformer and that the following data were obtained:

	Open-Circuit Test (on Primary)	Short-Circuit Test (on Primary)
Voltmeter	$V_{oc} = 7200\,\text{V}$	$V_{sc} = 250\,\text{V}$
Ammeter	$I_{oc} = 0.45\,\text{A}$	$I_{sc} = 13.8889\,\text{A}$
Wattmeter	$P_{oc} = 355\,\text{W}$	$P_{sc} = 1275\,\text{W}$

Determine the impedances of the approximate equivalent circuit referred to the primary side.

4.18 Consider a 75 kVA, 7500/480 V, 60 Hz distribution transformer. Resolve Example 4.6 using the following data:

	Open-Circuit Test (on Primary)	Short-Circuit Test (on Primary)
Voltmeter	$V_{oc} = 7500\,\text{V}$	$V_{sc} = 499\,\text{V}$
Ammeter	$I_{oc} = 0.35\,\text{A}$	$I_{sc} = 10\,\text{A}$
Wattmeter	$P_{oc} = 473\,\text{W}$	$P_{sc} = 1050\,\text{W}$

4.19 Consider a 25 kVA, 2400/240 V, 60 Hz distribution transformer. Resolve Example 4.6 using the following data:

	Open-Circuit Test (on Primary)	Short-Circuit Test (on Primary)
Voltmeter	$V_{oc} = 2400$ V	$V_{sc} = 214$ V
Ammeter	$I_{oc} = 0.125$ A	$I_{sc} = 10.417$ A
Wattmeter	$P_{oc} = 72$ W	$P_{sc} = 422$ W

4.20 Consider a 37.5 kVA, 7200/240 V, 60 Hz distribution transformer. Resolve Example 4.6 using the following data:

	Open-Circuit Test (on Primary)	Short-Circuit Test (on Primary)
Voltmeter	$V_{oc} = 7200$ V	$V_{sc} = 255$ V
Ammeter	$I_{oc} = 0.55$ A	$I_{sc} = 5.2083$ A
Wattmeter	$P_{oc} = 555$ W	$P_{sc} = 263$ W

4.21 A 50 kVA, 2400/240 V, 60 Hz transformer is to be tested to find out its excitation branch components and its series impedances. The following test data have been taken from the primary side of the transformer:

	Open-Circuit Test (on Primary)	Short-Circuit Test (on Primary)
Voltmeter	$V_{oc} = 2400$ V	$V_{sc} = 45$ V
Ammeter	$I_{oc} = 0.65$ A	$I_{sc} = 20.8333$ A
Wattmeter	$P_{oc} = 65$ W	$P_{sc} = 300$ W

(a) Find the values of R_c and X_m of the shunt (excitation) branch.
(b) Find the equivalent impedance of the transformer, referred to the primary side.
(c) Find the equivalent resistance and reactance of the transformer, referred to the primary side.
(d) Draw the equivalent circuit of the transformer, referred to the high-voltage side.

4.22 A 25 kVA, 2400/240 V, 60 Hz transformer is to be tested to find out its excitation branch components and its series impedances. The following test data have been taken from the primary side of the transformer:

	Open-Circuit Test (on Primary)	Short-Circuit Test (on Primary)
Voltmeter	$V_{oc} = 2400$ V	$V_{sc} = 70$ V
Ammeter	$I_{oc} = 0.4$ A	$I_{sc} = 10.41$ A
Wattmeter	$P_{oc} = 65$ W	$P_{sc} = 250$ W

(a) Find the values of R_c and X_m of the shunt (excitation) branch.
(b) Find the equivalent impedance of the transformer, referred to the primary side.
(c) Find the equivalent resistance and reactance of the transformer, referred to the primary side.
(d) Draw the equivalent circuit of the transformer, referred to the high-voltage side.

4.23 A 75 kVA, 2400/240 V, 60 Hz transformer is to be tested to find out its excitation branch components and its series impedances. The following test data have been taken from the primary side of the transformer:

	Open-Circuit Test (on Primary)	Short-Circuit Test (on Primary)
Voltmeter	$V_{oc} = 2400$ V	$V_{sc} = 185$ V
Ammeter	$I_{oc} = 0.7$ A	$I_{sc} = 31.25$ A
Wattmeter	$P_{oc} = 285$ W	$P_{sc} = 910$ W

(a) Find the equivalent impedance of the transformer, referred to the high-voltage side.

(b) Find the equivalent circuit parameters of the transformer, referred to the low-voltage side.

4.24 A 50 kVA, 2400/240 V, 60 Hz transformer is to be tested to find out its excitation branch components and its series impedances. The following test data have been taken from the primary side of the transformer:

	Open-Circuit Test (on Primary)	Short-Circuit Test (on Primary)
Voltmeter	$V_{oc} = 2400$ V	$V_{sc} = 195$ V
Ammeter	$I_{oc} = 0.9$ A	$I_{sc} = 20.8333$ A
Wattmeter	$P_{oc} = 395$ W	$P_{sc} = 950$ W

(a) Find the equivalent impedance of the transformer, referred to the high-voltage side.

(b) Find the equivalent circuit parameters of the transformer, referred to the low-voltage side.

4.25 Consider the power system given in Example 4.4 and determine the input impedance (i.e., the Thévenin's equivalent impedance) of the system looking into the system from bus 2. (Hint: Refer all impedances to the bus 2 side and then find the equivalent impedance.)

4.26 Consider the power system given in Example 4.10 and determine the input impedance (i.e., the Thévenin's equivalent impedance) of the system looking into the system from bus 2. (Hint: Refer all impedances to the bus 2 side and then find the equivalent impedance.)

4.27 Resolve Example 4.5 but assume that the transformer delivers the rated load current I_L at 0.9 leading PF.

4.28 Resolve Example 4.5 by using the following data: 100 kVA, 2400/240 V, 60 Hz distribution transformer with $Z_1 = 0.595 + j1.150 \, \Omega$, $Z_1 = 0.0059 + j0.011 \, \Omega$, and $Y_{el} = 5.55 \times 10^{-4} \angle -82° \, S$. Assume that the transformer delivers rated I_L at 0.85 lagging PF.

4.29 Resolve Example 4.7 but use the approximate equation, that is, Equation 4.84, for V_1.

4.30 A 50 kVA, 2400/240 V, 60 Hz distribution transformer has equivalent resistance and equivalent reactance, both referred to its secondary side, of 0.021888 and 0.09101 Ω, respectively. Use the exact equation for V_1 and determine the full-load voltage regulation:

(a) At a 0.8 lagging power factor

(b) At unity power factor

(c) At a 0.8 leading power factor

4.31 A 50 kVA, 2400/240 V, 60 Hz distribution transformer has equivalent resistance and equivalent reactance, both referred to its secondary side, of 0.021888 and 0.09101 Ω, respectively. Use the approximate equation for V_1 and determine the full-load voltage regulation:

(a) At a 0.9 lagging power factor

(b) At unity power factor

(c) At a 0.9 leading power factor

4.32 Consider a 75 kVA, 2400/240 V, 60 Hz transformer. Assume that the open-circuit and short-circuit tests were performed on the transformer and that the following data were obtained:

	Open-Circuit Test (on Primary)	Short-Circuit Test (on Primary)
Voltmeter	$V_{oc} = 2400\,V$	$V_{sc} = 200\,V$
Ammeter	$I_{oc} = 0.50\,A$	$I_{sc} = 31.25\,A$
Wattmeter	$P_{oc} = 75\,W$	$P_{sc} = 800\,W$

Also assume that the transformer operated at full load with a 0.92 lagging power factor. Note that the given power factor belongs to the load, not to the transformer. Determine the following:
   (a) The equivalent impedance, resistance, and reactance of the transformer, all referred to its primary side
   (b) Total loss, including the copper and core losses, at full load
   (c) The efficiency of the transformer
   (d) The percent voltage regulation of the transformer

4.33 Assume that a single-phase, two-winding transformer has a core loss of 1500 W. The secondary (i.e., output) voltage is 480 V and the output power is 40 kW at a 0.9 lagging power factor. The current in the primary winding is 9.2593 A. If its secondary and primary winding resistances are 0.03 and 3 Ω, respectively, determine the following:
   (a) The secondary current
   (b) The complex impedance of the load
   (c) The copper losses of the primary and secondary windings
   (d) The input power
   (e) The transformer efficiency

4.34 Consider the data given in Problem 4.24 and determine the following:
   (a) The efficiency of the transformer at full load, operating at a 0.8 lagging power factor
   (b) The efficiency of the transformer at half load, operating at the same power factor
   (c) The amount of load at which the transformer operates at its maximum efficiency

4.35 A 2400/2000 V autotransformer supplies a load of 100 kW at a power factor of 0.8. Find the current in each winding section and the kVA rating of the autotransformer.

4.36 A single-phase transformer has a core loss of 600 W. Its copper loss is 700 W, when the full-load secondary current is 20 A.
   (a) Determine the amount of the secondary current at which the transformer has its maximum efficiency.
   (b) If the transformer output is given in hp (Good grief!) as 20 hp, find its efficiency at full load.

4.37 Consider the data given in Problem 4.23 and the results of Example 4.7, and suppose that the transformer has 0.009318 Ω equivalent resistance referred to its secondary side. Determine the efficiency of the transformer at full load with these power factors:
   (a) PF = 0.85 lagging
   (b) PF = unity
   (c) PF = 0.85 leading

4.38 Consider the data given in Problem 4.24 and determine the efficiency of the transformer at full load with these power factors:
   (a) PF = 0.90 lagging
   (b) PF = unity
   (c) PF = 0.90 leading

4.39 Consider a 100 kVA, 7200/240 V, 60 Hz transformer. Assume that the open-circuit and short-circuit tests were performed on the transformer and that the following data were obtained:

	Open-Circuit Test (on Primary)	Short-Circuit Test (on Primary)
Voltmeter	$V_{oc}=7200\,V$	$V_{sc}=250\,V$
Ammeter	$I_{oc}=0.45\,A$	$I_{sc}=13.889\,A$
Wattmeter	$P_{oc}=355\,W$	$P_{sc}=1275\,W$

Also assume that the transformer operated at full load with a 0.85 lagging power factor. Note that the given power factor belongs to the load, not to the transformer. Determine the following:
(a) The equivalent impedance, resistance, and reactance of the transformer, all referred to its primary side
(b) Total loss, including the copper and core losses, at full load
(c) The efficiency of the transformer
(d) The percent voltage regulation of the transformer
(e) The phasor diagram of the transformer

4.40 Consider a 75 kVA, 7200/240 V, 60 Hz transformer. Assume that the open-circuit and short-circuit tests were performed on the transformer and that the following data were obtained:

	Open-Circuit Test (on Primary)	Short-Circuit Test (on Primary)
Voltmeter	$V_{oc}=7200\,V$	$V_{sc}=250\,V$
Ammeter	$I_{oc}=0.55\,A$	$I_{sc}=10.4167\,A$
Wattmeter	$P_{oc}=399\,W$	$P_{sc}=775\,W$

Also assume that the transformer operated at full load with a 0.90 lagging power factor. Note that the given power factor belongs to the load, not to the transformer. Determine the following:
(a) The equivalent impedance, resistance, and reactance of the transformer, all referred to its primary side
(b) Total loss, including the copper and core losses, at full load
(c) The efficiency of the transformer
(d) Percent voltage regulation of the transformer

4.41 Consider a 25 kVA, 2400/240 V, 60 Hz transformer with $Z_1=2.533+j2.995\,\Omega$ and $Z_2=(2.5333+j2.995)\times10^{-2}\,\Omega$, referred to the primary and secondary sides, respectively.
(a) Find $V_1, aV_2, I_1, I_2, R_{eq_1}, X_{eq_1}$.
(b) The transformer is connected at the receiving end of a feeder that has an impedance of $0.3+j1.8\,\Omega$. Let the sending end-voltage magnitude of the feeder be 2400 V. Also there is a load connected to the secondary side of the transformer that draws rated current from the transformer at a 0.85 lagging power factor. Neglect the excitation current of the transformer. Determine the secondary-side voltage of the transformer under such conditions.
(c) Draw the associated phasor diagram.

4.42 Consider a 75 kVA, 2400/240 V, 60 Hz transformer with $Z_1=0.52+j3.85\,\Omega$ and $Z_2=(0.52+j3.85)\times10^{-2}\,\Omega$, referred to the primary and secondary sides, respectively.
(a) Find $V_1, aV_2, I_1, I_2, R_{eq_1}, X_{eq_1}$.
(b) The transformer is connected at the receiving end of a feeder that has an impedance of $0.5+j2.3\,\Omega$. Let the sending end-voltage magnitude of the feeder be 2400 V. Also there is a load connected to the secondary side of the transformer that draws rated current from the transformer at a 0.92 lagging power factor. Neglect the excitation current of the transformer. Determine the secondary-side voltage of the transformer under such conditions.
(c) Draw the associated phasor diagram.

4.43 Repeat Example 4.9 assuming that the three-phase transformer is rated 45 MVA, 345/34.5 kV.

4.44 A three-phase, 150 kVA, 12,470/208 V distribution transformer bank supplies 150 kVA to a balanced, three-phase load connected to its secondary-side terminals. Find the kVA rating and turns ratio of each single-phase transformer unit as well as their primary- and secondary-side voltages and currents, if the transformer bank is connected in

(a) Wye–wye

(b) Delta–delta

4.45 A three-phase, 300 kVA, 480/4170 V distribution transformer bank supplies 300 kVA to a balanced, three-phase load connected to its secondary-side terminals. Find the kVA rating and turns ratio of each single-phase transformer unit as well as their primary- and secondary-side voltages and currents, if the transformer bank is connected in

(a) Wye–wye

(b) Delta–delta

4.46 Repeat Example 4.10 for a 2400/2640-V step-up connection, as shown in Figure 4.30b.

4.47 Redo Example 4.7 by using MATLAB and determine the following:

(a) Write the MATLAB program script.

(b) Give the MATLAB program output.

5 Electromechanical Energy Conversion Principles

Use all of our resources with maximum efficiency and effectiveness.

Lawrence J. Peter

Truth is the most valuable thing we have. Let us economize it.

Mark Twain, Henry Ward Beecher's Farm, 1885

If a man said all men are liars, would you believe him?

Author Unknown

5.1 INTRODUCTION

According to the **energy conversion principle**, *energy is neither created nor destroyed: it is simply changed in form.* The role of electromagnetic (or electromechanical) machines is to transmit energy or convert it from one type of energy to another. For example, the transformer transmits electrical energy, changing only the potential difference and current at which it exists. However, it also converts a small amount of electrical energy to heat. This is an unwanted result that is required to be minimized at the design stage.

However, a rotational or translational electromagnetic machine converts energy from mechanical to electrical form, or vice versa, that is, it operates as a *generator* or *motor*. In the process,* it also converts some electrical or mechanical energy to heat, which is also unwanted. In general, electric generators and motors of all kinds can be defined as *electromechanical energy converters*. Their main components are an electrical system, a mechanical system, and a coupling field, as shown in Figure 5.1.

5.2 FUNDAMENTAL CONCEPTS

In this section, some of the basic concepts involving electrical rotating machines are reviewed. Such concepts include angular velocity, angular acceleration, mechanical work, power, and torque. As explained in Chapter 1, most electrical machines rotate around an axis known as the shaft of the machine.

As can be seen in Figures 5.1 and 5.2, the input to a generator and the output of the motor are mechanical in nature. If the shaft rotates in *a counterclockwise* (CCW) direction rather than in *a clockwise* (CW) direction, the resultant rotational angle and direction are, by definition, considered positive; otherwise, they are considered negative.

The *angular position* θ of the shaft is the angle at which it is positioned measured from some arbitrarily selected reference point. Note that the angular position concept conforms to the linear

---

* Such a process occurs through the medium of the electric or magnetic field of the conversion device. In general, electromechanical devices can be classified as follows: (1) *transducers*, which are the devices used for measurement and control, such as torque motors, loudspeakers, and microphones; (2) *force-producing devices*, such as relays, electromagnets, and solenoid actuators; and (3) *continuous energy-conversion apparatus*, such as generators and motors.

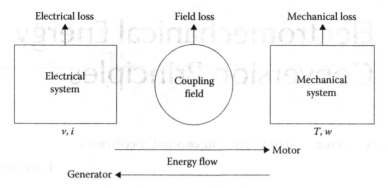

FIGURE 5.1 A representation of electromechanical energy conversion.

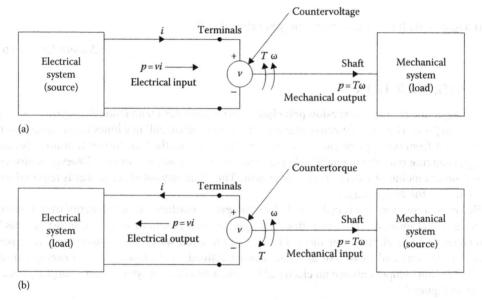

FIGURE 5.2 Illustration of some of the fundamental concepts associated with the operation of an electrical rotating machine: (a) motor action and (b) generator action.

distance concept along a line, and that it is measured in radians or degrees. However, the *angular velocity* (or *speed*) represents the rate of change in the angular position with respect to time. Thus, angular velocity can be expressed as

$$\omega = \frac{d\theta}{dt} \text{ rad/s} \tag{5.1}$$

as long as the angular position θ is measured in radians. Usually, the rotational speed n is given in revolutions per minute (i.e., rpm), so that

$$n = \left(\frac{60}{2\pi}\right)\omega \text{ rev/min} \tag{5.2}$$

The rate of change in angular velocity with respect to time is defined as the angular acceleration and is expressed as

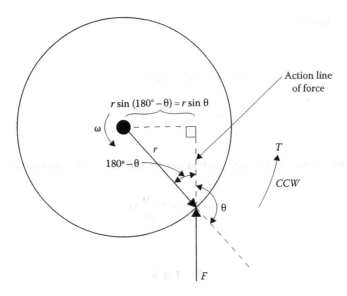

FIGURE 5.3 Illustration of the relationship between force and torque.

$$\alpha = \frac{d\omega}{dt} \text{ rad/s}^2 \tag{5.3}$$

In rotational mechanics, the **torque** (or twisting action on the cylinder), T, is defined as the tangential force times the radial distance at which it is applied, measured from the axis of rotation.

In other words, the torque is a function of the magnitude of the applied force F, and the distance between the axis of rotation and the line of action of the force. Hence, as illustrated in Figure 5.3, the rotational torque T can be expressed as

$$T = (\text{applied force})(\text{perpendicular distance})$$

$$= (F)(r\sin\theta) \tag{5.4}$$

$$= FR\sin\theta$$

In the SI system* of units, force is given in newtons (N) and distance is given in meters (m).

For linear motion, according to Isaac Newton, mechanical work W is defined as the *integral of force over distance*. Therefore,

$$W = f\int dx \text{ J} \tag{5.5}$$

The units of work are joules (J) on the SI system and foot pounds in the English system. However, for rotational motion, work is defined as the *integral of torque through an angle*. Thus,

$$W = T\int d\theta \text{ J} \tag{5.6}$$

---

* It is interesting to note that since 1954, the metric system known as the International System of Units (or Système International) (SI) has been in use all over the world with the exception of the United States. Today, the SI system is used even in England and Canada.

If the torque is unchanging,

$$W = T\theta \text{ J} \tag{5.7}$$

Power can be defined as the rate of doing work. Hence,

$$P = \frac{dW}{dt} = F\frac{dx}{dt} \text{ W} \tag{5.8}$$

However, for rotational motion having constant torque, power* can be expressed as

$$P = \frac{dW}{dt} = T\frac{d\theta}{dt} \text{ W} \tag{5.9}$$

or

$$P = T\omega \text{ W} \tag{5.10}$$

Also, from Equation 5.10 torque can be found as

$$T = \frac{P}{\omega} \text{ N} \cdot \text{m} \tag{5.11}$$

Since in the United States the English system of units is still in use, knowing the following conversion formulas may be useful:

$$T = \frac{7.04P \text{ (watts)}}{n} \text{ lb} \cdot \text{ft} \tag{5.12}$$

or

$$T = \frac{5252P \text{ (horse power)}}{n} \text{ lb} \cdot \text{ft} \tag{5.13}$$

where

$$n = \left(\frac{60}{2\pi}\right)\omega \text{ rev/min} \tag{5.2}$$

$$P \text{ (horse power)} = \frac{P \text{ (watts)}}{746} \tag{5.14}$$

$$T \text{ (lb} \cdot \text{ft)} = 0.738T \text{ (N} \cdot \text{m)} \tag{5.15}$$

Some of the fundamental concepts associated with the operation of an electrical rotating machine are illustrated in Figure 5.2. The magnetic field in such a machine establishes the necessary link between the electrical and mechanical systems, producing mechanical torque as well

---

* In the SI system of units, the work is in joules (since 1 J = 1 W/s) if the force is in newtons and the distance, *x*, is in meters. Thus, one watt equals one joule per second (i.e., 1 W = 1 J/s).

as inducing voltages in the coils. The magnetic field itself is developed by the current flowing through these coils.

Consider the *motor action* of the machine as illustrated in Figure 5.2a. The instantaneous power input to the motor is

$$p = v \times i \tag{5.16}$$

where v and i are the terminal voltage and current, respectively, as shown in Figure 5.2a. The magnetic field develops the output torque and induces a countervoltage (also called *counter-emf* since it opposes the current flow), which makes it possible for the machine to receive power from the electrical source and convert it into mechanical output. Here, the torque and the angular velocity are in the same direction.

In the generator action as shown in Figure 5.2b, the magnetic field induces the generated voltage and develops a countertorque (since it opposes the torque of the mechanical source) which makes it possible for the rotating machine to receive power from the mechanical source in order to convert it into electrical output. Note that the countertorque and the angular velocity are in the opposite direction.

Some large electromechanical machines that are used in different applications are shown in Figures 5.4 through 5.11. The dc motors, similar to the one shown in Figure 5.6, are typically used for continuous high-impact load duty on blooming and slabbing mills.

The ac generator shown in Figure 5.10 has a wound stator. Figure 5.11 shows the artist's rendition of one of the three ac generators that are in use at the Grand Coulee Dam. It is a water-cooled, low-speed hydro-electric generator with a combined thrust and guide bearing below the rotor. The stator's outside diameter is 23.9 m. The generator's complete weight is 2500 metric tons; its rotor weight is 1400 metric tons. The stator windings are water cooled. This is one of the world's most powerful hydroelectric generators.

FIGURE 5.4 The stator of a 9 MW, 250 rpm, 1650 V mill motor. (Courtesy of Simens, Munich, Germany.)

FIGURE 5.5   The rotor of a 5 MW, 60/120 rpm, 1750 V converter mill motor. (Courtesy of Siemens, Munich, Germany.)

FIGURE 5.6   A dc 8 MW, 50/100 rpm dc reversing mill motor operating at 899 V. (Courtesy of Siemens, Munich, Germany.)

Example 5.1

The rotational speed of a motor (i.e., its shaft speed) is 1800 rev/min. Determine its angular velocity (i.e., its shaft speed) in rad/s.

Solution

From Equation 5.2,

$$n = \left(\frac{60}{2\pi} \right) \omega \; \text{rev/min}$$

FIGURE 5.7 A large synchronous motor used in the mining industry. (Courtesy of General Electric Canada, Inc., Mississauga, Canada.)

or

$$\omega = \left(\frac{2\pi}{60}\right)n$$

$$= \left(\frac{2\pi}{60}\right)(1800\,\text{rev/min}) = 188.5 \text{ rad/s}$$

Example 5.2

Consider Figure 5.3 and assume that the rotational torque is 200 N · m and the radius of the rotor is 0.25 m. Determine the applied force in N:

    (a) If the angle θ is 15°
    (b) If the angle θ is 45°
    (c) If the angle θ is 90°
    (d) If the angle θ is 120°

FIGURE 5.8 A vertical synchronous generator used by a large industrial firm. (Courtesy of General Electric Canada, Inc., Mississauga, Canada.)

Solution

(a) From Equation 5.4,

$$T = Fr\sin\theta$$

from which

$$F = \frac{T}{r\sin\theta}$$

Thus, at $\theta = 15°$,

$$F = \frac{200\,\text{N}\cdot\text{m}}{(0.25\,\text{m})\sin 15°} = 3091\,\text{N}$$

(b) At $\theta = 45°$,

$$F = \frac{200\,\text{N}\cdot\text{m}}{(0.25\,\text{m})\sin 45°} = 1131.4\,\text{N}$$

FIGURE 5.9 Two large dc motors used in the metals industry. (Courtesy of General Electric Canada, Inc., Mississauga, Canada.)

(c) At $\theta = 90°$,

$$F = \frac{200 \text{ N} \cdot \text{m}}{(0.25 \text{ m})\sin 90°} = 800 \text{ N}$$

(d) At $\theta = 120°$,

$$F = \frac{200 \text{ N} \cdot \text{m}}{(0.25 \text{ m})\sin 120°} = 923.8 \text{ N}$$

Example 5.3

A coil, having a sectional area of 0.3 m^2 with $N = 20$ turns, is rotating around its horizontal axis with a constant speed of 3600 rpm in a uniform and vertical magnetic field of flux density $B = 0.8 \text{ T}$. If the total magnetic flux passing through the coil is given by $\Phi = AB \cos \omega t$ in Wb, where A is the sectional area of the coil, determine the maximum and effective values of the induced voltage in the coil.

Solution

Since the total magnetic flux passing through the coil is given as

$$\Phi = AB \cos \omega t$$

FIGURE 5.10   Installation of a 500 MVA, 200 rpm, ac generator rotor used at Churchill Falls hydroelectric plant. (Courtesy of General Electric Canada, Inc., Mississauga, Canada.)

According to Faraday's law, the induced voltage is

$$v = N \frac{d\phi}{dt}$$

$$= N \frac{d(AB\cos\omega t)}{dt}$$

$$= -N\omega AB \sin\omega t$$

Thus, its maximum voltage is

$$v_{max} = N \times \omega \times A \times B$$

where

$$\omega = \left(\frac{2\pi}{60}\right)(3600 \text{ rpm})$$

$$= 377 \text{ rad/s}$$

FIGURE 5.11 Cutaway view of a 718 MVA, 86 rpm, Grand Coulee generator. (Courtesy of General Electric Canada, Inc., Mississauga, Canada.)

Hence,

$$v_{max} = (20\,\text{turns})(377\,\text{rad/s})(0.3\,\text{m}^2)(0.8\,\text{T})$$

$$= 1809.6\,\text{V}$$

and its effective value is

$$v_{rms} = \frac{v_{max}}{\sqrt{2}}$$

$$= \frac{1809.6\,\text{V}}{\sqrt{2}} = 1279.6\,\text{V}$$

5.3 ELECTROMECHANICAL ENERGY CONVERSION

The energy conservation principle with regard to electromechanical systems can be expressed in various forms. For example, as shown in Figure 5.1, for *a sink* of electrical energy such as an electric motor, it can be expressed as

$$
\begin{pmatrix}
\text{Electrical} \\
\text{energy input} \\
\text{from source}
\end{pmatrix}
=
\begin{pmatrix}
\text{Mechanical} \\
\text{energy output} \\
\text{to load}
\end{pmatrix}
+
\begin{pmatrix}
\text{Increase in} \\
\text{stored energy} \\
\text{in coupling field}
\end{pmatrix}
+
\begin{pmatrix}
\text{Energy loss} \\
\text{converted} \\
\text{to heat}
\end{pmatrix}
\qquad (5.17)
$$

The last term of Equation 5.17 can be expressed as

$$
\begin{pmatrix}
\text{Energy loss} \\
\text{converted} \\
\text{to heat}
\end{pmatrix}
=
\begin{pmatrix}
\text{Resistance} \\
\text{loss of} \\
\text{winding}
\end{pmatrix}
+
\begin{pmatrix}
\text{Friction} \\
\text{and windage} \\
\text{losses}
\end{pmatrix}
+
\begin{pmatrix}
\text{Field} \\
\text{losses}
\end{pmatrix}
\qquad (5.18)
$$

The resistance loss is the $i^2 R$ loss in the resistance I of the winding. The friction and windage losses are associated with motion. Since the coupling field is the magnetic field, the field losses* are due to hysteresis and eddy-current losses, that is, the *core losses* are due to the changing magnetic field in the magnetic core.

If the energy losses[†] that are given in Equation 5.18 are substituted into Equation 5.17, the energy balance equation can be expressed as

$$
\begin{pmatrix}
\text{Electrical} \\
\text{energy input} \\
\text{from source} \\
\text{minus} \\
\text{resistance losses}
\end{pmatrix}
=
\begin{pmatrix}
\text{Mechanical} \\
\text{energy output} \\
\text{to load plus} \\
\text{friction and} \\
\text{windage losses}
\end{pmatrix}
+
\begin{pmatrix}
\text{Increase in} \\
\text{stored energy} \\
\text{in coupling field} \\
\text{plus core losses}
\end{pmatrix}
\qquad (5.19)
$$

Assume a differential time interval dt during which an increment of electrical energy dW_e (without including the $i^2 R$ *loss*) flows to the system. Then the net electrical input W_e can be equated to the increase in energy W_m so that, in incremental form,

$$
dW_e = dW_m + dW_f \ \text{J}
\qquad (5.20)
$$

where
dW_e is the differential electrical energy input[‡]
dW_m is the differential mechanical energy output[§]
dW_f is the differential increase in energy stored in the magnetic field

---

* They are also known as the *iron losses*, as previously stated.
† Furthermore, there are additional losses that arise from the nonuniform current distribution in the conductors and the core losses generated in the iron due to the distortion of the magnetic flux distribution from the load currents. Such losses are known as the *stray-load losses* and are very hard to determine precisely. Because of this, estimates that are based on tests, experience, and judgment are used. Typically, such stray losses range from 0.5% of the output in large machines to 5% of the output in medium-sized machines. In do machines, they are usually estimated to be about 1% of the output.
‡ It is also denoted by dW_i.
§ It is also called the *differential developed energy* and is denoted by dW_d.

Equation 5.20 is also known as the *incremental* (or *differential*) *energy-balance equation*. It provides a basis for the analysis of the operation of electromechanical machines. Since in time dt,

$$dW_e = v \times i \times dt \tag{5.21}$$

where v is the (reaction) voltage induced in the electric terminals by the changing magnetic stored energy. Therefore,

$$dW_e = v \times i \times dt = dW_m + dW_f \tag{5.22}$$

According to Faraday's law, the induced voltage v based on the flux linkages can be expressed as

$$v = \frac{d\lambda}{dt} \tag{5.23}$$

Then the net differential electrical energy input in time dt can be expressed as

$$
\begin{aligned}
dW_e &= v \times i \times dt \\
&= \left(\frac{d\lambda}{dt}\right) i \times dt \\
&= i \times d\lambda
\end{aligned}
\tag{5.24}
$$

The differential mechanical energy output for a *virtual displacement* (i.e., *linear motion*) dx when the force is $\mathcal{F}_f$ can be expressed as

$$dW_m = \mathcal{F}_f \times dx \tag{5.25}$$

Substituting Equations 5.24 and 5.25 into Equation 5.22,

$$dW_f = i \times d\lambda - \mathcal{F}_f \times dx \tag{5.26}$$

If the differential mechanical energy output is for a rotary motion, the force $\mathcal{F}_f$ is replaced by torque* T_f and the linear (differential) displacement dx *is* replaced by the angular (differential) displacement $d\theta$ so that

$$dW_m = T_f \times d\theta \tag{5.27}$$

and therefore

$$dW_f = i \times d\lambda - T_f \times d\theta \tag{5.28}$$

5.3.1 Field Energy

Suppose that the electromechanical system shown in Figure 5.12 has a movable part (i.e., *armature*) that can be kept in static equilibrium by a spring. If the movable part is kept stationary at some air gap and the current is increased from zero to a value I, a flux Φ will be maintained in the electro-magnetic system. Since no mechanical output can be produced,

$$dW_m = 0 \tag{5.29}$$

---

* It is also known as the *developed torque* and is denoted by T_d.

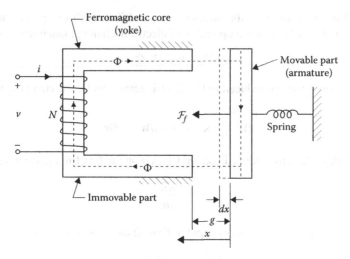

FIGURE 5.12   A simple electromechanical system.

and substituting Equation 5.29 into Equation 5.20,

$$dW_e = dW_f \tag{5.30}$$

Thus, if core loss is ignored, all the incremental electric energy input must be stored in the magnetic field. Since, from Equation 5.24,

$$dW_e = i \times d\lambda \tag{5.31}$$

then

$$dW_f = i \times d\lambda \tag{5.32}$$

$$dW_f = dW_e = v \times i \times dt = i \times d\lambda \tag{5.33}$$

Figure 5.13a shows the relationship between coil flux linkage λ and current I for a particular air-gap length. Since core loss is being ignored, the curve will be a single-valued curve passing

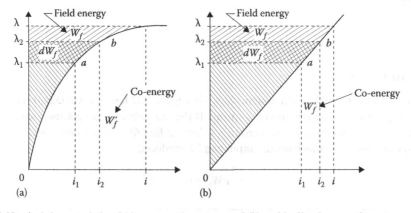

FIGURE 5.13 λ–i characteristic of (a) a magnetic system and (b) an idealized magnetic system.

through the origin. The incremental field energy dW_f is shown as the crosshatched area in Figure 5.13a. If the applied terminal voltage v is increased, causing a change in current from i_1 to i_2, there will be a matching change in flux linkage from λ_1 to λ_2. Therefore, the corresponding increase in stored energy is

$$dW_f = \int_{\lambda_1}^{\lambda_2} i \, d\lambda \tag{5.34}$$

as shown in Figure 5.13a. When the coil current and flux linkage are zero, the field energy is zero. Thus, if the flux linkage is increased from zero to ℓ, the total energy stored in the field is

$$dW_f = \int_0^{\lambda} i \, d\lambda \tag{5.35}$$

Such an integral represents the shaded area W_f between the A ordinate and the $\lambda-i$ characteristic, as shown in Figure 5.13a. Equation 5.35 can be used for any lossless electromagnetic system.

Also, if the leakage flux is negligibly small, then all flux Φ in the magnetic circuit links all N turns of the coil. Therefore,

$$\lambda = N\Phi \tag{5.36}$$

so that from Equations 5.33 and 5.36,

$$dW_f = i \times d\lambda = N \times i \times d\Phi = \mathcal{F} \times d\Phi \tag{5.37}$$

where

$$\mathcal{F} = N \times i \text{ A} \cdot \text{turns} \tag{5.38}$$

Thus, if the characteristic shown in Figure 5.13a is rescaled to show the relationship between Φ and $\mathcal{F}$ (so that the ordinate represents the Φ rather than the λ and the axis represents the $\mathcal{F}$ rather than the i), the shaded area again represents the stored energy.

Also, if the reluctance of the air gap makes up a considerably larger portion of the total reluctance of the magnetic circuit, then that of the magnetic material used may be ignored. The resultant $\lambda-i$ characteristic is represented by a straight line through the origin. Figure 5.13b shows such a characteristic of an idealized magnetic circuit. Hence, for this idealized system,

$$\lambda = L \times i \tag{5.39}$$

where the inductance of the coil is given by L. By substituting Equation 5.23 into Equation 5.35, the total energy stored in the field can be expressed as

$$\begin{aligned}
W_f &= \int_0^{\lambda} \frac{\lambda}{L} \, d\lambda \\
&= \frac{\lambda^2}{2L} \\
&= \frac{L \times i^2}{2} \\
&= \frac{i \times \lambda}{2}
\end{aligned} \tag{5.40}$$

On the other hand, if the reluctance of the magnetic system (i.e., of the air gap) as viewed from the coil is $\mathcal{R}$, then

$$\mathcal{F} = \mathcal{R} \times \Phi \text{ A} \cdot \text{turns} \tag{5.41}$$

and from Equation 5.37, the total energy stored in the field can be expressed as

$$W_f = \int_0^\Phi \mathcal{F} \times d\Phi = \mathcal{R}\frac{d\Phi^2}{2} = \frac{\mathcal{F}^2}{2\mathcal{R}} \tag{5.42}$$

Furthermore, if it is assumed that there is no fringing at the air gaps, and that the total field energy is distributed uniformly, the total energy stored in the field can be expressed as

$$W_f = \frac{i \times \lambda}{2}$$

$$= \frac{i(N \times \Phi)}{2}$$

$$= \frac{(N \times i)\Phi}{2}$$

$$= \mathcal{F}\left(\frac{\Phi}{2}\right) \tag{5.43}$$

or

$$W_f = \frac{H \times B \times \ell \times A}{2}$$

$$= \frac{B^2(vol)}{2\mu_0} \tag{5.44}$$

where
  $\ell = 2g$ represents the total length of the air gap in a flux path
  $vol = \ell \times A$ represents the total air gap volume
  A represents the cross-sectional area of the core
  B represents the flux density in the air gaps

$$\mu_0 = \frac{B}{A}$$

$$= \text{permeability of free space, H/m}$$

$$= 4\pi \times 10^{-7}$$

Since $\ell \times A$ is the total gap volume, the energy density W_f in the air gaps can be expressed as

$$w_f = \frac{W_f}{\ell \times A}$$

$$= \frac{B \times H}{2}$$

$$= \frac{\mu_0 H^2}{2}$$

$$= \frac{B^2}{2\mu_0} \tag{5.45}$$

The unit of the energy density is J/m³.

Example 5.4

Consider Example 3.7 and determine the following:

    (a) The mmf required by the air gap
    (b) The total mmf required
    (c) The mmf required by the ferromagnetic core
    (d) The energy density in the air gap
    (e) The energy stored in the air gap
    (f) The total energy stored in the magnetic system
    (g) The energy stored in the ferromagnetic core
    (h) The energy density in the ferromagnetic core

Solution

(a) The mmf required by the air gap is found from

$$\mathcal{F}_g = \Phi_g \times \mathcal{R}_g = \Phi_4 \times \mathcal{R}_4 = \Phi_2 \times \mathcal{R}_4$$

as

$$\mathcal{F}_g = (0.001197\,\text{Wb})\,(31{,}870.9886\,\text{A}\cdot\text{turns/Wb})$$

$$= 38.1017\,\text{A}\cdot\text{turns}$$

(b) The total mmf required is

$$\mathcal{F}_{tot} = \Phi_{tot} \times \mathcal{R}_{tot}$$

$$= (0.0015652\,\text{Wb})\,(255{,}565.4983\,\text{A}\cdot\text{turns/Wb})$$

$$= 400\,\text{A}\cdot\text{turns}$$

(c) The mmf required by the ferromagnetic core is

$$\mathcal{F}_{core} = \mathcal{F}_{tot} + \mathcal{F}_g$$

$$= 400 - 38.1017 = 361.8983\,\text{A}\cdot\text{turns}$$

(d) Since

$$B_g = B_2$$

$$= 0.2394 \, \text{Wb/m}^2$$

From Equation 5.45, the energy density in the air gap is

$$w_g = \frac{B_g^2}{2\mu_0}$$

$$= \frac{(0.2394 \text{ Wb/m}^2)^2}{2(4\pi \times 10 - 7)} = 0.0023 \times 10^7 \text{ J/m}^3$$

(e) The total air gap volume is

$$\text{vol}_g = \ell_g \times A_g$$

$$= (0.0002 \text{ m})(0.005 \text{ m}^2) = 1 \times 10^{-6} \text{ m}^3$$

Thus, the energy stored in the air gap is

$$W_g = w_g(vol_g)$$

$$= (0.0023 \times 10^7 \text{ J/m}^3)(1 \times 10^{-6} \text{ m}^3) = 0.023 \text{ J}$$

Alternatively, using Equation 5.42,

$$W_g = \mathcal{R}_g \frac{\Phi^{2/g}}{2}$$

$$= \frac{(31,830.9886 \text{ A} \cdot \text{turns/Wb})(0.001197 \text{ Wb})^2}{2}$$

$$= 0.023 \text{ J}$$

(f) The total energy stored in the magnetic system is

$$W_f = \mathcal{F}_{tot} \frac{\Phi_{tot}}{2}$$

$$= \frac{(400 \text{ A} \cdot \text{turns/Wb})(0.0015652 \text{ Wb})^2}{2}$$

$$= 0.313 \text{ J}$$

(g) Since

$$W_f = W_{core} + W_g$$

then the energy stored in the ferromagnetic core is

$$W_{core} = W_f - W_g$$

$$= 0.313 - 0.023$$

$$= 0.29 \text{ J}$$

(h) The volume of the ferromagnetic core is

$$vol_{core} = 2(0.05 \times 0.25)0.05 + 2(0.05 \times 0.4)\,0.05 + (0.15 - 0.0002)(0.10)0.05$$

$$= 4 \times 10^{-3}\,\text{m}^3$$

Since the energy stored in the ferromagnetic core is $W_{core} = w_{core}\,(vol_{core})$, then the energy density in the ferromagnetic core is

$$W_{core} = \frac{W_{core}}{vol_{core}}$$

$$= \frac{0.29\,\text{J}}{4 \times 10^{-3}\,\text{m}^3}$$

$$= 72.5\,\text{J/m}^3$$

5.3.2 Magnetic Force

The magnetic flux that crosses an air gap in a magnetic material produces a force $\mathcal{F}_f$ of attraction between the faces of the air gap, as shown in Figure 5.12. The core shown in the figure has an air gap of variable length g as dictated by the position of the movable part (i.e., the armature), which in turn is determined by the magnetic pulling force $\mathcal{F}_f$ and the spring. Note that, in Figure 5.12, the differential displacement can also be expressed as

$$dx = dg \tag{5.46}$$

Based on the symmetry involved (considering only one pole of the magnetic circuit), the differential change in volume can be found from

$$d(vol) = A\,dg \tag{5.47}$$

Ignoring leakage and fringing of the flux at the gaps

$$dW_f = \frac{BH\,d(vol)}{2}$$

$$= \frac{BHA\,dg}{2}$$

$$= \frac{B^2 A\,dg}{2\mu_0} \tag{5.48}$$

From the definition of work,

$$dW_m = \mathcal{F}_f dx = \mathcal{F}_f dg \tag{5.49}$$

When a magnetic pulling force is applied to the movable part (i.e., the armature), an energy dW equal to the magnetic energy dW_f stored in the magnetic field is expended. Therefore, at the equilibrium,

$$dW_f = dW_m \tag{5.50}$$

or substituting Equations 5.48 and 5.49 into Equation 5.50

$$\frac{B^2 A dg}{2\mu_0} = \mathcal{F}_f dg \tag{5.51}$$

from which the *magnetic pulling force per pole* on the movable part can be found as

$$\mathcal{F}_f = \frac{B^2 A}{2\mu_0} \text{ N} \tag{5.52}$$

Thus, the total magnetic pulling force on the movable part can be expressed as

$$\mathcal{F}_{f,total} = 2\frac{B^2 A}{2\mu_0} = \frac{B^2 A}{\mu_0} \text{ N} \tag{5.53}$$

It is important to understand that since the electrical input makes no contribution to the energy in the air gaps, due to the constant air-gap flux, the mechanical energy must be obtained from the stored energy in the air-gap fields (i.e., $id\ell = 0$). In other words, the air gaps give off energy by virtue of their decreased volume.

Example 5.5

Consider the linear electromechanical system shown in Figure 5.14. Assume that only the coil shown on the left is energized and that the core on the right acts as an armature (i.e., the movable part). The cross-sectional area of each air gap is 25×10^{-6} m². If the flux density is 1.1 Wb/m², determine the following:

(a) The magnetic pulling force per pole
(b) The total magnetic pulling force

Solution

(a) The magnetic pulling force per pole can be found from Equation 5.52 as

$$\mathcal{F}_f = \frac{B^2 A}{2\mu_0} = \frac{(1.1 \text{Wb/m}^2)^2 (25 \times 10^{-6} \text{ m}^2)}{2(4\pi \times 10^{-7})} = 12.04 \text{ N}$$

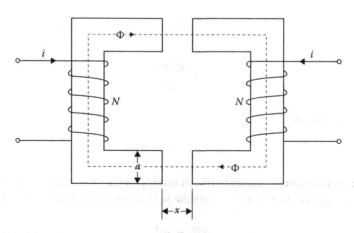

FIGURE 5.14   A doubly excited electromechanical (translational) system.

(b) The total magnetic pulling force can be found from Equation 5.53 as

$$\mathcal{F}_{f,total} = \frac{B^2 A}{\mu_0} = \frac{(1.1\,\text{Wb/m}^2)^2(25\times10^{-6}\,\text{m}^2)}{(4\pi\times10^{-7})} = 24.08\,\text{N}$$

Alternatively,

$$\mathcal{F}_{f,total} = 2\mathcal{F}_f = 2(12.04\,\text{N}) = 24.08\,\text{N}$$

5.3.3 ENERGY AND COENERGY

As previously stated, the shaded area in Figure 5.13a represents the total energy stored in a coil (which magnetizes the field) from zero to i A. Such energy can be determined by using Equation 5.35; that is, from

$$W_f = \int_0^\lambda i\,d\lambda \tag{5.35a}$$

In Figure 5.13a, the area between the i axis and $\lambda - i$ characteristic is defined as the *coenergy*, and can be determined from

$$W_f' = \int_0^i \lambda\,di \tag{5.54}$$

Such a magnetic coenergy has no physical meaning. However, it can be useful in determining force (or torque) developed in an electromagnetic system. From Figure 5.13a, for a coil current i and the resultant flux linkage λ

$$W_f + W_f' = \lambda \times i \tag{5.55}$$

$$\text{energy} + \text{coenergy} = \lambda \times i \tag{5.56}$$

Notice that W_f is greater than $W_{f'}$ if the $\lambda - i$ characteristic is nonlinear and that W_f is equal to $W_{f'}$ if the $\lambda - i$ characteristic is linear,* as shown in Figure 5.13b.

5.3.4 MAGNETIC FORCE IN A SATURABLE SYSTEM

Consider the electromechanical system shown in Figure 5.12 and assume that it is made up of saturable ferromagnetic material. It shows that when the air gap is large, the resultant, $\lambda - i$, characteristic is almost a straight line; when the air gap is very small, the characteristic is almost a straight line for small values of flux linkage. However, as flux linkage is increased, the curvature of the characteristic starts to appear because of the saturation of the magnetic core.

Assume that λ is a function of x and i and that there is a differential movement of the operating point corresponding to a differential displacement of dx of the armature (i.e., the movable part) made at a low speed (i.e., at a constant current), as shown in Figure 5.15a.

---

* In other words, if the magnetic core has a constant permeability, for example, as in the air, the energy and coenergy are equal.

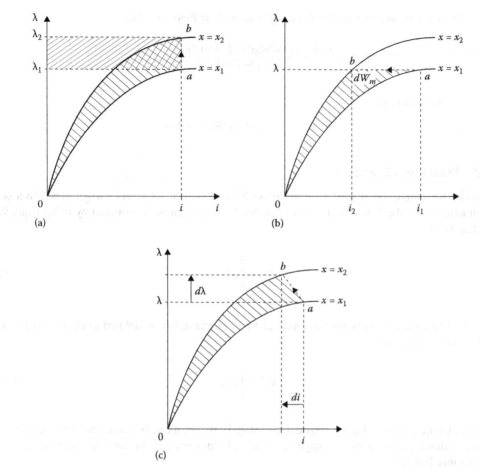

FIGURE 5.15 Energy balance in a saturable system: (a) constant current operation, (b) constant flux operation, and (c) a general case.

In other words, the armature of Figure 5.12 moves from the operating point a (where $x = x_1$) to a new operating point b (where $x = x_2$) so that at the end of the movement the air gap decreases. If the armature has moved slowly, the current i has stayed constant during the motion, causing the operating point to move upward from point a to b, as shown in Figure 5.15a. For this displacement,* during which the flux linkage changes, neither the emf nor the dW_e is zero. Therefore, from Equation 5.20,

$$dW_m = dW_e - dW_f = d(W_e - W_f) \tag{5.57}$$

Hence, for the displacement,

$$\mathcal{F}_f = \left. \frac{\partial (W_e - W_f)}{\partial x} \right|_{i=\text{constant}} \tag{5.58}$$

Since the motion has taken place under constant-current conditions, the mechanical work performed is depicted by the shaded area in Figure 5.15a. That area also represents the increase in the coenergy. Thus,

---

* Notice the increase in coenergy during the move from position a to b, as shown in Figure 5.15a.

$$dW_m = dW_f' \tag{5.59}$$

Substituting Equation 5.25 into Equation 5.59,

$$\mathcal{F}_f dx = dW_f' \tag{5.60}$$

The force on the armature is then

$$\mathcal{F}_f = \left. \frac{\partial W_f'(i, x)}{\partial x} \right|_{i=\text{constant}} \tag{5.61}$$

Since for any armature position,

$$W_f'(i, x) = \int_0^i \lambda di \tag{5.62}$$

and

$$\lambda = N\Phi \tag{5.63}$$

$$i = \frac{\mathcal{F}}{N} \tag{5.64}$$

Substituting Equations 5.63 and 5.64 into (5.62) gives the coenergy as a function of the mmf and displacement as

$$W_f'(i, x) = \int_0^{\mathcal{F}} \Phi d\mathcal{F} \tag{5.65}$$

and the force on the armature is then

$$\mathcal{F}_f = \left. \frac{\partial W_f'(\mathcal{F}, x)}{\partial x} \right|_{\mathcal{F}=\text{constant}} \tag{5.66}$$

Figure 5.15b illustrates a differential movement of the operating point in the λ–i diagram, corresponding to a differential displacement dx of the armature made at high speed, that is, at constant flux linkage. Here,

$$W_f = W_f(\lambda, x) \tag{5.67}$$

The electrical energy input for the movement is zero, since λ does not change and the emf is zero. The mechanical work done during the motion is represented by the shaded area, which depicts the decrease in the field energy. Since

$$dW_m = dW_f(\lambda, x) \tag{5.68}$$

and

$$\mathcal{F}_f dx = dW_m = -dW_f \tag{5.69}$$

Therefore,

$$\mathcal{F}_f = \frac{dW_m}{dx} = \left.\frac{\partial W_f(\lambda, x)}{\partial_x}\right|_{\lambda = \text{constant}} \tag{5.70}$$

It is interesting to see that at high-speed motion the electrical input is zero (i.e., $id\lambda = 0$) because the flux linkage has stayed constant and the mechanical output energy has been provided totally by the field energy. In the discussions so far, either i or λ has been kept constant.

In reality, however, neither condition is true. It is more likely that the change from position a to b follows a path such as the one shown in Figure 5.15c.

Also notice that for the linear case (i.e., when flux Φ *is* proportional to mmf $\mathcal{F}$), the energy and coenergy are equal. Thus,

$$W_f' = W_f \tag{5.71}$$

and

$$\mathcal{F}_f = \frac{\partial W_f'(\mathcal{F}, x)}{\partial_x} = \frac{\partial W_f(\mathcal{F}, x)}{\partial_x} \tag{5.72}$$

It is easier to use the inductance L *of* the excitation coil because L *is* independent *of* the current. Therefore,

$$W_f = \frac{Li^2}{2} \tag{5.73}$$

so that

$$\mathcal{F}_f = \frac{d}{dx}\left(\frac{Li^2}{2}\right) = \frac{i^2}{2}\frac{dL}{dx} \tag{5.74}$$

5.4   STUDY OF ROTATING MACHINES

In previous sections, the development *of translation motion* in an electromagnetic system has been reviewed extensively. However, most *of* the energy converters, especially the ones with higher power, develop *rotational motion*. Such a rotating electromagnetic system is made up *of* a fixed part known as the **stator** and a moving part known as the **rotor**, as previously explained in Section 2.2. In the following sections, singly excited and multiply excited rotating systems will be studied.

5.5   SINGLY EXCITED ROTATING SYSTEMS

To illustrate the application *of* electromechanical energy conversion principles to rotating systems, consider an elementary, singly excited two-pole rotating system, as shown in Figure 5.16. Such a system represents an elementary reluctance machine. Note that the *stator (pole) axis* is called the

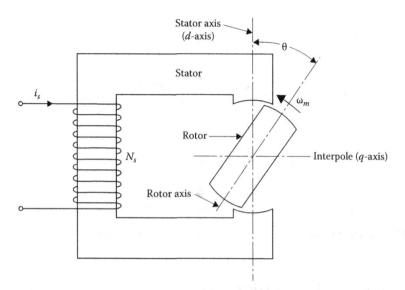

FIGURE 5.16 A singly excited rotating system.

direct axis or simply the *d-axis*, and that its **interpole axis** is also called the *quadrature axis* or simply the *q-axis*. Assume that a sinusoidal excitation is supplied to the stator winding, while the rotor is free to rotate on its shaft.

The variables are torque T and angle θ, and the differential mechanical energy output is $Td\theta$ when the torque and angle are assumed positive in the same direction (i.e., *motor action*). The developed torque can be expressed as

$$T_d = T_f = \frac{\partial W_f'}{\partial \theta} = \frac{\partial W_f'(i,\theta)}{\partial \theta} \tag{5.75}$$

For each revolution *of* the rotor, there are two cycles of reluctance, since the reluctance varies sinusoidally. Figure 5.16 shows the variation *of* inductance with rotor angular position θ as the rotor rotates with a uniform speed ω_m in a reluctance machine. Because the inductance is a periodic function *of* 2θ, it can be represented by a Fourier series as

$$L(\theta) = L_0 + L_2 \cos 2\theta \tag{5.76}$$

ignoring the higher-order terms. The variables used are defined as in Figure 5.17. The stator excitation current is

$$i = I \sin\omega_s t \tag{5.77}$$

which is a sinusoidal excitation whose angular frequency is ω_s. Since the air-gap region is linear, the coenergy in the magnetic field of the air-gap region can be expressed as

$$W_f' = W_f = \frac{1}{2} L(\theta) i^2 \tag{5.78}$$

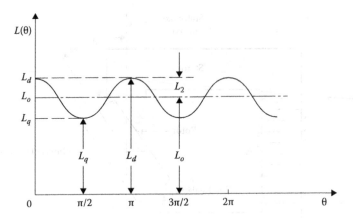

FIGURE 5.17 Variation of inductance with rotor angular position θ, as the rotor rotates in a reluctance machine.

and therefore the developed torque can be expressed as

$$T_d = \frac{\partial W_f'(i,\theta)}{\partial \theta}$$

$$= \frac{1}{2} i^2 \frac{\partial(\theta)}{\partial \theta} \tag{5.79}$$

which in terms of current and inductance variations can be reexpressed as

$$T_d = -I^2 L_2 \sin 2\theta \sin^2 \omega_s t \tag{5.80}$$

Here, it is assumed that the rotor rotates at an angular velocity ω_m; therefore, at any given time

$$\theta = \omega_m t - \delta \tag{5.81}$$

Note that at $t=0$ the current i is zero, thus the angular rotor position becomes

$$\theta = -\delta \tag{5.82}$$

The instantaneous torque expression given by Equation 5.80 can be expressed in terms of ω_m and ω_s by using the following trigonometric equations:

$$\sin^2 A = \frac{1 - \cos 2A}{2} \tag{5.83}$$

$$\sin A \cos B = \frac{1}{2} \sin (A+B) + \frac{1}{2} \sin (A-B) \tag{5.84}$$

Hence, the instantaneous (electromagnetic) developed torque becomes

$$T_d = -\frac{I^2 L_2}{2} \left\{ \sin 2(\omega_m t - \delta) - \frac{1}{2} \sin 2[(\omega_m + \omega_s)t - \delta] - \frac{1}{2} \sin 2[(\omega_m - \omega_s)t - \delta] \right\} \tag{5.85}$$

As can be seen from Equation 5.85, the torque equation is made up of the sum of sinusoids of various frequencies. Therefore, in most cases, the average torque over a period of time is zero, since the value of each term integrated over a period is zero. Under such conditions, the machine cannot operate as a motor to provide a load torque* to its shaft. The only case in which the average (*load*) *torque*[†] is nonzero is when

$$\omega_m = \omega_s \text{ rad/s} \tag{5.86}$$

so that

$$Td(ave) = -\frac{I^2 L_2}{4} \sin 2\delta \tag{5.87}$$

Also, it can be seen in Figure 5.17 that

$$L_2 = \frac{L_d - L_q}{2} \tag{5.88}$$

where L_d and L_q are defined as the *direct-axis inductance and quadrature-axis inductance*, representing the maximum and minimum values of inductance, respectively. Therefore, substituting Equation 5.88 into Equation 5.87, the average developed torque[‡] can be expressed as

$$BT_{d(ave)} = -\frac{I^2(L_d - L_q)}{8} \sin 2\delta \tag{5.89}$$

Based on the previous review, the following summary and conclusions can be made:

1. Only at a certain speed, given by Equation 5.86, can such a machine develop an average torque in either rotational direction. This speed is defined as the *synchronous speed*, at which the speed of mechanical rotation in radians per second is equal to the angular frequency of the electrical source.
2. Because the torque is a function of the reluctance variation with rotor position, such an apparatus is called a *synchronous reluctance machine*. Therefore, if there is no inductance or reluctance variation with rotor position (i.e., if $L_d = L_q$), the torque becomes zero. This can easily be concluded from Equation 5.89.
3. As can be concluded from Equation 5.89, the developed torque is a function of the angle δ, which is called the *torque angle*. The torque varies sinusoidally with the angle δ. Therefore, the angle δ can be used as a measure of the torque.
4. When $\delta < 0$ and $T_{d(ave)} > 0$, the developed torque is in the direction of rotation, and the machine operates as a *motor*, as can be seen in Figure 5.18. This torque maintains the speed of the rotor against friction, windage, and any external load torque applied to the rotor shaft.

---

* *Load torque* is defined as a torque in opposition to the rotor motion. Therefore, the total torque is equal to the difference between the magnetic torque and the load torque in the forward direction.
† It is also interesting to note that the total torque as a function of time has pulsating components even when $\omega_m = \omega_s$. However, because of the typical heavy steel rotor of a synchronous machine, it cannot significantly react to such pulsating components. Therefore, they cannot affect the average torque. Succinctly put, the rotor's mass functions as a *low-pass filter*.
‡ In general, the basic difference between various rotating machines is based on how the stator and rotor mmfs are kept displaced with respect to each other at all times so that they incline to align continuously and develop an average torque. This phenomenon is known as the *alignment principle*.

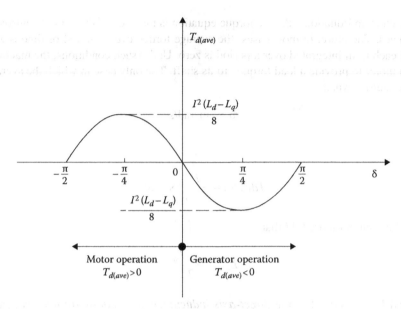

FIGURE 5.18   Variation of developed torque by a synchronous reluctance machine.

5. Ignoring the effects of friction and windage, the load torque can determine the angle δ. For example, a greater load torque can cause the rotor to operate at a larger negative δ. Since power and torque are proportional to constant speed, there is definitely a power limit, and at loads beyond the crest of the curve, shown in Figure 5.18, the motor will stall. The maximum torque for motor operation takes place at $\delta = -\pi/4$, and is called the **pull-out** *torque*. As previously stated, any load that requires a torque greater than the maximum torque causes an unstable operation of the machine; the machine pulls out of synchronism and comes to a standstill.

6. If the shaft of the same machine is driven by a prime mover, the angle δ will advance and the machine will absorb torque and power, and will supply electrical power as a generator. In other words, when $\delta > 0$ and $T_{d(ave)}$, the developed torque resists the rotation. Therefore, an external driving torque has to be applied to the rotor shaft to maintain the rotor at a synchronous speed. The mechanical energy supplied to the system after meeting the friction and windage losses is converted into electrical energy, that is, the machine operates as a generator. But this can happen only if the stator winding is already connected to an ac source, which acts as a sink when the external driving torque is applied and the machine begins to generate. As shown in Figure 5.18, the maximum torque for generator operation takes place at $\delta = \pi/4$.

7. If the driving torque provided by the prime mover is greater than the sum of the developed torque and that due to friction and windage, then the machine is driven above synchronous speed. It may therefore run away unless the prime mover speed is controlled and the continuous energy conversion process is stopped. In summary, a given machine can develop only a certain maximum power and is limited to the rate of energy conversion.

8. It is interesting that a mechanical speed $\omega_m = \omega_s$ will also provide a nonzero average developed torque. Therefore, such a reluctance motor cannot start by itself, but will continue to run in the direction in which it is started.

Because of the variation of reluctance with rotor position, the induced voltage in the stator coil will have a third-harmonic component. Such an unwanted characteristic makes reluctance machines useless as practical generators and restricts their size as motors.

However, small reluctance motors, when they are designed to develop starting torque, can be used to drive electric clocks, record players, and other devices, since they provide constant speed.

Example 5.6

Suppose that a two-pole reluctance motor operates at 60 Hz and 6 A. If its direct-axis inductance and quadrature-axis inductance are 0.8 H and 0.2 H, respectively, determine its maximum average developed torque.

Solution

From Equation 5.89,

$$T_{d(ave)} = -\frac{I2(L_d - L_q)}{8}\sin 2\delta$$

$$= -\frac{36(0.8 - 0.2)}{8}\sin 2\delta$$

$$= -2.7\sin 2\delta$$

Since $\sin 2\delta = 1$ when $\delta = 45°$,

$$T_{d(ave)} = -2.7\sin 90° = -2.7\,\text{N}\cdot\text{m}$$

5.6 MULTIPLY EXCITED ROTATING SYSTEMS

The general principles that were developed in the previous section also apply to multiply excited (i.e., multicoil) rotating systems. As an example, consider the doubly excited rotating system shown in Figure 5.19.

Notice that this system is the same as the one shown in Figure 5.16, except that the rotor also has a coil which is connected to its electrical source through *fixed (carbon) brushes* and rotor-mounted

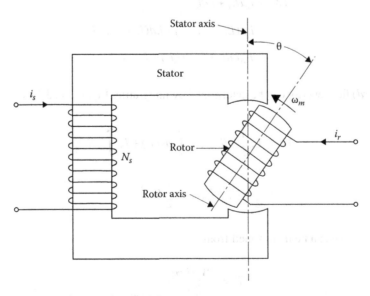

FIGURE 5.19 A doubly excited rotating system.

slip rings (or *collector rings*). The flux linkages of the stator and rotor windings, respectively, can be written as

$$\lambda_s = L_{ss}i_s + L_{sr}i_r \tag{5.90}$$

$$\lambda_r = L_{rs}i_s + L_{rr}i_r \tag{5.91}$$

where
  L_{ss} is the self-inductance of the stator winding
  L_{rr} is the self-inductance of the rotor winding
  $L_{sr}=L_{rs}$ is the mutual inductances between stator and rotor windings

Note that all these inductances depend on the position θ of the rotor (which is the angle between the magnetic axes of the stator and the rotor windings). Since for a linear magnetic system $L_{sr}=L_{rs}$, Equations 5.90 and 5.91 can be expressed in the matrix form as

$$\begin{bmatrix} \lambda_s \\ \lambda_r \end{bmatrix} = \begin{bmatrix} L_{ss} & L_{sr} \\ L_{sr} & L_{rr} \end{bmatrix} \begin{bmatrix} i_s \\ i_r \end{bmatrix} \tag{5.92}$$

If the system's rotor is prevented from rotating so that there is no mechanical output from its shaft, then the stored field energy $W_f W$ of the system can be found by establishing the currents i_s and i_r in its stator and rotor windings, respectively. Therefore,

$$dW_f = v_s i_s dt + v_r i_r dt$$

$$= i_s d\lambda_s + i_r d\lambda_r \tag{5.93}$$

Thus, for such a linear system, the differential field energy can be found by substituting Equations 5.90 and 5.91 into Equation 5.93 so that

$$dW_f = i_s d\lambda_s + i_r dl_r$$

$$= i_s d(L_{ss}i_s + L_{sr}i_r) + i_r d(L_{sr}i_s + L_{rr}i_r)$$

$$= L_{ss}i_s di_s + L_{sr} d(i_s i_r) + L_{rr}i_r di_r \tag{5.94}$$

The total (*stored*) field energy can be determined by integrating Equation 5.94 as

$$W_f = L_{ss}\int_0^{i_s} i_s di_s + L_{sr}\int_0^{i_s,i_r} d(i_s i_r) + L_{rr}\int_0^{i_r} i_r di_r$$

$$= \frac{1}{2}L_{SS}i_s^2 + L_{sr}i_s i_r + \frac{1}{2}L_{rr}i_r^2 \tag{5.95}$$

The developed torque can be determined from

$$T_d = \left. \frac{\partial W_f'(i,\theta)}{\partial \theta} \right|_{i=\text{constant}} \tag{5.96}$$

Since in a linear magnetic system, energy and coenergy are equal, that is

$$W_f = W'_f \tag{5.71}$$

the instantaneous (electromagnetic) developed torque can be expressed as

$$T_d = \frac{i_s^2}{2} \frac{dL_{ss}}{d\theta} + i_r i_r \frac{dL_{sr}}{d\theta} + \frac{i_r^2}{2} \frac{d\theta_{rr}}{d\theta} \tag{5.97}$$

Note that the first and third terms on the right-hand side of Equation 5.97 depict torques developed in the rotating machine due to variations of self-inductances as a function of rotor position. They represent the reluctance torque components of the torque; however, the second term represents the torque developed by the variations of the mutual inductance between the stator and rotor windings. Furthermore, multiply excited rotating systems, having more than two coils, are treated in a similar manner.

Consider the doubly excited rotating system shown in Figure 5.19 and assume that R_s and R_r are the resistances of the stator and rotor windings, respectively. The voltage–current relationships for the stator and rotor circuits can be written as

$$v_s = i_s R_s + \frac{d\lambda_s}{dt} \tag{5.98}$$

$$v_r = i_r R_r + \frac{d\lambda_r}{dt} \tag{5.99}$$

In general, the inductances L_{ss}, L_{rr}, and L_{sr} are functions of the angular position θ of the rotor, and the currents are time functions. Therefore, for the stator

$$\frac{d\lambda_s}{dt} = \frac{d}{dt}\left[L_{ss}(\theta)i_s(t) + L_{rs}(\theta)i_{ir}(t)\right]$$

$$\frac{d\lambda_s}{dt} = L_{ss}\frac{di_s}{dt} + i_s\frac{dL_{ss}}{dt}\frac{d\theta}{dt} + L_{rr}\frac{di_r}{dt} + i_r\frac{dL_{rs}}{d\theta}\frac{d\theta}{dt} \tag{5.100}$$

Similarly, for the rotor

$$\frac{d\lambda_r}{dt} = L_{rs}\frac{di_r}{dt} + i_s\frac{dL_{rs}}{dt}\frac{d\theta}{dt} + L_{rr}\frac{di_r}{dt} + i_r\frac{dL_{rr}}{d\theta}\frac{d\theta}{dt} \tag{5.101}$$

By substituting Equations 5.100 and 5.101 into Equations 5.98 and 5.99, respectively,

$$v_s = \left[i_s R_s + L_{ss}\frac{di_s}{dt}\right] + \left[\left(i_s\frac{dL_{ss}}{d\theta} + i_r\frac{dL_{rs}}{d\theta}\right)\frac{d\theta}{dt}\right] + \left[L_{rs}\frac{di_r}{dt}\right] \tag{5.102}$$

$$v_r = \left[i_r R_r + L_{rr}\frac{di_r}{dt}\right] + \left[\left(i_s\frac{dL_{rs}}{d\theta} + i_r\frac{dL_{rr}}{d\theta}\right)\frac{d\theta}{dt}\right] + \left[L_{rs}\frac{di_r}{dt}\right] \tag{5.103}$$

In Equations 5.102 and 5.103, the first terms on the right sides of the equations represent the *self-impedance voltage* v_z, the second terms represent the *speed voltage or motional voltage* v_m, and the third terms represent the *transformer voltage* v_t.

Therefore, the voltage equations for the stator and rotor can be expressed in the form

$$v = v_z + v_m + v_t \tag{5.104}$$

Note that in many cases, the self-inductances L_{ss} and L_{rr} are not dependent on the angular position of the rotor. Thus, Equations 5.97, 5.102, and 5.103 reduce to

$$T_d = i_s i_r \frac{dL_{sr}}{dt} \tag{5.105}$$

$$v_s = \left(i_s R_s + L_{ss} \frac{di_s}{dt} \right) + \left(i_r \frac{d\theta}{dt} \right) \frac{dL_{sr}}{d\theta} + L_{rs} \frac{di_r}{dt} \tag{5.106}$$

$$v_r = \left(i_r R_r + L_{rr} \frac{di_r}{dt} \right) + \left(i_s \frac{d\theta}{dt} \right) \frac{dL_{sr}}{d\theta} + L_{rs} \frac{di_s}{dt} \tag{5.107}$$

If the resistances of the stator and rotor are negligible, then Equations 5.106 and 5.107 further reduce to

$$v_s = L_{ss} \frac{di_s}{dt} + \left(i_r \frac{d\theta}{dt} \right) \frac{dL_{sr}}{d\theta} + L_{rs} \frac{di_r}{dt} \tag{5.108}$$

$$v_r = L_{rr} \frac{di_r}{dt} + \left(i_s \frac{d\theta}{dt} \right) \frac{dL_{sr}}{d\theta} + L_{rs} \frac{di_s}{dt} \tag{5.109}$$

In matrix notation, the total (stored) field energy, given by Equation 5.95 can be expressed as

$$W_f = \frac{1}{2} [i]^t [L][i] \tag{5.110}$$

where
   $[i]$ is the column matrix
   $[i]^t$ is the transpose of matrix $[i]$, that is, *a row matrix*
   $[L]$ is the inductance matrix of the system

Also, in matrix notation, the developed torque can be expressed as

$$T_d = \frac{1}{2} [i]^t \frac{\partial}{\partial \theta} ([L][i]) \tag{5.111}$$

and the voltage can be expressed as

$$[v] = [i][R] + \frac{d}{dt} ([L][i]) \tag{5.112}$$

Example 5.7

Consider the doubly excited rotating system shown in Figure 5.19. Assume that the self-inductances of the stator and rotor windings are 9 H and 1 H, respectively, and that the mutual inductance between its stator and rotor windings is 2 H. If its stator and rotor currents are 16 A and 8 A, respectively, determine the total stored magnetic field energy in the system.

Solution

From Equation 5.95, the total stored magnetic field energy in the system can be found as

$$W_f = \frac{1}{2}L_{ss}i_s^2 + L_{sr}i_s i_r + \frac{1}{2}L_{rr}i_r^2$$

$$= \frac{1}{2}(9\,\text{H})(16\,\text{A})^2 + 2(16\,\text{H})(8\,\text{A}) + \frac{1}{2}(1\,\text{H})(8\,\text{A})^2 = 1440\,\text{J}$$

5.7 CYLINDRICAL MACHINES

Figure 5.20 shows a cross-sectional view of a single-phase, two-pole cylindrical rotating machine with a uniform air gap. Such machines are also called *smooth-air-gap machines, uniform-air-gap machines,* or *round-rotor machines.* Note that previous sections dealt with rotating machines with salient poles.*

As shown in Figure 5.20, a cylindrical machine[†] has a cylindrical rotor in its cylindrical stator. The rotor is free to rotate, and its instantaneous angular position θ is defined as the displacement of the rotor's magnetic axis with respect to the stator's magnetic axis.

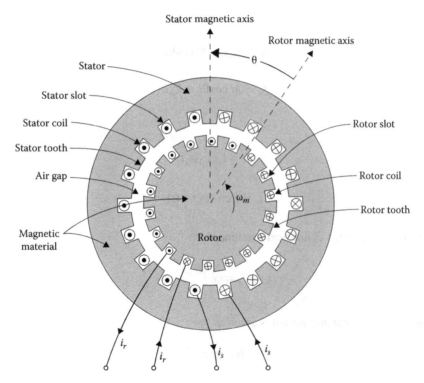

FIGURE 5.20 Cross-sectional view of a single-phase, smooth-air-gap machine having a cylindrical rotor in a cylindrical stator.

---

* The rotating machines can be classified based on their structures: (1) Those machines with salient stator but nonsalient rotor (i.e., the rotor is round or cylindrical), for example, dc commutator machines. (2) Those machines with nonsalient stator but salient rotor, for example, small reluctance machines and low-speed synchronous machines. (3) Those machines with salient stator and salient rotor, for example, some special rotating machines. (4) Those machines with nonsalient stator and nonsalient rotor, for example, induction motors and high-speed synchronous machines.

† Most electrical machines are of the cylindrical type because they develop greater torques even though their construction is more complex.

In a real rotating machine, the windings are distributed over a number of slots so that their mmf waves can be approximated by space sinusoids. The structure shown in Figure 5.20 is called *a smooth-air-gap* machine, because it can be accurately modeled mathematically by assuming that the reluctance of the magnetic path seen by each circuit is independent of rotor position. Also, such a model ignores the effects of slots and teeth on the magnetic path as the angle is changed. Of course, in an actual machine, the slots and teeth are relatively smaller than those shown in Figure 5.20.

Furthermore, special construction techniques, such as skewing the slots of one member slightly with respect to a line parallel to the axis, substantially minimize these effects. As a result of such construction, it can be assumed that the self-inductances L_{ss} and L_{rr} are constant and no reluctance torques are produced. The mutual inductance L_{sr} changes with rotor position. Therefore,

$$L_{sr} = M \cos \theta \tag{5.113}$$

where
  θ is the angle between the magnetic axis of the stator and rotor windings
  M is the peak value of the mutual inductance L_{sr}

Thus,

$$\lambda_s = L_{ss} i_s + M \cos \theta i_r \tag{5.114}$$

$$\lambda_r = M \cos \theta i_s + L_{rr} i_r \tag{5.115}$$

where

$$i_s = I_s \cos \omega_s t \tag{5.116}$$

$$i_r = I_r \cos(\omega_r t + \alpha) \tag{5.117}$$

The torque developed in the cylindrical machine is

$$T_d = i_s i_r \frac{dL_{sr}}{d\theta} \tag{5.118}$$

The position of the rotor at any instant is given as

$$\theta = \omega_m t + \delta \tag{5.119}$$

where
  ω_m is the angular velocity of the rotor in rad/s
  δ is the rotor position at $t=0$

Hence, by substituting Equations 5.113, 5.116, and 5.117 into Equation 5.118, the instantaneous electromagnetic torque developed by the machine can be expressed as

$$T_d = -I_s I_r M \cos \omega_s t \cos(\omega_r t + \alpha) \sin(\omega_m t + \delta) \tag{5.120}$$

Further, by using the trigonometric identities,

$$T_d = -\frac{I_sI_rM}{4}\{\sin\{[\omega_m+(\omega_s+\omega_r)]t+\alpha+\delta\}$$
$$+ \sin\{[\omega_m-(\omega_s+\omega_r)]t-\alpha+\delta\}$$
$$+ \sin\{[\omega_m-(\omega_s-\omega_r)]t-\alpha+\delta\}$$
$$+\sin\{[\omega_m-(\omega_s-\omega_r)]t+\alpha+\delta\}\} \tag{5.121}$$

Thus, the torque changes sinusoidally with time. As a result, the average value of each of the sinusoidal terms in Equation 5.121 is zero, except when the coefficient t is zero. Hence, the average developed torque will be nonzero if

$$\omega_m = \pm(\omega_s \pm \omega_r) \tag{5.122}$$

which may also be expressed as

$$|\omega_m| = |\omega_s \pm \omega_r| \tag{5.123}$$

In other words, the machine will develop average torque if it rotates in either direction, at a speed that is equal to the sum or difference of the angular speeds of the stator and rotor currents.

5.7.1 SINGLE-PHASE SYNCHRONOUS MACHINE

Assume that $\omega_r=0$, $\alpha=0$ and $\omega_m=\omega_s$. Here, the rotor excitation current is a direct current I_r and the machine rotates at synchronous speed. Therefore, from Equation 5.121 the developed torque can be expressed as

$$T_d = -\frac{I_sI_rM}{2}[\sin(2\omega_st+\delta)+\sin\delta] \tag{5.124}$$

This torque is a pulsating instantaneous torque. Thus, the average developed torque is

$$T_d = -\frac{I_sI_rM}{2}\sin\delta \tag{5.125}$$

The machine operates as an idealized *single-phase synchronous machine* and has an average (unidirectional) developed torque. It has dc excitation in the rotor and ac excitation in the stator. It is important to point out that when $\omega_m=0$, the machine cannot develop an average torque and hence is not self-starting.

Note that the pulsating torque can cause noise, speed fluctuation, and vibration, and therefore is waste of energy. Such pulsating torque can be avoided in a polyphase machine. All large synchronous machines are polyphase machines.

5.7.2 SINGLE-PHASE INDUCTION MACHINE

Assume that $\omega_m=\omega_s-\omega_r$ and that cos and ω_r are two different angular frequencies. Therefore, both stator and rotor windings have ac currents but at different frequencies. The motor operates at an *asynchronous speed* (i.e., $\omega_m \neq \omega_s$ or $\omega_r \neq \omega_s$). From Equation 5.121, the instantaneous developed torque can be expressed as

$$T_d = -\frac{I_s I_r M}{4}[\sin(2\omega_s t + \alpha - \delta) + \sin(-2\omega_r t - \alpha + \delta) + \sin(2\omega_s t - 2\omega_r t - \alpha + \delta) + \sin(\alpha + \delta)] \quad (5.126)$$

This instantaneous torque is a pulsating torque. The average developed torque is

$$T_d = -\frac{I_s I_r M}{4}\sin(\alpha + \delta) \tag{5.127}$$

The machine operates as a *single-phase induction machine*. Its stator winding is excited by an ac current, and an ac current is induced in the rotor winding. Such a single-phase induction machine *cannot self-start*, since when $\omega_m = 0$ no average unidirectional torque is developed. The machine has to be brought up to the speed of $\omega_m = \omega_s - \omega_r$ to achieve an average developed torque. In order to avoid pulsating torque, polyphase induction machines are used in most applications.

Example 5.8

Consider a two-pole cylindrical rotating machine as shown in Figure 5.20. If it operates with a speed of $\omega_s = \omega_r = \omega_m = 0$ and $\alpha = 0$, determine the following:

(a) The instantaneous developed torque
(b) The average developed torque

Solution

(a) Since $\omega_s = \omega_r = \omega_m = 0$ and $\alpha = 0$, the excitations are direct currents I_s and I_r. Therefore, from Equation 5.121, the instantaneous developed torque can be found as

$$T_d = -I_s I_r M\sin \delta$$

which is a constant.
(b) Thus, the average developed torque is

$$T_{d(ave)} = -I_s I_r M\sin \delta$$

Such a machine operates as a *do rotary actuator*, developing a constant torque against any displacement δ caused by an external torque placed on the rotor shaft.

5.8   FORCE PRODUCED ON A CONDUCTOR

According to field theory, the force on a differential length of conductor $d\mathbf{L}$, carrying i, and located in a field $\mathbf{B}$ can be expressed as

$$d\mathcal{F} = id\mathbf{L} \times \mathbf{B} \tag{5.128}$$

The direction of the force is determined from the cross product of the vectors $d\mathbf{L}$ and $\mathbf{B}$.

Assume that a current-carrying conductor, having a length of $\mathbf{L}$, is within a uniform magnetic field of flux density $\mathbf{B}$, as shown in Figure 5.21a and b. Figure 5.21b shows the flux density $\mathbf{B}$, pointing into the page. The developed force on the conductor will make the conductor move, and the induced force can be expressed as

$$\mathcal{F} = i(L \times B)\mathrm{N} \tag{5.129}$$

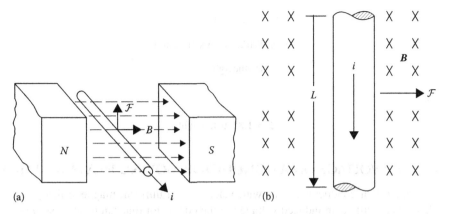

FIGURE 5.21   (a) A current-carrying straight conductor in a uniform magnetic field and (b) without a flux density B, pointing into the page.

where
    i is the magnitude of the current in the conductor
    L is the length of the conductor, given as a vector and with the same direction as the current flow
    B is the magnetic flux density vector

The direction of the force produced on the conductor is found by *Flemming's right-hand rule*. Therefore, if the index finger of the right-hand points in the direction of the vector L, and the middle finger points in the direction of the flux density vector B, then the thumb will point in the direction of the developed force on the conductor. The resulting electromechanical force* can be expressed as

$$\mathcal{F} = B \times i \times L \times \sin\theta \qquad (5.130)$$

where θ is the angle between the conductor and the flux density vector. It is important to note that as the current-carrying conductor is placed in the field B, *the field itself will change due to the effect of current i*. Therefore, the field B in the above equation is the magnetic field that exists *before the presence of the current i*. The maximum value of the force takes place when $\theta = 90°$. Thus,

$$\mathcal{F}_{max} = B \times i \times L \qquad (5.131)$$

In summary, the induction of such a mechanical force caused by a current flowing through the conductor in a magnetic field produces *motor action*.

Example 5.9

Consider a current-carrying conductor that is within a uniform magnetic field, as shown in Figure 5.21a. Assume that the magnetic flux density is 0.3 Wb/m², pointing into the page, and that the current flowing through the 2 m-long conductor is 3 A. Determine the magnitude and direction of the developed force on the conductor.

Solution

Based on the right-hand rule, the direction of the force is to the right, as shown in Figure 5.21b. Its magnitude is

---

* According to the electromagnetic force law, the interaction between a magnetic field and a current-carrying conductor produces a mechanical force.

$$\mathcal{F} = B \times i \times L \times \sin\theta$$

$$= (0.3\,\text{Wb/m}^2)(3\,\text{A})(2\,\text{m})\sin 90°$$

$$= 1.8\,\text{N to the right}$$

Thus,

$$\mathcal{F} = 1.8\,\text{N to the right}$$

5.9 INDUCED VOLTAGE ON A CONDUCTOR MOVING IN A MAGNETIC FIELD

Suppose that a straight conductor moves with velocity in a uniform magnetic field, as shown in Figure 5.22a. There will be an induced voltage* in the conductor that can be expressed as

$$e_{ind} = (v \times B) \cdot L \tag{5.132}$$

where
  v is the velocity of the conductor
  B is the magnetic flux density
  L is the length of the conductor

Assume that the vector L is in the same direction as the conductor's positive end.[†] The voltage induced in the conductor builds up so that the positive end is in the direction of the vector $(v \times B)$, as shown in Figure 5.22a. In summary, the induction of voltages in a conductor moving in a magnetic field causes *generator action*.

Note that mathematically the vector cross product $v \times B$ has a magnitude that is equal to the product of the magnitudes of v and B and the sine of the angle between them. Its direction can be found from the right-hand rule, which states that when the thumb of the right-hand points in the direction of v and the index finger points in the direction of B, $v \times B$ will be parallel to L. *If the conductor*

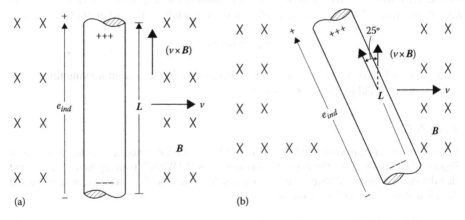

(a)   (b)

FIGURE 5.22 (a) A straight and vertical conductor moving in a uniform magnetic field and (b) a non-vertical conductor moving in a uniform field.

---

* In 1831, Faraday called this voltage an *induced voltage* because it occurred only when there was relative motion between the conductor and a magnetic field without any actual "physical" contact between them.

† The selection of the positive end is totally arbitrary, because if the selection is wrong, the resultant computed voltage value will be negative, indicating that a wrong assumption has been made in selecting the positive end.

is not oriented on a vertical line, the direction *of* **L** must be selected to make the smallest possible angle with the direction *of* $v \times B$.

Example 5.10

Consider a 2 m-long conductor that is moving with a velocity *of* 4 m/s to the right, within a uniform magnetic field, as shown in Figure 5.22a. The magnetic field density is 0.3 Wb/m², pointing into the page, and the conductor length is oriented from the bottom toward the top. Determine the following:

(a) The magnitude *of* the resulting induced voltage
(b) The polarity *of* the resulting induced voltage

Solution

(a) The velocity vector v is perpendicular to the magnetic field density vector **B**, and therefore $v \times B$ *is* parallel to the conductor length vector **L**. The magnitude *of* the resulting induced voltage is

$$e_{ind} = (v \times B) \cdot L$$

$$= (vB \sin 90°)L \cos 0°$$

$$= v \times B \times L$$

$$= (4\,\text{m/s})(0.3\,\text{Wb/m}^2)(2\,\text{m})$$

$$= 2.4\,\text{V}$$

(b) The polarity *of* the resulting induced voltage is positive at the top *of* the conductor and negative at the bottom *of* the conductor, as shown in Figure 5.22a.

Example 5.11

Consider a 1.5 m long conductor that is moving with a velocity *of* 5 m/s to the right within a uniform magnetic field, as shown in Figure 5.22b. *If* the magnetic field density is 0.8 Wb/m², pointing into the page, and the conductor length is oriented from the bottom toward the top, determine the following:

(a) The magnitude *of* the resulting induced voltage
(b) The polarity of the resulting induced voltage

Solution

(a) The magnitude *of* the resulting induced voltage is

$$e_{ind} = (v \times B) \cdot L$$

$$= (vB \sin 90°)L \cos 0°$$

$$= v \times B \times L$$

$$= [(5\,\text{m/s})(0.8\,\text{Wb/m}^2)\sin 90°](1.5\,\text{m})\cos 25°$$

$$= 5.44\,\text{V}$$

(b) The polarity of the resulting induced voltage is positive at the top of the conductor and negative at the bottom of the conductor, as shown in Figure 5.22a.

PROBLEMS

5.1   The rotational speed of a motor (i.e., its shaft speed) is 3600 rev/min. Determine its angular velocity (i.e., its shaft speed) in rad/s.

5.2   If the motor in Problem 5.1 is operating at 50 Hz frequency, instead of 60 Hz, determine its new angular velocity in rad/s.

5.3   A special-purpose motor is operating at 25 Hz frequency. Determine the following:
(a) Its angular velocity (i.e., its shaft speed) in rad/s
(b) Its rotational speed (i.e., its shaft speed) in rev/min

5.4   If a motor is delivering 200 N m of torque to its mechanical load at a shaft speed of 3600 rpm, determine the following:
(a) The power supplied to the load in watts
(b) The power supplied to the load in horsepower

5.5   Assume that a coil, having a sectional area of 0.25 m² with $N = 15$ turns, is rotating around its horizontal axis with a constant speed of 1800 rpm in a uniform and vertical magnetic field of flux density $B = 0.75$ T. If the total magnetic flux passing through the coil is given by $\Phi = AB \cos \omega t$ Wb, where A is the sectional area of the coil, determine the maximum and effective values of the induced voltage in the coil.

5.6   Consider the linear electromechanical system shown in Figure 5.14. Since the system is considered to be linear, its core reluctance is negligibly small. The core depth is given as b. If the system is excited by two identical current sources, determine the following:
(a) The force of attraction between the poles in terms of current i and the geometry involved
(b) The force between the poles, if the current is reversed in one coil

5.7   Assume that there is a two-pole cylindrical rotating machine, as shown in Figure 5.20, and that it operates with $\omega_s = \omega_r$ and $\omega_m = 0$. Determine the following:
(a) The instantaneous developed torque
(b) The average developed torque

5.8   Consider the current-carrying conductor that is within a uniform magnetic field, as shown in Figure 5.21b. Assume that the magnetic field density is 0.25 Wb/m², pointing out of the page, and that the current flowing through the 0.3 m long conductor is 1.5 A. Determine the following:
(a) The magnitude of the developed force in N
(b) The direction of the developed force, if the current is flowing from the top toward the bottom
(c) The direction of the developed force, if the current is flowing from the bottom toward the top

5.9   Consider a current-carrying conductor that is within a uniform magnetic field, as shown in Figure 5.21b. Assume that the magnetic field density is 0. 5 Wb/m² and that the current flowing through the 0.6 in long conductor (in the direction as shown, i.e., from the top toward the bottom) is 2 A. Determine the following:
(a) The magnitude of the developed force in N
(b) The direction of the developed force, if the magnetic flux density vector is pointing into the page
(c) The direction of the developed force, if the magnetic flux density vector is pointing out of the page
(d) The direction of the developed force, if the magnetic flux density vector is pointing into the page and the direction of the current flow is reversed (i.e., from the bottom toward the top)
(e) The direction of the developed force, if the magnetic flux density is pointing out of the page and the direction of the current flow is reversed (i.e., from the bottom toward the top)

5.10 Consider a 0.5 m long conductor that is moving with a velocity of 2 m/s to the right within a uniform magnetic field, as shown in Figure 5.22a. Assume that the magnetic field density is 0.25 Wb/m², pointing into the page, and that the conductor length is oriented from the bottom toward the top. Determine the following:

(a) The magnitude of the resulting induced voltage

(b) The polarity of the resulting induced voltage

(c) The polarity of the resulting induced voltage, if the conductor length is oriented from the top toward the bottom

5.11 Consider a 0.5 m long conductor that is moving with a velocity of 4 m/s to the right within a uniform magnetic field, as shown in Figure 5.22b. Assume that the magnetic field density is 0.75 Wb/m², pointing into the page, and that the conductor length is oriented from the bottom toward the top. Determine the following:

(a) The magnitude of the resulting induced voltage

(b) The polarity of the resulting induced voltage

(c) The polarity of the resulting induced voltage, if the magnetic flux density vector is pointing out of the page

(d) The polarity of the resulting induced voltage, if the magnetic flux density vector is pointing into the page and the conductor length is oriented from the bottom toward the top

(e) The polarity of the resulting induced voltage, if the magnetic flux density vector is pointing out of the page, the conductor length is oriented from the top toward the bottom, and the conductor is moving to the left

6 Induction Machines

> Any man may make a mistake; none but a fool will stick to it.
>
> **M.T. Cicero, 51 BC**

> Time is the wisest counselor.
>
> **Pericles, 450 BC**

> When others agree with me, I wonder what is wrong!
>
> **Author Unknown**

6.1 INTRODUCTION

Because of its relatively low cost, simple and rugged construction, minimal maintenance requirements, and good operating characteristics that satisfy a wide variety of loads, the induction motor is the most commonly used type of ac motor. Induction motors range in size from a few watts to about 40,000 hp. Small fractional-horsepower motors are usually single phase and are used extensively for domestic appliances, such as refrigerators, washers, dryers, and blenders.

Large induction motors (usually above 5 hp) are always designed for three-phase operation to achieve a constant torque and balanced network loading. In particular, where very large machinery is to be operated, the three-phase induction motor* is the *workhorse* of the industry. In contrast to dc motors, induction motors can operate from supplies in excess of 10 kV. Figure 6.1a shows a typical three-phase induction motor, while Figure 6.1b shows a large three-phase induction motor. Figure 6.2 shows a totally enclosed, fan-cooled, three-phase induction motor. Figure 6.3 shows a totally enclosed, fan-cooled, explosion-proof, three-phase induction motor for use in hazardous environments.

In typical induction motors, the stator winding (the *field winding*) is connected to the source, and the rotor winding (the *armature winding*) is short-circuited for many applications, or may be closed through external resistances. In such a motor, alternating current passing through a fixed stator winding sets up a rotating magnetic field.

Thus, an induction motor is *a singly excited* motor (as opposed to a *doubly* excited synchronous motor). In such motor, alternating current passing through a fixed stator winding sets up a rotating magnetic field. This moving field induces† currents in closed loops of wire mounted on the rotor. These currents set up magnetic fields around the wires and cause them to follow the main magnetic field as it rotates.

---

* The whole concept of polyphase ac, including the induction motor, was developed by Nikola Tesla and patented in 1888. In 1895, the Niagara Falls hydroplant, using the Tesla polyphase ac system concept, went into operation. This was the first large-scale application of the polyphase ac system. However, the first paper written on the induction machine was authored by Galileo Ferraris, an Italian, who also developed a new per-phase equivalent circuit for the new motor. The circuit had a primary and secondary in much the same manner as the present per-phase circuit. Unfortunately, he did not recognize the need for slip s in the circuit, instead he used R_2 rather than R_2 1s as the secondary resistor. After some mathematical analysis, Ferraris concluded that this new motor was not practical since from the maximum power transfer theorem only a maximum efficiency of 50% can be attained when $R_2 = R_1$! Thus, Ferraris promptly gave up the development of the motor as impractical and went on to become famous in other areas, but not in electrical machines. Needless to say, Tesla, being an experimentalist, was never bothered with such niceties, but simply proceeded directly from the concept to the implementation and physically demonstrated that it worked.

† An *induction motor* is so called, because the driving force is provided by an electric current induced in a rotor due to its interaction with a magnetic field.

(a)

(b)

FIGURE 6.1   (a) A small three-phase induction motor and (b) a large three-phase induction motor. (Courtesy of Reliance Electric Company, Beachwood, OH.)

Therefore, the operation of the induction motor depends on the rotating field passing through the loops on the rotor, which must always turn more slowly than the rotating field. Since no current has to be supplied to the rotor, the induction motor is simple to construct and reliable in operation.

This class of rotating machines derives its name from the fact that the rotor current results from *induction*, rather than *conduction*.* A given induction machine can be operated in the motor region, generator region, or braking region, as shown in Figure 6.4.

In the *motor mode*, its operating speed is slightly less than its synchronous speed,[†] but in the *generator mode*, its operating speed is slightly greater than its synchronous speed and it needs magnetizing reactive power from the system that it is connected to in order to supply power. The

---

* Since an induction motor runs below synchronous speed, it is also known as an *asynchronous* (i.e., not synchronous) machine.
[†] As the winding loops of the rotor turns faster they try to catch up to the rotating magnetic field, and the difference between the two speeds gets smaller. The size of the induced currents, and therefore the size of the driving force, also gets smaller. The rotor thus settles down to a steady speed, which is slower than that of the rotating magnetic field.

FIGURE 6.2 A totally enclosed, fan-cooled, three-phase induction motor. (Courtesy of General Electric Canada, Inc., Mississauga, Canada.)

full-load speed of three-phase induction motors is often within 7% of the synchronous speed, even though full-load speeds of 1% below the synchronous speed are not uncommon.

In the *braking mode* of operation, a three-phase induction motor running at a steady-state speed can be brought to a quick stop by interchanging two of its stator leads. By doing this, the phase sequence, and therefore the direction of rotation of the magnetic field, is suddenly reversed; the motor comes to a stop under the influence of torque and is immediately disconnected from the line before it can start in the other direction. This is also known as the *plugging operation*.

Since the induction motor cannot produce its excitation, it needs reactive power; it draws a lagging current from the source and operates at a power factor that is less than unity (usually, above 0.85). However, it runs at lower lagging power factors when lightly loaded.

To limit the reactive power, the magnetizing reactance has to be high, and, thus, the air gap is shorter than in synchronous motors of the same size and rating (with the exception of small motors).

Also, the starting current of an induction motor is usually five to seven times its full-load (i.e., rated) current. In general, the speed of an induction motor is not easily controlled.

Even though the induction machine, with a wound rotor, can be used as a generator, its performance characteristics (especially in comparison to a synchronous generator) have not been found satisfactory for most applications.

However, induction generators are occasionally used at hydroelectric power plants. For example, they are presently in use as generators at the Folsom Dam in Northern California. Also, the induction machine with a wound rotor can be used as a *frequency changer*.

FIGURE 6.3 A totally enclosed, fan-cooled, explosion-proof, three-phase induction motor for use in hazardous environments. (Courtesy of General Electric Canada, Inc., Mississauga, Canada.)

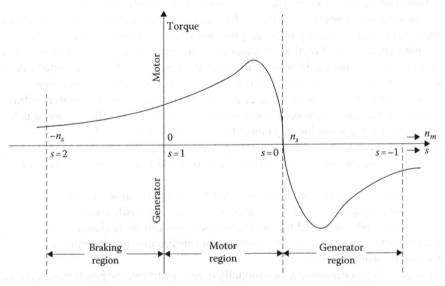

FIGURE 6.4 An induction machine's torque–speed characteristic curve showing braking-, motor-, and generator regions.

6.2 CONSTRUCTION OF INDUCTION MOTORS

In general, the stator construction of a three-phase induction machine is the same as that for a synchronous machine.* However, the same cannot be said for their rotors. In fact, the three-phase induction motors are classified based on their rotor types as wound-rotor or squirrel-cage motors.

Figure 6.5a shows a cross-sectional view of the magnetic circuit of an induction motor that has a wound rotor. The rotor iron is laminated and slotted to contain the insulated windings. The wound-rotor motor has a three-phase symmetrical winding similar to that in the stator and is wound for the same number of poles as the stator winding. These rotor phase windings are wye-connected with the open end of each phase connected to a slip ring mounted on the rotor shaft. Figure 6.6 shows that three equal external variable resistors used for speed control are connected to the slip rings by carbon brushes.[†]

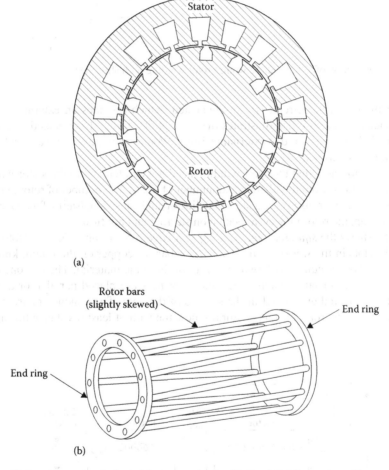

FIGURE 6.5 (a) A cross-sectional view of the magnetic circuit of an induction motor with a wound rotor and (b) the squirrel-cage winding of a cage rotor of an induction motor.

---

* That is, the stator core is built of sheet-steel laminations that are supported in a stator frame of cast iron or fabricated steel plate. Its windings, quite similar to those of the revolving field synchronous machine, are spaced in the stator slots 120 electrical degrees apart. The stator-phase windings can be either wye- or delta connected. The stator windings constitute the armature windings.

† The rotor winding is not connected to a supply. The slip rings and brushes simply provide a means of connecting an external variable-control resistance (called a slip-ring rheostat) to the rotor circuit.

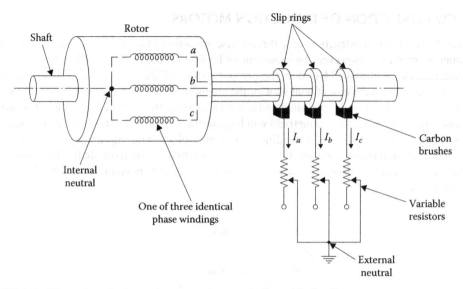

FIGURE 6.6   Illustration of a three-phase wound-rotor winding with slip rings.

Note that the total rotor circuit is wye-connected, which provides an external neutral that is usually grounded. Figure 6.7 also illustrates the concept of a three-phase induction motor that has wound-rotor windings connected to external resistors. Wound-rotor motors are also called **slip-ring motors**, for obvious reasons.

It is important to know that the rotor winding need not be identical to the stator winding; however, the two have to be wound with an equal number of poles. The number of rotor and stator slots should not be equal, otherwise several slots may line up and cause a pulsating flux. Occasionally, if the slots do line up, the rotor may even lock up on starting and not turn.

Figure 6.5b shows the *squirrel-cage winding of a cage rotor of* an induction motor. Instead *of* a winding, the slots in the *squirrel-cage** rotor have bars of copper or aluminum, known as rotor bars, which are short-circuited with two *end rings of* the same material. There is one ring at each end *of* the stack *of* rotor laminations. The solid rotor bars are placed parallel, or approximately parallel, to the shaft and embedded in the surface *of* the core. The conductors are not insulated from the core, since the rotor currents naturally flow the path of least resistance through the rotor conductors.

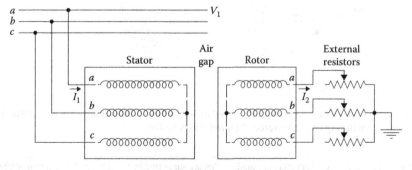

FIGURE 6.7   Illustration of a three-phase induction motor with its wound-rotor windings connected to external resistors.

---

* It is called this because of its appearance, since it resembles the exercise wheel used for hamsters and gerbils.

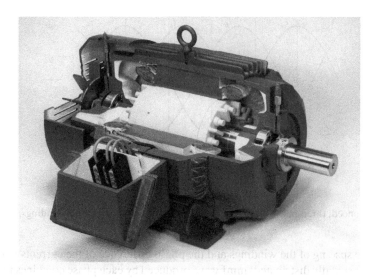

FIGURE 6.8 A cutaway view of a squirrel-cage, three-phase induction motor (Omega XL). (Courtesy of Reliance Electric Company, Beachwood, OH.)

In large motors, the rotor bars that may be copper alloy are driven into the slots and brazed to the end rings. In small motors, the rotors usually have die-cast aluminum bars, and their core laminations are stacked in a mold which is then filled with molten aluminum, as shown in Figure 6.8. Note that in such a rotor, the rotor bars, end rings, and cooling-fan blades are cast at the same time, for economical reasons.

Squirrel-cage rotor bars are not always placed parallel to the motor shaft, but are sometimes skewed, as shown in Figure 6.5b. This provides a more uniform torque and also reduces the magnetic humming noise and mechanical vibrations when the motor is running.

The induction motor is basically a fixed drive. Therefore, in order to function efficiently, its rotor has to rotate at a speed near the synchronous speed. The synchronous speed itself is a function of the frequency of the applied stator voltages and the number of poles of the motor.

Thus, efficient variable-speed operation basically requires changing the frequency of the power supply. Recent developments in solid-state technology have resulted in more efficient variable-frequency power sources and have therefore substantially increased the possible applications of induction motors.

Squirrel-cage rotor bars are not always placed parallel to the motor shaft, but are sometimes skewed as shown in Figure 6.4b. This provides a more uniform torque and also reduces the magnetic humming noise and mechanical vibrations when the motor is running.

The induction motor is basically a fixed drive. Therefore, in order to function efficiently, its rotor has to rotate at a speed near the synchronous speed. The synchronous speed itself is a function of the frequency of the applied stator voltages and the number of poles of the motor. Thus, efficient variable-speed operation basically requires changing the frequency of the power supply. Recent developments in solid-state technology have resulted in more efficient variable-frequency power sources and have therefore substantially increased the possible applications of induction motors.

6.3   ROTATING MAGNETIC FIELD CONCEPT

When the three-phase stator windings of an induction motor are supplied by three-phase voltages, currents will flow in each phase. These currents are time displaced from each other by 120 electrical degrees in a two-pole machine, as shown in Figure 6.9. An induction motor's operation depends on a rotating magnetic field established by the stator currents in the air gap of the motor.

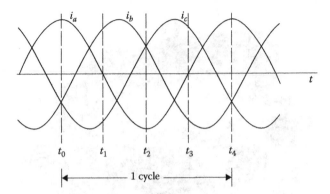

FIGURE 6.9 Balanced, three-phase alternating currents applied to three-phase windings.

Because of the spacing of the windings and the phase difference of the currents in the windings, the pulsating (sinusoidally distributed) mmf wave produced by each phase combines to form a resultant mmf $\mathcal{F}$, which moves around the inner circumference of the stator surface (i.e., in the air gap) at a constant speed. The resultant flux is called the *rotating magnetic field*. If balanced, three-phase excitation is applied with *abc* phase sequence and then the currents can be expressed as

$$i_a = I_m \cos\omega t \tag{6.1}$$

$$i_b = I_m \cos(\omega t - 120°) \tag{6.2}$$

$$i_b = I_m \cos(\omega t - 240°) \tag{6.3}$$

where I_m is the maximum value of the current and the time origin is arbitrarily assumed to be the instant when the phase-a current is at its positive maximum. Figure 6.9 shows such instantaneous currents. The resultant mmf wave is a function of the three component mmf waves caused by these currents. It can be determined either graphically or analytically.

6.3.1 GRAPHICAL METHOD

Since the rotating magnetic field is produced by the mmf contribution of space-displaced phase windings with appropriate time-displaced currents, one has to take into account various instants of time and determine the magnitude and direction of the resultant mmf wave. For example, consider the instant of time (indicated in Figure 6.9) $t = t_0$ and notice that the currents in the phase windings *a*, *b*, and *c*, respectively, are

$$i_a = I_m \tag{6.4}$$

$$i_b = -\frac{I_m}{2} \tag{6.5}$$

$$i_b = -\frac{I_m}{2} \tag{6.6}$$

Note that each phase in Figure 6.10a, for the sake of convenience and simplicity, is represented by a single coil. For example, coil $a - a'$ represents the entire phase a winding (normally distributed over 60 electrical degrees), with its mmf axis directed along the horizontal. The right-hand rule readily confirms this statement. Similarly, the mmf axis of the phase b winding is 120 electrical degrees apart from phase a, and that of phase c is 120 electrical degrees displaced from phase b. Obviously, the unprimed and primed letters refer to the beginning and end terminals of each phase, respectively. Also, notice that the current directions in the corresponding coils are indicated by dots and crosses, as shown in Figure 6.10.

The current in the phase-a winding is at its maximum at $t = t_0$, and is represented by a phasor $\mathcal{F}_a = \mathcal{F}_m$ along the axis of phase a, as shown in Figure 6.10a. The mmfs of phases b and c are represented by phasors $\mathcal{F}_b$ and $\mathcal{F}_c$, respectively, each with a magnitude of $\mathcal{F}_m/2$ and located in the negative direction along their corresponding axes. The sum of the three phasors is a phasor $\mathcal{F} = 1.5\mathcal{F}_m$ affecting in the positive direction along the phase a axis, as shown in Figure 6.10. Figure 6.11 shows the corresponding component mmf waves and the resultant mmf wave at the instant $t = t_0$.

Now consider a later instant of time t_1, as shown in Figure 6.10b. The currents and mmf associated with the phase winding can be expressed as

$$i_a \text{ and } \mathcal{F}_a = 0 \qquad\qquad \text{w(6.7)}$$

$$i_b = \frac{\sqrt{3}}{2} I_m \quad \text{and} \quad \mathcal{F}_b = \frac{\sqrt{3}}{2} \mathcal{F}_{max} \qquad\qquad (6.8)$$

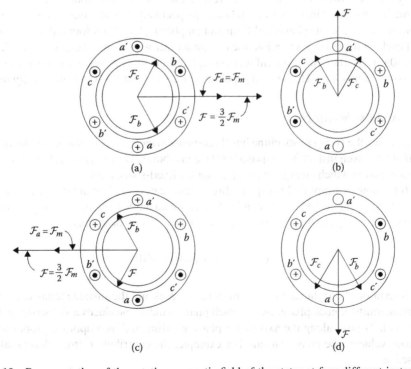

FIGURE 6.10 Representation of the rotating magnetic field of the stator at four different instants of time (indicated in Figure 6.9): (a) time $t = t_0 = t_4$, (b) time $t = t_1$, (c) time $t = t_2$, and (d) time $t = t_3$.

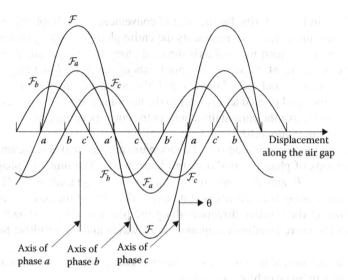

FIGURE 6.11   Component and resultant mmf field distributions corresponding to time $t = t_0$ in Figure 6.10a.

$$i_c = -\frac{\sqrt{3}}{2} I_m \quad \text{and} \quad \mathcal{F}_c = -\frac{\sqrt{3}}{2} \mathcal{F}_{max} \tag{6.9}$$

The figure shows the current directions, the component mmfs, and the resultant mmf at $t = t_1$. Note that the resultant mmf has now rotated counterclockwise 90 electrical degrees in space.

Similarly, Figure 6.10c and d shows the corresponding current directions, component mmfs, and resultant mmfs at the other instants $t = t_2$ and $t = t_3$, respectively. It is obvious that as time passes, the resultant mmf wave keeps its sinusoidal form and amplitude, but shifts forward around the air gap. In one full cycle of the current variation, the resultant mmf wave comes back to the position shown in Figure 6.10a. Thus, the resultant mmf wave completes one revolution per cycle of the current variation in a two-pole machine. Hence, in a p-pole machine, the mmf wave rotates by $2/p$ revolutions.

6.3.2 ANALYTICAL METHOD

Assume again that the two-pole machine has three-phase windings on its stator, so that the resultant stator mmf at any given instant is composed of the contributions of each phase. Each phase winding makes a contribution which changes with time along a fixed-space axis.

Figure 6.12 shows a simplified two-pole, three-phase stator winding arrangement. The resultant mmf wave, at any point in the air gap, can be defined by an angle θ. Notice the origin of the axis of phase a, as shown in Figure 6.12. The resultant mmf along θ can be expressed as

$$\mathcal{F}(\theta) = \mathcal{F}_a(\theta) + \mathcal{F}_b(\theta) + \mathcal{F}_c(\theta) \tag{6.10}$$

where each term on the right side of Equation 6.10 represents the instantaneous contributions of the alternating mmfs of each phase. Hence, each phase winding produces a sinusoidally distributed mmf wave with its peak along the axis of the phase winding and its amplitude proportional to the instantaneous value of the phase current. For example, the contribution from phase a along θ can be expressed as

$$\mathcal{F}_a(\theta) = \mathcal{F}_m \cos\theta \tag{6.11}$$

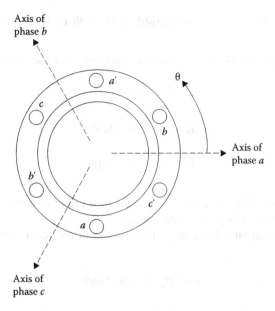

Axis of phase b

Axis of phase a

Axis of phase c

FIGURE 6.12 Simplified two-pole, three-phase stator winding arrangement.

where $\mathcal{F}_m$ is the maximum instantaneous value of the phase a mmf wave. Therefore, Equation 6.11 can be rewritten as

$$\mathcal{F}_a(\theta) = Ni_a \cos\theta \tag{6.12}$$

where

N is the effective number of turns in the phase a winding

i_a is the instantaneous value of the current in the phase a winding

Since the phase axes shown in Figure 6.12 are shifted from each other by 120° electrical degrees, the mmf contributions from phase b and phase c can be expressed, respectively, as

$$\mathcal{F}_b(\theta) = \mathcal{F}_m \cos(\theta - 120°) \tag{6.13}$$

$$\mathcal{F}_c(\theta) = \mathcal{F}_m \cos(\theta - 240°) \tag{6.14}$$

or

$$\mathcal{F}_b(\theta) = Ni_b \cos(\theta - 120°) \tag{6.15}$$

$$\mathcal{F}_c(\theta) = Ni_c \cos(\theta - 240°) \tag{6.16}$$

Hence, the resultant mmf at point θ is

$$\mathcal{F}(\theta) = \mathcal{F}_m \cos\theta + \mathcal{F}_m \cos(\theta - 120°) + \mathcal{F}_m \cos(\theta - 240°) \tag{6.17}$$

or, alternatively,

$$\mathcal{F}(\theta) = Ni_a\cos\theta + Ni_b\cos(\theta - 120°) + Ni_c\cos(\theta - 240°) \tag{6.18}$$

However, the instantaneous currents i_a, i_b, and i_c are functions of time and are expressed as

$$i_a = I_m\cos\omega t \tag{6.19}$$

$$i_b = I_m\cos(\omega t - 120°) \tag{6.20}$$

$$i_c = I_m\cos(\omega t - 240°) \tag{6.21}$$

where I_m is the maximum value of the current, and the time origin is arbitrarily taken as the instant when the phase a current is at its positive maximum.

The quantity w is the angular frequency of oscillation of the stator currents, which by definition is

$$\omega = 2\pi f \text{ electrical rad/s} \tag{6.22}$$

where f is the frequency of the stator currents in hertz. Therefore, Equation 6.18 can be expressed as

$$\mathcal{F}(\theta,t) = NI_m\cos\omega t \cos\theta$$
$$+ NI_m\cos(\omega t - 120°)\cos(\theta - 120°)$$
$$+ NI_m\cos(\omega t - 240°)\cos(\theta - 240°) \tag{6.23}$$

By use of the identity

$$\cos x\cos y = \frac{1}{2}\cos(x+y) + \frac{1}{2}\cos(x-y) \tag{6.24}$$

each term on the right side of Equation 6.23 can be rewritten as the sum of two cosine functions, one involving the difference and the other the sum of the two angles. The resultant mmf of the total three-phase winding can be expressed as

$$\mathcal{F}(\theta,t) = \underbrace{\frac{1}{2}NI_m\cos(\omega t - \theta)} + \overbrace{\frac{1}{2}NI_m\cos(\omega t + \theta)}$$

$$+ \underbrace{\frac{1}{2}NI_m\cos(\omega t - \theta)} + \overbrace{\frac{1}{2}NI_m\cos(\omega t + \theta - 240°)}$$

$$+ \underbrace{\frac{1}{2}NI_m\pi\cos(\omega t - \theta)}_{\text{Forward-rotating components}} + \overbrace{\frac{1}{2}NI_m\cos(\omega t + \theta + 240°)}^{\text{Backward-rotating components}} \tag{6.25}$$

However, this expression defines a *space* field. Therefore, the second, fourth, and sixth terms, being equal in amplitude and 120° apart, yield a net value of zero. Thus, Equation 6.25 simplifies to

$$\mathcal{F}(\theta,t) = \frac{3}{2}NI_m\cos(\omega t - \theta) \tag{6.26}$$

or

$$\mathcal{F}(\theta, t) = \frac{3}{2} \mathcal{F}_m \cos(\omega t - \theta) \tag{6.27}$$

which represents the resultant field mmf wave rotating counterclockwise with an angular velocity of ω rad/s in the air gap. The speed of such a revolving field is usually denoted by ω_s and is referred to as synchronous speed ($\omega_s = \omega$). Suppose that at a given time t_1, the resultant mmf wave is distributed sinusoidally around the air gap with its positive peak occurring along $\theta = \omega t_1$. If, at a later time t_2, the positive peak of the sinusoidally distributed wave is along $\theta = \omega t_2$, then the resultant mmf wave has moved by $\omega(t_2 - t_1)$ around the air gap. Therefore, polyphase currents cause a rotating magnetic field* to develop in the air gap as if there were a physically rotating permanent magnet present within the stator of the machine.

6.4 INDUCED VOLTAGES

Assume that the rotor winding is wound-type, wye-connected, and open-circuited. Since the rotor winding is open-circuited, no torque can develop. This represents the *standstill operation* of a three-phase induction motor. The application of a three-phase voltage to the three-phase stator winding results in a rotating magnetic field that "cuts" both the stator and rotor windings at the supply frequency f_1. Hence, the rms value of the induced voltage per phase of the rotor winding can be expressed as

$$E_2 = \frac{2\pi}{\sqrt{2}} f_1 N_2 \phi k_{\omega 2} \tag{6.28}$$

$$E_2 = 4.44 f_1 N_2 \phi k_{\omega 2} \tag{6.29}$$

where the subscripts 1 and 2 are used to denote stator- and rotor-winding quantities, respectively. Since the rotor is at standstill, the stator frequency f_1 is used in Equations 6.28 and 6.29. Here, the flux ϕ is the mutual flux per pole involving both the stator and rotor windings. Similarly, the rms value of the induced voltage per phase of the stator winding can be expressed as

$$E_1 = 4.44 f_1 N_1 \phi k_{\omega 1} \tag{6.30}$$

Thus, it can be shown that

$$\frac{E_1}{E_2} = \frac{N_1 k_{\omega 1}}{N_2 k_{\omega 2}} \tag{6.31}$$

where $k_{\omega 1}$ and $k_{\omega 2}$ are the winding factors for the stator and rotor windings, respectively. Since usually they are the same, turns ratio a can be found from

---

* It is interesting to note that a reversal of the phase sequence of the currents on the stator windings causes the rotating mmf (as well as the shaft of the motor) to rotate in the opposite direction. For example, if current i_s flows through the phase a winding as before, but the currents i_b and i, now flow through the phase c and phase b windings, respectively, the rotating mmf (as well as the shaft of the motor) will rotate in a clockwise direction. In summary, the direction of the rotation of a three-phase motor may be reversed by interchanging any of the three motor supply lines.

$$\frac{E_1}{E_2} = \frac{N_1}{N_2} = a \qquad (6.32)$$

Notice the similarities between the induction motor at standstill and a transformer. Also note that the stator and rotor windings are represented by the primary and secondary, respectively.*

6.5 CONCEPT OF ROTOR SLIP

In the event that the stator windings are connected to a three-phase supply and the rotor circuit is closed, the induced voltages in the rotor windings produce three-phase rotor currents. These currents in turn cause another rotating magnetic field to develop in the air gap. This induced rotor magnetic field also rotates at the same synchronous speed, n_s. In other words, the stator magnetic field and the rotor magnetic field are stationary with respect to each other. As a result, the rotor develops a torque according to the principle of alignment of magnetic fields.

Thus, the rotor starts to rotate in the direction of the rotating field of the stator, due to Lenz's law. Here, the stator magnetic field can be considered as dragging the rotor magnetic field. The torque is maintained as long as the rotating magnetic field and the induced rotor currents exist. Also, the voltage induced in the rotor windings depends on the speed of the rotor *relative* to the magnetic fields. At steady-state operation, the rotor's shaft speed[†] n_m is less than the synchronous speed n_s at which the stator rotating field rotates in the air gap. The synchronous speed is determined by the applied stator frequency[‡] f_1, in hertz, and the number of poles, p, of the stator winding. Therefore,

$$n_s = \frac{120 f_1}{p} \text{ rev/min} \qquad (6.33)$$

Of course, at $n_m = n_s$, there would be no induced voltages or currents in the rotor windings and, therefore, no torque. Thus, *the shaft speed of the rotor can never be equal to the synchronous speed*, but has to be at some value below that speed.

The **slip speed** (also called the slip rpm) is defined as the difference between synchronous speed and rotor speed and indicates how much the rotor slips[§] behind the synchronous speed. Hence,

$$n_{slip} = n_s - n_m \qquad (6.34)$$

where

n_{slip} is the slip speed of motor in rpm
n_s is the synchronous speed (i.e., speed of magnetic fields) in rpm
n_m is the mechanical shaft speed of rotor in rpm

Therefore, the term **slip** describes this relative motion in per unit or in percent. Thus, the slip in per unit is

$$s = \frac{n_s - n_m}{n_s} \qquad (6.35)$$

---

* Because of such similarities, the induction motor has also been called a "rotating transformer."
† It is also called the mechanical shaft *speed of the rotor.*
‡ In other words, the frequency of the applied three-phase supply system.
§ The term "slip" is used because it describes what an observer riding with the stator field sees when looking at the rotor; it appears to be slipping backward.

and the slip in percent is

$$s = \frac{n_s - n_m}{n_s} \times 100 \qquad (6.36)$$

Alternatively, the slip can be defined in terms of angular velocity w (rad/s) as

$$s = \frac{\omega_s - \omega_m}{\omega_s} \times 100 \qquad (6.37)$$

By closely inspecting Equation 6.35 and Figure 6.4 and simply applying *deductive reasoning,** one can observe the following:

1. If $s=0$, it means that $n_m=n_s$, that is, the rotor turns at synchronous speed. (In practice, it can only occur if the direct current is injected into the rotor winding.)
2. If $s=1$, it indicates that $n_m=0$, that is, the rotor is stationary. In other words, the rotor is at standstill.
3. If $1>s>0$, it signals that the rotor turns at a speed somewhere between standstill and synchronous speed. In other words, the motor runs at an asynchronous speed as it should, as illustrated in Figure 6.13a.
4. If $s>1$, it signifies that the rotor rotates in a direction opposite of the stator rotating field, as shown in Figure 6.13c. Therefore, in addition to electrical power, mechanical power (i.e., shaft power) must be provided.
   Since power comes in from both sides, the copper losses of the rotor increase tremendously. The rotor develops a braking torque that forces the motor to stop. This mode of induction machine operation is called **braking** (or *plugging*) **mode**.
5. If $s<0$, it means that the machine operates as a generator with a shaft speed that is greater than the synchronous speed, as shown in Figure 6.13b. This mode of operation is called *generating mode*.

Also note that the mechanical shaft speed of the rotor can be obtained from the following two equations, which involve only slip and synchronous speed:

$$n_m = (1-s)n_s \text{ rpm} \qquad (6.38)$$

$$\omega_m = (1-s)\omega_s \text{ rad/s} \qquad (6.39)$$

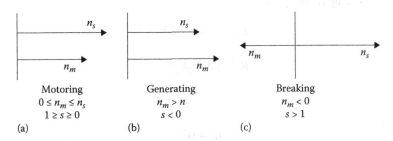

FIGURE 6.13 Three operation modes of an induction machine: (a) motoring, (b) generating, and (c) plugging.

---

* Thanks to Sherlock Holmes, who more than once said: "You see, but you do not observe!" See the *Adventures of Sherlock Holmes* by Arthur Conan Doyle, 1891.

6.6 EFFECTS OF SLIP ON THE FREQUENCY AND MAGNITUDE OF INDUCED VOLTAGE OF THE ROTOR

If the rotor of an induction motor is rotating, the frequency of the induced voltages (as well as the induced currents) in the rotor circuit is no longer the same as the frequency of its stator. Under such a running operation, the frequency of the induced voltages (and the currents) in the rotor is directly related to the slip rpm (i.e., the relative speed between the rotating field and the shaft speed of the rotor). Therefore,

$$f_2 = \frac{p \times n_{slip}}{120} \tag{6.40a}$$

$$f_2 = \frac{p \times n_{slip}}{120} \tag{6.40b}$$

where f_2 is the frequency of the voltage and current in the rotor winding. Using Equation 6.35, Equation 6.40 can be expressed as

$$f_2 = \frac{p \times s \times n_s}{120} \tag{6.41a}$$

or

$$f_2 = \frac{s \times p \times n_s}{120} \tag{6.41b}$$

By substituting Equation 6.33 into Equation 6.41b,

$$f_2 = s \times f_1 \tag{6.42a}$$

or

$$f_r = s \times f_1 \tag{6.42b}$$

That is, the rotor frequency f_2 or f_r is found simply by multiplying the stator's frequency f_1 by the per unit value of the slip. Because of this,* f_2 or f_r is also called the slip frequency.

Therefore, the voltage induced in the rotor circuit at a given slip s can be found from Equation 6.29 simply by replacing f_1 with f_2 as

$$E_r = 4.44 f_2 N_2 \phi k_{\omega 2} \tag{6.43a}$$

$$E_r = 4.44 s f_1 N_2 \phi k_{\omega 2} \tag{6.43b}$$

or

$$E_r = s E_2 \tag{6.43c}$$

where E_2 is the induced voltage in the rotor circuit at standstill, that is, at the stator frequency f_1.

---

* An induction machine with a wound rotor can also be used as a frequency changer.

The induced currents in the three-phase rotor windings also develop a rotating field. The speed of this rotating magnetic field of the rotor with respect to rotor itself can be found from

$$n_2 = \frac{120 \times f_2}{p} \tag{6.44a}$$

or

$$n_r = sn_s$$

$$n_2 = \frac{120 \times s \times f_1}{p} \tag{6.44b}$$

$$n_2 = s \times n_s \tag{6.44c}$$

or

$$n_r = s \times n_s \tag{6.44d}$$

where

n_r is the speed of the rotating magnetic field of the rotor

n_s is the speed of the rotating magnetic field of the stator

However, since the rotor itself is rotating at n_m, the developed rotor field rotates in the air gap at a speed of

$$n_m + n_2 = (1 - s)n_s + sn_s = n_s \tag{6.45}$$

Thus, one can prove that both the stator field and the rotor field rotate in the air gap at the same synchronous speed n_s. In other words, the stator and rotor fields are stationary with respect to each other, producing a steady torque and maintaining rotation.

Example 6.1

A three-phase, 60 Hz, 25 hp, wye-connected induction motor operates at a shaft speed of almost 1800 rpm at no load and 1650 rpm at full load. Determine the following:

(a) The number of poles of the motor
(b) The per-unit and percent slip at full load
(c) The slip frequency of the motor
(d) The speed of the rotor field with respect to the rotor itself
(e) The speed of the rotor field with respect to the stator
(f) The speed of the rotor field with respect to the stator field
(g) The output torque of the motor at the full load

Solution

(a) From Equation 6.33,

$$n_s = \frac{120 f_1}{p}$$

from which

$$p = \frac{120 f_1}{n_s}$$

$$= \frac{120 \times 60}{1800} = 4\,\text{poles}$$

(b) Since

$$n_m = n_s(1 - s)$$

Then

$$s = \frac{n_s - n_m}{n_s}$$

$$= \frac{1800 - 1650}{1800}$$

$$= 0.08333\,\text{pu or } 8.33\%$$

(c) The slip frequency is

$$f_2 = s f_1 = 0.0833 \times 60 = 5\,\text{Hz}$$

(d) The speed of the rotor field with respect to the rotor itself can be determined from

$$n_2 = \frac{120 f_2}{p}$$

$$= \frac{120 \times 5}{4}$$

$$= 150\,\text{rpm}$$

or

$$n_2 = s \times n_s$$

$$= 0.08333 \times 1800$$

$$= 150\,\text{rpm}$$

(e) The speed of the rotor field with respect to the stator can be found from

$$n_m + n_2 = 1650 + 150 = 1800\,\text{rpm}$$

or

$$n_m + n_2 = n_s = 1800\,\text{rpm}$$

(f) The speed of the rotor field with respect to the stator field can be determined from

$$(n_m + n_2) - n_s = 1800 - 1800 = 0$$

or since

$$n_m + n_2 = n_s$$

then

$$n_s - n_s = 0$$

(g) The output torque of the motor at the full load can be determined from

$$T_{out} = T_{shaft} = \frac{P_{out}}{\omega_m}$$

$$= \frac{(25 \text{ hp})(746 \text{ W/hp})}{(1650 \text{ rev/min})(2\pi \text{ rad/rev})(1 \text{ min}/60 \text{ s})}$$

$$= 108 \text{ N} \cdot \text{m}$$

or in English units,

$$T_{out} = T_{shaft} = \frac{5252P}{n}$$

$$= \frac{5252(25 \text{ hp})}{1650 \text{ rev/min}}$$

$$= 79.6 \text{ lb} \cdot \text{ft}$$

6.7   EQUIVALENT CIRCUIT OF AN INDUCTION MOTOR

Assume that a three-phase, wound-rotor* induction motor has a balanced wye connection, as shown in Figure 6.7, so that the currents are always line values and the voltages are always line-to-neutral values. If the currents flow in both the stator and rotor windings, there will be rotating magnetic fields in the air gap. Since these magnetic fields rotate at the same speed in the air gap, they will develop a resultant air-gap field rotating at synchronous speed.

Because of this air-gap field, voltages will be induced in the stator windings at the supply frequency f_1 and in the rotor windings at the slip frequency f_2. As with a balanced polyphase transformer, only one phase of the circuit model need be considered.

6.7.1   STATOR CIRCUIT MODEL

Figure 6.14a shows the equivalent circuit of the stator. The stator terminal voltage differs from the induced voltage (i.e., the counter-emf) in the stator winding because of the voltage drop in the stator leakage impedance. Therefore,

$$V_1 = E_1 + I_1(R_1 + jX_1) \qquad (6.46)$$

where
   V_1 is the per-phase stator terminal voltage
   E_1 is the per-phase induced voltage (counter-emf) in the stator winding
   I_1 is the stator current
   R_1 is the per-phase stator winding resistance
   X_1 is the per-phase stator leakage reactance

---

* In the case of a squirrel-cage rotor, the rotor circuit can be represented by an equivalent three-phase rotor winding.

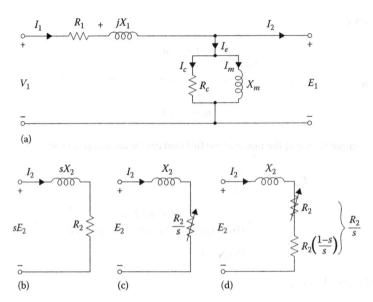

FIGURE 6.14 Development of the per-phase stator and rotor-equivalent circuits of an induction motor: (a) stator-equivalent circuit, (b) actual rotor circuit, (c) rotor-equivalent circuit, and (d) modified equivalent rotor circuit.

One can easily observe that the equivalent circuit of the stator winding is the same as the equivalent circuit of the transformer winding. As is the case in the transformer model, the stator current I_1 can be separated into two components, that is, a load component I_2 and an excitation component I_e. Here, the load component I_2 produces an mmf that exactly counteracts the mmf of the rotor current. The excitation component I_e is the extra stator current needed to create the resultant air-gap flux. In the shunt branch of the model, R_c and X_m represent per-phase stator core-loss resistance and per-phase stator magnetizing reactance, respectively, as is the case in transformer theory. However, the magnitudes of the parameters are considerably different. For example, I_e is much larger in the induction machine due to the air gap. It can be as high as 30%–50% of the rated current in an induction machine versus 1%–5% in a transformer.

Due to the air gap, the value of magnetizing reactance X_m is relatively small in comparison to that of a transformer; but the leakage reactance X_1 is greater than the magnetizing reactance than in transformers. Another reason for this is that the stator and rotor windings are distributed along the periphery of the air gap instead of being stacked on a core as they are in transformers.

6.7.2 Rotor-Circuit Model

Figure 6.14b shows the actual rotor circuit of an induction motor operating under load at a slip s. The rotor current per phase can be expressed as

$$I_2 = \frac{sE_2}{R_2 + jsX_2} \tag{6.47}$$

where
E_2 is the per-phase induced voltage in the rotor at standstill (i.e., at stator frequency f_1)
R_2 is the per-phase rotor-circuit resistance
X_2 is the per-phase rotor leakage inductive reactance

The figure illustrates that I_2 is a slip-frequency current produced by the slip frequency–induced emf sE_2 acting in a rotor circuit with an impedance per phase of $R_2 + jsX_2$. Therefore, the total rotor copper loss can be expressed as

$$P_{2,cu} = 3I_2^2 R_2 \tag{6.48}$$

which represents the amount of real power involved in the rotor circuit. Equation 6.47 can be rewritten by dividing both the numerator and the denominator by the slip s so that

$$I_2 = \frac{E_2}{(R_2/s) + jX_2} \tag{6.49}$$

This equation suggests the rotor-equivalent circuit shown in Figure 6.14c. Of course, the magnitude and phase angle of I_2 remain unchanged by this process, but there is a significant difference between these two equations and the circuits they represent. The current I_2 given by Equation 6.47 is at slip frequency f_2, whereas I_2 given by Equation 6.49 is at line frequency f_1.

Also in Equation 6.47, the rotor leakage reactance sX_2 changes with speed, but the resistance R_2 remains unchanged; whereas in Equation 6.49, the resistance R_2/s changes with speed, but the leakage reactance X_2 remains unchanged. The total rotor copper loss associated with the equivalent rotor circuit shown in Figure 6.14c is

$$P = 3I_2^2 \left(\frac{R_2}{s} \right)$$

$$= \frac{P_{2,cu}}{s} \tag{6.50}$$

Since induction machines are run at low slips, the power associated with Figure 6.14c is substantially greater. The equivalent circuit given in Figure 6.14c is at the stator frequency and therefore is the rotor-equivalent circuit as seen from the stator. Thus, the power determined by using Equation 6.50 is the power transferred across the air gap (i.e., P_g) from the stator to the rotor which includes the rotor copper loss as well as the developed mechanical power. Here, the equation can be expressed in a manner that stresses this fact. Therefore,

$$P = P_g = 3I_2^2 \frac{R_2}{2} = 3I^2 \left[R_2 + \frac{R_2}{s}(1-s) \right] \tag{6.51}$$

The corresponding equivalent circuit is shown in Figure 6.14d. The speed-dependent resistance* $R_2(1 - s)/s$ represents the mechanical power developed by the induction machine to overcome the mechanical shaft load. Therefore, the total developed mechanical power can be found from

$$P_d = P_{mech} = 3I_2^2 \frac{R_2}{s}(1-s) \tag{6.52a}$$

$$P_d = P_{mech} = (1-s)P_g \tag{6.52b}$$

---

* It is known as the *dynamic resistance* or *load* resistance. Note that in the braking mode of the operation, this resistance is negative and represents a source of energy.

or

$$P_d = P_{mech} = \frac{1-s}{s} P_{2,cu} \qquad (6.52c)$$

where

$$P_{2,cu} = 3I_2^2 R_2 = sP_g \qquad (6.53)$$

A small portion of the developed mechanical power is also lost due to windage and friction. The rest of the mechanical power is defined as the *output shaft power*.

6.7.3 COMPLETE EQUIVALENT CIRCUIT

If the stator-equivalent circuit shown in Figure 6.14a and c and the rotor-equivalent circuit shown in Figure 6.14d are at the same line frequency f_1, they can be joined together. However, if the turns in the stator winding and the rotor winding are different, then E_1 and E_2 can be different, as shown in Figure 6.15a. Because of this, the turns ratio $(a = N_1/N_2)$ needs to be taken into account. Figure 6.15c shows the resultant equivalent circuit of the induction machine.

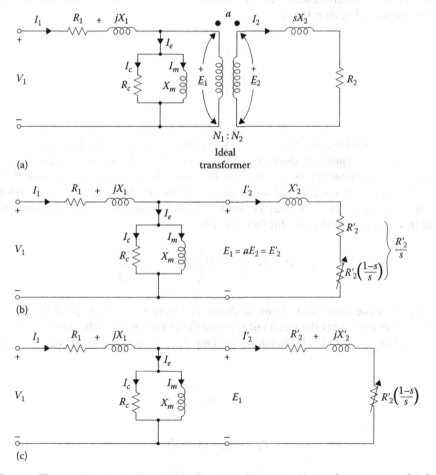

FIGURE 6.15 The per-phase equivalent circuit of an induction motor: (a) transformer model of an induction motor, (b) exact equivalent circuit, and (c) alternative form of the equivalent circuit.

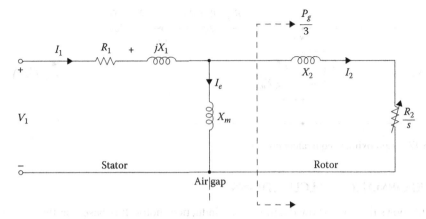

FIGURE 6.16   Exact equivalent circuit with the core-loss resistor omitted.

Notice that such an equivalent circuit form is identical to that of the two-winding transformer. Also note that the prime notation is used to denote *stator-referred* rotor quantities. Therefore,

$$I'_2 = \frac{I_2}{2} \tag{6.54}$$

$$E_1 = aE_2 = E'_2 \tag{6.55}$$

$$R'_2 = a^2 R_2 \tag{6.56}$$

$$X'_2 = a^2 X_2 \tag{6.57}$$

$$R'_2\left(\frac{1-s}{s}\right) = a^2 R_2\left(\frac{1-s}{s}\right) \tag{6.58}$$

Due to the presence of the air gap in the induction machine, the magnetizing impedance is low and therefore the exciting current I_e is high (about 30%–50% of full-load current). The leakage reactance X_1 is also high.

The equivalent circuit of the induction machine can be simplified by omitting resistance R_c and lumping the corresponding core loss with the friction and windage losses. The error involved is negligible. Figure 6.16 *shows* the resultant equivalent circuit.* Therefore, if core loss is assumed to be constant, then such an equivalent circuit should be used. Note that all stator-referred rotor quantities are shown without prime notation as is customary. However, from now on, *it should be understood that they are stator referred.*

6.7.4   Approximate Equivalent Circuit

The computation can be simplified with very little loss of accuracy, by moving the magnetizing (shunt) branch (i.e., R_c and X_m) to the machine terminals, as shown in Figure 6.17. This modification is based mainly on the assumption that $V_1 = E_1 = E_2$. A further simplification can be achieved by also omitting the resistance R_c.

---

* It is recommended by IEEE and known as the *Steinmetz model of one phase of a three*-phase induction machine.

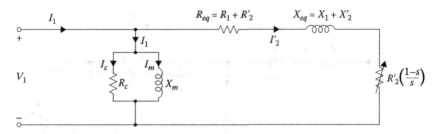

FIGURE 6.17   Approximate equivalent circuit.

6.8   PERFORMANCE CALCULATIONS

Figure 6.18 shows the power-flow diagram of an induction motor. It is based on the equivalent circuit shown in Figure 6.15c. The input power is the electrical power input to the stator of the motor. Therefore,

$$P_{in} = P_1 = \sqrt{3}V_L I_L \cos\theta \tag{6.59}$$

The total stator copper losses are

$$P_{1,cu} = 3I_1^2 R_1 \tag{6.60}$$

The total core losses can be found from

$$P_{core} = 3E_1^2 G_c \tag{6.61a}$$

or

$$P_{core} = \frac{3E_1^2}{R_c} \tag{6.61b}$$

Therefore, the total air-gap power can be given as

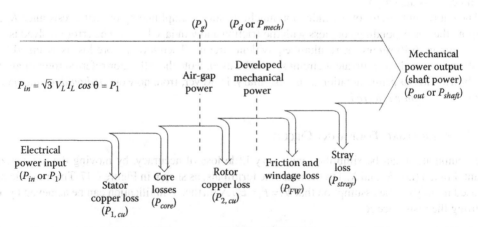

FIGURE 6.18   Power-flow diagram of an induction motor.

$$P_g = P_{in} - P_{1,cu} - P_{core} \tag{6.62a}$$

or

$$P_g = 3I_2^2 \frac{R_2}{s} \tag{6.62b}$$

The total rotor copper losses are

$$P_{2,cu} = 3I_2^2 R_2 \tag{6.63a}$$

or

$$P_{2,cu} = sP_g \tag{6.63b}$$

Thus, the total mechanical power developed can be found by

$$P_d = P_{mech} = P_g - P_{2,cu} \tag{6.64a}$$

or

$$P_d = P_{mech} = P_g(1-s) \tag{6.64b}$$

or

$$P_d = P_{mech} = \left(\frac{1-s}{s} \right) P_{2,cu} \tag{6.64c}$$

If the friction and windage losses and the stray losses* are known, the output power (or shaft power) can be determined from

$$P_{out} = P_{shaft} = P_d - P_{FW} - P_{stray} \tag{6.65}$$

If the core losses are assumed to be constant, they can be lumped in with the friction and windage losses, and the stray losses. Their sum is called the *rotational losses*. Thus, the rotational loss is given as

$$P_{rot} = P_{core} + P_{FW} + P_{stray} \tag{6.66}$$

Therefore, the corresponding output power can be found from

$$P_{out} = P_d - P_{rot} \tag{6.67a}$$

---

* Stray *losses* consist of all losses not otherwise included above, for example, the losses due to nonuniform current distribution in the copper, and additional core losses developed in the iron core as a result of distortion in the magnetic flux by the load current. They may also include losses due to harmonic fields. The stray losses are also called *miscellaneous losses*.

$$P_{out} = P_d - (P_{core} + P_{FW} + P_{stray})$$ (6.67b)

The corresponding equivalent circuit is shown in Figure 6.16.

The developed torque is defined as the mechanical torque developed by the electromagnetic energy conversion process. It can be found by dividing the developed power by the shaft speed. Therefore, the developed torque* can be expressed as

$$T_d = \frac{P_d}{\omega_m}$$ (6.68a)

or

$$T_d = \frac{P_g(1-s)}{\omega_s(1-s)}$$ (6.68b)

or

$$T_d = \frac{P_g}{\omega_s}$$ (6.68c)

$$T_d = \frac{3I_2^2 R_2}{s\omega_s}$$ (6.68d)

The output torque (or shaft torque) is

$$T_{out} = \frac{P_{out}}{\omega_m}$$ (6.69)

The efficiency of the induction motor can be determined from

$$\eta = \frac{P_{out}}{P_{in}}$$ (6.70a)

$$\eta = \frac{P_{out}}{P_{out} + P_{loss}}$$ (6.70b)

where P_{loss} represents the total losses.

Example 6.2

A three-phase, 480 V, 50 hp induction motor is supplied 70 A at a 0.8 lagging power factor. Its stator and rotor copper losses are 4257.53 and 1000 W, respectively. Its core losses are 3000 W, the friction and windage losses are 800 W, and the stray losses are 200 W. Determine the following:

  (a) The air-gap power
  (b) The mechanical power developed

---

* Because the developed torque can be expressed by Equation 6.68c, the air-gap power is also called the *torque in synchronous watts*.

(c) The shaft output power
(d) The efficiency of the motor

Solution

(a) Since

$$P_{in} = P_1 = \sqrt{3}V_L I_L \cos\theta$$

$$= \sqrt{3}(480 \text{ V})(70 \text{ A})0.8 = 46,557.53 \text{ W}$$

therefore, the air-gap power is

$$P_g = P_{in} - P_{1,cu} - P_{core}$$

$$= 46,557.53 - 4,257.53 - 3,000$$

$$= 39,300 \text{ W}$$

(b) The developed mechanical power is the same as the developed power. Thus,

$$P_d = P_{mech}$$

$$= P_g - P_{2,cu}$$

$$= 39,300 - 1,000$$

$$= 38,300 \text{ W}$$

(c) The shaft output power can be found as

$$P_{out} = P_d - P_{FW} - P_{stray}$$

$$= 38,300 - 800 - 200$$

$$= 37,7,300 \text{ W}$$

or in horsepower,

$$P_{out} = (37,300 \text{ W})\left(\frac{1\text{hp}}{746 \text{ W}}\right) = 50 \text{ hp}$$

(d) The efficiency of the motor is

$$\eta = \frac{37,300 \text{ W}}{46,557.53 \text{ W}} \times 100 = 80.1\%$$

Example 6.3

A three-phase, two-pole, 35 hp, 480 V, 60 Hz, wye-connected induction motor has the following constants in ohms per phase referred to the stator:

$$R_1 = 0.322\,\Omega \quad R_2 = 0.196\,\Omega$$

$$X_1 = 0.675\,\Omega \quad X_2 = 0.510\,\Omega$$

$$X_m = 12.5\,\Omega$$

The total rotational losses are 1850 W and are assumed to be constant. The core loss is lumped in with the rotational losses. For a rotor slip of 3% at the rated voltage and rated frequency, determine the following:

(a) The speed in rpm and in rad/s
(b) The stator current
(c) The power factor
(d) The developed power and output power
(e) The developed torque and output torque
(f) The efficiency

Solution

(a) The synchronous speed is

$$n_s = \frac{120 f_1}{p}$$

$$= \frac{120(60\,\text{Hz})}{2}$$

$$= 3600 \text{ rev/min}$$

or

$$\omega_s = (3600 \text{ rev/min})\left(\frac{2\pi\,\text{rad}}{1\,\text{rev}}\right)\left(\frac{1\,\text{min}}{60\,\text{s}}\right)$$

$$= 376.99 \,\text{rad/s}$$

Thus, the rotor's mechanical shaft speed is

$$n_m = (1-s)n_s$$

$$= (1-0.03)\,3600$$

$$= 3492 \,\text{rpm}$$

or

$$\omega_m = (1-s)\omega_s$$

$$= (1-0.03)376.99$$

$$= 365.68 \,\text{rad/s}$$

(b) Since the core loss is assumed to be constant, the appropriate equivalent circuit for the motor is the one shown in Figure 6.16.
To determine the stator current, the equivalent impedance of the circuit has to be found. Therefore, the referred rotor impedance is found from

$$Z_2 = \frac{R_2}{s} + jX_2 s$$

$$= \frac{0.196}{0.03} + j0.510$$

$$= 6.55\angle 4.46°$$

(c) Since this rotor impedance is in parallel with the magnetization branch, the corresponding impedance is

$$Z_{eq} = \frac{1}{\frac{1}{jX_m} + \frac{1}{Z_2}}$$

$$= \frac{1}{\frac{1}{j12.5} + \frac{1}{6.55\angle 4.46°}}$$

$$= 5.63\angle 31.13°$$

Therefore, the total impedance is

$$Z_{tot} = (R_1 + jX_1) + Z_{eq}$$

$$= (0.322 + j0.675) + 5.63\angle 31.13°$$

$$= 6.265\angle 34.89° \ \Omega$$

Thus, the stator current is

$$I_1 = \frac{V_1}{Z_{tot}}$$

$$= \frac{\left(480/\sqrt{3}\right)\angle 0°}{6.265\angle 34.89°}$$

$$= 44.24\angle -34.89° \ A$$

The power factor of the motor is

$$PF = \cos 34.89° = 0.82 \text{ lagging}$$

(d) The input power to the motor is

$$P_{in} = P_1 = 3V_1 I_1 \cos\theta$$

$$= 3\frac{(480\,\text{V})}{\sqrt{3}}(44.24\,\text{A})(0.82)$$

$$= 30,166.38 \ W$$

The stator copper losses are

$$P_{1,Cu} = 3I_1^2 R_1$$

$$= 3(44.24\,\text{A})^2(0.322\,\Omega)$$

$$= 1890.27\,W$$

The air-gap power is

$$P_g = P_{in} - P_{1,Cu}$$

$$= 30{,}186.33 - 1{,}890.27$$

$$= 28{,}276.2\,\text{W}$$

Thus, the developed power is

$$P_d = (1-s)P_g$$

$$= (1-0.03)\,28{,}276.1$$

$$= 27{,}427.82\,\text{W}$$

Therefore, the output power is

$$P_{out} = P_d - P_{rot}$$

$$= 27{,}427.82 - 1{,}850$$

$$= 25{,}577.82\,\text{W}$$

(e) The developed torque is

$$T_d = \frac{P_g}{\omega_s}$$

$$= \frac{28{,}276.1\,\text{W}}{376.99\,\text{rad/s}}$$

$$= 75\,\text{N} \cdot \text{m}$$

The output torque *is*

$$T_{out} = \frac{P_{out}}{\omega_m}$$

$$= \frac{25{,}577.82\,\text{W}}{365.58\,\text{rad/s}}$$

$$= 69.96\,\text{N} \cdot \text{m}$$

(f) The motor's efficiency at this operating condition is

$$\eta = \frac{P_{out}}{P_{in}} \times 100$$

$$= \frac{25{,}577.82\,\text{W}}{30{,}166.38\,\text{W}} \times 100$$

$$= 84.79\,\%$$

6.9 EQUIVALENT CIRCUIT AT START-UP

At start-up, the rotor is at standstill and therefore the slip of the motor is 1.0. The corresponding equivalent circuit is the same as the one shown in Figure 6.16, except that all the values of slip are set to a value of 1.0. All powers and torques can be found as shown before, except for the output quantities. Since the motor is at standstill, there are no windage and friction losses.

Furthermore, since ω_m is zero, T_{out} is undefined. Similarly, if Equation 6.68a is used, T_d is also undefined. However, T_d can be found by using Equations 6.68c and d and by setting s equal to 1.0. Therefore, the starting torque can be determined from

$$T_{d,start} = \frac{P_g}{\omega_s} \tag{6.71}$$

or

$$T_{d,start} = \frac{3I_2^2 R_2}{\omega_s} \tag{6.72}$$

Example 6.4

Consider the induction motor given in Example 6.3 and assume that the rotor is at standstill. Determine the following:

(a) The speed at start-up
(b) The stator current at start-up
(c) The power factor at start-up
(d) The developed power and output power at start-up
(e) The developed torque and output torque at start-up

Solution

(a) The synchronous speed is

$$n_s = 3600 \, \text{rev/min}$$

or

$$\omega_s = 376.99 \, \text{rad/s}$$

However, the rotor's mechanical shaft speed is

$$n_m = (1-s)n_s = (1-1)3600 = 0$$

(b) The referred rotor impedance is

$$Z_2 = \frac{R_2}{s} + jX_2$$

$$= \frac{0.196}{1.0} + j0.510$$

$$= 0.5464 \angle 68.98° \, \Omega$$

Also,

$$Z_{eq} = \frac{1}{1/jX_m + 1/Z_2}$$

$$= \frac{1}{1/j12.5 + 1/0.5464\angle 68.98°}$$

$$= 0.525 \angle 69.84° \, \Omega$$

Thus, the total impedance is

$$Z_{tot} = (R_1 + jX_1) + Z_{eq}$$

$$= (0.322 + j0.675) + 0.525\angle 69.84°$$

$$= 1.275\angle 66.29° \, \Omega$$

Therefore, the stator current is

$$I_1 = \frac{V_1}{Z_{tot}}$$

$$= \frac{\left(480/\sqrt{3}\right)\angle 0°}{1.275\angle 66.29°}$$

$$= 217.28\angle - 66.29° \, A$$

Notice that the starting current is almost five times the load current found in Example 6.3. *Such a starting current would blow the fuses.*

(c) The power factor of the motor is

$$PF = \cos 66.29° = 0.4 \text{ lagging}$$

(d) The input power to the motor is

$$P_{in} = P_1 = \sqrt{3}V_1 I_1 \cos\theta$$

$$= 3\left(\frac{480\,V}{\sqrt{3}}\right)(217.28\,A)\,0.4$$

$$= \sqrt{3}(480\,V)(217.28\,A)\,0.4$$

$$= 72,171.72\,W$$

The stator copper losses are

$$P_{1,cu} = 3I_1^2 R_1$$

$$= 3(217.28)^2 \times 0.322$$

$$= 45,605.44\,W$$

The air-gap power is

$$P_g = P_{in} - P_{1,cu}$$

$$= 72,171.72 - 45,605.44$$

$$= 26,566.28\,W$$

Therefore, the developed power is

$$P_d = (1-s)P_g$$

$$= (1-1)26,566.28$$

$$= 0$$

Thus, the output power is

$$P_{out} = P_d - P_{rot}$$

$$= 0$$

(e) The developed torque is

$$T_{d,start} = \frac{P_g}{\omega_s}$$

$$= \frac{26,566.28}{376.99}$$

$$= 70.5 \, \text{N} \cdot \text{m}$$

Example 6.5

A three-phase, two-pole, 60 Hz induction motor provides 25 hp to a load at a speed of 3420 rpm. If the mechanical losses are zero, determine the following:

(a) The slip of the motor in percent
(b) The developed torque
(c) The shaft speed of the motor, if its torque is doubled
(d) The output power of the motor, if its torque is doubled

Solution

(a) Since

$$n_s = \frac{120 f_1}{p}$$

$$= \frac{120 \times 60}{2}$$

$$= 3600 \, \text{rpm}$$

the slip is

$$s = \frac{n_s - n_m}{n_s} \times 100$$

$$= \frac{3600 - 3420}{3600} \times 100$$

$$= 5\%$$

(b) Since the mechanical losses are zero,

$$T_d = T_{load} = T_{out}$$

and

$$P_d = P_{load} = P_{out}$$

the developed torque is

$$T_d = \frac{P_d}{\omega_m}$$

$$= \frac{(25\,\text{hp})(746\,\text{W/hp})}{(3420\,\text{rpm})(2\pi\,\text{rad/rev})(1\,\text{min/60s})}$$

$$= 52.07\,\text{N} \cdot \text{m}$$

or in English units,

$$T_d = \frac{5252 P_d}{n_m}$$

$$= \frac{5252(25\,\text{hp})}{4320\,\text{rpm}}$$

$$= 38.4\,\text{lb} \cdot \text{ft}$$

Alternatively, the torque in lb · ft can be found directly from

$$T_d = \left(\frac{550}{746}\right)(T_d\,\text{N} \cdot \text{m})$$

$$= \left(\frac{550}{746}\right)(52.07\,\text{N} \cdot \text{m})$$

$$= 38.4\,\text{lb} \cdot \text{ft}$$

(c) The developed torque is proportional to the slip. If the developed torque is doubled, then the slip also doubles and the new slip is

$$s = 2 \times 0.05 = 0.10$$

Hence, the shaft speed becomes

$$n_m = (1-s)n_s$$

$$= (1-0.10)3600$$

$$= 3240\,\text{rpm}$$

(d) Since

$$P_d = T_d \times \omega_m$$

the power supplied by the motor is

$$P_d = (2 \times 52.07)\left[(3{,}240\,\text{rpm})(2\pi\,\text{rad/rev})(1\,\text{min}/60\text{s})\right]$$

$$= 35{,}333.9\,\text{W}$$

or in English units,

$$P_d = \frac{T_d \times n_m}{5252}$$

$$= \frac{(2 \times 38.4)(3,240\,\text{rpm})}{5252}$$

$$= 47.4\,\text{hp}$$

6.10 DETERMINATION OF POWER AND TORQUE BY USE OF THÉVENIN'S EQUIVALENT CIRCUIT

According to Thévenin's theorem, a network of linear impedances and voltage sources can be represented by a single-voltage source and a single impedance as viewed from two terminals. The equivalent voltage source is the voltage that appears across these terminals when the terminals are open-circuited. The equivalent impedance is the impedance that can be found by looking into the network from the terminals with all voltage sources short-circuited.

Therefore, to find the current I_2 in Figure 6.19a, Thévenin's theorem can be applied to the induction-motor equivalent circuit. The Thévenin voltage can be found by separating the stator circuit from the rotor circuit, as indicated in the figure. Thus, by voltage division,

$$V_{th} = V_1 \left(\frac{jX_m}{R_1 + jX_1 + jX_m} \right) \tag{6.73}$$

The magnitude of the Thévenin voltage is

$$\mathbf{V}_{th} = \mathbf{V}_1 \left(\frac{X_m}{[R_1^2 + (X_1 + jX_m)^2]^{1/2}} \right) \tag{6.74}$$

However, since $R_1^2 \ll (X_1 + X_m)^2$, the voltage is approximately

$$\mathbf{V}_{th} = \mathbf{V}_1 \left(\frac{X_m}{X_1 + X_m} \right) \tag{6.75}$$

$$\mathbf{Z}_{th} = R_{th} + jX_{th}$$

$$= \frac{jX_m(R_1 + jX_1)}{R_1 + j(X_1 + X_m)} \tag{6.76}$$

Since $X_1 \ll X_m$ and $R_1^2 \ll \left(X_1 + X_m^2 \right)$, the Thévenin resistance and reactance are approximately

$$R_{th} \cong R_1 \left(\frac{X_m}{X_1 + X_m} \right)^2 \tag{6.77}$$

and

$$X_{th} \cong X_1 \tag{6.78}$$

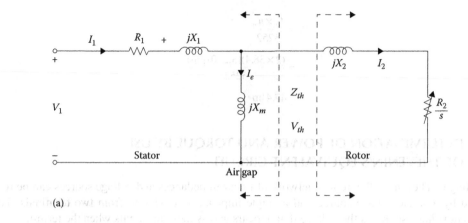

(a)

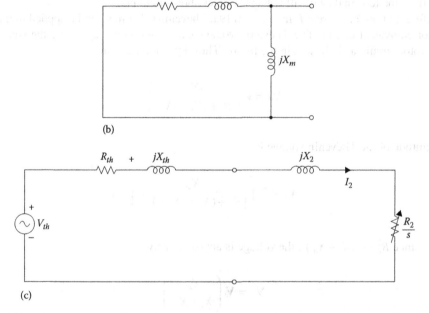

(b)

(c)

FIGURE 6.19 (a) Application of Thévenin's theorem to the induction motor circuit model, (b) the stator circuit used to determine the Thévenin-equivalent impedance of the stator circuit, and (c) the resultant induction-motor equivalent circuit simplified by Thévenin's theorem.

Figure 6.19c shows the resultant equivalent circuit of the induction motor. Here, the rotor current can be found from

$$I_2 = \frac{V_{th}}{Z_{th} + Z_2} \tag{6.79a}$$

or

$$I_2 = \frac{V_{th}}{R_{th} + R_2 / s + j(X_{th} + jX_2)} \tag{6.79b}$$

The magnitude of the rotor current is

$$I_2 = \frac{V_{th}}{\left[(R_{th} + R_2 / s)^2 + (X_{th} + X_2)^2 \right]^{1/2}} \tag{6.80}$$

Thus, the corresponding air-gap power is

$$P_g = 3I_2^2 \left(\frac{R_2}{s} \right) \tag{6.81a}$$

or

$$P_g = \frac{3V_{th}(R_2 / s)}{\left[(R_{th} + R_2 / s)^2 + (X_{th} + X_2)^2 \right]} \tag{6.81b}$$

Therefore, the developed torque is

$$T_d = \frac{P_g}{\omega_s}$$

$$= \frac{3I_2^2(R_2 / s)}{\omega_s} \tag{6.82a}$$

or

$$T_d = \frac{3V_{th}^2(R_2 / s)}{\omega_s \left[(R_{th} + R_2 / s)^2 + (X_{th} + X_2)^2 \right]} \tag{6.82b}$$

Since at start-up the slip is unity, the developed starting torque is

$$T_{start} = \frac{3V_{th}^2 R_2}{\omega_s \left[(R_{th} + R_2)^2 + (X_{th} + X_2)^2 \right]} \tag{6.83}$$

6.11   PERFORMANCE CHARACTERISTICS

The performance characteristics of the induction machine include starting torque, maximum (or pull-out) torque,* maximum power, current, power factor, and efficiency. The maximum torque can be determined by using Thévenin's equivalent circuit. Since

$$T_d = \frac{P_g}{\omega_s} \tag{6.84}$$

The developed torque will be maximum when the air-gap power is a maximum. The air-gap power can be found from Equation 6.81. To find at what value of the variable R_2/s is the maximum P_g takes

---

* It is also *known* as *the maximum internal or breakdown torque.*

place, the derivative of the right side of Equation 6.81 with respect to R_2/s must be determined and set equal to zero. Thus,

$$\frac{3V_{th}^2\left[R_{th}^2-(R_2/s)^2+(X_{th}+X_2)^2\right]}{\left[(R_{th}+R_2/s)^2+(X_{th}+X_2)^2\right]^2}=0 \tag{6.85}$$

by setting the numerator of this equation equal to zero,

$$R_{th}^2-\left(\frac{R_2}{s}\right)^2+(X_{th}+X_2)^2=0 \tag{6.86}$$

from which

$$\frac{R_2}{s}=\left[R_{th}^2+(X_{th}-X_2)^2\right]^{1/2} \tag{6.87}$$

That is, the maximum power is transferred to the air-gap power resistor $R_2/2$ when this resistor is equal to the impedance looking back into the source. Therefore, the slip $s_{max\,T}$ at which the maximum (or pull-out) torque is developed is

$$s_{maxT}=\frac{R_2}{\left[R_{th}^2+(X_{th}+X_2)^2\right]^{1/2}} \tag{6.88}$$

The slip at which maximum torque takes place is directly proportional to the rotor resistance and may be increased by using a larger rotor resistance. Thus, the **maximum** or **pull-out torque** can be found, by inserting Equation 6.88 into Equation 6.82, as

$$T_{max}=\frac{3V_{th}^2}{2\omega_s\left\{R_{th}+\left[R_{th}^2+(X_{th}+X_2)^2\right]^{1/2}\right\}} \tag{6.89}$$

The maximum torque is proportional to the square of the supply voltage and is also inversely related to the size of the stator resistance and reactance, and the rotor reactance.

Figure 6.20 shows the torque–speed or torque–slip characteristic curve of an induction motor. As shown in the figure, the full-load torque of an induction motor is less than its starting torque. If the value of the supply voltage is halved, both the starting torque and the maximum torque become one-fourth of their respective full-voltage values. Figure 6.21 shows the torque versus speed (or slip) characteristic curves of an induction motor at full and half of the supply voltage.

According to Equation 6.89, the maximum torque developed by the induction motor is independent of the rotor-winding resistance. Note that the value of the rotor-winding resistance R_2 determines the speed at which the maximum torque will take place, as suggested by Equation 6.88.

In other words, increasing the rotor-winding resistance by inserting external resistance increases the slip at which the maximum (or *pull-out*) torque occurs, but leaves its magnitude unchanged. Figure 6.22 shows the effect of increasing the rotor resistance on the torque–speed characteristics of a wound-rotor induction motor. Notice that as the rotor resistance increases, the curve becomes flatter. Also notice that the starting torque and the maximum torque are the same at the given rotor resistance value.

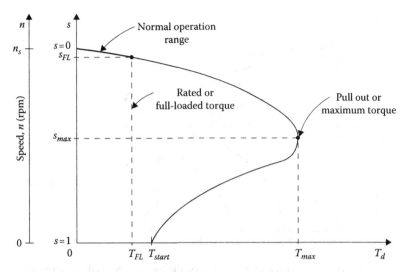

FIGURE 6.20   Torque–speed or torque–slip characteristic curve of an induction motor.

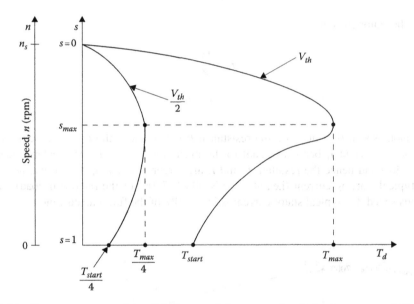

FIGURE 6.21   Torque versus speed (or slip) characteristic curves of an induction motor at full and halved supply voltage.

As can be observed in Figure 6.16, the input impedance of an induction motor is

$$\mathbf{Z}_1 = R_1 + jX_1 + \frac{jX_m(R_2 \, / \, s + jX_2)}{R_2 \, / \, s + j(X_m + X_2)} \tag{6.90a}$$

or

$$\mathbf{Z}_1 = Z_1 \angle \theta_1 \tag{6.90b}$$

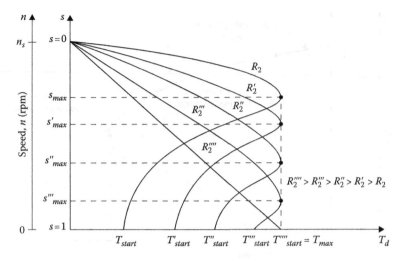

FIGURE 6.22 The effect of increased rotor resistance on the torque–speed characteristics of a wound-rotor induction motor.

Therefore, the stator current is

$$I_1 = \frac{V_1}{Z_1}$$

$$= I_e + I_2 \tag{6.91}$$

At synchronous speed (i.e., at $s=0$), the resistance R_2/s becomes infinite and therefore I_2 is zero. Thus, the stator current I_1 becomes equal to the excitation current I_e. At greater values of slip, $R_2/s + jX_2$ is low and hence the resultant I_2 and I_1 are larger. For example, as illustrated in Figure 6.23, the typical starting current (i.e., at $s=1$) is 500%–700% of the rated (full-load) current. At synchronous speed, the typical stator current is 25%–50% of the full-load current.

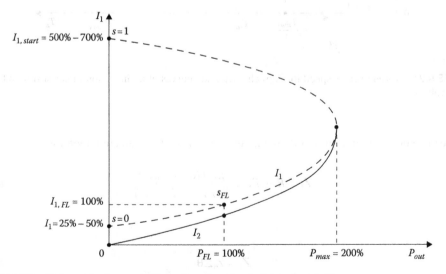

FIGURE 6.23 Stator and rotor currents as a function of output power and slip.

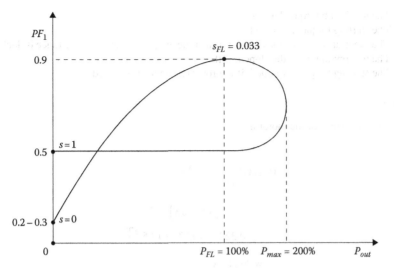

FIGURE 6.24   Power factor of a typical induction motor as a function of output power and slip.

The power factor of an induction motor is $\cos \theta_1$ where θ_1 is the phase angle of the stator current I_1. This phase angle θ_1 is the same as the input impedance angle of the equivalent circuit shown in Figure 6.16 and given by Equation 6.90b. Figure 6.24 shows the typical power factor variation as a function of output power and slip. Note that the figure is not drawn to scale.

Figure 6.25 shows the efficiency of a typical induction motor as a function of output power and slip. The full-load efficiency of a large induction motor may be as high as 95%. As an induction motor is loaded beyond its rated output power, its efficiency decreases considerably.

Example 6.6

The induction motor given in Example 6.3 has a wound rotor. Determine the following:

(a) The slip at which the maximum torque is developed
(b) The speed at which the maximum torque is developed

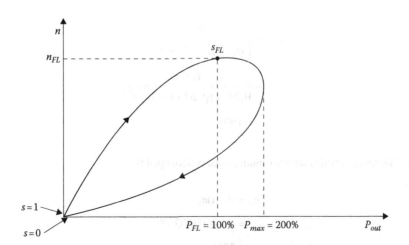

FIGURE 6.25   The efficiency of a typical induction motor as a function of output power and slip.

(c) The maximum torque developed
(d) The starting torque developed
(e) The speed at which the maximum torque is developed, if the rotor resistance is doubled
(f) The maximum torque developed, if the rotor resistance is doubled
(g) The starting torque developed, if the rotor resistance is doubled

Solution

The Thévenin voltage of the motor is

$$V_{th} = V_1 \left(\frac{X_m}{[R_1^2 + (X_1 + X_m)^2]^{1/2}} \right)$$

$$= \frac{\left(480 / \sqrt{3}\right)12.5}{\left[0.322^2 + (0.675 + 12.5)^2\right]^{1/2}}$$

$$= 262.85\,\text{V}$$

The Thévenin resistance is

$$R_{th} \cong R_1 \left(\frac{X_m}{X_1 + X_m} \right)^2$$

$$= 0.322 \left(\frac{12.5}{0.675 + 12.5} \right)^2$$

$$= 0.29\,\Omega$$

The Thévenin reactance is

$$X_{th} \cong X_1 = 0.675\,\Omega$$

(a) The slip at which the maximum torque is developed is

$$s_{maxT} = \frac{R_2}{\left[R_{th}^2 + (X_{th} + X_2)^2\right]^{1/2}}$$

$$= \frac{0.196}{0.29^2 + [(0.675 + 0.51)^2]^{1/2}}$$

$$= 0.1607$$

(b) The speed at which the maximum torque is developed is

$$n_m = (1 - s)n_s$$

$$= (1 - 0.1607)3600$$

$$= 3021.6\,\text{rpm}$$

(c) The maximum torque developed is

$$T_{max} = \frac{3V_{th}^2}{2\omega_s\{R_{th} + [R_{th} + (X_{th} + X_2)^2]^{1/2}\}}$$

$$= \frac{3(262.85)^2}{2(377\,\text{rad/s})\left\{0.29 + \left[0.29^2 + (0.306^2 + (0.675+).51)^2\right]^{1/2}\right\}}$$

$$= 182\,\text{N} \cdot \text{m}$$

(d) The starting torque developed is

$$T_{start} = \frac{3V_{th}^2 R_2}{\omega_s[(R_{th} + R_2)^2 + (X_{th} + X_2)^2]^2}$$

$$= \frac{3(262.85)^2 \times 0.196}{(377\,\text{rad/s})[(0.29 + 0.196)^2 + (0.675 + 0.51)^2]^{1/2}]}$$

$$= 107.3\,\text{N} \cdot \text{m}$$

(e) Since the rotor resistance is doubled, the slip at which the maximum torque occurs also doubles. Thus,

$$s_{max} = 2(0.1607) = 0.3214$$

and the speed at maximum torque is

$$n_m = (1 - s)n_s$$

$$= (1 - 0.3214)3600$$

$$= 2443\,\text{rpm}$$

(f) The maximum torque still remains at

$$T_{max} = 182\,\text{N} \cdot \text{m}$$

(g) If the rotor resistance is doubled, the developed starting torque becomes

$$T_{start} = \frac{3(262.85)^2\ 0.392}{377\left[(0.29 + 0.196)^2 + (0.675 + 0.51)^2\right]}$$

$$= 131.29\,\text{N} \cdot \text{m}$$

6.12 CONTROL OF MOTOR CHARACTERISTICS BY SQUIRREL-CAGE ROTOR DESIGN

In principle, an increase in rotor resistance decreases the speed at which a given torque is found, increases the starting torque, and lowers the motor efficiency. Also, it decreases the starting current and increases the power factor. If the increase in the power factor is greater than the decrease in the starting current, it results in a better starting torque.

In general, in a squirrel-cage rotor design, the rotor resistance value is determined by a compromise between conflicting requirements of good speed regulation and good starting torque. Usually, the resistance of a squirrel-cage motor, referred to the stator, is less than that of a wound-rotor machine of the same size. The rotor resistance of a squirrel-cage motor can be increased by decreasing the cross-sectional area of the end rings. The resistance of the stator windings should be minimal to reduce both the stator's copper losses and its internal voltage drop. Also, an increase in the stator resistance causes the maximum torque to decrease.

The leakage reactances are affected by changes in the air gap and slot openings. As the air gap is increased, these reactances decrease, causing a greater excitation current to flow at a lesser power factor. Open slots cause the same thing. Also, an increase in reactance decreases the pull-out torque. Therefore, induction motors are designed with as small an air gap as possible to reduce the excitation current. The rotor frequency changes with speed and at standstill is the same as the stator frequency. As the motor speeds up, the rotor frequency decreases in value to 1 or 3 Hz at full load in a typical 60 Hz motor.

By using suitable shapes and arrangements for rotor bars, it is possible to design squirrel-cage rotors so that their effective resistance at 60 Hz is several times their resistance at 1 or 3 Hz. This results from the inductive effect of the slot-leakage flux on the current distribution in the rotor bars. The change in the resistance of the rotor bars is due to what is commonly known as the **skin effect**.

Various slot shapes for squirrel-cage induction motor rotors which produce the National Electrical Manufacturers Association (NEMA) design characteristics* are shown in Figure 6.26. The rotor bars of the NEMA design class A motor shown in Figure 6.26a are quite large and located near the surface of the rotor. They have low resistance (because of their large cross sections) and a low leakage reactance X_2 (because the bars are located near the stator). Such motors are also called *class* A motors. They have low slip at full load, high running efficiency, and high pull-out torque (due to the low rotor resistance). However, because R_2 is small, the starting torque of the motor is small, but its starting current is large. These motors are usually used in constant-speed drives to drive pumps, fans, lathes, blowers, and other devices.

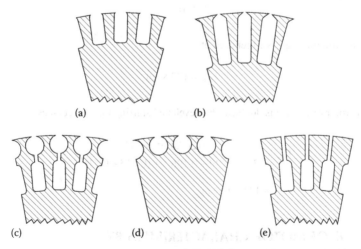

FIGURE 6.26 Various slot shapes for squirrel-cage induction motor rotors which produce NEMA design-class characteristics: (a) NEMA design class A, (b) NEMA design class B, (c) NEMA design class C, (d) NEMA design class D, and (e) old NEMA design class F.

---

* These standard designs are commonly called *design classes*. Each of them provides different torque–speed curves. Recently, the International Electrotechnical Commission (IEC) in Europe adopted similar design classes for motors.

Figure 6.26b shows the deep-bar rotor slots of the *NEMA design class B* motor. Rotor bars embedded in deep slots provide a high effective resistance and a large torque at start-up. Due to the skin effect, the current has a tendency to concentrate at the top of the bars at start-up, when the frequency of the rotor currents is high.

Under normal operating conditions with low slips (since the frequency of the rotor currents is much smaller), the skin effect is negligible and the current tends to distribute almost uniformly throughout the entire rotor-bar cross section. Thus, the rotor resistance decreases to a small value causing a higher efficiency. Such motors are used in applications that are comparable to those for design class A. Due to their lower starting current requirements, they are usually preferred to design class A motors.

Figure 6.26c shows a NEMA *design class C* rotor slot shape, which is an alternative design for the deep-bar rotor. Such a double-cage arrangement is used to attain greater starting torque and better running efficiency. The squirrel-cage winding is made up of two layers of bars short-circuited by end rings. The inner cage, consisting of low-resistance bottom bars, is deeply embedded in the rotor's iron core. However, the outer cage has relatively high-resistance bars located close to the inner stator surface. At start-up, the frequency of the rotor currents is relatively high (almost equal to stator frequency) and the leakage reactance of the cage made up of the larger (inner) rotor bars is also high, suppressing the current in that cage. Therefore, the outer cage with the smaller bars, because of its higher resistance and lower leakage inductance (because of skin effect), predominates during start-up, producing high start-up torque. At the steady-state (i.e., *normal*) operation of the motor, its speed is normal. Thus, the rotor frequency is so low that the leakage reactance of the low-resistance cage is substantially lower than its resistance, and the current densities in the two cages are practically equal. Therefore, during the normal running period, because of the negligible skin effect, the current penetrates the full depth of the lower cage causing an efficient steady-state operation.

In summary, such a design results in high rotor resistance on start-up and low resistance at normal speed. These motors are more expensive than the others. They are used in applications involving high-starting-torque loads, such as compressors and conveyors that are fully loaded when started, and loaded pumps.

Figure 6.26d shows *NEMA design class D* rotor slot shapes. Class D motors are characterized by high starting torque, low starting current, and high operating slip. The rotor cage bars are made of higher-resistance material such as brass. The maximum torque takes place at a slip of 0.5 or higher. Because of high rotor resistance, the full-load slip for theses motors is high (about 7%–17% or more). Thus, the running efficiency is low. The high losses in the rotor circuit dictate that the machine be large and is therefore expensive for a given power. They are ideal for loads with rapid acceleration or high impact such as punch presses and shears.

In addition to the four design classes reviewed so far, NEMA used to also have design classes E and F. They were called *soft-start* induction motors, with very low starting currents and torques, but are no longer in use. Figure 6.26e shows old NEMA design class F. It was once used in large motors designed for very easily started loads such as industrial fans. Figure 6.27 shows torque–speed characteristics for the four design classes.

6.13   STARTING OF INDUCTION MOTORS

A wound-rotor induction motor can be connected directly to the line with its rotor open-circuited, or with relatively high resistances connected to the slip-ring brushes. At full operating speed, the brushes are shorted together so that there is zero external resistance in each phase. However, induction motors with squirrel-cage rotors can be started by a number of different methods, depending on the size and type of the motor involved.

In general, the basic methods of starting can be classified as *direct-on-line starting, reduced-voltage starting,* and *current limiting by series resistance or impedance.* Other methods of starting

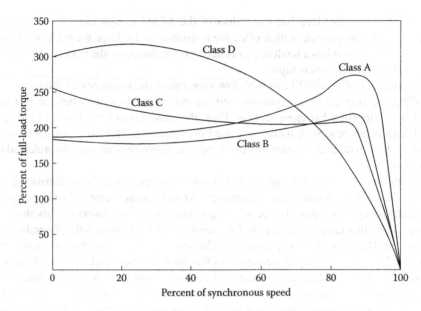

FIGURE 6.27   Typical torque–speed characteristics of cage motors.

include *part-winding starting* and *multicircuit starting*. In these methods, the motor is connected asymmetrically during the starting period.

6.13.1 DIRECT-ON-LINE STARTING

In general, most induction motors are rugged enough so that they can be started across the line without any resultant damage to the motor windings, even though about five to seven times the rated current flows through the stator at rated voltage at standstill. However, such across-the-line starting of large motors is not advisable for two reasons: (1) the lines supplying the induction motor may not have enough capacity and (2) the large starting current may cause a large *voltage dip*,* resulting in reduced voltage across the motor. Since the torque changes approximately with the square of the voltage, the starting torque can become so small at the reduced line voltage that the motor may not even start on load.

To estimate the starting current, all squirrel-cage motors now have a starting *code letter* (not to be mixed up with their *design-class letter*) on their nameplates. The code letter gives the maximum limit of current for the motor at starting conditions. Table 6.1 gives the starting kVA per hp at starting conditions for the machine. To find the starting current for an induction motor, its rated voltage, horsepower, and code letter are read from its nameplate. Therefore, its starting apparent power is

$$S_{start} = (\text{rated horsepower})(\text{code letter factor})$$

and its starting current is

$$I_L = \frac{S_{start}}{\sqrt{3}V_t}$$

---

* For further information, see Gönen (2008).

TABLE 6.1
The NEMA Code Letters

Code Letter	Locked Rotor kVA/h	Code Letter	Locked Rotor kVA/hp
A	0–3.14	L	9.00–9.99
B	3.15–3.54	M	10.00–11.19
C	3.55–3.99	N	11.20–12.49
D	4.00–4.49	P	12.50–13.99
E	4.50–4.99	R	14.00–15.99
F	5.00–5.59	S	16.00–17.99
G	5.60–6.29	T	18.00–19.99
H	6.30–7.09	U	20.00–22.39
J	7.10–7.99	V	22.40 and up
K	8.00–8.99		

Source: Reproduced by permission from NEMA *Motors and Generators Standards*, NEMA Publications MG-1 1978, Copyright 1982 by NEMA, National Electrical Manufacturers Association, New York.

6.13.2 REDUCED-VOLTAGE STARTING

At the time of starting, a reduced voltage is supplied to the stator and slowly increased to the rated value when the motor is within approximately 25% of its rated speed. Such reduced-voltage starting can be achieved by means of *wye–delta starting, autotransformer starting,* and *solid-state voltage controller starting.*

1. *Wye–delta starting.* In this starting, both ends of each phase of the stator winding must be brought out to the terminals of a wye–delta switch so that at start the stator windings are connected in wye. As the motor approaches full speed, the switch is operated and the stator windings are connected in delta, and the motor runs at full speed. Here, the motor used has to be designed for delta operation and is connected in wye only during the starting period. Since the impedance between the line terminals for the wye connection is three times that of the delta connection for the same line voltage, the line current is reduced to one-third of its value for the delta connection. That is, when the motor is connected in wye, it takes one-third as much starting current and develops one-third as much torque. A wye–delta starter is equivalent to an autotransformer with a ratio of $n = 1/\sqrt{3}$.

2. *Autotransformer starting.* Here, the setting of the autotransformer can be predetermined to limit the starting current to any given value. Therefore, an autotransformer, which reduces the voltage applied to the motor to x times the normal voltage, will reduce the starting current in the supply system as well as the starting torque of the motor to x^2 times the normal values. Here, the x is known as the *compensator turns ratio.*

   When the motor reaches about 80% of normal speed, the connections are changed so that the autotransformers are de-energized and the motor is connected to full line voltage. Most compensators are provided with three sets of standard taps in order to apply 80%, 65%, or 50% of the line voltage to the motor. To achieve a satisfactory starting, the lowest tap is normally used. Figure 6.28 shows an autotransformer starter for squirrel-cage motors. Such starters are also called *starting compensators.*

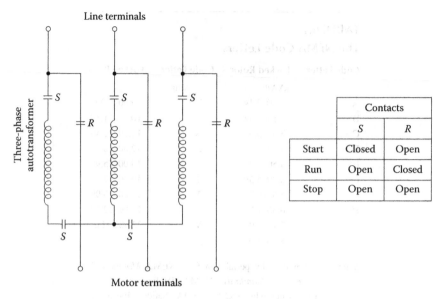

FIGURE 6.28   An autotransformer starter for an induction motor.

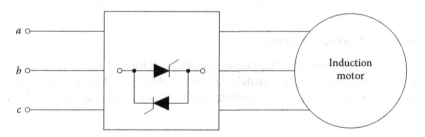

FIGURE 6.29   Solid-state voltage controller starting.

3. *Solid-state voltage controller starting.* In this starting, a solid-state voltage controller is used as a reduced-voltage starter, as shown in Figure 6.29. The advantage of this method is that the solid-state voltage controller provides a smooth starting and also controls the speed of the induction motor during running.

6.13.3 Current Limiting by Series Resistance or Impedance

Such a technique may be used if the starting torque requirement is not too great. In this method, series resistances, or impedances, are inserted in the three lines to limit the starting current. These resistors, or impedances, are shorted out when the motor gains speed. Because of the extra power losses in the external resistances during start-up, it is an inefficient method.

Example 6.7

A 20 hp, 480 V, three-phase induction motor has a code letter *H* on its nameplate. Determine its starting current.

Solution

From Table 6.1, the maximum kVA per hp is 7.09. Thus, the maximum starting kVA of this motor is

$$S_{start} = (20\,hp)(7.09)$$

$$= 141.9\,kVA$$

Therefore, the starting current is

$$I_L = \frac{S_{start}}{\sqrt{3}V_t}$$

$$= \frac{141,800\,VA}{\sqrt{3}\,(480\,V)}$$

$$= 170.6\,A$$

Example 6.8

A three-phase, eight-pole, 100 hp, 440 V, 60 Hz, wye-connected induction motor has the following constant in ohms per phase referred to the stator: $R_1 = 0.085\,\Omega$, $X_1 = 0.196\,\Omega$, $R_2 = 0.067\,\Omega$, $X_2 = 0.161\,\Omega$, and $X_m = 6.65\,\Omega$. The motor has a wound rotor with a turns ratio of 2.0. To produce maximum torque at starting (i.e., $T_{start} = T_{max}$), the wound-rotor windings are connected to external resistors. The Thévenin resistance and reactance of the motor are 0.0802 and 0.191 Ω, respectively. Determine the value of the external rotor resistance under the following conditions:

(a) If they are referred to the stator and connected in wye
(b) If they are referred to the rotor and connected in wye
(c) If they are referred to the stator and connected in delta
(d) If they are referred to the rotor and connected in delta

Solution

(a) The external rotor-circuit resistance is adjusted until the motor produces its maximum torque at start so that $T_{start} = T_{max}$. Since T_{max} occurs at starting $s_{maxT} = 1.0$, from Equation 6.88

$$\frac{R_{2,tot}}{s_{maxT}} = \left[R_{th}^2 + (X_{th} + X_2)^2 \right]^{1/2}$$

or

$$\frac{R_{2,tot}}{1.0} = \left[0.0802^2 + (0.191 + 0.161)^2 \right]^{1/2}$$

$$= 0.361$$

Therefore, $R_{2,tot} = 0.361\,\Omega$ referred to the stator is found. Since this value represents the total resistance in the rotor circuit, the value of the external rotor resistance referred to the stator and connected in wye is

$$R'_{2,ext} = R_{2,tot} - R_2$$

$$= 0.361 - 0.067$$

$$= 0.294\,\Omega \text{ per phase}$$

(b) The value of the external resistance referred to the rotor (i.e., in terms of physical resistance) and connected in wye is

$$R_{2,ext} = \frac{R'_{2,ext}}{a^2}$$

$$= \frac{0.294}{2^2}$$

$$= 0.0735 \ \Omega \text{ per phase}$$

(c) The value of the external rotor resistance *referred to the stator* and connected in delta is

$$R'_{2,ext} = 3(0.294 \ \Omega)$$

$$= 0.882 \ \Omega$$

(d) The value of the external rotor resistance *referred to the rotor* and connected in delta is

$$R_{2,ext} = 3(0.0735\Omega)$$

$$= 0.2205\Omega$$

or

$$R_{2,ext} = \frac{R'_{2,ext}}{a^2} = \frac{0.882}{2^2}$$

$$= 0.2205\Omega$$

Example 6.9

Consider the induction motor given in Example 6.8 and assume that either its wound-rotor slip rings have been short-circuited or its wound rotor has been replaced by an equivalent squirrel-cage rotor. Investigate the line-voltage starting of the motor. Neglect jX_m as a reasonable simplifying assumption because of the large slip (i.e., $s_{start} = 1.0$) at starting. Determine the following:

(a) The stator current at starting
(b) The rotor current at starting
(c) The air-gap power at starting
(d) The developed power at starting
(e) The developed torque at starting

Solution

(a) The stator current at starting is

$$I_{1,start} = \frac{V_1}{(R_1 + R_2/s) + j(X_1 + X_2)}$$

$$= \frac{254\angle 0°}{(0.085 + 0.067/1.0) + j(0.196 + 0.161)}$$

$$= 655\angle -66.9° \text{A}$$

(b) Since

$$I_{2,start} \cong I_{1,start} = 655 \angle -66.9° \text{ A}$$

(c) The air-gap power at starting is

$$P_{g,start} = 3I_{2,start}^2 \left(\frac{R_2}{s} \right)$$

$$= 3(655)^2 0.067$$

$$\cong 86 \text{ kW}$$

(d) The developed power at starting is

$$P_{d,start} = 3I_{2,start}^2 R_2 \left(\frac{1-s}{s} \right)$$

$$= 3(655)^2 \left(\frac{1-1}{1} \right) = 0$$

(e) The developed torque at starting is

$$T_{d,start} = \frac{P_d}{\omega_s}$$

$$= \frac{86,139 \text{ W}}{94.24 \text{ rad/s}}$$

$$= 914 \text{ N} \cdot \text{m}$$

where

$$\omega_s = \frac{2\pi}{60} \left(\frac{120 f_1}{p} \right)$$

$$= \frac{2\pi}{60} \left(\frac{120 \times 60}{8} \right)$$

$$= 94.24 \text{ rad/s}$$

Since $P_{d,\,star\,t} = 0$ and $\omega_m = 0$, the torque Equation 6.68a cannot be used to determine the developed torque at starting.

Example 6.10

Consider autotransformer-type reduced-voltage starting with either the squirrel-cage rotor or the equivalent wound rotor given in Example 6.9, ignoring the shunt branch X_m. Apply ideal autotransformer theory, find and tabulate the motor's stator current $I_{1,start}$, the line current into the autotransformer starter I_L, and the developed starting torque. Determine the aforementioned values as a function of the autotransformer taps, which are 50%, 65%, and 80% of the line voltage.

Solution

Figure 6.30 shows the application of the autotransformer-type reduced-voltage starting. The results are given in Table 6.2 and are based on the results of Example 6.9. Note that such a starting technique is the most suitable for centrifugal-type loads. However, it fails in applications such as elevators where constant load torque is required.

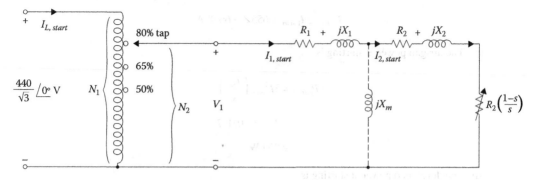

FIGURE 6.30 The autotransformer starting used for Example 6.9.

TABLE 6.2
Results of Example 6.10

V_1	N_2/N_1	$(N_2/N_1)^2$	$I_{1,start}=655(N_2/N_1)$	$I_{1,start}=655(N_2/N_1)^2$	$T_{d,start}=914(N_2/N_1)^2$
100[a]	1.0	1.0	655	655	914
80	0.80	0.64	524	419	585
65	0.65	0.4225	426	277	386
50	0.50	0.25	328	164	229

[a] From Example 6.8.

Example 6.11

Consider the induction motor given in Example 6.10 and replace the autotransformer-type reduced-voltage starting either by primary resistor starting or by primary reactor starting, as shown in Figure 6.31. The existence of the resistor and the reactor are mutually exclusive, that is, there is either R or X in the box shown but both cannot be present. It is required that the starting torque at such reduced voltage be 25% of the starting torque at full voltage. Determine the following:

(a) The starting torque at such a reduced voltage
(b) The primary voltage at starting
(c) The stator current at starting
(d) The total input impedance at starting
(e) The value of the resistor R in ohms
(f) The value of the reactor X in ohms

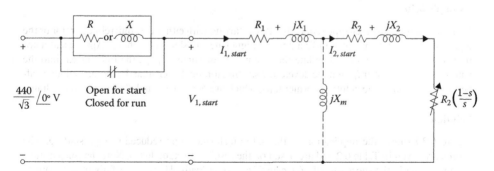

FIGURE 6.31 The primary resistor or primary reactor starting used for Example 6.10.

Solution

(a) In Example 6.10, the starting torque at full voltage was $914\,\text{N}\cdot\text{m}$. Therefore, the new starting torque will be

$$T_{starting} = \frac{914\,\text{N}\cdot\text{m}}{4}$$

$$= 228.5\,\text{N}\cdot\text{m}$$

(b) If the value of the supply voltage is halved, the starting torque becomes one-forth of its full-voltage value. Therefore, the primary voltage at starting is

$$V_{1,starting} = \frac{1}{2}\left(\frac{440\,\text{V}}{\sqrt{3}}\right) = 127\,\text{V}$$

(c) Therefore, the stator current at starting is

$$I_{1,starting} = \frac{1}{2}(655\,\text{A})$$

$$= 327.5\,\text{A}$$

(d) Thus, the total input impedance at starting is

$$Z_{input} = \frac{V_{1,starting}}{I_{1,starting}} = \frac{127\,\text{V}}{327.5\,\text{A}} = 0.388\,\Omega$$

(e) Since the shunt branch jX_m is neglected, the input impedance, for the primary resistor starting, can be expressed as

$$\left|Z_{input}\right| = (R_1 + R_2 + R) + j(X_1 + X_2)$$

Since the magnitude of this input impedance is $0.388\,\Omega$, the value of R is about $0.4 \times 10^{-4}\,\Omega$.

(f) Since the input impedance for the primary reactor starting can be expressed as

$$\left|Z_{input}\right| = (R_1 + R_2) + j(X_1 + X_2 + X)$$

where the magnitude of this input impedance is $0.388\,\Omega$, the value of X from the aforementioned equation can be approximately found as $-0.196 \times 10^{-4}\,\Omega$. This technique is often used to reduce the acceleration torque T_a or the developed torque T_d. However, it should not be used to reduce the stator current I_1.

6.14 SPEED CONTROL

An induction motor is basically a constant-speed motor when it is connected to a constant-voltage and constant-frequency power supply. Even though a large number of industrial drives run at constant speed, there are many applications in which variable speed is a requirement. Examples include elevators, conveyors, and hoists. Traditionally, dc motors have been used in such adjustable-speed drive systems.

However, dc motors are expensive, require frequent maintenance of commutators and brushes, and should not be used in hazardous environments. The synchronous speed of an induction motor can be changed by changing the number of poles or varying the line frequency. The operating slip can be changed by varying the line voltage, varying the rotor resistance, or applying voltages with appropriate frequency to the rotor circuits.

1. *Pole-changing method.* In this method, the stator winding of the motor can be designed so that by simple changes in coil connections the number of poles can be changed by the ratio of 2 to 1. In this way, two synchronous speeds can be obtained. This method is not suitable for wound-rotor motors, since the rotor windings would also have to be reconnected to have the same number of poles as the stator.

   However, a squirrel-cage rotor automatically develops a number of magnetic poles equal to those of the air-gap field. With two independent sets of stator windings, each arranged for pole changing, as many as four synchronous speeds can be achieved in a squirrel-cage motor. For example, 600, 1200, 1800, and 3600 rev/min can be attained for a 60 Hz operation. In addition, the motor phases can be connected either in wye or delta, resulting in eight possible combinations.

2. *Variable-frequency method.* The synchronous speed of an induction motor can be controlled by changing the line frequency. The change in speed is continuous or discrete depending upon whether or not supply frequency is continuous or discrete. To maintain approximately constant flux density, the line voltage must also be changed with the frequency. Therefore, the maximum torque remains nearly constant.

   This type of control is known as *constant volts per hertz* and is possible only if a variable-frequency supply is available. A wound-rotor induction machine can be used as a frequency changer. The arrival of solid-state devices with relatively large power ratings has made it possible to use solid-state frequency converters.*

3. *Variable line-voltage method.* The torque developed by an induction motor is proportional to the square of the applied voltage. Therefore, the speed of the motor can be controlled over a limited range by changing the line voltage. If the voltage can be varied continuously from V_1 to V_2, the speed of the motor can also be varied continuously from speeds n_1 to n_2 for a given load. This method is used for small squirrel-cage motors driving fans and pumps.

4. *Variable rotor-resistance method.* This method can only be used with wound-rotor motors. By varying the external resistance connected to the rotor through the slip rings, the torque–speed characteristics of a wound-rotor induction motor can be controlled. The high available torque permits reduced starting voltage to be used while maintaining a sufficiently high starting torque.

   In addition, the maximum torque and the starting torque may be made the same. By continuous variation of rotor-circuit resistance, continuous variation of speed can also be achieved. The disadvantages of this method include low efficiency at reduced speeds and poor speed regulation with respect to changes in the load.

5. *Variable-slip method.* Without sacrificing efficiency at low-speed operation or affecting the speed with load variation, the induction motor speed can be controlled by using semiconductor converters.

6. *Speed control by solid-state switching.* With the exception of the cycloconverter- or inverter-driven motor, the speed of a wound-rotor motor is controlled by the inverter in the rotor circuit or by regulating the stator voltage with solid-state switching devices such as power transistors or silicon-controlled rectifiers (SCRs or thyristors). In general, SCR-based control provides a wider range of operation and is more efficient than other slip-control methods.

6.15   TESTS TO DETERMINE EQUIVALENT-CIRCUIT PARAMETERS

The parameters of the equivalent circuit of the induction motor can be found from the *no-load* and *blocked-rotor tests.*[†] These tests correspond to the no-load and short-circuit tests done on the transformer. The stator resistance can be determined from the *dc test.*

---

* One such arrangement is a SCR supplying do voltage to a static inverter, with solid-state components which in turn supplies the variable frequency to the motor. This arrangement is called a *cycloconverter.*
† For further information, see IEEE, Std, 112–1978 (1984).

6.15.1 NO-LOAD TEST

Rated balanced voltage at rated frequency is applied to the stator, and the motor is permitted to run without a load. The voltage, current, and power input to the stator are measured. At no load, R_2 is very small with respect to $R_2(1 - s)/s$; therefore, the no-load rotor copper loss is negligible. The no-load input power is the sum of the stator copper loss and the rotational losses. Thus,

$$P_{n\ell} = P_{1,cu} + P_{rot} \tag{6.92a}$$

$$P_{n\ell} = 3I_{1,n\ell}^2 R_1 + P_{rot} \tag{6.92b}$$

where

$$P_{n\ell} = P_{core} + P_{FW} \tag{6.93}$$

assuming that the stray losses are negligible. Hence, the rotational losses can be found from

$$P_{rot} = P_{n\ell} - 3I_{1,n\ell}^2 R_1 \tag{6.94}$$

where
  $P_{n\ell}$ is the total three-phase power input to the machine at rated voltage and frequency
  $I_{1,n\ell}$ is the average of the three line currents

Under no load conditions, R_1 is small with respect to X_m, and the overall input power factor is very small, about 0.1. The equivalent input impedance is

$$\left| \boldsymbol{Z}_{eq} \right| = \left| \boldsymbol{Z}_{n\ell} \right| = \frac{V_{n\ell}}{\sqrt{3}I_{n\ell}} \cong X_1 + X_m \tag{6.95}$$

where $V_{n\ell}$ is the line-to-line terminal voltage. Thus,

$$X_m = \frac{V_{n\ell}}{\sqrt{3}I_{n\ell}} - X_1 \tag{6.96}$$

6.15.2 DC TEST

The stator resistance R_1 can be considered equal to its dc value. Thus, it can be measured independently of the rotor impedance. In such a test, a dc power supply is connected to two of the three terminals of a wye-connected induction motor. The current in the stator windings is adjusted to the rated value, and the voltage between the terminals is measured. Since the current flows through two of the wye-connected windings, the total resistance in the path is $2R_1$. Hence,

$$2R_1 = \frac{V_{dc}}{I_{dc}} \tag{6.97}$$

or

$$R_1 = \frac{V_{dc}}{2I_{dc}} \tag{6.98}$$

This value of R_1 may now be used in Equation 6.94 to determine the stator copper loss as well as the rotational losses. Usually, the calculated R_1 has to be corrected for the skin effect and the temperature of the windings during the short-circuit test.* Therefore, the *ac* resistance is found by multiplying the *dc* resistance by a factor which varies from 1.2 to about 1.8, depending on the frequency and other factors.

6.15.3 BLOCKED-ROTOR TEST

This test corresponds to the short-circuit test of a transformer and is also called the *locked-rotor test*. Here, the rotor of the machine is blocked ($s = 1.0$) to prevent it from moving. A reduced voltage[†] is applied to the machine so that the rated current flows through the stator windings. This input power, voltage, and current suggest a blocked-rotor test frequency of 25% of the rated frequency. The input power to the motor is

$$P_{br} = \sqrt{3}V_{br}I_{br}\cos\theta \tag{6.99}$$

so that the blocked-rotor power factor can be expressed as

$$\cos\theta = \frac{P_{br}}{\sqrt{3}V_{br}I_{br}} \tag{6.100}$$

The magnitude of the total impedance in the motor can be expressed as

$$|Z_{br}| = \frac{V_\phi}{I_1} = \frac{V_{br}}{\sqrt{3}V_{br}I_{br}} \tag{6.101}$$

Since the impedance angle is θ,

$$Z_{br} = R_{br} + jX_{br} \tag{6.102a}$$

$$Z_{br} = Z_{br}\cos\theta + jZ_{br}\sin\theta \tag{6.102b}$$

Since R_1 is found by the dc test, the blocked-rotor resistance is

$$R_{br} = R_1 + R_2 \tag{6.103}$$

and the blocked-rotor reactance is

$$X'_{br} = X'_{br} + X'_2 \tag{6.104}$$

where X'_1 and X'_2 represent the stator and rotor reactances *at the test frequency*, respectively.
    Alternatively, this blocked-rotor reactance can be expressed as

---

* For further information, see IEEE, Std, 112–1978 (1984).
† If full voltage at the rated frequency were applied, the current would be five to eight or more times the rated value. Because of this, blocked-rotor tests are *not done at full voltage* except for small motors. Even then, such tests are made as rapidly as possible to prevent overheating of the windings.

$$X'_{br} = Z_{br} \sin\theta \qquad (6.105)$$

or

$$X'_{br} = \left(Z^2_{br} - R^2_{br}\right)^{1/2} \qquad (6.105b)$$

or

$$X'_{br} = \left[\left(\frac{V_{br}}{\sqrt{3}I_{br}}\right)^2 - \left(\frac{P_{br}}{3I^2_{br}}\right)^2\right]^{1/2} \qquad (6.105c)$$

The rotor resistance R_2 can be found from

$$R_2 = R_{br} - R_1 \qquad (6.106)$$

where R_1 is found from the dc test. Therefore,

$$R_2 = \frac{P_{br}}{3I^2_{br}} - \frac{V_{dc}}{2I_{dc}} \qquad (6.107)$$

Since the reactance is directly proportional to the frequency, the total equivalent reactance at the normal operating frequency can be expressed as

$$X_{br} = \frac{f_{rated}}{f_{test}} X'_{br} = X_1 + X_2 \qquad (6.108)$$

Unfortunately, there is no simple way to determine the stator and rotor reactances. This information is known only by the designer of the machine. Table 6.3 gives the approximate values of X_1 and X_2 as fractions of X_{br}. In general, how X_{br} is divided between X_1 and X_2 is not that important, but what is important is the amount of X_{br} since it affects the breakdown torque.

Also note that Design B and C rotors are designed so that their rotor resistances change with frequency.

TABLE 6.3

Determination of X_1 and X_2 from Blocked-Rotor Reactance at the Rated Frequency

Rotor Design	X_1 and X_2 as Fractions of X_{br}	
—	X_1	X_2
Wound rotor	$0.5X_{br}$	$0.5Xb_r$
Design A	$0.5X_{br}$	$0.5X_{br}$
Design B	$0.4X_{br}$	$0.6X_{br}$
Design C	$0.3Xb_r$	$0.7Xb_r$
Design D	$0.5X_{br}$	$0.5X_{br}$

Example 6.12

The following test data were taken on a three-phase, four-pole, 150 hp, 480 V, 60 Hz, Design B wye-connected induction motor with a rated current of 101.3 A.

DC test

$$V_{dc} = 20.26 \quad I_{dc} = 101.3\,\text{A}$$

No-load test

$$V_{n\ell} = 480\,\text{V} \quad f = 60\,\text{Hz}$$

$$I_a = 34.8\,\text{A} \quad P_{n\ell} = 3617.5\,\text{W}$$

$$I_b = 35\,\text{A}$$

$$I_c = 35.2\,\text{A}$$

Blocked-rotor test

$$V_{br} = 51.3\,\text{V} \quad f_{test} = 15\,\text{Hz}$$

$$I_a = 101.3\,\text{A} \quad P_{br} = 5200\,\text{W}$$

$$I_b = 100.4\,\text{A}$$

$$I_c = 102.5\,\text{A}$$

Determine the following:
The R_1 and R_2 of the motor. Use a factor of 1.5 in computing the effective ac armature resistance per phase.
The X_1, X_2, and X_m of the motor.

Solution

(a) From the dc test,

$$R_1 = \frac{V_{dc}}{2I_{dc}}$$

$$= \frac{20.26\,\text{V}}{2(101.3\,\text{A})}$$

$$= 0.1\,\Omega$$

Thus, this resistance in ac is $(0.1\,\Omega)1.5 = 0.15\,\Omega$. From the no-load test,

$$I_{1,n\ell} = \frac{I_a + I_b + I_c}{3}$$

$$= \frac{34.8 + 35 + 35.2}{3}$$

$$= 35\,\text{A}$$

and since

$$|\mathbf{Z}_{n\ell}| = \frac{V_{n\ell}}{\sqrt{3}(35\,\text{A})}$$

$$\cong X_1 + X_m$$

so that when X_l is known, X_m can be determined. The stator copper losses are

$$P_{1,cu} = 3I_{1,n\ell}^2 R_1$$

$$= 3(35\,\text{A})^2(0.15\,\Omega)$$

$$= 551.25\,\text{W}$$

Hence, the no-load rotational losses are

$$P_{rot} = P_{n\ell} - P_{1,cu}$$

$$= 3617.5 - 551.25$$

$$= 3066.25\,\text{W}$$

From the blocked-rotor test,

$$I_L = \frac{101.3 + 100.4 + 102.5}{3}$$

$$= 101.4\,\text{A}$$

The blocked-rotor impedance can be found from

$$|\mathbf{Z}_{br}| = \frac{V_\phi}{I_1}$$

$$= \frac{V_{br}}{\sqrt{3}I_{br}}$$

as

$$|\mathbf{Z}_{br}| = \frac{51.3\,\text{V}}{\sqrt{3}(101.4\,\text{A})}$$

$$= 0.292\,\Omega$$

and the impedance angle θ is

$$\theta = \cos^{-1}\left(\frac{P_{br}}{\sqrt{3}V_{br}I_{br}}\right)$$

$$= \cos^{-1}\left(\frac{5200\,\text{W}}{\sqrt{3}(51.3\,\text{V})(101.4\,\text{A})}\right)$$

$$= 54.75°$$

Thus,

$$R_{br} = Z_{br} \cos \theta$$
$$= 0.292 \cos 54.75°$$
$$= 0.169 \,\Omega$$

Since

$$R_{br} = R_1 + R_2$$

then

$$R_2 = R_{br} - R_1$$
$$= 0.169 - 0.15$$
$$= 0.019 \,\Omega$$

(b) The reactance at 15 Hz is

$$X'_{br} = Z_{br} \sin \theta$$
$$= 0.292 \sin 54.75°$$
$$= 0.239 \,\Omega$$

The equivalent reactance at 60 Hz is

$$X_{br} = f_{rated} X'_{br}$$
$$= \left(\frac{60 \,\text{Hz}}{15 \,\text{Hz}} \right)(0.239 \,\Omega)$$
$$= 0.956 \,\Omega$$

Since it is a design class B induction motor,

$$X_1 = 0.4 X_{br}$$
$$= 0.4(0.956)$$
$$= 0.382 \,\Omega$$

and

$$X_2 = 0.6 X_{br}$$
$$= 0.6(0.956)$$
$$= 0.574 \,\Omega$$

Also,

$$X_m = 7.92 - X_1$$
$$= 7.92 - 0.382$$
$$= 7.538 \,\Omega$$

PROBLEMS

6.1 A 50 hp, three-phase, 60 Hz, wye-connected induction motor operates at a shaft speed of almost 900 rpm at no load and 873 rpm at full load. Determine the following:
(a) The number of poles of the motor
(b) The per-unit and percent slip at full load
(c) The slip frequency of the motor
(d) The speed of the rotor field with respect to the rotor itself
(e) The speed of the rotor field with respect to the stator
(f) The speed of the rotor field with respect to the stator field
(g) The shaft torque of the motor at full load

6.2 A three-phase, 60 Hz, wye-connected induction motor operates at a shaft speed of almost 3600 rpm at no load and 3420 rpm at full load. Determine the following:
(a) The number of poles of the motor
(b) The per-unit and percent slip at full load
(c) The slip frequency of the motor
(d) The speed of the rotor field with respect to the rotor itself
(e) The speed of the rotor field with respect to the stator
(f) The speed of the rotor field with respect to the stator field

6.3 Solve Problem 6.1 but assume that the 10 hp motor runs at a shaft speed of almost 1800 rpm at no load and 1761 rpm at full load.

6.4 Solve Problem 6.2 but assume that the motor runs at a shaft speed of almost 120 rpm at no load and 114 rpm at full load.

6.5 A three-phase, 480 V, 25 hp, two-pole 60 Hz induction motor has a full-load slip of 5%. Determine the following:
(a) The synchronous speed of the motor
(b) The rotor speed at full load.
(c) The slip frequency at full load
(d) The shaft torque at full load

6.6 A three-phase, 480 V, 50 hp, four-pole, 60 Hz induction motor has a full-load slip of 3.5%. Determine the following:
(a) The synchronous speed of the motor
(b) The rotor speed at full load
(c) The slip frequency at full load
(d) The shaft torque at full load

6.7 A three-phase, 208 V, 25 hp induction motor is supplied with 75 A at a 0.85 PF lagging. Its stator and rotor copper losses are 1867 and 650 W, respectively. Its core losses are 1500 W, the friction and windage losses are 300 W, and the stray losses are negligible. Determine the following:
(a) The air-gap power
(b) The mechanical power developed
(c) The shaft output power
(d) The efficiency of the motor

6.8 An induction motor draws 50 A from a 480 V, three-phase line at a lagging power factor of 0.85. Its stator and rotor copper losses are 1000 and 500 W, respectively. Its core losses are 500 W, the friction and windage losses are 250 W, and the stray losses are 250 W. Determine the following:
(a) The air-gap power
(b) The mechanical power developed
(c) The shaft output power in horsepower
(d) The efficiency of the motor

6.9 Consider the data given in Problem 6.8. If the frequency of the power source is 60 Hz and the induction motor has two poles, determine the following:
(a) The slip in percent
(b) The operating speed in rad/s and in rpm
(c) The developed torque
(d) The output torque

6.10 Show (i.e., prove by derivation) that if rotor copper loss were the only loss in an induction motor, the efficiency of the machine would be $\eta = 1 - s$, where s represents per-unit slip (i.e., slip given as a fraction).

6.11 An induction motor draws 50 A from a 380 V, three-phase line at a lagging power factor of 0.90. Its stator and rotor copper losses are 1000 and 500 W, respectively. Its core losses are 650 W, the friction and windage losses are 200 W, and the stray losses are 250 W. Determine the following:
(a) The air-gap power
(b) The mechanical power developed
(c) The shaft output power in W and hp
(d) The efficiency of the motor

6.12 Consider the data given in Problem 6.11. If the frequency of the power source is 50 Hz, and the machine has four poles, determine the following:
(a) The slip in percent
(b) The operating speed in rad/s and rpm
(c) The developed torque
(d) The output torque

6.13 A two-pole, 60 Hz induction motor has its full-load torque at a speed of 3492 rpm. Determine the following:
(a) Its speed at half rated torque
(b) Its speed at half rated torque and half rated voltage, if rotor resistance per phase is doubled

6.14 Consider a 10 hp, 120 V, 60 Hz induction motor. If the motor is operated at 120 Hz, determine the following:
(a) The amount of voltage that should be applied to the motor to maintain the normal degree of iron saturation
(b) The approximate value of the rated horsepower at such a frequency

6.15 The input to the rotor of a 208 V, three-phase, 60 Hz, 24-pole induction motor is 20 kW. Determine the following:
(a) The developed (i.e., electromagnetic) torque in $N \cdot m$ and $lb \cdot ft$
(b) The speed in rpm and rad/s, and the hp output of the motor, if the rotor current is 63.25 A per phase and the rotor resistance is 0.05 Ω per phase. Ignore the rotational losses

6.16 A three-phase, four-pole, 100 hp, 480 V, 60 Hz, wye-connected induction motor has the following constants in ohms per phase referred to the stator:

$$R_1 = 0.1\,\Omega \quad R_2 = 0.079\,\Omega$$

$$X_1 = 0.205\,\Omega \quad X_2 = 0.186\,\Omega$$

$$X_m = 7.15\,\Omega$$

The total rotational losses are 2950 W and are assumed to be constant. The core loss is lumped in with the rotational losses. For a rotor slip of 3.33% at the rated voltage and rated frequency, determine the following:
(a) The speed in rpm and in rad/s
(b) The stator current

    (c) The power factor
    (d) The developed power and output power
    (e) The developed torque and output torque
    (f) The efficiency of the motor

6.17 A three-phase, two-pole, 25 hp, 380 V, 60 Hz, wye-connected induction motor has the following constants in ohms per phase referred to the stator:

$$R_1 = 0.525\,\Omega \quad R_2 = 0.295\,\Omega$$

$$X_1 = 1.75\,\Omega \quad X_2 = 0.8\,\Omega$$

$$X_m = 20.5\,\Omega$$

The total rotational losses are 1850 W and are assumed to be constant. The core loss is lumped in with the rotational losses. For a rotor slip of 3.33% at the rated voltage and rated frequency, determine the following:
    (a) The speed in rpm and in rad/s
    (b) The stator current
    (c) The power factor
    (d) The developed power and output power
    (e) The developed torque and output torque
    (f) The efficiency of the motor

6.18 A three-phase, four-pole, 150 hp, 480 V, 60 Hz, wye-connected induction motor has the following constants in ohms per phase referred to the stator:

$$R_1 = 0.1\,\Omega \quad R_2 = 0.085\,\Omega$$

$$X_1 = 0.25\,\Omega \quad X_2 = 0.175\,\Omega$$

$$X_m = 6.25\,\Omega$$

The total rotational losses are 3250 W and are assumed to be constant. The core loss is lumped in with the rotational losses. For a rotor slip of 3% at the rated voltage and rated frequency, determine the following:
    (a) The speed in rpm and in rad/s
    (b) The stator current
    (c) The power factor
    (d) The developed power and output power
    (e) The developed torque and output torque
    (f) The efficiency of the motor

6.19 A three-phase, four-pole, 60 Hz induction motor supplies 50 hp to a load at a speed of 1701 rpm. Assume that the mechanical losses are zero and determine the following:
    (a) The slip of the motor in percent
    (b) The developed torque in N · m and lb · ft
    (c) The shaft speed of the motor, if its torque is doubled
    (d) The output power of the motor in W and hp, if its torque is doubled

6.20 A three-phase, 60-pole, 60 Hz induction motor supplies 30 hp to a load at a speed of 114 rpm. Assume that the mechanical losses are zero and determine the following:
    (a) The slip of the motor in percent
    (b) The developed torque in N · m and lb · ft

(c) The shaft speed of the motor, if its torque is doubled

(d) The output power of the motor in W and hp, if its torque is doubled

6.21 A three-phase, four-pole, 60 Hz induction motor supplies 100 hp to a load at a speed of 1701 rpm. Assume that the mechanical losses are zero and determine the following:

(a) The slip of the motor in percent

(b) The developed torque in N · m and lb · ft

(c) The shaft speed of the motor, if its torque is doubled

(d) The output power of the motor in W and hp, if its torque is doubled

6.22 A three-phase, 30-pole, 60 Hz induction motor supplies 50 hp to a load at a speed of 233 rpm. Assume that the mechanical losses are zero and determine the following:

(a) The slip of the motor in percent

(b) The developed torque in N · m and lb · ft

(c) The shaft speed of the motor, if its torque is doubled

(d) The output power of the motor in W and hp, if its torque is doubled

6.23 Consider the induction motor given in Problem 6.16 and assume that it has a wound rotor. Determine the following:

(a) The slip at which the maximum torque is developed

(b) The speed at which the maximum torque is developed

(c) The maximum torque developed

(d) The starting torque developed

(e) The speed at which the maximum torque is developed, if the rotor resistance is doubled

(f) The maximum torque developed, if the rotor resistance is doubled

(g) The starting torque developed, if the rotor resistance is doubled

6.24 A three-phase, six-pole, 75 hp, 480 V, 60 Hz, wye-connected induction motor has the following constants in ohms per phase referred to the stator:

$$R_1 = 0.245\,\Omega \quad R_2 = 0.198\,\Omega$$

$$X_1 = 0.975\,\Omega \quad X_2 = 0.72\,\Omega$$

$$X_m = 14.5\,\Omega$$

Determine the following:

(a) The slip at which the maximum torque is developed

(b) The speed at which the maximum torque developed

(c) The maximum torque developed

(d) The starting torque developed

(e) The speed at which the maximum torque is doubled, if the rotor resistance is doubled

(f) The maximum torque developed, if the rotor resistance is doubled

(g) The starting torque developed, if the rotor resistance is doubled

6.25 A three-phase, six-pole, 150 hp, 380 V, 50 Hz, wye-connected induction motor has the following constants in ohms per phase referred to the stator:

$$R_1 = 0.09\,\Omega \quad R_2 = 0.07\,\Omega$$

$$X_1 = 0.195\,\Omega \quad X_2 = 0.172\,\Omega$$

$$X_m = 6.83\,\Omega$$

Determine the following:

(a) The slip at which the maximum torque is developed

(b) The speed at which the maximum torque is developed

(c) The maximum torque developed

(d) The starting torque developed

(e) The speed at which the maximum torque is doubled, if the rotor resistance is doubled

(f) The maximum torque developed, if the rotor resistance is doubled

(g) The starting torque developed, if the rotor resistance is doubled

6.26 A three-phase, eight-pole, 100 hp, 440 V, 60 Hz, wye-connected induction motor has the following constants in ohms per phase referred to the stator: $R_1 = 0.085\,\Omega$, $X_1 = 0.196\,\Omega$, $R_2 = 0.067\,\Omega$, $X_2 = 0.161\,\Omega$, and $X_m = 6.65\,\Omega$. The total rotational losses are 3200 W and are assumed to be constant. The core losses are lumped in with the rotational losses. For a rotor slip of 3.33% at the rated voltage and rated frequency, determine the following:

(a) The equivalent (input) impedance of the motor

(b) The stator (or starting) current

(c) The power factor of the motor

(d) The induced voltage in the stator winding

(e) The rotor current

(f) The developed torque in N · m and lb · ft

(g) The total losses of the motor

(h) The output power in W and hp

(i) Efficiency

6.27 Consider the data given in Problem 6.26 and determine the following:

(a) The maximum torque and slip at which the maximum torque is developed.

(b) Neglect the jX_m and determine the maximum torque and the slip at which the maximum torque is developed. (Note that ignoring X_m at the maximum slip is reasonable because at the maximum *slip* I_2 is much greater than I_e.)

6.28 Consider the induction motor given in Example 6.8 and assume that the motor is now supplied from a power line that has an impedance of $Z_L = R_L + jX_L = 0.02 + j0.05\,\Omega$ per phase and that the constant voltage source connected at the sending end of the line has a voltage of $V_\phi = 440/\sqrt{3}$ V . Determine the following:

(a) The slip at which the maximum torque is developed, if the jX_m is neglected

(b) The developed torque at starting

(c) The developed torque at starting, if the jX_m is not neglected

6.29 A wound-rotor induction motor is operating from a constant line voltage V_1 and with variable external rotor-circuit resistance $R_{2,ext}$. The rotor-winding resistance per phase is R_2 so that $R_{2,tot} = R_2 + R_{2,ext}$. If the motor is mechanically loaded with a constant torque load, that is, $T_d = $ constant, then explain analytically the reasons why the following are true:

(a) Rotor current I_2 is constant as $R_{2,ext}$ is varied

(b) $R_{2,tot}/s$ is constant

6.30 Consider the induction motor given in Example 6.8 and assume that the motor is uncoupled from its mechanical load and operated at the rated stator terminal voltage. Determine the following:

(a) The slip at which the motor is operating

(b) The stator phasor current I_1

(c) The three-phase power input to the stator, that is, P_1

6.31 Figure P6.31 shows a system that can be used to convert balanced, 60 Hz voltages to other frequencies. The synchronous motor has four poles and drives the interconnecting shaft in a clockwise direction. The induction motor has eight poles, and its stator windings are connected to the lines to produce a counterclockwise rotating field (i.e., in the opposite direction to the synchronous motor). As shown in the figure, the induction machine has a wound rotor with terminals brought out through slip rings. Determine the following:

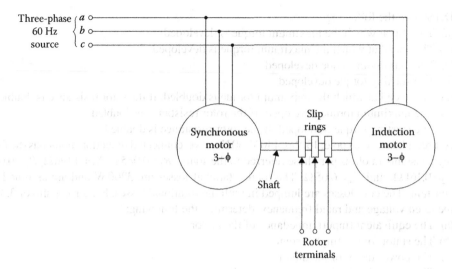

FIGURE P6.31 Figure for problem 6.31.

(a) The speed at which the motor runs

(b) The frequency of the rotor voltages in the induction machine

6.32 The following test data were taken on a three-phase, four-pole, 100 hp, 480 V, 60 Hz, design A wye-connected induction motor with a rated current of 116.3 A:

Dc test

$$V_{dc} = 23.26 \text{ V} \quad I_{dc} = 116.3 \text{ A}$$

No-load test

$$V_{nl} = 480 \text{ V} \quad I_a = 24.5 \text{ A}$$

$$f = 60 \text{ Hz} \quad I_b = 24.1 \text{ A}$$

$$P_{nl} = 4900 \text{ W} \quad I_c = 24.7 \text{ A}$$

Blocked-rotor test

$$V_{br} = 42.2 \text{ V} \quad I_a = 116.4 \text{ A}$$

$$f_{test} = 15 \text{ Hz} \quad I_b = 116.4 \text{ A}$$

$$P_{br} = 6100 \text{ W} \quad I_c = 116.2 \text{ A}$$

Determine the following:

(a) The resistances R_1 and R_2 of the motor. Use a factor of 1.2 in computing the effective (ac) armature resistance per phase.

(b) The reactances X_1, X_2, and X_m of the motor.

7 Synchronous Machines

Who neglects learning in his youth, loses the past and is dead for the future.

Euripides, 438 BC

You cannot teach a crab to walk straight.

Aristophanes, 421 BC

But you can teach a man to walk like a crab.

Turan Gönen

7.1 INTRODUCTION

Almost all three-phase power is generated by three-phase synchronous machines operated as generators. Synchronous generators are also called **alternators** and are normally large machines producing electrical power at hydro, nuclear, or thermal power plants. Efficiency and economy of scale dictate the use of very large generators. Because of this, synchronous generators rated in excess of 1000 MVA (mega-volt-amperes) are quite commonly used in generating stations. Large synchronous generators have a high efficiency which at ratings greater than 50 MVA usually exceeds 98%. The term *synchronous* refers to the fact that these machines operate at constant speeds and frequencies under steady-state operations.

A given synchronous machine can operate as a generator or as a motor. Such machines are used as motors in constant-speed drives in industrial applications and also for pumped-storage stations. In small sizes with only fractional horsepower, they are used in electric clocks, timers, record players, and in other applications which require constant speed.

Synchronous motors with frequency changers such as inverters or cycloconverters can also be used in variable-speed drive applications. An overexcited synchronous motor with no load can be used as a *synchronous capacitor* or *synchronous condenser** to correct power factors. A linear version of a synchronous motor can develop linear or translational motion. Presently, such a *linear synchronous motor* (LSM) is being developed for future high-speed public transportation systems in Japan. However, in general, the LSM is not being used as much as the *linear induction motor* (LIM).

7.2 CONSTRUCTION OF SYNCHRONOUS MACHINES

In a synchronous machine, the *armature*[†] winding is on the stator and the field winding is on the rotor. In normal operation, the three-phase stator currents (in the three-phase distributed stator winding) set up a rotating magnetic field. The synchronous machine *rotors* are simply rotating electromagnets, which have the same number of poles as the stator winding. The rotor winding is supplied from an external dc source through slip rings and brushes; therefore, it produces a rotor magnetic field. Since the rotor rotates in synchronism with the stator magnetic field, the total magnetic field is the result of these two fields.

---

* *Condenser* is an old name for capacitor.
† In rotating machinery, the term *armature* refers to the machine part in which an alternating voltage is generated due to relative motion with respect to a magnetic flux field.

A synchronous machine is a constant-speed (i.e., synchronous speed) machine. Its rotor structure therefore depends on its speed rating. For this reason, high-speed machines have cylindrical (or non-salient pole) rotors, whereas low-speed machines have salient*-pole rotors. With a cylindrical rotor, the reluctance of the magnetic circuit of the field is independent of its actual direction and relative to the direct axis.

However, with salient poles, the reluctance is lowest when the field is along the direct axis where the air gap is the minimum. It is highest when the field is directly halfway between the poles, that is, along the quadrature axis.

Since the rotor field structure depends upon the speed rating of the synchronous machine, turbo-generators (also known as turbo-alternators or turbine-generators), which are high-speed machines, have cylindrical rotors with two or four poles. Figure 7.1 shows a four-pole cylindrical rotor. Hydroelectric and diesel-electric generators are low-speed machines that have salient-pole rotors with four or more poles. Figure 7.2 shows a salient-pole rotor of a hydroelectric generator. This type of rotor structure typically has a relatively short axial length and a relatively large diameter.

The stator of a synchronous machine is basically similar to that of a three-phase induction machine. The stator winding is the source of voltage and electric power when the machine is operating as a generator, and the input winding when it is operating as a motor. It is usually made of preformed stator coils in a double-layer winding.

The winding itself is distributed and chorded to reduce the harmonic content of the output voltages and currents. Figure 7.3 shows the stator of a 650 MW synchronous generator. Figure 7.4 shows the stator winding of a synchronous generator that is being wound.

In salient-pole synchronous machines with laminated rotor construction, where induced currents are not allowed to flow in the rotor body, heavy copper bars are installed in slots in the pole faces. These bars are all shorted together at both ends of the rotor similar to the squirrel-cage rotor of an induction motor. Such a winding is known as the *amortisseur or damper winding*. Damper windings are installed in almost all synchronous machines that have *salient* poles.

When the load on a synchronous machine changes, the load angle also changes. As a result, oscillations in the load angle and corresponding mechanical oscillations in the synchronous rotation

FIGURE 7.1 Insertion of a four-pole cylindrical rotor into a 1200 MW nuclear generator unit. (Courtesy of ABB Corporation, Zurich, Switzerland.)

---

* The word *salient* means "protruding" or "sticking out." Thus, in a salient-pole rotor, a magnetic pole protrudes from the surface of the rotor, whereas a non-salient pole is built flush with the surface of the rotor; and its winding laid in slots in the rotor periphery.

FIGURE 7.2   A salient-pole type hydroelectric generator rotor being lowered into a turbine pit at an electric utility site. (Courtesy of General Electric Canada, Inc., Mississauga, Canada.)

of the shaft take place. These rotor oscillations are known as the *hunting*. The damper windings produce damping torques to eliminate these rotor oscillations caused by such transients and starting torques in synchronous motors.

Cylindrical-rotor machines are formed from solid-steel forgings. Because transient rotor currents can be induced in the solid-rotor body itself, there is no need for a damper winding in such a machine.

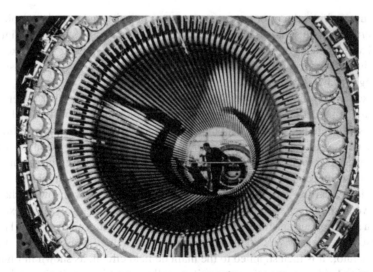

FIGURE 7.3   Internal measurement of the stator of a 650 MW synchronous generator. (Courtesy of ABB Corporation, Zurich, Switzerland.)

FIGURE 7.4   Winding the stator of a synchronous generator. (Courtesy of MagneTex, Nashville, TN.)

7.3   FIELD EXCITATION OF SYNCHRONOUS MACHINES

In a synchronous machine, the rotor poles have constant polarity and must be supplied with direct current. The current can be supplied by an external dc generator or by a rectifier. Figure 7.5a and b shows physical arrangements for a shaft-mounted exciter, and for a shaft-mounted exciter and a pilot exciter, respectively. The arrangement shown in Figure 7.5b is usually used in slow-speed machines with large ratings, such as hydrogenerators.

Here, the exciter may not be self-excited; instead, a self-excited or permanent-magnet type *pilot exciter* may be used to activate the exciter. Figure 7.6a shows a conventional shaft-mounted exciter that is a self-excited dc generator mounted on the same shaft as the rotor of the synchronous machine. In such an arrangement, the generator is the *exciter*. Stationary contacts called *brushes* ride on the slip rings to provide current from the dc source to the rotating field windings. The slip rings are metal rings that completely encircle the shaft of a machine, but are insulated from it. The brushes are made of a carbon compound which provides good contact with low mechanical friction.

An alternative form of excitation is to mount the armature of a relatively small exciter alternator on the shaft of a synchronous machine with a stationary field mounted on the stator. The three-phase output of the exciter generator is rectified to direct current by a three-phase rectifier circuit mounted on the shaft generator. It is then supplied to the main dc field circuit, as shown in Figure 7.6b. This means of supplying field current to the rotor coils is called *brushless excitation* and is now used in most large synchronous machines.

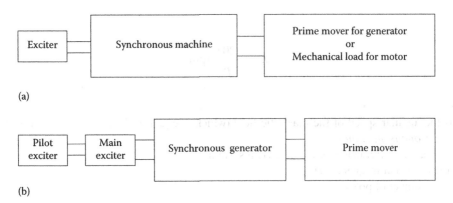

(a)

(b)

FIGURE 7.5 Conventional excitation systems for synchronous machines: (a) physical arrangement for a shaft-mounted exciter and (b) physical arrangement for a shaft-mounted exciter and pilot exciter.

7.4 SYNCHRONOUS SPEED

A synchronous machine operates only at *synchronous speed*, a constant speed that can be determined by the number of poles and the frequency of alternation of the armature-winding voltage. Synchronous machines are called *synchronous* because their speed is directly related to the stator electrical frequency. Therefore, the synchronous speed can be expressed as

$$\omega_s = \frac{\omega}{p/2} = \frac{2\pi f}{p/2} = \frac{4\pi f}{p} \text{ rad/s} \tag{7.1}$$

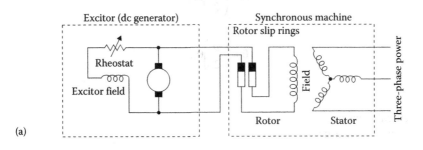

(a)

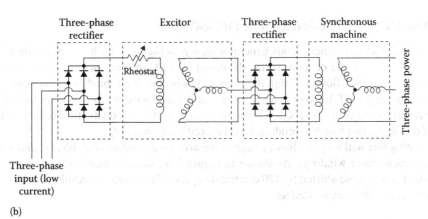

(b)

FIGURE 7.6 Circuit diagram for (a) conventional shaft-mounted exciter and (b) brushless exciter.

or

$$n_s = \frac{120f}{p} \text{ rpm} \tag{7.2}$$

where
  ω_s is the angular speed of the magnetic field (which is equal to the *angular rotor speed of the synchronous machine*)
  ω is the angular frequency of the electrical system
  f is the electrical frequency, Hz
  p is the number of poles

Since the rotor rotates at the same speed as the magnetic field, the stator electrical frequency can be expressed as

$$f = \frac{p \times n_s}{120} \tag{7.3}$$

Note that the frequency in hertz for a two-pole machine is the same as the speed of the rotor in revolutions per second; that is, the electrical frequency is synchronized with the mechanical speed of rotation. Therefore, a two-pole synchronous machine *must* rotate at 60 rps or 3600 rpm to produce a 60 Hz voltage. Alternatively, the radian frequency ω of the voltage wave in terms of ω_m, the mechanical speed in radians per second, is given as

$$\omega = \left(\frac{p}{2}\right)\omega_m \tag{7.4}$$

$$\theta = \left(\frac{p}{2}\right)\theta_m \tag{7.5}$$

where
  θ is in electrical measure
  θ_m is in mechanical measure

7.5   SYNCHRONOUS GENERATOR OPERATION

Consider the elementary synchronous generator shown in Figure 2.1a. It has three identical stator coils (*aa'*, *bb'*, *cc'*), of one or more turns, displaced by 120° in space with respect to each other. When the field current I_f flows through the rotor field winding, it establishes a sinusoidally distributed flux in the air gap. If the rotor is now driven counterclockwise at a constant speed by the prime mover, a revolving magnetic field is developed in the air gap. This magnetic field is called the *excitation field* due to the fact that it is produced by the excitation current I_f.

The rotating flux will vary the flux linkage of the armature windings *aa'*, *bb'*, *cc'* and will induce voltages in these stator windings. As shown in Figure 2.1b, these induced voltages have the same magnitudes but are phase-shifted by 120 electrical degrees. Therefore, the resultant voltages in each of the three coils can be expressed as

$$e_{aa'}(t) = E_{max} \sin\omega t \tag{7.6a}$$

$$e_{bb'}(t) = E_{max} \sin(\omega t - 120°) \tag{7.6b}$$

$$e_{cc'}(t) = E_{max} \sin(\omega t - 240°) \tag{7.6c}$$

The peak voltage in any phase of a three-phase stator is

$$E_{max} = \omega \times N \times \Phi \tag{7.7}$$

However, if the winding is distributed over several slots, the induced voltage is less and is given as

$$E_{max} = \omega \times N \times \Phi \times k_w \tag{7.8}$$

since $\omega = 2\pi f$, then

$$E_{max} = 2\pi \times f \times N \times \Phi \times k_w \tag{7.9}$$

where
  N is the number of turns in each phase winding
  Φ is the flux per pole due to the excitation current I_f
  k_w is the winding factor*

Thus, the rms voltage of any phase of this three-phase stator is

$$E_{max} = \frac{2\pi}{\sqrt{2}} f \times N \times \Phi \times k_w \tag{7.10}$$

$$E_a = 4.44 f \times N \times \Phi \times k_w \tag{7.11}$$

This voltage is a function of the frequency or speed of rotation, the flux that exists in the machine, and, of course, the construction of the machine itself. Therefore, it is possible to rewrite Equation 7.10 as

$$E_a = K \times \Phi \times \omega \tag{7.12}$$

where K is a constant representing the construction of the machine. Thus,

$$K = \frac{N \times k_w}{\sqrt{2}} \tag{7.13}$$

if ω is given in electrical radians per second. Alternatively,

$$K = \frac{N \times p \times k_w}{2\sqrt{2}} \tag{7.14}$$

---

* Its value is less than unity and depends on the winding arrangement.

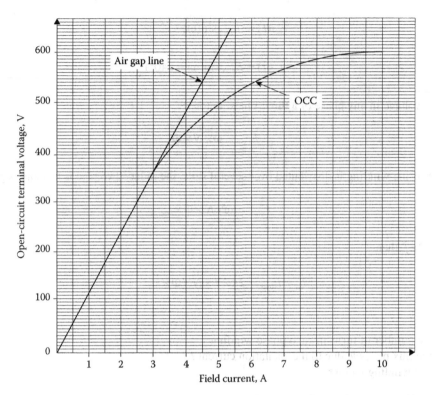

FIGURE 7.7   Open-circuit characteristic (OCC) or magnetization curve of a synchronous machine.

if ω is given in *mechanical* radians per second. Note that E_a is the *internal generated voltage**
or simply the generated voltage. Its value depends upon the flux and the speed of the machine.
However, the flux itself depends upon the current I_f flowing in the rotor field circuit. Therefore, for a
synchronous generator operating at a constant synchronous speed, E_a is a function of the field current, as shown in Figure 7.9.

Such a curve is known as the *open-circuit characteristic* (OCC) or *magnetization curve* of the
synchronous machine. Contrary to the plot shown in Figure 7.7, at $I_f = 0$ the internal generated voltage (i.e., the induced voltage) is not zero due to the residual magnetism.

At the beginning, the voltage rises linearly with the field current. As the field current is increased
further, the flux Φ does not increase linearly with I_f (as suggested by the air gap line) due to saturation of the magnetic circuit and E_a levels off. If the machine terminals are kept open, the internal
generated voltage E_a is the same as the terminal voltage V_t and can be determined using a voltmeter.

Example 7.1

An elementary two-pole three-phase 50 Hz alternator has a rotating flux of 0.0516 Wb. The number of turns in each phase coil is 20. Its shaft speed is 3000 rev/min and its stator winding factor
is 0.96. Determine the following:

   (a)   The angular speed of the rotor
   (b)   The three phase voltages as a function of time
   (c)   The rms phase voltage of this generator if the stator windings are connected in delta
   (d)   The rms terminal voltage if the stator windings are connected in wye

---

* However, the *internal generated voltage* E_a is also known as the *excitation voltage* E_f. Since the excitation voltage (sometimes called the *field voltage*) can be confused with the dc voltage across the field winding, it is preferable to use the
former.

Solution

(a) The angular speed of the rotor is

$$\omega = (3000 \text{ rev/min})(2\pi \text{ rad/rev})(1 \text{ min/60 s})$$

$$= 314.16 \text{ rad/s}$$

(b) The magnitudes of the peak phase voltages are

$$E_{max} = \omega \times N \times \Phi \times k_w$$

$$= (314.16 \text{ rad/s})(20)(0.0516 \text{ Wb})(0.96)$$

$$= 311.13 \text{ V}$$

Thus, the three phase voltages are

$$e_{aa'}(t) = 311.13 \sin 314.16t \text{ V}$$

$$e_{bb'}(t) = 311.13 \sin(314.16t - 120°) \text{ V}$$

$$e_{cc'}(t) = 311.13 \sin(314.16t - 240°) \text{ V}$$

(c) If the stator windings are delta-connected, the rms phase voltage of the generator is

$$E_a = \frac{E_{max}}{\sqrt{2}} = \frac{311.13 \text{ V}}{\sqrt{2}} = 220 \text{ V}$$

(d) If the stator windings are wye-connected,

$$V_t = \sqrt{3}E_a = \sqrt{3}(220 \text{ V}) = 380 \text{ V}$$

Example 7.2

Determine the value of the K constant of the generator given in Example 7.1.
(a) If ω is in *electrical* radians per second.
(b) If ω is in *mechanical* radians per second.
(c) Determine the value of the internal generated voltage E_a.

Solution

(a) If ω is in *electrical* radians per second, then

$$K = \frac{N \times k_w}{\sqrt{2}} = \frac{(20)(0.96)}{\sqrt{2}} = 13.58$$

(b) If ω is in *mechanical* radians per second, then

$$K = \frac{N \times p \times k_w}{2\sqrt{2}} = \frac{(20)(2)(0.96)}{2\sqrt{2}} = 13.58$$

(c) The value of the internal generated voltage is

$$E_a = K \times \Phi \times \omega$$

$$= (13.58)(0.0516 \text{ Wb})(314.16)$$

$$= 220 \text{ V}$$

7.6 EQUIVALENT CIRCUITS

Figure 7.8 shows the complete equivalent circuit representation of a three-phase synchronous generator. A dc power source supplies the rotor field circuit. The field current I_f is controlled by a rheostat connected in series with the field winding. Each phase has an internal generated voltage with series resistance R_a and series reactance X_s.

Assuming balanced operation of the machine, the rms phase currents are equal to each other and 120° apart in phase. The same thing is also true for the voltages. In other respects, the three phases are identical to each other. Therefore, the armature (stator) winding can be analyzed on *a per-phase* basis.

Figure 7.9a, b, and c shows the per-phase equivalent circuits of a synchronous generator. Even though the internal generated voltage E_a is induced in the armature (stator) winding, the voltage that exists at the terminal of the winding is V_ϕ.

The reasons for the difference are (1) the resistance of the armature winding, (2) the leakage reactance of the armature winding, (3) the distortion of the air-gap magnetic field caused by the load current flowing in the armature winding, and (4) the effect of salient-pole rotor shapes if the machine has a salient-pole rotor.

The resistance R_a is the *effective resistance** of the armature winding and is about 1.6 times the dc resistance of the stator winding. It includes the effects of the operating temperature and the **skin effect** caused by the alternating current flowing through the armature winding.

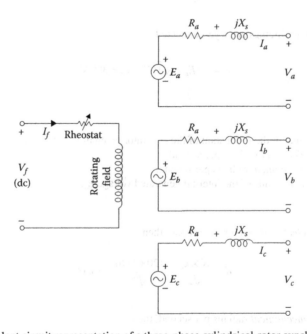

FIGURE 7.8 Equivalent circuit representation of a three-phase cylindrical-rotor synchronous generator.

---

* It is also called the *ac resistance*.

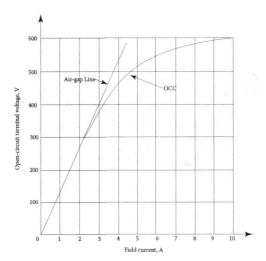

FIGURE 7.9 The per-phase equivalent circuits of a cylindrical-rotor synchronous generator: (a) for phase a, (b) for phase b, and (c) for phase c.

The leakage reactance X_a of the armature winding is caused by the leakage fluxes linking the armature windings due to the currents in the windings. These fluxes do not link with the field winding. For easy calculations, the leakage reactance can be divided into (1) end-connection leakage reactance, (2) slot-leakage reactance, (3) tooth-top and zigzag (or differential) leakage reactance, and (4) belt-leakage reactance. However, in most large machines, the last two reactances are a small portion of the total leakage reactance.

The air-gap magnetic field (caused by the rotor magnetic field) is distorted by the armature (stator) magnetic field because of the load current flowing in the stator. This effect is known as the *armature reaction*, and the resultant reactance X_{ar} is called the *armature reactance*.*

The two reactances X_{ar} and X_a are combined into one reactance and called the *synchronous reactance*[†] X_s, which can be expressed as

$$X_s = X_a + X_{ar} \qquad (7.15)$$

Therefore, as shown in Figure 7.9b, the *synchronous impedance* becomes

$$\mathbf{Z}_s = R_a + jX_s \qquad (7.16)$$

In general, as the machine size increases, the per-unit resistance decreases but the per-unit synchronous reactance increases. Thus, the magnitude of the synchronous impedance becomes

$$\mathbf{Z}_s = (R_a^2 + X_s^2)^{1/2} \cong X_s \qquad (7.17)$$

Because of this, R_a is omitted[‡] from many analyses of synchronous machine operations.

Figures 7.10a and b and 7.11 show the phasor diagrams[§] of a cylindrical-rotor synchronous generator operating at a lagging, leading, and unity power factor, respectively. Note that the dc current

---

* It is also called as the magnetizing reactance.
[†] It is also called the direct-axis synchronous reactance and denoted by Xd.
[‡] The magnitude of R_a is about 0.5%–2% of X_s, and therefore can be ignored except inefficiency computations.
[§] In the phasor diagrams, the length of the armature-resistance voltage drop phasor has been shown larger than it should be in order to make it noticeable.

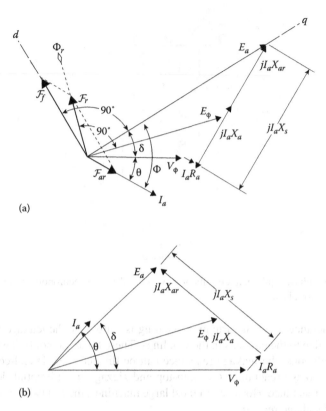

(a)

(b)

FIGURE 7.10 Phasor diagrams of a cylindrical-rotor synchronous generator operating at (a) lagging (overexcited) and (b) leading (underexcited) power factor. (The diagrams shown are not drawn to scale.)

I_f in the field winding produces the mmf $\mathcal{F}_a$ in the air gap and that the ac load current flowing in the stator produces the mm $\mathcal{F}_{ar}$ due to the armature reaction.

The vector sum of the two mmfs gives the resultant mmf of $\mathcal{F}_r$. The flux produced by an mmf is in phase with the mmf and the voltage induced* by a certain flux is behind the corresponding mmf by 90°. Thus, as shown in Figure 7.10a, the mmf $\mathcal{F}_f$ *is* ahead of E_a by 90° and the mmf $\mathcal{F}_{ar}$ is in

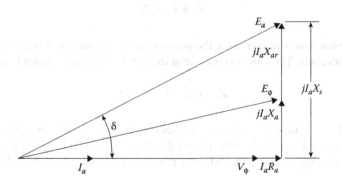

FIGURE 7.11 Phasor diagram of a cylindrical-rotor synchronous generator operating at unity power factor. (The diagram shown is not drawn to scale.)

---

* According to Lenz's Law, $e = -N(d\phi/dt)$.

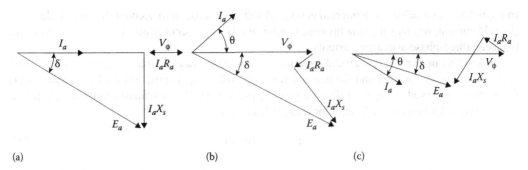

(a) (b) (c)

FIGURE 7.12 Phasor diagram of a synchronous motor operating at (a) unity, (b) leading (overexcited), and (c) lagging (underexcited) power factor.

phase with I_a. The resultant mmf $\mathcal{F}_r$ is ahead of E_ϕ by 90°. The *armature* reaction *voltage* E_{ar} can be determined from

$$E_{ar} = -jX_{ar}I_a \tag{7.18}$$

From Figure 7.12a,

$$E_\phi = E_a + E_{ar} \tag{7.19}$$

or

$$E_a = E_\phi - E_{ar} \tag{7.20}$$

Thus,

$$E_a = E_\phi + jX_{ar}I_a \tag{7.21}$$

The phase voltage V_ϕ is found from

$$I_\phi = E_a - R_aI_a - j(X_a + X_{ar})I_a \tag{7.22a}$$

$$= E_a - R_aI_a - jX_s I_a \tag{7.22b}$$

$$= E_a - (R_a + jX_s)I_a \tag{7.22c}$$

$$= E_a - Z_s I_a \tag{7.22d}$$

Alternatively, the air-gap voltage and the internal generated voltage* can be expressed, respectively, as

$$E_\phi = V_\phi + (R_a + jX_a)I_a \tag{7.23}$$

$$E_a = V_\phi + (R_a + jX_s)I_a \tag{7.24}$$

---

* It is interesting to note that even though synchronous motors, almost without exception, have salient poles, they are often treated as cylindrical-rotor machines. Therefore, Equation 7.33 can also be used for motors as long as the sign of current I_a is made negative to yield $V_\phi = E_a + (R_a + jX_s)I_a$.

In a synchronous machine, the internal generated voltage E_a takes into account the flux produced by the field current, whereas the synchronous reactance takes into account all the flux produced by the balanced three-phase armature currents.

For a given unsaturated cylindrical-rotor machine operating at a constant frequency, the synchronous reactance is a constant and the internal generated voltage E_a is proportional to the field current I_f. As can be observed in Figure 7.10, for a given phase voltage V_f and armature current I_a, a greater E_a is required for lagging loads than for leading loads. Since

$$E_a = K \times \Phi \times \omega \tag{7.12}$$

then a greater field current (*overexcitation*) is required with lagging loads to keep the V_ϕ constant. (Here, ω has to be constant to hold the frequency constant.)

However, for leading loads, a smaller E_a is needed, and, thus, a smaller field current (*underexcitation*) is required. Succinctly put, the synchronous machine is said to be *overexcited* when voltage E_a exceeds voltage V_ϕ; otherwise, if $V_\phi > E_a$ it is said to be *underexcited*. The angle δ between voltage E_a and voltage V_ϕ is called the *torque angle* or the *power angle* of the synchronous machine.

The **voltage regulation** of a synchronous generator at full load, power factor, and rated speed is defined as

$$V\ Reg = \frac{E_a - V_\phi}{V_\phi} \tag{7.25}$$

It is often expressed as percent voltage regulation. Thus,

$$\%V\ Reg = \frac{E_a - V_\phi}{V_\phi} \times 100 \tag{7.26}$$

where
$\quad V_\phi$ is the voltage at full load
$\quad E_a$ is the internal generated voltage (i.e., the V_ϕ voltage) at no load

The voltage regulation is a useful measure in comparing the voltage behavior of generators. It is positive for an inductive load since the voltage rises when the load is removed, and is negative for a capacitive load if the load angle is large enough for the voltage to drop.

Example 7.3

A three-phase, 13.2 kV, 60 Hz, 50 MVA, wye-connected cylindrical-rotor synchronous generator has an armature reactance of 2.19 Ω per phase. The leakage reactance is 0.137 times the armature reactance. The armature resistance is small enough to be negligible. Also ignore the saturation. Assume that the generator delivers full-load current at the rated voltage and 0.8 lagging power factor. Determine the following:

(a) The synchronous reactance in ohms per phase
(b) The rated load current
(c) The air gap voltage
(d) The internal generated voltage
(e) The power angle
(f) The voltage regulation

Solution

(a) The leakage reactance per phase is

$$X_a = 0.137X_{ar}$$

$$= 0.137(2.19\ \Omega)$$

$$\cong 0.3\ \Omega$$

Therefore, the synchronous reactance per phase is

$$X_s = X_a + X_{ar}$$

$$= 0.3 + 2.19$$

$$= 2.49\ \Omega$$

(b) The rated load (or full load) current is

$$I_a = \frac{S}{\sqrt{3}V_t}$$

$$= \frac{50 \times 10^6}{\sqrt{3}(13{,}200)}$$

$$= 2{,}186.93\ \text{A}$$

and when expressed as a phasor,

$$\mathbf{I}_a = I_a(\cos\theta - j\sin\theta)$$

$$= 2{,}186.93(0.8 - j0.6)$$

$$= 1{,}749.55 - j1{,}312.16$$

$$= 2{,}186.93\angle -36.87°\ \text{A}$$

(c) The air-gap voltage (also known as the *voltage behind the leakage reactance*) is

$$\mathbf{E}_\phi = \mathbf{V}_\phi + jX_a\mathbf{I}_a$$

$$= 13{,}200\angle 0° + j0.3(2{,}186.93\angle -36.87°)$$

$$= 8{,}031.84\angle 3.75°\ \text{V}$$

(d) The internal generated voltage (also known as the *voltage behind the synchronous reactance*) is

$$\mathbf{E}_a = \mathbf{V}_\phi + jX_s\mathbf{I}_a$$

$$= 13{,}200\angle 0° + j2.49(2{,}186.93\angle -36.87°)$$

$$= 11{,}727.44\angle 21.81°\ \text{V}$$

(e) The power angle (also called the *torque angle*) is

$$\delta = 21.81°$$

(f) The voltage regulation at full load is

$$V\ Reg = \frac{E_a - V_\phi}{V_\phi} = \frac{11,727.44 - 7,621.02}{7,621.02} \times 100 = 53.88\%$$

7.7 SYNCHRONOUS MOTOR OPERATION

A given synchronous machine can also operate as a motor. However, when the synchronous machine makes the transition from generator to motor action, reversal of power flow takes place. Instead of current flowing *out of* the armature (stator) terminals, it flows *into* the armature terminals.

The speed of the synchronous motor is constant as long as the source frequency is constant. Thus, the equivalent circuit of a synchronous motor is exactly the same as the equivalent circuit of a synchronous generator (as shown in Figure 7.9), with one exception: the direction of the current I_a is reversed. The corresponding KVL equations for the motor are

$$V_\phi = E_a + (R_a + jX_s)I_a \tag{7.27}$$

$$V_\phi = E_\phi + (R_a + jX_a)I_a \tag{7.28}$$

$$E_a = V_\phi - (R_a + jX_s)I_a \tag{7.29}$$

$$E_\phi = V_\phi - (R_a + jX_a)I_a \tag{7.30}$$

The power output of a synchronous motor depends totally on the mechanical load on the shaft. As previously stated, its speed depends on the source frequency. Since the speed does not vary as the field current I_f is changed, varying I_f has no effect on the output power.

However, changing I_f affects the E_a (i.e., E_a increases when I_f increases) and the power factor of the current I_a drawn from the three-phase source. Figure 7.12a, b, and c shows phasor diagrams of a synchronous motor which is operating at unity, leading, and lagging power factors, respectively.

When the magnitude of E_a is equal to V_ϕ, this condition is referred to as 100% *excitation* (i.e., operating at *unity* power factor). When $E_a > V_\phi$, it is called *overexcitation* (i.e., operating at *leading* power factor). Finally, when $E_a < V_\phi$, it is called *underexcitation* (i.e., operating at *lagging* power factor).

7.8 POWER AND TORQUE CHARACTERISTICS

In a synchronous generator, the input power is provided by a prime mover in terms of shaft power. The input mechanical power of the generator can be expressed as

$$P_{in} = P_{shaft} = T_{in} \times \omega_m \tag{7.31}$$

where
   T_{in} represents the applied torque to the shaft by the prime mover
   ω_m denotes the mechanical speed of the shaft rotation

In contrast, the power internally developed from mechanical to electrical form can be expressed as

$$P_d = T_d \times \omega_m \tag{7.32}$$

or

$$P_d = 3E_a I_a \cos \Phi \tag{7.33}$$

where Φ is the angle between E_a and I_a as shown in Figure 7.12a. Therefore,

$$\Phi = \theta + \delta \tag{7.34}$$

The difference between output power and input power gives the losses of the machine, whereas the difference between the input power and the developed power gives the mechanical and core losses of the generator. The electrical output power of the generator can be found in terms of line quantities as

$$P_{out} = \sqrt{3} V_t I_L \cos \theta \tag{7.35}$$

or in phase quantities as

$$P_{out} = 3 V_\phi I_a \cos \theta \tag{7.36}$$

Similarly, the reactive power can be found in terms of line quantities as

$$Q_{out} = \sqrt{3} V_t I_L \sin \theta \tag{7.37}$$

or in phase quantities as

$$Q_{out} = 3 V_\phi I_a \sin \theta \tag{7.38}$$

The real and reactive power output of a synchronous generator can also be expressed as a function of the terminal voltage, the internal generated voltage, the synchronous impedance, and the *power angle* or *torque angle* δ.

This is also true for the real and reactive power received by a synchronous motor. Since X_s is much greater than R_a, then it can be proven easily that

$$E_a \sin \delta = X_s I_a \cos \theta \tag{7.39}$$

Thus,

$$I_a \cos \theta = \frac{E_a \sin \delta}{X_s} \tag{7.40}$$

and substituting this equation into Equation 7.36,

$$P = \left(\frac{3 E_a V_\phi}{X_s} \right) \sin \delta \tag{7.41}$$

Because the stator losses are ignored (i.e., R_a is assumed to be zero), P represents both the developed power P_d (or the air-gap power) and the output power P_{out}.

The power output of the synchronous generator depends on the angle between E_a and V_ϕ. If the angle δ is increased gradually, the real power output increases, reaching a maximum when δ is 90°. Therefore, the *maximum power* becomes

$$P_{max} = \frac{3E_a V_\phi}{X_s} \qquad (7.42)$$

This is also known as the *steady-state power limit* or the *static stability* limit. From Equation 7.32, the developed torque of the synchronous machine can be found as

$$T_d = \frac{P}{\omega_m} \qquad (7.43a)$$

But since $\omega_s = \omega_m$

$$= \frac{3E_a V_\phi}{\omega_m X_s} \sin \delta \qquad (7.43b)$$

$$= T_{max} \sin \delta \qquad (7.43c)$$

where the maximum torque is

$$T_{max} = \frac{P_{max}}{\omega_m} \qquad (7.44a)$$

$$= \frac{3E_a V_\phi}{\omega_m X_s} \qquad (7.44b)$$

Therefore, any increase in the mechanical power to the synchronous generator or in the mechanical output of the synchronous motor* after δ has reached 90° produces a decrease in real electrical power. The generator accelerates while the motor decelerates, and either way the result is a *loss of synchronism*.[†] The maximum torque T_{max} is also known as the **pull-out torque**.[‡]

Figure 7.13 shows the steady-state power-angle or torque-angle characteristic of a synchronous machine with negligible armature resistance. Note that when δ becomes negative, the power flow reverses. In other words, when power flows into the electrical terminals, the machine starts acting as a motor with a negative δ. In the generator mode, the power flows out of the electrical terminals and the angle δ becomes positive. This behavior can be explained by Equation 7.42. Similarly, the torque reverses direction (sign) when the machine goes from generator operation to motor operation according to Equation 7.44b. In generator mode the torque is positive, that is, a *counter torque*, and therefore the T_d is opposite to ω_m.

In motor mode, the torque is negative which means that it is in the same direction as ω_m. Figure 7.14 shows the superimposed power and torque-angle characteristics of a synchronous machine operating in generator mode.

---

* Even though the rotors of three-phase synchronous motors are salient-pole rather than cylindrical type, applying cylindrical-rotor theory yields a good degree of approximation.

† It is also known as *pulling out of step*.

‡ More precisely, the pull-out torque is the maximum sustained torque that the motor will develop at synchronous speed for 1 min, with rated voltage applied at rated frequency and with normal excitation.

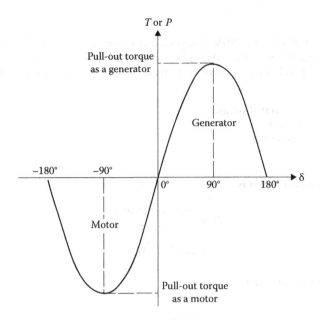

FIGURE 7.13 Synchronous machine power or torque as a function of power angle, δ.

Note that both the maximum power and the maximum torque take place when δ is 90°. If the prime mover tends to drive the generator to *supersynchronous* speed by excessive driving torque, the field current can be increased to develop more counter torque to overcome such a tendency.

Similarly, if a synchronous motor is apt to pull out of synchronism due to excessive load torque, the field current can be increased to produce greater torque and prevent a loss of synchronism.

The reactive power of a synchronous machine can be expressed as

$$Q = \frac{3\left(E_a V_\phi \cos\delta - V_\phi^2\right)}{X_s} \tag{7.45}$$

Here, *positive Q* means *supplying inductive vars* in the generator mode or *receiving inductive vars* in the motor mode, and *negative Q* means *supplying capacitive vars* in the generator mode or *receiving capacitive vars* in the motor mode.

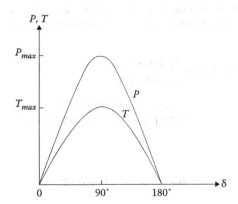

FIGURE 7.14 The superimposed power and torque-angle characteristics of a synchronous generator.

Example 7.4

A three-phase, 100 hp, 60 Hz, 480 V, four-pole, wye-connected, cylindrical-rotor synchronous motor has an armature resistance of $0.15\,\Omega$ and a synchronous reactance of $2\,\Omega$ per phase, respectively. At the rated load and a leading power factor of 0.8, the motor efficiency is 0.95. Determine the following:

(a) The internal generated voltage
(b) The torque angle δ
(c) The maximum torque

Solution

(a) The motor input power is

$$P_{in} = \frac{(100 \text{ hp})(746 \text{ W/hp})}{0.95}$$

$$= 78,526.32 \text{ W}$$

The rated load current is

$$I_a = \frac{P_{in}}{\sqrt{3}V_L \cos\theta}$$

$$= \frac{78,526.32 \text{ W}}{\sqrt{3}\,(480 \text{ V})0.8}$$

$$= 118.07 \text{ A}$$

The voltage per phase is

$$V_\phi = \frac{480 \text{ V}}{\sqrt{3}} = 277.13 \text{ V}$$

and $\theta = \cos^{-1} 0.8 = 36.87°$ leading

$$\boldsymbol{E}_a = \boldsymbol{V}_\phi - (R_a + jX_s)\boldsymbol{I}_a$$

$$= 277.13\angle 0° - (0.15 + j2)(118.07\angle 36.87°)$$

$$= 451.17\angle - 26.25° \text{ V}$$

(b) The negative sign indicates that $\boldsymbol{E}_a$ lags V_ϕ (which is used as the reference phasor). The $26.25°$ also represents the torque angle δ. Alternatively, this torque angle can be found from

$$\tan(\theta + \delta) = \frac{V_\phi \sin\theta + I_a X_s}{V_\phi \cos\theta - I_a R_a}$$

$$= \frac{277.13 \sin 36.87° + (118.07)2}{277.13 \cos 36.87° + (118.07)0.15} = 1.97$$

Thus,

$$\theta + \delta = \tan^{-1}(1.97) = 63.12°$$

$$\delta = 63.12° - 36.87° = 26.25°$$

(c) Since the machine has four poles, its speed is 1800 rpm or 188.495 rad/s. Thus, the maximum torque is

$$T_{max} = \frac{P_{max}}{\omega_m} = \frac{3E_aV_\phi}{\omega_m X_s} = \frac{3(451.17)277.13}{(188.495)2} = 994.98 \text{ N m}$$

7.9 STIFFNESS OF SYNCHRONOUS MACHINES

The ability of a synchronous machine to endure the forces that tend to pull it out of synchronism is called *stiffness*. Stiffness, which represents the slope of the power–angle curve at a given operating point, can be determined by taking the partial derivative of the power delivered with respect to the torque angle. The unit of such a rate of power is W per radian. Since

$$P = \frac{3E_aV_\phi}{X_s}\sin\delta \qquad (7.46)$$

For small displacements of $\Delta\delta$, the change in power is ΔP. Also,

$$K_s = \frac{\Delta P}{\Delta\delta} \cong \frac{dP}{d\delta} \qquad (7.47)$$

Thus, the stiffness can be expressed as

$$K_s = \frac{3E_aV_\phi}{X_s}\cos\delta \qquad (7.48)$$

The maximum stiffness is referred to as *synchronizing power*. Of course, at pull-out, the stiffness of the machine is zero.

Example 7.5

Consider the synchronous generator given in Example 7.4 and assume that the machine has eight poles. Determine the following:

(a) The synchronizing power in MW per electrical radian and in MW per electrical degree
(b) The synchronizing power in MW per mechanical degree
(c) The synchronizing torque in MW per mechanical degree

Solution

(a) From Example 7.4, the synchronizing power is

$$P_s = \frac{3E_aV_\phi}{X_s}\cos\delta$$

$$= \left(\frac{3(11,727)(7,621.02)}{2.49}\right)\cos 21.8°$$

$$= 99,076 \text{ MW per electrical radian}$$

$$= \frac{99,076 \text{ MW per electrical radian}}{57.3}$$

$$= 1.745 \text{ MW per electrical degree}$$

(b) Since the machine has four pole pairs, there are four electrical cycles for each mechanical revolution. Thus,

$$P_s = 4 \times 1.745 = 6.98 \text{ MW per mechanical degree}$$

(c) Since the synchronous speed is

$$n_s = \frac{60}{4} = 15 \text{ rev/s}$$

the synchronizing torque is

$$T_s = \frac{P_s}{\omega_m} = \frac{6,979,656.6 \text{ W}}{2\pi(15 \text{ rev/s})} = 74,056.4 \text{ N} \cdot \text{m per mechanical degree}$$

7.10 EFFECT OF CHANGES IN EXCITATION

One of the important characteristics of the synchronous machine is that its power factor can be controlled by the field current. In other words, *the power factor of the stator (or line) current can be controlled by changing the field excitation.*

However, the behavior of a synchronous generator (alternator) connected to an infinite bus (large system) is quite different from that of one operating alone.

7.10.1 SYNCHRONOUS MACHINE CONNECTED TO AN INFINITE BUS

Assume a constant-power operation of a synchronous machine connected to an infinite bus so that it operates at constant frequency and terminal voltage. Under such circumstances, the power factor is determined by the field current.

In the generator mode, the amount of power generated and the frequency (or speed) are determined by the prime mover. In the motor mode, the speed is determined by the line frequency, and the output depends on the mechanical load on the shaft.

Figure 7.15a, b, and c shows the phasor diagrams of a synchronous machine *operating as an overexcited generator,* a normal excited generator, and an underexcited motor, respectively. The figures show that the locus of the current I_a for constant real power is a vertical line, while the locus of the internal generated voltage E_a is a horizontal line. Notice that the variation in the power-factor angle θ is very significant, but the variation in the torque angle δ is almost insignificant.

As shown in Figure 7.15a, when the machine is operating *as an overexcited generator,* it has a lagging power factor due to a high field current. The maximum power P_{max} is large, and, therefore, the machine operation is stable.

As shown in Figure 7.15b, when the machine is operating *as a normal excited generator,* it has a unity power factor as a result of normal field current. Finally, as shown in Figure 7.15c, when the machine is operating as an underexcited generator, it has leading power factor due to a low field current. Therefore, the maximum power P_{max} is small and hence the machine operation is less stable.

Figure 7.15d, e, and f shows the phasor diagrams of a synchronous machine operating as an overexcited motor, normal excited motor, and underexcited motor, respectively. The figures show that the locus of the current I_a for constant real power is a vertical line while the locus of the internal generated voltage E_a is a horizontal line.

As shown in Figure 7.15d, *when the machine is operating as an overexcited motor,* it has leading power factor due to a high field current. Therefore, the maximum power P_{max} is large and the machine operation is stable.

As shown in Figure 7.15e *when the machine is operating as a normal excited motor,* it has a unity power factor due to a normal field current. Under such conditions, the motor draws the minimum stator current I_a.

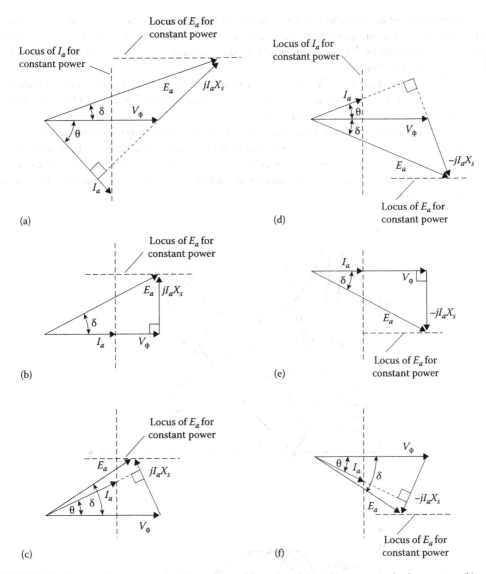

FIGURE 7.15   Phasor diagram of a synchronous machine operating as (a) an overexcited generator, (b) a normally excited generator, (c) an underexcited generator, (d) an overexcited motor, (e) a normally excited motor, and (f) an underexcited motor.

Finally, as shown in Figure 7.15f, *when the machine is operating as an underexcited motor*, it has lagging power factor due to a low field current. Here, the maximum power P_{max} is small and hence the machine operation is less stable.

From the phasor diagrams shown in Figure 7.15, one can observe that the voltage E_a leads the voltage V_ϕ when the synchronous machine operates as a generator, and *lags* when it operates as a motor. Also, note that the torque angle or the power angle δ is positive when generating and negative when motoring.

Succinctly put, the *power factor at which a synchronous machine operates and its stator (arma-ture) current can be controlled by changing its field excitation.**

---

* In the event that the synchronous machine is not transferring any power but is simply *floating* on the infinite bus, the machine power factor is zero. In other words, the armature current either lags or leads the terminal voltage by 90°.

The curve showing the relationship between the stator current and the field current at a constant terminal voltage with a constant real power is called a *synchronous machine V curve* because of its shape. The V curves can be developed for synchronous generators as well as for synchronous motors and will be almost identical.*

Figure 7.16a shows *a family of V curves for a synchronous motor.* Note that there are three V curves in the figure corresponding to full load, half load, and no load. The dashed lines are loci for constant power and are called *compounding curves.*

Notice that *minimum armature* (i.e., stator or line) *current is always associated with a unity power factor.* The corresponding field current is indicated as *normal* excitation. Also notice that the region to the right of the unity-power-factor compounding curve is associated with overexcitation and a leading power factor and that the region to the left is associated with underexcitation and a lagging power factor.

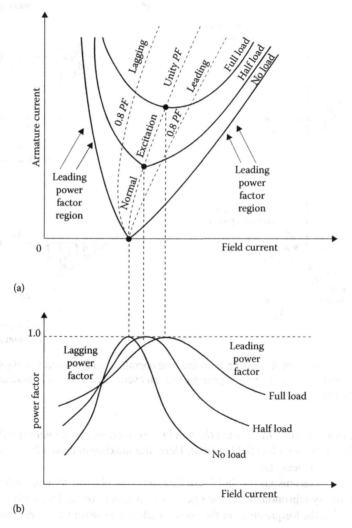

(a)

(b)

FIGURE 7.16 Synchronous-motor V curves: (a) stator current versus field current and (b) power factor versus field current.

---

* If it were not for the small effects of armature resistance.

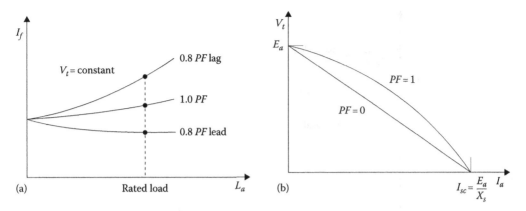

FIGURE 7.17 Characteristics of a generator operating alone: (a) compounding curves and (b) terminal voltage versus load current at constant-field current.

Figure 7.16b shows the correlation between the power factor and the field current. These curves show that a synchronous motor can be overexcited and carry a substantial leading power factor.

Also notice that both curves, in Figure 7.16a and b, show that a slightly increased field current is needed to produce normal excitation as the load increases.

7.10.2 SYNCHRONOUS GENERATOR OPERATING ALONE

In general, synchronous machines operating in generator mode or in motor mode are connected to an infinite bus. However, there are many applications in which synchronous generators may be used to supply an isolated (independent) power system.

For such applications, the *infinite bus* theory (i.e., having constant voltage and constant frequency) cannot be used, since there are no other generators connected in parallel to compensate for changes in field excitation and prime-mover output in order to keep the terminal voltage and frequency constant. Here, the prime mover is most likely a diesel or gasoline engine.

The frequency depends totally on the speed of the prime mover. Thus, a *governor is needed to maintain the constant frequency*. The power factor is the load power factor and changes as the load changes. Hence, the power factor and armature current cannot be controlled at the generator site. In fact, the only control that can be used at the generator site is that of the field current. Thus, at constant speed, if the field current is increased, the terminal voltage will increase, as shown in Figure 7.17a.

Since the terminal voltage changes drastically as the load changes, an automatic voltage regulator is required to control I_f so that the terminal voltage can be kept constant with a changing load. Otherwise, as shown in Figure 7.17b, as the load current I_a is increased, the terminal voltage drops sharply with a drop in the load power factor.

From the study of characteristics of a synchronous generator, given in Figure 7.17, one can conclude that (1) the addition of inductive loads causes the terminal voltage to drop drastically, (2) the addition of purely resistive loads causes the terminal voltage to drop very little (almost insignificantly), and (3) the addition of capacitive loads causes the terminal voltage to rise drastically. Figure 7.18 shows the constant-field current volt-ampere characteristic curves of a synchronous generator operating alone. Note that the curves shown are for three different values of constant-field current and power factors.

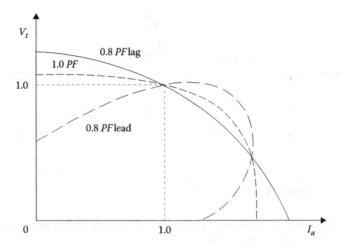

FIGURE 7.18 The constant-field current volt-ampere characteristic curves of a synchronous generator operating alone.

7.11 USE OF DAMPER WINDINGS TO OVERCOME MECHANICAL OSCILLATIONS

If the load on a synchronous machine varies, the load angle changes from one steady value to another. *During such transient phenomena, oscillations in the load angle and resultant mechanical oscillations* of the rotor take place.* To dampen out such oscillations, damper windings are used in most salient-pole synchronous machines. The damper windings[†] are made of copper or brass bars located in pole-face slots on the pole shoes of a salient-pole rotor and their ends are *connected together*.

When the rotor speed is different from the synchronous speed, currents are induced in the damper windings. The damper windings behave like the squirrel-cage rotor of an induction motor, developing a torque to eliminate mechanical oscillations and restore the synchronous speed.

Note that cylindrical-rotor machines do not have damper windings because the eddy currents that exist in the solid rotor during such transients play the same role as the currents in damper windings in a salient-rotor machine.

The load angle and resultant mechanical oscillations[‡] of the rotor take place. To dampen out such oscillations, damper windings are used in most salient-pole synchronous machines. The damper windings[§] are made of copper or brass bars located in pole-face slots on the pole shoes of a salient-pole rotor and their ends are connected together.

7.12 STARTING OF SYNCHRONOUS MOTORS

A synchronous motor is not a self-starter. In other words, if its rotor winding is connected to a dc source and its stator winding is supplied by an ac source, the motor will not start,[¶] but simply

---

[*] In other words, any variance in load causes an oscillatory motion superimposed on the normal (i.e., synchronous) motion of the machine shaft. This motion is also called *hunting*.

[†] They are also called the *amortisseur (killer) windings*.

[‡] In other words, any variance in load causes an oscillatory motion superimposed on the normal (i.e., synchronous) motion of the machine shaft. This motion is also called *hunting*.

[§] They are also called the *amortisseur (killer) windings*.

[¶] To produce the required torque, the rotor must be rotating at the same speed as the armature (stator) field. Therefore, if the rotor is turning (or not turning, in this case) at some other speed, the rotating armature-field poles will be moving past the rotor poles first attracting, then repelling them. Thus, the average torque is zero and the motor cannot start. Such a synchronous motor has no starting torque.

vibrates. The methods that can be used to start a synchronous motor include the following: (1) starting the motor as an induction motor, (2) starting it with a variable-frequency supply, and (3) starting it with the help of a dc motor.

The *first method* is the most practical and is therefore the most popular. When the field windings are disconnected from the dc source and the stator windings are connected to its ac source, the motor acts like an induction motor because of its damper windings. Such an induction-motor start brings the machine almost up to synchronous speed and when the dc field windings are excited, the rotor *falls into step*, that is, starts to rotate at the synchronous speed. At synchronous speed, there is no current induced in the damper windings and therefore there is no torque produced by them.

The *second method* involves starting the motor with low-frequency ac voltage by employing a frequency converter. As a result, the armature field rotates slowly to make the rotor poles follow the armature poles. Later, the motor can start operating at its synchronous speed by slowly increasing the supply frequency to its nominal value. The third method involves bringing the motor to its synchronous speed by using a dc motor before connecting the motor to the ac supply.

7.13 OPERATING A SYNCHRONOUS MOTOR AS A SYNCHRONOUS CONDENSER

As previously stated, overexcited* synchronous motors can generate reactive power. When synchronous motors are used as synchronous condensers they are manufactured without a shaft extension, since they are operated with no mechanical load. The ac input power supplied to such a motor can only provide for its losses. These losses are very small and the power factor of the motor is almost zero.

Therefore, the armature current leads the terminal voltage by close to 90°, as shown in Figure 7.19a, and the power network perceives the motor as a capacitor bank. As can be seen in Figure 7.19b, when this motor is overexcited it behaves like a *capacitor* (i.e., *synchronous condenser*), with $E_a > V_\phi$, whereas when it is underexcited, it behaves like an inductor (i.e., a synchronous reactor), with $E_a < V_\phi$.

Synchronous condensers are used to correct power factors at load points, or to reduce line voltage drops and thereby improve the voltages at these points, as well as to control reactive power flow. Large synchronous condensers are usually more economical than static capacitors.

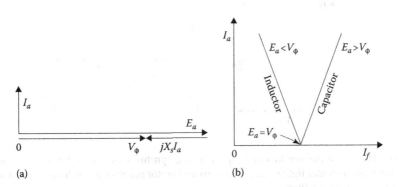

FIGURE 7.19 Synchronous condenser operation: (a) overexcited motor operation and (b) overexcited versus underexcited operation of the motor.

---

* In fact, an overexcited synchronous machine produces reactive power whether or not it is operating as a motor or as a generator.

7.14 OPERATING A SYNCHRONOUS MOTOR AS A SYNCHRONOUS REACTOR

In general, it is not economical to correct the full-load power factor to unity. Therefore, a transmission line usually operates at a lagging power factor. Assume that an overexcited synchronous motor is being used as a synchronous capacitor to correct the power factor of a transmission line which is supplying a load with a lagging power factor. Figure 7.20a shows the phasor diagram of such a compensated transmission line, with a sending-end voltage of V_s and a receiving-end voltage of V_R, operating under peak-load conditions with $V_s > V_R$.

However, *under no-load* or *light-load conditions* due to the reactive current I_r, the receiving-end voltage V_R becomes much greater than V_s, that is, $V_s < V_R$ as shown in Figure 7.20b. This condition is known as the *Ferranti effect.**

To prevent this, the dc field excitation of the synchronous motor can be controlled by a voltage regulator and reduced as the load decreases and the V_R increases. As shown in Figure 7.20d, when the synchronous motor[†] is underexcited, it becomes a synchronous reactor and starts to provide an inductive voltage drop by means of the inductive current I_r to counteract the capacitive line drop. Figure 7.20c and d shows the corresponding full-load and no-load corrections, respectively. Observe that the relationship between the two receiving-end voltages is about the same. The use of a synchronous condenser provides a constant voltage at the receiving end despite changes in the load current and power factor.

Example 7.6

A three-phase, 750 hp, 4160 V, wye-connected induction motor has a full-load efficiency of 90%, a lagging power factor of 0.75, and is connected to a power line. To correct the power factor of such a load to a lagging power factor of 0.85, a synchronous condenser is connected at the load. Determine the reactive power provided by the synchronous capacitor.

Solution

The input power of the induction motor is

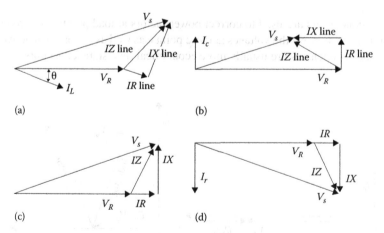

(a)   (b)

(c)   (d)

FIGURE 7.20 The use of the synchronous capacitor as a synchronous reactor for a transmission line: (a) under maximum load, (b) under light load, (c) underexcited motor becomes a synchronous reactor under full-load, and (d) under no-load conditions.

---

* See Gönen (1988) for further information.

† The synchronous motors used for this purpose are usually equipped with salient rotors, since a synchronous motor with a cylindrical rotor may step out of synchronism and stop when it is operating with underexcitation. This is due to the fact that decreasing the field current too far can cause the developed torque to be less than the rotational torque required.

$$P = P_{in}(0.746 \text{ kW/hp})$$

$$= \frac{P_{out}(0.746 \text{ kW/hp})}{\eta}$$

$$= \frac{(750 \text{ hp})(0.746 \text{ kW/hp})}{0.90}$$

$$= 621.67 \text{ kW}$$

The reactive power of the motor at the uncorrected power factor is

$$Q_1 = P \tan \theta_1$$

$$= 621.67 \tan(\cos^{-1} 0.75)$$

$$= 621.67 \tan(41.4096°)$$

$$= 548.26 \text{ kvar}$$

The reactive power of the motor at the corrected power factor is

$$Q_2 = P \tan \theta_2$$

$$= 621.67 \tan(\cos^{-1} 0.85)$$

$$= 621.67 \tan(31.788°)$$

$$= 385.27 \text{ kvar}$$

Thus, the reactive power provided by the synchronous capacitor is

$$Q_c = Q_1 - Q_2 = 548.26 - 385.27 = 162.99 \text{ kvar}$$

7.15 TESTS TO DETERMINE EQUIVALENT-CIRCUIT PARAMETERS

The equivalent-circuit parameters of a synchronous machine can be determined from three tests,*
namely, the open-circuit test, the short-circuit test, and the dc test.

7.15.1 OPEN-CIRCUIT TEST

As discussed in Section 7.5, the OCC of a synchronous machine can be developed based on the
open-circuit test. As shown in Figure 7.21a, the machine is driven at synchronous speed with its
armature terminals open and its field current set at zero. The open-circuit (line-to-line) terminal
voltage V_{oc} is measured as the field current I_f is increased.

Since the terminals are open $V_{oc} = E_a = V_t$, assuming that the armature windings are connected in
wye. The plot of this voltage with respect to the field excitation current I_f gives the OCC, as shown
in Figure 7.21c.

Therefore, the internal generated voltage E_a at any given field current I_f can be found from the
OCC characteristic. Observe that as the field current is increased, the OCC starts to separate from

---

* It is applicable to both cylindrical-rotor and salient-rotor synchronous machines.

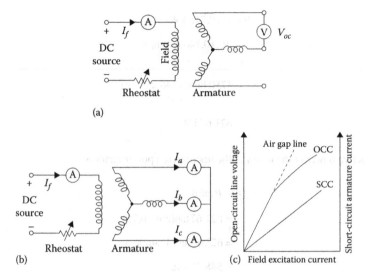

FIGURE 7.21 Open-circuit and short-circuit tests: (a) connection diagram *for open*-circuit test, (b) connection diagram for short-circuit test, and (c) plots of open-circuit and short-circuit characteristics.

the air-gap line due to the saturation of the magnetic core. The no-load rotational losses (i.e., friction, windage, and core losses) can be found by measuring the mechanical power input. While the friction and windage losses remain constant, the core loss is proportional to the open-circuit voltage.

7.15.2 SHORT-CIRCUIT TEST

As shown in Figure 7.21b, the armature terminals are short-circuited through suitable ammeters* and the field current is set at zero. While the synchronous machine is driven at synchronous speed, its armature current I_a is measured as the field current gradually increases until the armature current is about 150% of the rated current.

The plot of the average armature current I_a versus the field current I_f gives the *short-circuit characteristic* (SCC) of the machine, as shown in Figure 7.21c. The SCC is a linear line since the magnetic-circuit iron is unsaturated.[†]

7.15.3 DC TEST

If it is necessary, the resistance R_a of the armature winding can be found by applying a dc voltage to two of the three terminals of a wye-connected synchronous machine while it is stationary. Since current flows through two of the wye-connected armature windings, the total resistance of the path is $2R_a$. Thus,

$$2R_a = \frac{V_{dc}}{I_{dc}} \tag{7.49}$$

$$R_a = \frac{V_{dc}}{2I_{dc}} \tag{7.50}$$

---

* If necessary, an instrument current transformer can be used with an ammeter in its secondary.
† When the short-circuit current is equal to the rated current, the voltage E_a will only be about 20% of its rated value. Therefore, the magnetic-circuit iron is unsaturated.

Usually, the calculated R_a has to be corrected for the skin effect and the temperature of the winding during the short-circuit test. However, the resistance R_a of synchronous machines with ratings greater than even a few hundred kVA is generally very small and is often ignored except in efficiency computations.

7.15.4 Unsaturated Synchronous Reactance

As can be observed in Figure 7.22, if the synchronous machine is unsaturated, the open-circuit line voltage will increase linearly with the field current along the air-gap line. As a result, the short-circuit armature current is directly proportional to the field current. Therefore, the unsaturated synchronous impedance for a specific value of the field current can be found from Figure 7.22 as

$$\mathbf{Z}_{s,un} = \frac{E_{ac}}{\sqrt{3}I_{ab}} = R_a + jX_{s,un} \tag{7.51}$$

If R_a is small enough to be ignored,

$$X_{s,un} \cong \frac{E_{ac}}{\sqrt{3}I_{ab}} \tag{7.52}$$

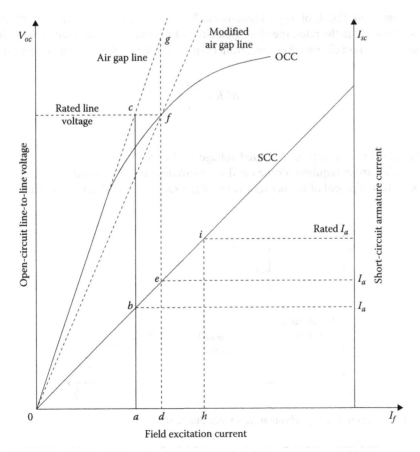

FIGURE 7.22 Open-circuit and short-circuit characteristics of a synchronous machine.

7.15.5 SATURATED SYNCHRONOUS REACTANCE

Under normal operating conditions, the magnetic circuit is saturated. Therefore, if the field current is changed, the internal generated voltage will vary along the *modified air-gap line*, as shown in Figure 7.22. Thus, the saturated synchronous impedance at the rated voltage is given by

$$Z_s = \frac{E_{df}}{\sqrt{3}I_{de}} = R_a + jX_s \qquad (7.53)$$

If R_a is small enough to be ignored,

$$X_s = \frac{E_{df}}{\sqrt{3}I_{de}} \qquad (7.54)$$

However, the machine is unsaturated in the short-circuit test.

Therefore, the determination of the synchronous reactance based on short-circuit test data and open-circuit test data is only an approximation* at best. Figure 7.23 shows the variation of the synchronous reactance due to saturation.

7.15.6 SHORT-CIRCUIT RATIO

The *short-circuit ratio* (SCR) of a synchronous machine is the ratio of the field current required to generate rated voltage at the rated speed at open circuit to the field current needed to produce rated armature current at short circuit. Therefore, from Figure 7.22, the *SCR* of the synchronous machine is

$$SCR = \frac{od}{oh} \qquad (7.55)$$

where
  od is the field current that produces rated voltage on the OCC
  oh is the field current required for the rated short-circuit armature current
  the *SCR* is the reciprocal of the per-unit value of the saturated synchronous reactance

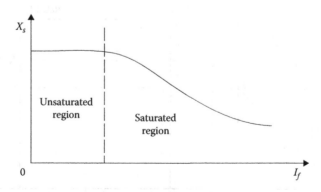

FIGURE 7.23   Variation of the synchronous reactance due to saturation.

---

* See McPherson (1981) for a more accurate determination of the saturated synchronous reactance.

Example 7.7

The following data are taken from the open-circuit and short-circuit characteristics of a 100 kVA, three-phase, wye-connected, 480 V, 60 Hz synchronous machine with negligible armature resistance:

From the OCC
   Line-to-line voltage = 480 V
   Field current = 3.2 A

From the air-gap line
   Line-to-line voltage = 480 V
   Field current = 2.94 A

From the SCC

Armature current 90.35 A	120.28 A
Field current 3.2 A	4.26 A

Determine the following:
   (a) The unsaturated synchronous reactance
   (b) The saturated synchronous reactance at rated voltage
   (c) The SCR

Solution

   (a) The field current of 2.94 A needed for rated line-to-line voltage of 480 V on the air-gap line produces a short-circuit armature current of 2.94(90.35/3.2) = 83.01 A. Therefore, the unsaturated synchronous reactance is

$$X_{s,un} = \frac{480 \text{ V}}{\sqrt{3}(83.01 \text{ A})} = 3.34 \text{ }\Omega/\text{phase}$$

   (b) The field current of 3.32 A produces the rated voltage on the OCC and a short-circuit armature current of 90.35 A. Therefore, the saturated synchronous reactance at the rated voltage is

$$X_s = \frac{480 \text{ V}}{\sqrt{3}(90.35 \text{ A})} = 3.07 \text{ }\Omega/\text{phase}$$

   (c) The SCR is

$$SCR = \frac{3.2}{4.26} = 0.75$$

7.16   CAPABILITY CURVE OF SYNCHRONOUS MACHINE

Figure 7.24a shows the capability curve of a synchronous machine. The rated MVA of the synchronous machine is dictated by stator heating in terms of maximum allowable stator current. The upper and lower portions of the area inside the circle, with a radius of maximum S, represent the generator and motor operation, respectively. The maximum allowable field current is limited by the rotor heating. The maximum permissible torque angle is dictated by the steady-state stability limits that exist in the generator and motor modes of operation and further restricts the operation area of the synchronous machine. In the generator mode, the power limit is determined by the prime-mover rating.

Figure 7.24b shows the capability curve of a synchronous generator. Any point that lies within the area is a safe operating point for the generator from the standpoint of heating and stability. Assume that the operating point S is chosen as shown in Figure 7.24b, and that the corresponding real and reactive powers are P and Q, respectively.

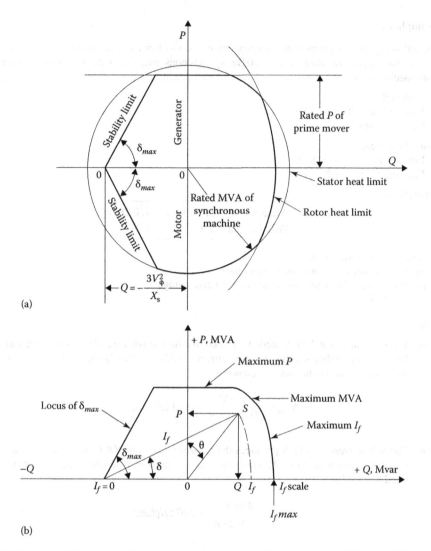

FIGURE 7.24 Capability curves of a synchronous machine: (a) construction of the capability curve and (b) the capability curve of a synchronous generator.

For this operation, the power-factor angle can be readily determined from the diagram as θ by drawing a line from the operating point S to the origin. A line drawn from the operating point S to the origin of the I_f axis facilitates finding the power or torque angle δ from that axis.

7.17 PARALLEL OPERATION OF SYNCHRONOUS GENERATORS

The process of connecting a synchronous generator to an infinite bus is called *paralleling with the infinite bus*. The generator to be added to the system is referred to as the one *to be put on line*. As shown in Figure 7.25, to connect the incoming generator to the infinite bus, a definite procedure called the *synchronizing procedure* must be followed before closing the circuit breaker CB to prevent any damage to the generator or generators.

Accordingly, the following conditions must be met: (1) the rms voltages of the generator must be the same as the rms voltages of the infinite bus, (2) the phase sequence of the voltages of the generator must be the same as the phase sequence of the infinite bus, (3) the phase voltages of the generator

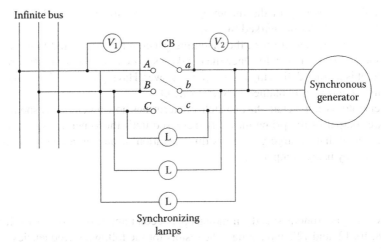

FIGURE 7.25   Parallel operation of synchronous generators.

must be in phase with the phase voltages of the infinite bus, and (4) the frequency of the generator must be almost equal to that of the infinite bus.

By using voltmeters, the field current of the incoming generator is increased to a level at which the voltages of the generator are the same as the voltages of the infinite bus. The phase sequence of the incoming generator has to be compared to the phase sequence of the infinite system. This can be done in various ways. One *method* is to connect three light bulbs as shown in Figure 7.25. If the phase sequence is correct, all three light bulbs will have the same brightness.

A *second method* is to connect a small induction motor first to the terminals of the infinite bus and then to the terminals of the incoming generator. If the motor rotates in the same direction each time, then the phase sequences are the same.

A *third method* is to use an instrument known as *a phase-sequence indicator*. In any case, if the phase sequences are different, then two of the terminals of the incoming generator have to be reversed. The two sets of voltages must be in phase and the three phase voltages have to be almost equal to the voltages of the infinite bus.

The frequency of the incoming generator has to be a little higher than the frequency of the infinite bus. The reason for this is that when it is connected, it will come on line providing power as a generator rather than being used as a motor. The frequency (or speed) of the incoming generator can also be compared to the frequency of the infinite bus by using an instrument known as a *synchroscope*. When the two frequencies are identical, the pointer of the synchroscope locks into a vertical position, as shown in Figure 7.26.

On the other hand, if the radian frequency of the incoming generator is, for example, 381 rad/s and that of the infinite bus is 377 rad/s, then the pointer rotates at 4 rad/s in the direction marked

FIGURE 7.26   The face of a synchroscope.

FAST. If the radian frequency of the incoming generator is, for example, 372 rad/s, the pointer rotates at 5 rad/s in the direction marked SLOW.

When all of the conditions are met, the incoming generator is connected to the infinite bus by closing the circuit breaker (or switch). Once the breaker is closed, the incoming generator is on line. At this moment, it is neither delivering nor receiving power. Having $E_a = V_\phi$, $I_a = 0$, and $\delta = 0$ in each phase, it is simply *floating* as explained in Section 7.10.1.

After the generator is connected, the dispatcher determines how much power should be produced by it. The power output of the prime mover is increased until the generator starts to produce the required power. Note that in large generators this operation of putting a new generator on line is done automatically by using computers.

PROBLEMS

7.1   Compute the synchronous speeds in radians per second and revolutions per minute for poles of 2, 4, 8, 10, 12, and 120 and tabulate the results for the following frequencies:
(a) 60 Hz
(b) 50 Hz

7.2   An elementary four-pole, three-phase, 60 Hz alternator has a rotating flux of 0.0875 Wb. The number of turns in each phase coil is 25. Its shaft speed is 1800 rev/min. Let its stator winding factor be 0.95 and find the following:
(a) The angular speed of the rotor
(b) The three phase voltages as a function of time
(c) The rms phase voltage of this generator if the stator windings are delta-connected
(d) The rms terminal voltage if the stator windings are wye-connected

7.3   Determine by derivation the value of the K constant given in Equations 7.13 and 7.14.
(a) If ω is given in *electrical* radians per second
(b) If ω is given in *mechanical* radians per second

7.4   Consider the synchronous generator given in Example 7.3 and assume that the generator delivers full-load current at the rated voltage and 0.8 leading power factor. Determine the following:
(a) The rated load current
(b) The air-gap voltage
(c) The internal generated voltage
(d) The power angle
(e) The voltage regulation

7.5   A three-phase, 4.17 kV, 60 Hz, wye-connected, cylindrical-rotor synchronous generator has a leakage reactance of 2.2 Ω per phase and a winding resistance of 0.15 Ω per phase. If the load connected to the generator is 950 kVA at 0.85 lagging power factor, determine the air-gap voltage.

7.6   A three-phase synchronous motor is supplied by 100 A at unity power factor from a bus. Determine the following:
(a) The current at a leading power factor of 0.85, if the bus voltage is constant.
(b) The current at a lagging power factor of 0.85, if the bus voltage is constant.
(c) What usually happens to the bus voltage as the power factor becomes more leading?

7.7   A three-phase, wye-connected generator supplies a unity power factor load at 4.16 kV. If the synchronous reactance voltage drop is 262 V per phase and the resistance voltage drop is 25 V per phase, what is the percent regulation?

7.8   A 25 hp synchronous motor has a full-load efficiency of 92% and operates at a leading power factor of 0.8. Determine the following:
(a) The power input to the motor
(b) The kVA input to the motor

7.9 A three-phase, four-pole, 12 kV, 60 Hz, 40 MVA, wye-connected, cylindrical-rotor synchronous generator has a synchronous reactance of 0.3 per unit. Ignore the armature resistance and the effects of saturation. Also assume that the generator delivers full-load current at 0.85 lagging power factor. Determine the following:
  (a) The rated (full-load) current
  (b) The internal generated voltage
  (c) The torque angle
  (d) The synchronizing power in W per electrical radian and in W per electrical degree
  (e) The synchronizing power in W per mechanical degree
  (f) The synchronizing torque in W per mechanical degree

7.10 A three-phase, 100 kVA, 60 Hz, 480 V, four-pole, wye-connected, cylindrical-rotor synchronous motor has an armature resistance and a synchronous reactance of 0.09 and 1.5 Ω per phase, respectively. Its combined friction and windage losses are 2.46 kW and its core losses are 1.941 kW. Ignore its dc field losses. If the motor is operating at unity power factor, determine the following:
  (a) The internal generated voltage E_a
  (b) The torque angle δ
  (c) The efficiency at full load
  (d) The output torque at full load

7.11 Solve Problem 7.10 but assume that the machine is operating at 0.85 leading power factor.

7.12 Solve Problem 7.10 but assume that the machine is operating at 0.85 lagging power factor.

7.13 Assume that the following data are obtained from the open-circuit and short-circuit characteristics of a 500 kVA, three-phase, wye-connected, 480 V, 60 Hz synchronous machine with negligible armature resistance:

From the OCC
  Line-to-line voltage = 480 V
  Field current = 16 A

From the air-gap line
  Line-to-line voltage = 480 V
  Field current = 14.7 A

From the SCC

Armature current	451.76 A	601.41 A
Field current	16.0 A	21.3 A

Determine the following:
  (a) The unsaturated synchronous reactance
  (b) The saturated synchronous reactance at the rated voltage
  (c) The SCR

7.14 A three-phase, 60 Hz, 480 V, two-pole, delta-connected, cylindrical-rotor synchronous generator has a synchronous reactance of 0.12 Ω and an armature resistance of 0.010 Ω. Its OCC is shown in Figure 7.9. Its combined friction and windage losses are 30 kW and its core losses are 20 kW. Neglect its dc field losses and determine the following:
  (a) The amount of field current for the rated voltage of 480 V at no load
  (b) The amount of field current required for the rated terminal voltage of 480 V when the rated load current is 1000 A at a lagging power factor of 0.85
  (c) The efficiency of the generator under the rated load conditions
  (d) The terminal voltage if the load is suddenly disconnected
  (e) The amount of field current required to have the rated terminal voltage of 480 V when the rated load current is 1000 A at a leading power factor of 0.85

7.15 A three-phase, 60 Hz, 480 V, four-pole, delta-connected, cylindrical-rotor synchronous generator has an armature resistance and a synchronous reactance of 0.012 and 0.15 Ω, respectively.

Its OCC is shown in Figure 7.7. Its combined friction and windage losses are 35 kW and its core losses are 25 kW. Neglect its dc field losses and determine the following:

(a) The field current for the rated terminal voltage of 480 V at no load
(b) The field current for the rated terminal voltage of 480 V when the rated load current is 1100 A at a lagging power factor of 0.90
(c) The efficiency of the generator under the rated load conditions
(d) The terminal voltage if the load is suddenly disconnected
(e) The field current for the rated terminal voltage of 480 V when the rated load current is 1100 A at a leading power factor of 0.90

7.16 Solve Problem 7.15 but assume that the generator is wye-connected and the rated load current is 500 A.

7.17 A three-phase, 60 Hz, 480 V, two-pole, delta-connected, cylindrical-rotor synchronous generator has an armature resistance and a synchronous reactance of 0.012 and 0.15 Ω, respectively. Its OCC is shown in Figure 7.7. Its combined friction and windage losses are 25 kW and its core losses are 25 kW. Neglect its dc field losses and determine the following:

(a) The field current for the rated terminal voltage of 480 V at no load
(b) The field current for the rated terminal voltage of 480 V when the rated load current is 1100A at a lagging power factor of 0.90
(c) The efficiency of the generator under the rated load conditions
(d) The terminal voltage if the load is suddenly disconnected
(e) The field current for the rated terminal voltage of 480 V when the rated load current is 1100 A at a leading power factor of 0.90

7.18 A three-phase, 60 Hz, 480 V, four-pole, delta-connected, cylindrical-rotor synchronous generator has an armature resistance and a synchronous reactance of 0.012 and 0.15 Ω, respectively. Its OCC is shown in Figure 7.7. Its combined friction and windage losses are 25 kW and its core losses are 30 kW. The field current is constant at no load. Neglect its dc field losses and determine:

(a) The field current for the rated terminal voltage of 480 V at no load
(b) The field current for the rated terminal voltage of 480 V when the rated load current is 1000 A at a lagging power factor of 0.85
(c) The efficiency of the generator under the rated load conditions
(d) The terminal voltage if the load is suddenly disconnected
(e) The field current for the rated terminal voltage of 480 V when the rated load current is 1000 A at a leading power factor of 0.85

7.19 A three-phase, 60 Hz, 480 V, wye-connected, two-pole cylindrical-rotor synchronous generator has a synchronous reactance of 1.2 Ω, per phase. Its armature resistance is small enough to be neglected. Its full-load armature current is 75 A at 0.85 lagging power factor. Its combined friction and windage losses are 1.8 kW and its core losses are 1.1 kW. The field current is constant at no load. Neglect its dc field losses and determine the following:

(a) Its terminal voltage when it is loaded with the rated current having a power factor of (1) 0.85 lagging, (2) unity, and (3) 0.85 leading
(b) Its efficiency when it is loaded with the rated current at a lagging power factor of 0.85
(c) Its input torque and induced counter torque when it is operating at full load
(d) Its voltage regulation when it is operating under full load with a power factor of (1) 0.85 lagging, (2) unity, and (3) 0.85 leading

7.20 A three-phase, 60 Hz, 480 V, Y-connected, two-pole cylindrical-rotor synchronous generator has a synchronous reactance of 0.95 Ω per phase. Its armature resistance is negligible. Its combined friction and windage losses are 1.3 kW and its core losses are 0.95 kW. Neglect its dc field losses. The field current is constant at no load. Its full-load armature current is 55 A at 0.9 PF lagging. Determine the following:

(a) Its terminal voltage when it is loaded with the rated current at a power factor of (1) 0.9 lagging, (2) unity, and (3) 0.9 leading

(b) Its efficiency when it is loaded with the rated current at a power factor of 0.9

(c) Its input torque and induced counter torque when it is operating at full load

(d) Its voltage regulation when it is operating under full load with a power facto of (1) 0.9 lagging, (2) unity, and (3) 0.9 leading

7.21 A three-phase, 60 Hz, 480 V, delta-connected, four-pole cylindrical-rotor synchronous generator has a synchronous reactance of 1.5 Ω per phase. Its armature resistance is negligible. Its combined friction and windage losses are 2.1 kW and its core losses are 1.2 kW. Neglect its dc field losses. The field current I_f is constant at no load. Its full-load armature current is 100 A at 0.80 PF lagging. Determine the following:

(a) Its terminal voltage when it is loaded with the rated current at a power factor of (1) 0.80 lagging, (2) unity, and (3) 0.80 leading

(b) Its efficiency when it is loaded with the rated current at a lagging power factor of 0.80

(c) Its input torque and induced counter torque when it is operating at full load

(d) Its voltage regulation when it is operating under full load with a power factor of (1) 0.80 lagging, (2) unity, and (3) 0.80 leading

7.22 Solve Problem 7.21 but assume that the generator is wye-connected.

7.23 A three-phase, 60 Hz, 480 V, delta-connected, six-pole cylindrical-rotor synchronous generator has a synchronous reactance of 0.95 Ω, per phase. Its armature resistance is negligible. Its combined friction and windage losses are 1.5 kW and its core losses are 1.25 kW. Neglect its dc field losses. The field current is constant at no load. Its full-load armature current is 55 A at 0.85 lagging PF. Determine the following:

(a) Its terminal voltage when it is delivering the rated current at a power factor of (1) 0.85 lagging, (2) unity, and (3) 0.85 leading

(b) Its efficiency when it is loaded with the rated current at a lagging power factor of 0.85

(c) Its input torque and induced counter torque when it is operating at full load

(d) Its voltage regulation when it is operating under full load with a power factor of (1) 0.85 lagging, (2) unity, and (3) 0.85 leading

7.24 Solve Problem 7.23 but assume that the generator is wye-connected.

7.25 A three-phase, 60 Hz, $(480/\sqrt{3})$ V, delta-connected, cylindrical-rotor synchronous motor has a synchronous reactance of 3.5 Ω. Its armature resistance is negligible. Its combined friction and windage losses are 2 kW and its core losses are 1.45 kW. The motor is connected to a 25 hp mechanical load and is operating at a leading power factor of 0.85.

(a) Find the values of I_a, I_L, and E_a of the motor and draw its phasor diagram.

(b) If the mechanical load is increased to 50 hp, draw the new phasor diagram.

(c) Find the values of I_a, I_L, and E_a and the PF of the motor in Part (b).

7.26 A three-phase, 60 Hz, 277.1281-V, delta-connected, cylindrical-rotor synchronous motor has a synchronous reactance of 4 Ω. Its armature resistance is negligible. Its combined friction and windage losses are 2.5 kW and its core losses are 2.25 kW. The motor is connected to a 20 hp mechanical load and is operating at a leading power factor of 0.80.

(a) Find the values of I_a, I_L, and E_a of the motor and draw its phasor diagram.

(b) If the mechanical load is increased to 75 hp, draw the new phasor diagram.

(c) Find the values of I_a, I_L, and E_a and the PF of the motor in Part (b).

7.27 Assume that two three-phase induction motors and a three-phase synchronous motor are connected to the same bus. The first induction motor is 150 kW and operating at 0.85 lagging power factor. The second induction motor is 250 kW and operating at 0.70 lagging power factor. The real power of the synchronous motor is 200 kW. If the bus voltage is 480 V and the synchronous motor is operating at 0.90 lagging power factor, determine the following:

(a) The total real and reactive power at the bus.

(b) The total bus current and its power factor.

(c) If the synchronous motor is operating at 0.90 leading power factor, the new total bus current and its power factor.

(d) Consider the results of Parts (b) and (c), and determine the percent reduction in the power line losses.

7.28 Suppose that two three-phase induction motors and a three-phase synchronous motor are connected to the same bus. The first induction motor is 250 kW and operating at 0.90 lagging power factor. The second induction motor is 350 kW and operating at 0.75 lagging power factor. The real power of the synchronous motor is 300 kW. If the bus voltage is 480 V and the synchronous motor is operating at 0.90 lagging power factor, determine the following:

(a) The total real and reactive power at the bus.

(b) The total bus current and its power factor.

(c) If the synchronous motor is operating at 0.90 leading power factor, the new total bus current and its power factor.

(d) Consider the results of Parts (b) and (c), and determine the percent reduction in the power line losses.

7.29 A manufacturing plant has a load of 500 kW at 0.78 lagging power factor. If a 75 hp synchronous motor is added and operated at 0.85 leading power factor, determine the new total load and new power factor of the plant. Neglect the losses of the synchronous motor.

7.30 A 75 MVA, 20 kV, 60 Hz, wye-connected, three-phase synchronous generator has $X_d = 2.1\,\Omega$ and $X_q = 1.0\,\Omega$. Ignore its resistance. Assume that it operates at full load at 0.8 lagging power factor and determine the following:

(a) The phase voltage and phase current at full load

(b) The internal generated voltage E_a if it has a cylindrical rotor

(c) The internal generated voltage E_a if it has a salient-pole rotor

7.31 Suppose that two three-phase induction motors and a three-phase synchronous motor are connected to the same bus. The first induction motor is 275 kW and operating at 0.8 lagging power factor. The second induction motor is 125 kW and operating at 0.8 lagging power factor. If the bus voltage is 480 V and the 350 kW synchronous motor is operating at 0.95 lagging power factor, determine the following:

(a) The total real and reactive power at the bus

(b) The total bus current and its power factor

(c) If the synchronous motor is operating at 0.95 leading power factor, the new total bus current and its power factor

7.32 A three-phase, 60 Hz, 480 V, wye-connected, cylindrical-rotor synchronous motor has a synchronous reactance of $3\,\Omega$. Its armature resistance is negligible. Its friction and windage losses are 4 kW and its core losses are 3 kW. The motor is connected to a 75 hp mechanical load and is operating at a leading power factor of 0.85.

(a) Find the values of I_a, I_L, and of the motor and draw its phasor diagram.

(b) If the mechanical load is increased to 100 hp, draw the new phasor diagram.

(c) Find the values of I_a, I_L, and E_a and the PF of the motor in Part (b).

7.33 A three-phase, 50 Hz, 380 V, wye-connected, four-pole cylindrical-rotor synchronous generator has a synchronous reactance of $0.9\,\Omega$ per phase. Its armature resistance is negligible. Its friction and windage losses are 1.0 kW and its core losses are 1.0 kW. Neglect its armature resistance and dc field losses. The field current is constant at no load. Its full-load armature current is 50 A at 0.9 PF lagging. Determine the following:

(a) Its terminal voltage when it is delivering the rated current at a unity power factor

(b) Its efficiency when it is loaded with the rated current at a lagging power factor of 0.9

(c) Its input torque and induced counter torque when it is operating at full load

(d) Its voltage regulation when it is operating under full load at a unity power factor

8 Direct-Current Machines

Nothing is so firmly believed as what we least know.

M.E. De Montaigne, Essays, 1580

Talk sense to a fool and he calls you foolish. Euripides

The Bacchae, 407 BC

But talk nonsense to a fool and he calls you a genius.

Turan Gönen

8.1 INTRODUCTION

A direct-current (dc) machine is a versatile machine, that is, the same machine can be used as a generator to convert mechanical energy to dc electrical energy or as a motor to convert dc electrical energy into mechanical energy. However, the use of dc machines as dc generators to produce bulk power has rapidly disappeared due to the economic advantages involved in the use of alternating-current generation, transmission, and distribution. This is partly due to the high efficiency and relative simplicity with which transformers convert voltages from one level to another.

Today, the need for dc power is often met by the use of solid state–controlled rectifiers. However, dc motors are used extensively in many industrial applications because they provide constant mechanical power output or constant torque, adjustable motor speed over wide ranges, precise speed or position control, efficient operation over a wide speed range, rapid acceleration and deceleration, and responsiveness to feedback signals.

Such machines can vary in size from miniature permanent-magnet motors to machines rated for continuous operation at several thousand horsepower. Examples of small dc motors include those used for small control devices, windshield-wiper motors, fan motors, starter motors, and various servomotors. Application examples for larger dc motors include industrial drive motors in conveyors, pumps, hoists, overhead cranes, forklifts, fans, steel and aluminum rolling mills, paper mills, textile mills, various other rolling mills, golf carts, electrical cars, street cars or trolleys, electric trains, electric elevators, and large earth-moving equipment.

Obviously, dc machine applications are very significant, but the advantages of the dc machine must be weighed against its greater initial investment cost and the maintenance problems associated with its brush-commutator system. Figure 8.1 shows the housing of a dc motor being mounted in an industrial mill.

8.2 CONSTRUCTIONAL FEATURES

The schematic diagram of the construction of a dc machine is shown in Figure 8.2. The construction has two basic parts, namely, the stator (which stands still) and the rotor (which rotates). The stator has salient poles that are excited by one or more field windings. The armature winding of a dc machine is located on the rotor with current flowing through it by carbon brushes making contact with copper commutator segments.

FIGURE 8.1   Mounting the housing of a dc mill motor. (Courtesy of Siemens, Munich, Germany.)

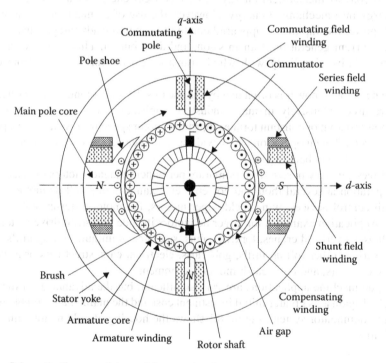

FIGURE 8.2   Schematic diagram of dc machine construction.

Both the main poles and armature core are made up of laminated materials to reduce core losses. With the exception of a few small machines, the dc machines also have *commutating poles**
between the main poles of the stator. Each commutating pole has its own winding which is known as the *commutating winding.*

---

* They are known as the interpoles or compoles.

The main (or field) poles are located on the stator and are attached to the stator yoke (or frame). The stator yoke also serves as a return path for the pole flux. Because of this, the yokes are being built with laminations to decrease core losses in solid state–driven motors. The ends of the poles are called the *pole shoes*. The surface of the pole shoe opposite the rotor is called the pole face. The distance between the *pole face* and the rotor surface is called the **air gap**. As shown in the figure, there is a special winding located in the slots of the pole faces called the *compensating winding*.*

The field windings are located around the pole cores and are connected in series and/or in shunt (i.e., in parallel) with the armature circuit. The shunt winding is made up of many turns of relatively thin wires, whereas the series winding has only a few turns and is made up of thicker wires. As shown in the figure, if the field has both windings, the series winding is located on top of the shunt winding. The two windings are separated by extra insulating material, which is usually paper. Figure 8.3 shows the schematic connection diagram for a dc machine with commutating and compensating windings in addition to series and shunt windings. The series and shunt windings are located on the *d*-axis. This axis is called the *field axis*, or **direct axis**, because the air-gap flux distribution due to the field windings is symmetric at the center line of the field poles. Both the compensating and commutating winding brushes are located on the *q*-axis. This axis is called the *quadrature axis* because it is 90 electrical degrees from the *d*-axis and represents the neutral zone. Figures 8.4 and 8.5 show the cutaway views of two different dc machines.

The *commutator* is located on the armature and consists of a number of radial segments assembled into a cylinder which is attached to and insulated from the shaft, as shown in Figures 8.6 and 8.7. These segments are well insulated from each other by mica. The leads of the armature coils are connected to these commutator segments. Current is conducted to the armature coils by carbon brushes that ride on the commutator segments. The brushes are fitted to the surface of the commutator and are held in *brush holders*. These brush holders use springs to push the brushes against the commutator surface to maintain constant pressure and problem-free riding. The connection between the brush and brush holder is by a flexible copper cable called a *pigtail*. The rotor itself is mounted on a shaft that rides in the bearings.

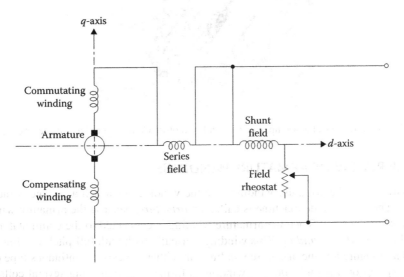

FIGURE 8.3   Schematic connection diagram for a dc machine having commutating poles and compensating winding in addition to series and shunt windings.

---

* Sometimes it is called the pole-face winding, for obvious reasons.

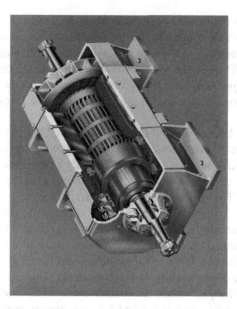

FIGURE 8.4 (See color insert.) Cutaway view of a mill duty dc motor. (Courtesy of General Electric, Fairfield, CT.)

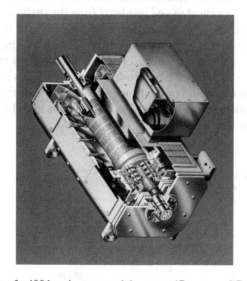

FIGURE 8.5 Cutaway view of a 400 hp, shunt-wound dc motor. (Courtesy of General Electric, Fairfield, CT.)

8.3 BRIEF REVIEW OF ARMATURE WINDINGS

As previously stated, the armature windings are the windings in which a voltage is induced in a dc machine. The rotor of a dc machine is called an *armature*, because the armature windings are placed in slots of the rotor. Since the armature winding is connected to the commutator, it is also known as the *commutator winding*. This winding is usually built with full-pitch windings.

As shown in Figure 8.8, the armature windings are either the closed continuous type of double-layer *lap windings* or *wave windings*. A winding is formed by connecting several coils in series, and a coil is formed by connecting several turns (loops) in series. Each turn is made up of two conductors connected to one end by an end connection. In other words, each side of a turn is called a **conductor**.

FIGURE 8.6 A close-up view of the commutator of a 500 hp dc motor. (Courtesy of MagneTex, Inc., Menomonee Falls, WI.)

FIGURE 8.7 (See color insert.)   The armature of a 500 V, 150 hp dc motor. (Courtesy of General Electric, Fairfield, CT.)

In a *lap winding*, there are always as many paths in parallel through the armature winding as there are a number of poles (or brushes). Each path is made up of a series connection between a number of terminal coils that are approximately equal to the total number of armature coils divided by the number of poles. In such a lap winding, the current in each armature coil is equal to the armature terminal current divided by the number of poles.

In a *wave winding*, there are always two paths in parallel through the armature winding from one terminal to the other. At any given time, each path is made up of a series connection of approximately

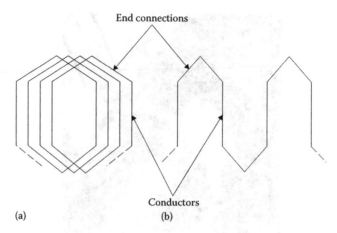

FIGURE 8.8   Basic armature winding types: (a) lap winding and (b) wave winding.

one-half of the total armature coils between the terminals. The current in each armature coil is one-half of the armature terminal current.

   The internal generated voltages in each of the parallel paths of a lap winding are equal if the machine is geometrically and magnetically symmetrical. If complete symmetry does not exist, the internal generated voltage of the different paths will not be exactly the same. Therefore, there will be a circulating current flowing between brush sets of the same polarity. To prevent this, equalizers are used. *Equalizers* are bars located on the rotor of a lap-wound dc machine that short together the points in the winding at the same voltage level in different paths. Figure 8.7 shows a close-up view of the commutator of a dc motor. Notice that the equalizers are mounted in the small ring just in front of the commutator segments.

8.4   ELEMENTARY DC MACHINE

Figure 8.9 shows an elementary two-pole dc generator. The armature winding consists of a single coil of N turns. The voltage induced in this rotating armature is alternating. However, by using a commutator this voltage is rectified mechanically into dc voltage for the external circuit. Here, the commutator has two half rings that are made up of two copper segments insulated from each other and from the shaft.

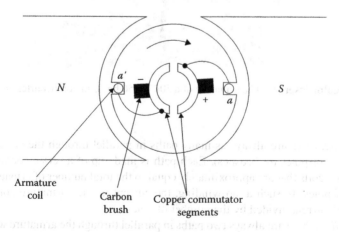

FIGURE 8.9   Simple representation of a dc machine.

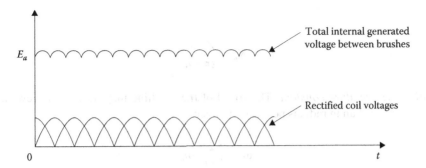

FIGURE 8.10   The total internal generated voltage between brushes in a dc machine as a function of time.

Each end of the armature coil is connected to a segment. Stationary carbon brushes held against the commutator surface connect the coil to the external armature terminals. Since the brushes remain in the same position as the coil rotates, each fixed terminal is always connected to the side of the coil where the relative motion between the coil side and the field is the same.

In other words, the action of the commutator is to reverse the armature coil connections to the external circuit when the current reverses in the armature coil. Therefore, the commutator at all times connects the coil side under the south pole to the positive brush and the one under the north pole to the negative pole. Thus, the polarity of the voltage difference between the two fixed brushes is always the same and the voltage is now unidirectional.

However, a pulsating dc, like the one produced by this type of single-coil generator, is not suitable for most commercial uses. As shown in Figure 8.10, the total internal generated voltage between brushes (i.e., simply the brush voltage) can be made practically constant by using a large number of coils and commutator segments with the coils evenly distributed around the armature surface.

8.5   ARMATURE VOLTAGE

In a dc machine, the armature voltage is the internal generated voltage. By applying Faraday's law of electromagnetic induction, the armature voltage* can be expressed as

$$E_a = \left(\frac{Z \times p}{2\pi \times a} \right) \Phi_d \times \omega_m \tag{8.1}$$

where
   Z is the total number of conductors in the armature winding
   p is the number of poles (of field or stator)
   a is the number of parallel paths in the armature winding
   Φ_d is the direct-axis air-gap flux per pole in webers
   ω_m is the angular velocity of the armature (or shaft) in mechanical radians per second

The armature voltage can also be expressed as

$$E_a = K_a \Phi_d \omega_m \tag{8.2}$$

---

* It is also known as the speed voltage. Some authors define this voltage as the internal source voltage when the machine is operating as a generator, and as the countervoltage (or back emf when the machine is operating as a motor).

where

$$K_a = \frac{Z \times p}{2\pi \times a} \tag{8.3}$$

and is called the *armature constant*. The speed of the machine may be given in revolutions per minute (rpm) rather than in radians per second. Since

$$\omega_m = \left(\frac{2\pi}{60}\right) n_m \tag{8.4}$$

the armature voltage can be expressed as

$$E_a = K_{a1} \times \Phi_d \times n_m \tag{8.5}$$

where

$$K_{a1} = \frac{Z \times p}{60 \times a} \tag{8.6}$$

Therefore, the armature voltage is a function of the flux in the machine, the speed of its rotor, and a constant that depends on the machine.

The armature voltage, or more precisely the internal generated voltage, is not the terminal voltage. Consider the circuit representation of a separately excited dc generator and motor as shown in Figure 8.11a and b, respectively. The armature voltage E_a can be expressed as

$$E_a = V_t \pm I_a R_a \tag{8.7}$$

where

the plus sign is used for a generator and the minus sign for a motor
V_t is the terminal voltage
R_a is the armature resistance*

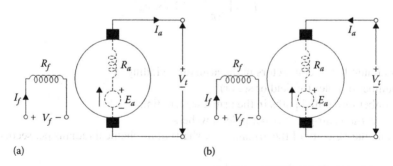

(a)   (b)

FIGURE 8.11   Simple representation of a dc machine: (a) circuit representation of a dc generator and (b) circuit representation of a dc motor.

---

* Even though the armature resistance R_a actually exists between the brushes in the armature, in some books it is not explicitly represented in the armature circuits. Such representation agrees with the dc machine panels of an electromechanical laboratory since only armature terminals A_1 and A_2 can possibly be reached, as shown in Figure 8.12. However, in the rest of the book, the armature resistance is explicitly represented to avoid confusing the beginner.

Therefore, in the case of a generator, the armature voltage is always greater than the terminal voltage. In a motor, the armature voltage is less than the terminal voltage. Regardless of whether the machine is used as a generator or as a motor, a *brush-contact* **voltage drop**, usually assumed to be 2 V, exists due to the resistive voltage drop between the brushes and commutator.

Also, the term *armature winding circuit resistance* can include not only the resistance of the armature winding R_a, but also the resistances of the series-field winding R_{se}, commutating winding R_{cw}, compensating winding R_{cp}, as well as the resistance of any external wires (used in laboratories to make the necessary connections) $R_{a,\,ext}$. Therefore, the general expression for the armature voltage becomes

$$E_a = V_t \pm I_a \left(\sum R_a \right) \pm 2.0 \tag{8.8}$$

where

$$\sum R_a = R_a + R_{se} + R_{cw} + R_{cp} + R_{a,\,ext} \tag{8.9}$$

and represents the total armature winding circuit resistance. In Equation 8.8, the plus sign is used for a generator and the minus sign for a motor. Note that the voltage polarity of the brushes is a function of the rotational direction and the magnetic polarity of the stator field poles.

Example 8.1

Assume that a four-pole dc machine has an armature with a radius of 15 cm and an effective length of 30 cm, and that the poles cover 70% of the armature periphery. The armature winding has 50 coils of five turns each and is wave-wound with $a = 2$ paths. Assume that the average flux density in the air gap under the pole faces is 0.7 T and determine the following:

(a) The total number of conductors in the armature winding
(b) The flux per pole
(c) The armature constant K_a
(d) The induced armature voltage if the speed of the armature is 900 rpm
(e) The current in each coil if the armature current is 200 A

Solution

(a) The total number of conductors in the armature winding is
(b) Since the pole area is

$$A_p = \frac{2\pi(0.15 \text{ m})(0.30 \text{ m})(0.7)}{6}$$

$$= 0.033 \text{ m}^2$$

the flux per pole is

$$\Phi_d = A_P \times B$$

$$= (0.033 \text{ m}^2)(0.7 \text{ Wb/m}^2)$$

$$= 0.023 \text{ Wb}$$

(c) The armature constant is

$$K_a = \frac{Z \times p}{2\pi \times a}$$

$$= \frac{(500)(6)}{2\pi(2)}$$

$$= 238.73$$

(d) The speed of the armature is

$$\omega_m = n_m \left(\frac{2\pi}{60} \right)$$

$$= (900 \text{ rpm}) \left(\frac{2\pi}{60} \right)$$

$$= 94.25 \text{ rad/s}$$

Therefore, the induced armature voltage is

$$E_a = K_a \times \Phi_d \times \omega_m$$

$$= (238.73)(0.023 \text{ Wb})(94.25 \text{ rad/s})$$

$$= 517.5 \text{ V}$$

(e) The current in each coil is

$$I_{coil} = \frac{I_a}{2}$$

$$= \frac{200 \text{ A}}{2}$$

$$= 100 \text{ A}$$

Example 8.2

Assume that a separately excited shunt dc machine has a rated terminal voltage of 230 V and a rated armature current of 100 A. Its armature winding resistance, commutating winding resistance, and compensating winding resistance are 0.08, 0.01, and 0.008 Ω, respectively. The resistance of external wires (i.e., $R_{a,ext}$) is 0.002 Ω. Determine the following:

(a) The induced armature voltage if the machine is operating as a generator at full load
(b) The induced armature voltage if the machine is operating as a motor at full load

Solution

(a) Since the total armature winding circuit resistance is

$$\sum R_a = R_a + R_{se} + R_{cw} + R_{cp} + R_{a,ext}$$

$$= 0.08 + 0 + 0.01 + 0.008 + 0.002$$

$$= 0.10 \ \Omega$$

When the machine is operating as a generator, its induced armature voltage is

$$E_a = V_t + I_a \left(\sum R_a \right) + 2.0$$

$$= 230 + (100 \text{ A})(0.10 \ \Omega) + 2.0$$

$$= 242 \text{ V}$$

(b) When the machine is operating as a motor, its induced armature voltage is

$$E_a = V_t - I_a \left(\sum R_a \right) - 2.0$$

$$= 230 - (100 \text{ A})(0.10 \ \Omega) - 2.0$$

$$= 218 \text{ V}$$

8.6   METHODS OF FIELD EXCITATION

The field circuit and the armature circuit of a dc machine can be interconnected in several different ways to produce various operating characteristics. There are basically two types of field windings, namely, *shunt-field winding* and *series-field winding*.

The shunt windings have a great many turns and are built from thinner wires. Therefore, the required field current is a very small portion (less than 5%) of the rated armature current. On the other hand, the series windings have relatively less number of turns and are built from thicker wires. The series windings are connected in series with the armature and therefore its field current is the armature current.

The shunt winding can be separately excited from a separate source. In a *separately excited* machine, there is no electrical interconnection between the field and the armature windings, as shown in Figure 8.12a. When the field is interconnected with the armature winding, the machine is said to be *self-excited*.

The self-excited machines may be shunt, series, or compound, as shown in Figure 8.12b through 8.12g. Notice that a compound machine has both shunt- and series-field windings in addition to the armature winding.

If the relative polarities of the shunt and series-field windings are additive, the machine is called *cumulative compound*. If they oppose each other, the machine is called *differential compound*. A compound machine may be connected *short shunt* with the shunt field in parallel with the armature only or *long shunt* with the shunt field in parallel with both the armature and series field. The circuits shown in Figure 8.12 are labeled according to the NEMA standards.*

8.7   ARMATURE REACTION

Armature reaction is defined as the effect of the armature mmf field upon the flux distribution of the machine. Figure 8.13a shows the main field flux Φ_f that is established by the mmf produced by the field current when there is no current flowing in the armature. Figure 8.13b shows the armature flux Φ_a that is established by the armature mmf produced by the current flowing in the armature when there is no current flowing in the field winding of the machine.

The brushes are located on the magnetic neutral axis. Figure 8.13c shows the situation when both the main field flux and the armature flux exist at the same time. It is clear that the armature flux

---

* Notice that currents flowing into terminals F_1 and S_1 result in a cumulative compound effect; whereas currents flowing into terminal F_1 and out of terminal S_1 result in a differential compound effect. Also notice that if F_1 were connected to A_1 in each case, it would become a short-shunt connection.

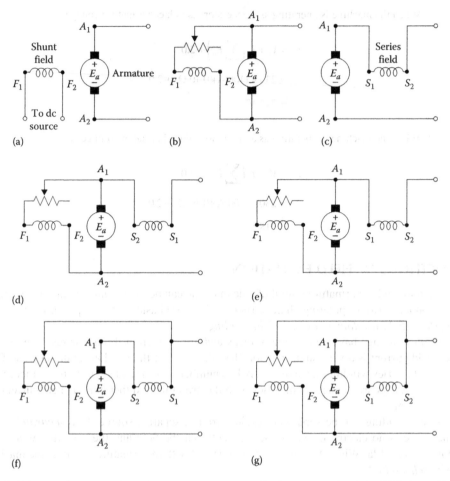

FIGURE 8.12 Typical field-excitation methods for a dc machine: (a) separately excited, (b) shunt, (c) series, (d) short-shunt connection for cumulative compound motor or differential compound generator, (e) short-shunt connection for cumulative compound generator or differential compound motor, (f) long-shunt connection for cumulative compound motor or differential compound generator, and (g) long-shunt connection for cumulative compound generator or differential compound motor.

causes a distortion in the distribution of the main field flux. As shown in the figure, the phasor sum of the two mmfs produces a resultant flux Φ_r.

Notice that the flux produced by the armature mmf opposes the flux under one-half of the pole and aids under the other half of the pole. As a result, flux density under the pole increases in one-half of the pole and decreases under the other half of the pole.

As shown in Figure 8.14d, the magnetic neutral axis is shifted from the geometric neutral axis. The shift is forward in the direction of rotation for a generator and backward against rotation for a motor. The magnitude of flux shift is a function of saturation in the pole tips and the amount of armature (load) current.

If there is no saturation, the increase of flux in one pole tip is canceled by a corresponding decrease in the others.[*] With saturation, on the other hand, there is a net decrease in total flux,[†] causing a decrease in the terminal voltage of the generator and an increase in the speed of the motor.

---

[*] It is called the cross-magnetizing armature reaction.
[†] It is called the demagnetizing effect of the cross-magnetizing armature reaction.

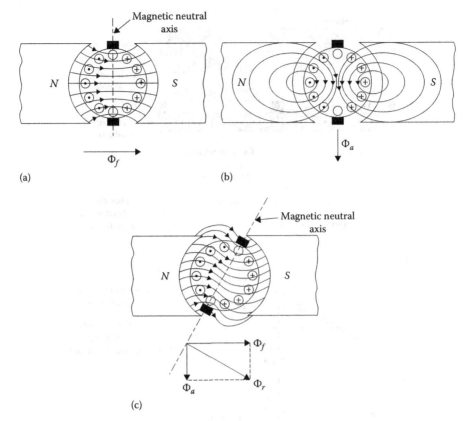

FIGURE 8.13 The effect of armature reaction in a dc machine: (a) the effect of the air gap on the pole flux distribution, (b) the armature flux alone, and (c) resultant distortion of the field flux produced by the armature flux.

8.8 COMMUTATION

Commutation is the process of reversing the direction of the current in an armature coil as the commutator segments to which the coil is connected move from the magnetic field of one polarity to the influence of the magnetic field of the opposite polarity, as shown in Figure 8.15. The time interval required for this reversal is called the *commutation period*.

As the commutator sweeps past the brushes, any given coil connected to one of the segments has current in a particular direction. The current in the coil is reversed as the commutator segment approaches and passes the brush. Consider the coil *b* connected to commutator segments 2 and 3, as shown at the top of Figure 8.15, and notice that the current in the coil is flowing from left to right.

The middle diagram of Figure 8.15 shows that there is no current flow in the coil since it is short-circuited by the brush. The bottom of Figure 8.15 shows the moment at which brush contact with commutator segment 3 is interrupted. Notice that the direction of current flow in coil b is reversed and that it is now flowing from right to left.

As shown in Figure 8.16, during the commutation period Δ*t*, the commutator segments to which a coil is connected are passing under the brush; the current *I* has to be totally reversed as the commutator segments pass from under the brush to prevent the formation of an arc.

The reversal of the coil induces a self-inductance voltage which opposes the change of current that can cause a spark to appear at the trailing edge of the brush. Figure 8.17 shows the ideal process of commutation as well as under commutation* due to the reactance voltage. If the reactance

---

* It is also called the incomplete or delayed commutation.

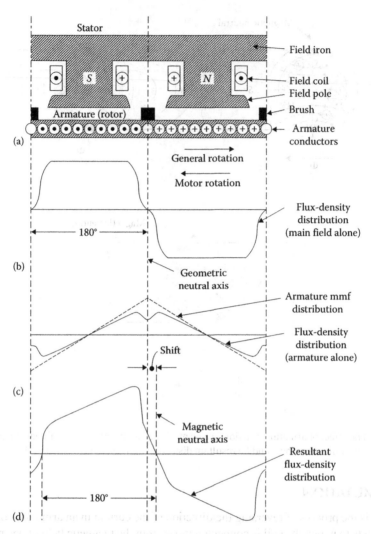

FIGURE 8.14 Main field, armature, and resultant flux-density distributions with brushes on the geometric neutral axes: (a) linear representation of stator and armature-magnetic circuits, (b) flux-density distribution due to main field alone, (c) flux-density distribution due to armature mmf alone, and (d) resultant flux-density distribution.

voltage is large enough, it may cause sparking at the trailing edge of the brush. Excessive sparking burns the brushes and the commutator surface, but sparking can be prevented by inducing a voltage in the coil undergoing commutation.

This can be accomplished by the use of commutating poles, which are small poles placed between the main (Figure 8.15 on pp. 387 of *Electrical Machines* by Gönen.) poles of a dc machine. The commutating windings are interconnected in such a way that they have the same polarity as the following main pole in the direction of rotation. Almost all dc machines of more than 1 hp are furnished with commutating poles (or interpoles).

As previously stated, the commutating windings are permanently connected in series with the armature and their leads are not brought out to the terminal box. In small machines (with 1 hp or less), the commutation can be improved by shifting the brushes. As a result, the coils undergoing commutation can have current reversals supported by flux from the main poles.

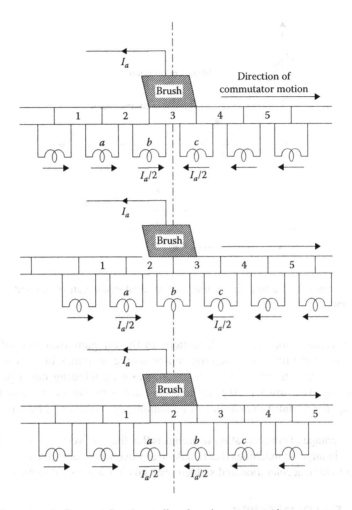

FIGURE 8.15 The reversal of current flow in a coil undergoing commutation.

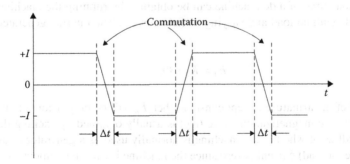

FIGURE 8.16 Waveform of a current in an armature coil during linear commutation.

8.9 COMPENSATING WINDINGS

As the armature current increases due to the armature reaction, the corresponding flux density distortion also increases, which in turn causes the commutator flashover probability to increase. The commutating windings located on the commutating poles can neutralize the effect of armature reaction in the interpolar areas.

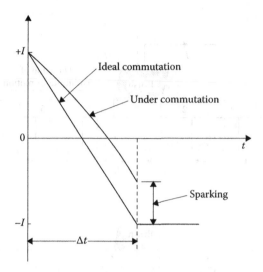

FIGURE 8.17 Coil current as a function of time during ideal commutation and under commutation (or delayed commutation).

However, they cannot stop the flux distortions in the air gaps over the pole faces. These flux distortions can be eliminated by placing compensating windings in slots distributed along the pole faces, as shown in Figure 8.2. Each compensating winding has a polarity opposite that of the adjoining armature winding. By allowing armature current to flow through such a pole-face winding, the armature reaction can be completely neutralized by a proper amount of ampere-turns.

The only disadvantage of compensating windings is that they are very expensive. For this reason, they are only used in large dc machines that handle heavy overloads, or suddenly changing loads, or in motors subject to high acceleration and sudden reversals in rotational directions.

8.10 MAGNETIZATION CURVE

The magnetization curve of a dc machine can be obtained by running the machine as a generator at its rated speed* with no load and varying its field current. The internal generated voltage of the machine is

$$E_a = K_a \times \Phi_d \times \omega_m \tag{8.2}$$

and if the speed of the armature is kept constant, then E_a will be proportional to the flux setup by the field winding. The magnetization curve (mc) is usually obtained by exciting the field winding separately, regardless of whether the machine is normally used as a generator or as a motor. Note that the armature (or load) current is zero since the machine is running without a load, and that the terminal voltage is equal to the internal generated voltage E_a.

As the magnetizing flux per pole is increased by raising the current in the field winding, the voltage E_a also increases. However, above a certain point called the *saturation point*, it becomes increasingly difficult to further magnetize the core. Due to the saturation of the magnetic core above the saturation point (i.e., the knee of the curve), the relationship between the voltage E_a and the field current I_f becomes nonlinear, as shown in Figures 8.18a and b. The resultant mc curve shows the relationship between the voltage E_a and the field mmf or the field current I_f.

---

* The rated speed is the speed at which the machine is designed to operate to produce the rated voltage.

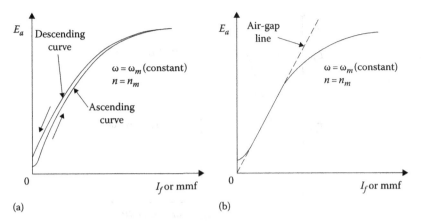

FIGURE 8.18 Magnetization curve: (a) explicitly showing the effect of hysteresis and (b) average magnetization curve.

However, the shape of the curve is determined mainly by the characteristics of the magnetic circuit. Because of a small residual magnetism that exists in the field structure, the voltage does not start from zero, except in a new machine.*

As shown in Figure 8.18a, if the field current is increased from zero to a value that yields an armature voltage well above the rated voltage of the machine, the resultant curve is the *ascending curve*. If the field current I_f is progressively decreased to zero again, this curve is the *descending curve*. The reason that the descending curve is above the ascending curve can be explained by hysteresis. The magnetization curve[†] shown in Figure 8.18b is the average of the two curves. Note that ω_m and n_m represent the rated speed[‡] of the machine at which the magnetization curve is developed given in rad/s and rpm, respectively.

Consider the magnetization curves shown in Figure 8.19. Assume that the top and the bottom curves are obtained at constant speeds of ω_{m1} and ω_{m2}, respectively. Therefore, the associated internal generated voltages are

$$E_{a1} = K_a \times \Phi_d \times \omega_{m1} \tag{8.10}$$

and

$$E_{a1} = K_a \times \Phi_d \times \omega_{m1} \tag{8.11}$$

Since

$$K_a \times \Phi_d = \frac{E_{a1}}{\omega_{m1}} \tag{8.12}$$

and

---

* Just as a new machine may not have residual magnetic flux, in any machine it may be lost as a result of conditions such as mechanical jarring during transportation, excessive vibrations, inactivity for long periods of time, extreme heat, or having its alternating current unintentionally connected across the field winding. In such cases, the field winding must receive an initial do excitation to provide the machine with a suitable level of residual flux. This process is known as flashing the field.

† It is also called the saturation curve, the open-circuit characteristic, or the no-load characteristic.

‡ Hence, it is called the magnetization-curve speed.

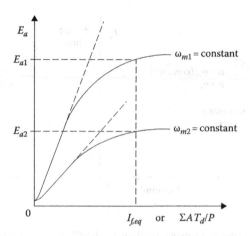

FIGURE 8.19   Magnetization curves developed at two separate speeds with the excitation held constant.

$$K_a \times \Phi_d = \frac{E_{a2}}{\omega_{m2}} \tag{8.13}$$

then

$$E_{a2} = \left(\frac{\omega_{m2}}{\omega_{m1}} \right) E_{a1} \tag{8.14}$$

where E_{a1} and E_{a2} are generated at constant speeds ω_{m1} and ω_{m2}, respectively. Alternatively, if the constant speeds are given in rpm, then

$$E_{a2} = \left(\frac{n_{m2}}{n_{m1}} \right) E_{a1} \tag{8.15}$$

Up to now, it has been assumed that the machine involved is a separately excited shunt machine. However, it is possible to apply this approach to other types of dc machines. This can be accomplished by considering the direct-axis air-gap flux produced by the combined mmf.

For example, suppose that the machine involved is a compound machine. Its net excitation per pole on the d-axis (i.e., the total ampere turns per pole on the d-axis) can be expressed as

$$\sum \frac{AT_d}{p} = N_f I_f \pm N_{se} I_a \tag{8.16}$$

where
   N_f is the number of turns per pole of the shunt-field winding
   I_f is the current in the shunt-field winding
   N_{se} is the number of turns per pole of the series winding
   I_a is the current in the series-field winding

This equation can be modified so that the effects of the armature reaction can be taken into account. Thus,

$$\sum \frac{AT_d}{p} = \frac{V_t N_f}{\sum R_f} \pm N_{se} I_a - K_d I_a \tag{8.17}$$

where the term $K_d I_a$ is a simplified linear approximation to account for the demagnetization of the d-axis (i.e., the armature reaction) caused by the armature mmf. In Equations 8.16 and 8.17, the plus sign is used for a cumulative-compounded machine and the minus sign is used for a differential compounded machine. The K_d is the armature reaction constant for the machine involved. Hence, Equation 8.17 gives the total effective mmf per pole.

If the machine has a self-excited shunt field, the net excitation per pole on the d-axis can be expressed as

$$\sum \frac{AT_d}{p} = \frac{V_t N_f}{\sum R_f} \pm N_{se} I_a - K_d I_a \tag{8.18}$$

where the total shunt-field circuit resistance is

$$\sum R_f = R_f + R_{rheo} \tag{8.19}$$

where
V_t is the terminal voltage
R_f is the resistance of the shunt-field winding
R_{rheo} is the resistance of the shunt-field rheostat*

However, if the machine has a separately excited shunt field, the net excitation per pole on the d-axis can be expressed as

$$\sum AT_d I_p = \frac{V_f N_f}{\sum R_f} \pm N_{se} I_a - K_d I_a \tag{8.20}$$

where V_f is the voltage across the shunt-field winding.

If the magnetization curve is given in terms of E_a versus I_f, it is necessary to define an equivalent if that would produce the same voltage E_a as the combination of all the mmfs in the machine.[†] Since

$$\sum \frac{AT_d}{p} = N_f I_{f,eq} \tag{8.14}$$

such equivalent shunt-field current can be found from

$$I_{f,eq} = \frac{\sum AT_d / p}{N_f} \tag{8.21}$$

---

* Usually, a rheostat is included in the circuit of the shunt winding to control the field current and to vary the shunt-field mmf.

† In other words, it is as if the machine were replaced by an equivalent machine with a shunt field in order to find the corresponding equivalent I_f current.

Once the equivalent shunt-field current is found, the corresponding voltage E_a can be found from the magnetization curve. Some magnetization curves are plotted in terms of E_a versus $\sum AT_d / p$ or $I_{f, eq}$, as shown in Figure 8.19. As can be observed from Equation 8.20, when the I_a is zero, the net excitation per pole on the d-axis is produced only by the shunt-field winding. Since

$$\sum \frac{AT_d}{p} = N_f I_f \tag{8.22}$$

then

$$I_f = \frac{\sum AT_d / p}{N_f} \tag{8.23}$$

8.11 DC GENERATORS

As stated in Section 8.1, dc generators are generally not used to produce bulk power today. Instead, solid state–controlled rectifiers are preferred for many applications. However, there are still some dc generators that are used to provide dc power to isolated loads and for special applications. In such use, the dc machine operating as a generator is driven by a prime mover at a constant speed with the armature terminals connected to the load. It can be used as a separately excited generator, a self-excited shunt generator, a series generator, or a compound generator.

8.12 SEPARATELY EXCITED GENERATOR

In the separately excited dc generator, the shunt-field winding is supplied by a separate external dc power source, as shown in Figure 8.20. The external dc power source can be a small dc generator, a solid-state dc power supply, or a battery.

In the equivalent circuit, E_a is the internal generated voltage, V_t is the terminal voltage, I_a is the armature current, which is also the load current, I_f is the field current, R_a is the resistance of the armature winding, R_f is the resistance of the field winding, R_{rheo} is the resistance of the shunt-field rheostat, and V_f is the voltage of a separate source. According to Kirchhoff's voltage law, the terminal voltage is

$$V_t = E_a - I_a R_a - 2.0 \tag{8.24}$$

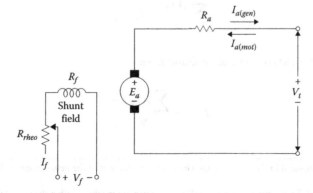

FIGURE 8.20 The separately excited shunt generator or motor schematic diagram with current directions.

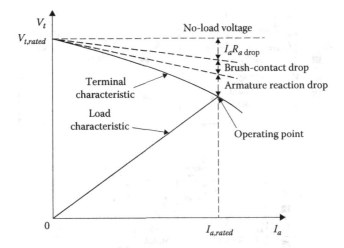

FIGURE 8.21 Terminal and load characteristics of a separately excited generator.

and the field voltage is

$$V_f = I_f R_f \tag{8.25}$$

Equation 8.24 represents the *terminal characteristic** of the separately excited dc generator as shown in Figure 8.21. Notice that the terminal voltage differs from the no-load voltage by the three voltage drops representing the armature resistance voltage drop, the brush-contact voltage drop, and the armature reaction voltage drop. The *load characteristic* is determined by

$$V_t = I_a R_L$$

where R_L represents the load resistance. As shown in Figure 8.21, the intersection of the terminal characteristic and the load characteristic is the *operating point* for the generator. The operation of the separately excited generator is stable with any field excitation. Therefore, a wide range of output voltages are available.

8.13 SELF-EXCITED SHUNT GENERATOR

As shown in Figure 8.22a, in the self-excited shunt generator, the field winding is connected directly across the armature winding. Therefore, the armature voltage can provide the field current. However, any change in the armature current results in a change in the $I_a R_a$ voltage drop.

Because of this, both the terminal voltage and the field current[†] must vary. Thus, the internal generated voltage E_a in a self-excited generator is a function of the armature current I_a. Accordingly, the terminal voltage V_t changes as the I_a changes. Hence,

$$V_t = E_a - I_a R_a - 2.0 \tag{8.26}$$

and the field voltage is

$$V_t = I_f R_f \tag{8.27}$$

---

* It is also called the external characteristic.
† Typically, the field current I_f is about 5% of the rated armature current.

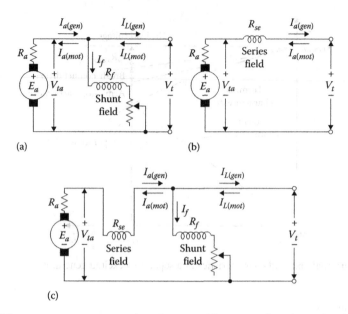

FIGURE 8.22 Motor or generator connection diagram with current directions of a dc machine having (a) shunt winding, (b) series winding, and (c) compound (long-shunt) winding.

The terminal characteristic of a self-excited generator is similar to that of the separately excited generator except that its terminal voltage falls off faster as the load current I_L increases. Notice that

$$I_a = I_L + I_f \tag{8.28}$$

and

$$I_f = \frac{V_f}{R_f} \tag{8.29}$$

The decrease in the terminal voltage, due to increased $I_a R_a$ and armature reaction voltage drops, also causes the field current I_f to decrease and the terminal voltage to drop further.

The field current can be adjusted by using the shunt-field rheostat, which is connected in series with the shunt-field winding. Under no-load conditions, the armature current is equal to the field current.

The operation of the self-excited generator depends on the existence of some residual magnetism in its magnetic circuit. As shown in Figure 8.23, when such a generator is brought up to its constant speed because of the presence of a *residual flux* $\Phi_{d,res}$ in its field poles, there will be a small internal generated voltage $E_{a,res}$ even at zero field current since

$$E_{a,res} = K_a \times \Phi_d \times \omega_m \tag{8.30}$$

Once this voltage appears at the terminals, there will be a small amount of field current flowing in the shunt-field winding since it is connected directly across the brushes. This field current produces an mmf in the field, causing the flux to increase and inducing a higher E_a, which in turn causes more current to flow through the field windings.

This *voltage buildup* is depicted on the mc curve in Figure 8.23. Notice that at no-load field current $I_{f,nl}$, the corresponding no-load terminal voltage is $V_{t,nl}$. Also shown in this figure is the field resistance line which is a plot of $R_f I_f$ versus I_f. *Field resistance* is governed by

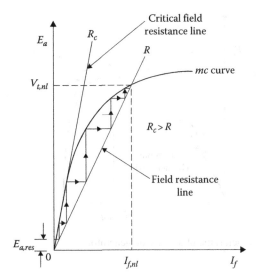

FIGURE 8.23   Voltage built up in a self-excited shunt dc generator.

$$R_f = \frac{V_f}{I_f} \tag{8.31}$$

As long as there is some residual magnetism in the magnetic poles of the generator, the voltage buildup will take place if the following conditions are satisfied: (1) there must be a residual flux in the magnetic circuit and (2) the field winding mmf must act to aid this residual flux where (3) the total field-circuit resistance must be less than the critical field-circuit resistance.

As shown in Figure 8.23, the *critical field resistance* is the value that makes the resistance line coincides with the linear portion of the mc curve. In other words, it represents the resistance value of the shunt-field circuit below which the voltage buildup takes place.

Therefore, the slope of the mc curve in the linear region is the *critical resistance*. Since a shunt generator maintains approximately constant voltage on load, it is widely used as an exciter to provide the field current for a large generator.

8.14   SERIES GENERATOR

The series generator is a self-excited generator that has its field winding connected in series with its armature, as shown in Figure 8.22b. Therefore, its armature current I_a, field current I_{se}, and load current I_L are all equal to each other. Thus,

$$I_a = I_{se} = I_L \tag{8.32}$$

and

$$V_t = E_a - I_a(R_a + R_{se}) \tag{8.33}$$

where R_{se} is the resistance of the series-field winding. Under no-load conditions, the internal generated voltage is due to the residual magnetism. As the load increases, so does the field current ($I_f = I_a$) and the voltage E_a, as shown in Figure 8.24.

The terminal voltage continues to increase until the magnetic circuit of the machine becomes saturated. Series generators are used as *voltage boosters* and as *constant-current generators* in arc welding.

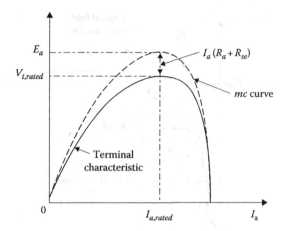

FIGURE 8.24 Terminal characteristic of a series dc generator.

8.15 COMPOUND GENERATOR

As shown in Figure 8.22c, a compound generator has both series and shunt field windings. If the mmf of the series field aids the mmf of the shunt field, it is called a *cumulative compound genera-tor*. On the other hand, if the mmf of the series field opposes the mmf of the shunt field, it is called a *differential compound generator*. If the shunt-field winding is connected across the armature, as shown in Figures 8.12d and e, this type of compound generator is called a *short-shunt genera-tor*. If the shunt-field winding is connected across the series combination *of* armature and series windings, as shown in Figure 8.12c, it is called a *long-shunt generator*. For the short-shunt com-pound generator,

$$V_t = E_a - I_a R_a - I_L R_{se} \tag{8.34}$$

and

$$I_a = I_f + I_L \tag{8.35}$$

where
R_{se} is the resistance of the series-field winding
I_L is the load current

For the long-shunt compound generator,

$$V_t = E_a - I_a(R_a + R_{se}) \tag{8.36}$$

$$I_a = I_f + I_L \tag{8.37}$$

where

$$I_f = \frac{V_t}{R_f} \tag{8.38}$$

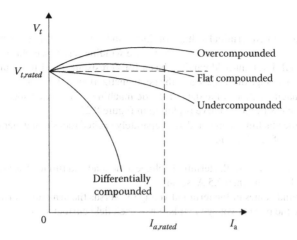

FIGURE 8.25 Terminal characteristics of compound generators operating at constant speed.

or

$$I_f = \frac{V_t}{R_a + R_{se}}$$

As shown in Figure 8.25, *a cumulatively compounded generator* may be *flat-compounded, overcompounded*, or *undercompounded*, depending on the strength of the series field.

If the terminal voltage at the rated load (i.e., full load) is equal to the rated voltage (i.e., no-load voltage), the generator is called a *flat-compounded generator*. If the terminal voltage at the rated load is greater than the no-load voltage, the generator is called an *overcompounded generator*. In the event that the terminal voltage at the rated load is less than the no-load voltage, the generator is called an *undercompounded generator*.

Overcompounding can be used to compensate for line drop when the load served is located far from the generator. It may also be used to counteract the effect of a drop in the prime-mover speed as the load increases.

8.16 VOLTAGE REGULATION

The terminal voltage of a generator normally changes as the load current changes. This voltage variation is described by voltage regulation. The percent voltage regulation (%V Reg) of a generator is defined as

$$\%\text{V Reg} = \frac{V_{t,nl} - V_{t,fl}}{V_{t,fl}} \times 100 \tag{8.39}$$

where
$V_{t,nl}$ is the no-load terminal voltage
$V_{t,fl}$ is the full-load terminal voltage

Also,

$$\%\text{V Reg} = \frac{E_a - V_{t,fl}}{V_{t,fl}} \times 100 \tag{8.40}$$

since E_a is equal to $V_{t,\,nl}$.

Example 8.3

A dc machine has a rated terminal voltage of 250 V and a rated (full-load) armature current of 100 A. Its armature-circuit resistance (i.e., armature winding resistance plus commutating field resistance) is $0.10\,\Omega$. Its shunt-field winding has a $100\,\Omega$ resistance and 1000 turns per pole. The total brush-contact voltage drop is 2 V and the demagnetization of the d-axis by armature mmf is neglected. The magnetization curve data of the dc machine for the rated speed n_{mc} of 1200 rpm is tabulated in Table 8.1 and the curve is plotted in Figure 8.26.

Assume that the machine is operated as a separately excited (dc) shunt generator and is driven at a constant speed of 1200 rev/min.

(a) Determine the values of the terminal voltage at no load and full load as the field current is set at 1.0, 1.5, and then 2.5 A, separately.
(b) Plot the found values of the terminal voltage V_t versus the armature current I_a.
(c) Determine the percent voltage regulation at each field current setting.

Solution

(a) When the field current is set at 1.0 A at no load, the terminal voltage $V_{t(0\,A)}$ is equal to the internal generated voltage $E_{a(mc)}$ found from the magnetization curve given in Table 8.1. Thus,

$$V_{t(0\,A)} = E_{a(mc)} = E_a = 140 \text{ V}$$

At full load, the internal voltage drop due to the armature-circuit resistance is

$$I_a\left(\sum R_a\right) = (100 \text{ A})(0.1\ \Omega) = 10 \text{ V}$$

Hence, at full load, the terminal voltage of the generator is

$$V_{t(100\,A)} = E_a - I_a\left(\sum R_a\right) - 2.0$$

$$= 140 - (100 \text{ A})(0.10\ \Omega) - 2.0$$

$$= 128 \text{ V}$$

Similarly, when the field current is set at 1.5 and then at 2.5 A, the corresponding full-load terminal voltages can be found in the same way. The results are presented in Table 8.2. Figure 8.27 shows the plot of terminal voltage at no load versus the field current I_f.

(b) The plot of the terminal voltage V_t versus the armature current I_a is shown in Figure 8.28.

TABLE 8.1

The Magnetization Curve Data for $n_{me} = 1200$ rpm

$E_{a(mc)}$ (V)	70	140	195	235	260	276
I_f (A)	0.5	1.0	1.5	2.0	2.5	3.0

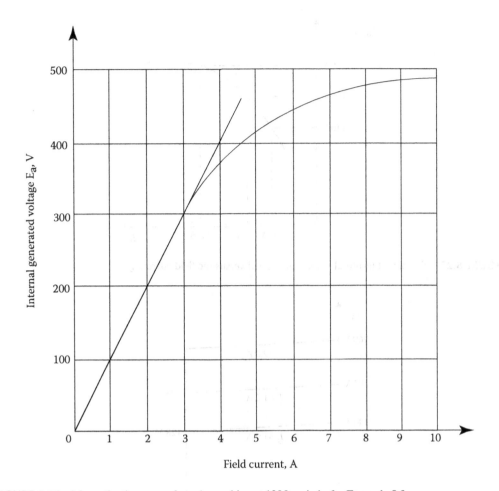

FIGURE 8.26   Magnetization curve for a dc machine at 1200 rev/min for Example 8.3.

TABLE 8.2
For the Results of Example 8.3

I_f	At $I_a =$ 0 $V_{t(0\,A)} = E_{a(mc)}$ = 0	At $I_a = 100$ A $I_a\left(\sum R_a\right)$	$V_{t(100\,A)}$
1.0 A	140 V	10 V	128 V
1.5	195	10	183
2.5	260	10	248

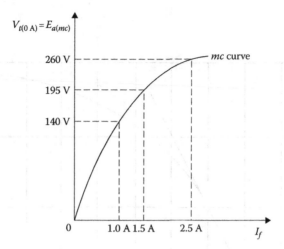

FIGURE 8.27   The plot of terminal voltage at no load versus the field current If.

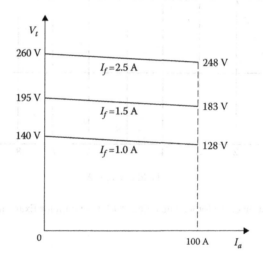

FIGURE 8.28   The plot of terminal voltage versus the armature current.

(c) At $I_f = 1.0$ A:

$$\%\text{V Reg} = \frac{V_{t,\,nl} - V_{t,\,fl}}{V_{t,\,fl}} \times 100$$

$$= \frac{140 - 128}{128} \times 100$$

$$= 9.4$$

At $I_f = 1.5$ A:

$$\%\text{V Reg} = \frac{195 - 183}{183} \times 100$$

$$= 6.6$$

At $I_f = 2.5$ A:

$$\%V \text{ Reg} = \frac{260 - 248}{248} \times 100$$

$$= 4.8$$

8.17 DEVELOPED POWER

As previously stated, a dc machine is a versatile machine, which can be used as a generator or a motor. In the generator mode, the input is the mechanical power provided by a prime mover (a diesel engine, a gas turbine, or an electrical motor) and the output is the electrical power. Conversely, in the motor mode, the input is the electrical power and the output is the mechanical power.

However, as illustrated in Figure 8.29, in both modes of operation, a dc excitation current must be provided to establish the magnetic field. The developed power of a separately excited dc machine is

$$P_d = E_a I_a \tag{8.41}$$

The power at the armature terminals is

$$P_{ta} = V_{ta} \times I_a \tag{8.42}$$

where V_{ta} is the voltage at the armature terminals. Thus, from Equation 8.8,

$$V_{ta} = E_a \pm I_a \left(\sum R_a \right) \pm 2.0 \tag{8.43}$$

where the total armature winding circuit resistance is given by Equation 8.9 as

$$\sum R_a = R_a + R_{se} + R_{cp} + R_{cw} + R_{a,\,ext} \tag{8.9}$$

Observe that in Equation 8.43, the minus sign is used for a generator and the plus sign for a motor. By substituting Equation 8.43 into Equation 8.42, the power at the armature terminals can be expressed as

$$P_{ta} = \left[E_a \mp I_a \left(\sum R_a \right) \mp 2.0 \right] I_a \tag{8.44}$$

$$P_{ta} = E_a I_a \mp I_a^2 \left(\sum R_a \right) \mp 2.0 I_a \tag{8.45}$$

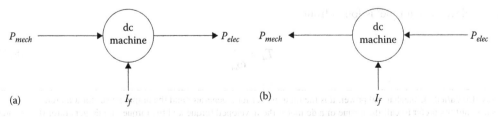

FIGURE 8.29 Block diagram of a dc machine operating in (a) generator mode and (b) motor mode.

or

$$P_{ta} = P_d \mp P_{cu} \mp P_{brush} \qquad (8.46)$$

where

P_d is the developed power
P_{cu} is the armature-circuit copper losses
P_{brush} is the brush-contact loss

Therefore,

$$P_{cu} = I_a^2 \left(\sum R_a \right) \qquad (8.47)$$

and

$$P_{brush} = 2.0 I_a \qquad (8.48)$$

The developed power P_d is greater than the armature terminal power P_{ta} for a generator, but it is less than P_{ta} for a motor.

The *shaft power** can be expressed as

$$P_{shaft} = P_d \pm P_{rot} \qquad (8.49)$$

where P_{rot} represents the rotational losses and is the sum of the friction and windage losses P_{FW} and the core losses P_{core}. Thus,

$$P_{rot} = P_{FW} + P_{core} \qquad (8.50)$$

In Equation 8.49, the plus sign is used for a generator and the minus sign for a motor because the shaft power P_{shaft} is greater than the developed power P_d for a generator, but it is less than P_d for a motor.

In addition to the rotational losses, there may be a *stray-load loss* P_{stray} for those losses that cannot be easily accounted for. It is usually ignored in small machines, but in large machines above 100 hp it is generally assumed to be about 1% of the output power.

8.18 DEVELOPED TORQUE

Assume that a dc machine has an armature voltage of E_a and armature current of I_a. Its developed power can be expressed as

$$P_d = E_a I_a \qquad (8.51)$$

and its developed torque[†] is found from

$$T_d = \frac{P_d}{\omega_m} \qquad (8.52)$$

---

* It is also called the mechanical power. It is the input power for a generator and the output power for a motor.
† Some authors prefer to call the torque of a dc motor the developed torque and the torque of a dc generator the counter-torque or the induced torque since it opposes the torque applied to the shaft by the prime mover.

or

$$T_d = \frac{E_a I_a}{\omega_m} \tag{8.53}$$

Using Equation 8.2,

$$T_d = \frac{(K_a \times \Phi_d \times \omega_m) I_a}{\omega_m} \tag{8.54}$$

$$T_d = K_a \times \Phi_d \times I_a \tag{8.55}$$

where

$$K_a = \frac{Z \times p}{2\pi \times a} \tag{8.3}$$

and is defined as the *winding constant* since it is fixed by the design of the winding. Hence, the developed torque of a dc machine is a function of the flux in the machine, the armature current in the rotor, and a constant that depends on the machine.

In the above equations, the torque is in N·m. If it is requested in English units, it must be multiplied by 0.7373. Note that the developed torque can also be determined from

$$T_d = \frac{P_d}{\omega_m} = \frac{E_a \times I_a}{\omega_m}$$

Since

$$\omega_m = \frac{2\pi n_m}{60}$$

then

$$T_d = \frac{E_a \times I_a}{2\pi n_m / 60} \tag{8.56}$$

where the speed n_m is in rpm.

8.19 POWER FLOW AND EFFICIENCY

Consider the equivalent-circuit diagram of the self-excited compound machine shown in Figure 8.22c. Figure 8.30a shows the power flow of this type of dc generator. The input power is the mechanical power or the shaft power.

Depending on the machine size, the rotational losses are between 3% and 15%, the stray-load loss is about 1% for machines larger than 100 hp (otherwise, it is usually ignored), the armature-circuit copper loss is between 3% and 6%, the shunt-field loss is between 1% and 5%, and the brush-contact loss is about $2I_a$ as explained before.

If the shunt field is separately excited, its losses are not supplied by the prime mover through the shaft and therefore, must be handled separately. Figure 8.30b shows the power flow of a self-excited

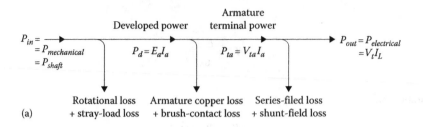

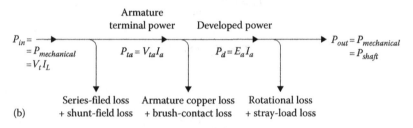

FIGURE 8.30 Power flow in a dc machine: (a) generator and (b) motor.

compound motor. Notice that here, the input is in electrical power and the output is in mechanical power. The efficiency* of a dc machine can be determined from

$$\text{Efficiency} = \frac{P_{out}}{P_{in}} \tag{8.57}$$

but since

$$P_{out} = P_{in} - \sum P_{loss} \tag{8.58}$$

then

$$\text{Efficiency} = 1 - \frac{\sum P_{loss}}{P_{in}} \tag{8.59}$$

Thus, the percent efficiency can be expressed as

$$\% \, \eta = \left(1 - \frac{\sum P_{loss}}{P_{in}}\right) \times 100 \tag{8.60}$$

The maximum efficiency at a given constant speed is obtained when the sum of the rotational loss and the shunt-field copper loss is equal to the armature copper loss. The percent efficiency can also be found from

$$\% \, \eta = \frac{P_{out}}{P_{out} + \sum P_{loss}} \times 100 \tag{8.61}$$

---

* The efficiency of a dc machine can be determined more accurately by a test using the Kapp–Hopkinson method. In such a test, two similar machines are mechanically coupled and electrically connected back to back. For further information, see Daniels (1968).

Example 8.4

Assume that a separately excited shunt motor operating at 1000 rev/min has a load current of 100 A and a terminal voltage of 240 V. If the armature winding resistance is 0.1 Ω, determine the following:

(a) The developed torque
(b) The shaft speed and the load current if the torque is doubled at the same excitation

Solution

(a) From Equation 8.8,

$$E_a = V_t - I_a \left(\sum R_a \right) - 2.0$$

$$= 240 - (100)(0.1) - 2.0$$

$$= 228 \text{ V}$$

at a speed of

$$\omega_m = (1000 \text{ rev/min}) \left(\frac{2\pi}{60} \right)$$

$$= 104.72 \text{ rad/s}$$

From Equation 8.2,

$$E_a = K_a \times \Phi_d \times \omega_m$$

or

$$K_a \times \Phi_d = \frac{E_a}{\omega_m}$$

$$= \frac{228 \text{ V}}{104.72 \text{ rad/s}}$$

$$= 2.1772$$

Thus, the developed torque is

$$T_d = K_a \times \Phi_d \times I_a$$

$$= (2.1772)(100)$$

$$= 217.72 \text{ N} \cdot \text{m}$$

(b) When $T_d = 2(217.72) = 435.44 \text{ N·m}$

$$I_a = \frac{T_d}{K_a \times \Phi_d}$$

$$= \frac{435.44}{2.1772}$$

$$= 200 \text{ A}$$

Therefore, the corresponding speed is

$$\omega_m = \frac{E_a}{K_a \times \Phi_d}$$

$$= \frac{V_t - I_a R_a}{K_a \times \Phi_d}$$

$$= \frac{240 - (200)(0.1)}{2.1772}$$

$$= 101.05 \text{ rad/s}$$

or

$$n = (101.05 \text{ rad/s})\left(\frac{60}{2\pi}\right)$$

$$= 964.96 \text{ rev/min}$$

Example 8.5

Assume that a 25 hp, 250 V, self-excited shunt motor is supplied by a full-load line current of 83 A. The armature and field resistances are 0.1 and 108 Ω, respectively. If the total brush-contact voltage drop is 2 V and the friction and core losses are 650 W, determine the following:

(a) The shunt-field winding loss
(b) The armature winding loss
(c) The total loss of the motor
(d) The percent efficiency of the motor

Solution

The input power is

$$P_{in} = V_t I_L$$

$$= (250 \text{ V})(83 \text{ A})$$

$$= 20,750 \text{ W}$$

(a) The shunt-field winding loss is

$$P_f = I_f^2 R_f$$

$$= V_f I_f$$

$$= \frac{V_f^2}{R_f}$$

$$= 579 \text{ W}$$

Thus, the field current is

$$I_f = \frac{P_f}{V_f}$$

$$= \frac{579 \text{ W}}{250 \text{ V}}$$

$$= 2.316 \text{ A}$$

(b) Since the full-load armature current is

$$I_a = I_L - I_f$$

$$= 83 - 2.316$$

$$= 80.684 \text{ A}$$

the armature winding loss is

$$P_a = I_a R_a$$

$$= (80.684 \text{ A})^2 (0.10 \text{ }\Omega)$$

$$= 651 \text{ W}$$

and the brush-contact loss is

$$P_{brush} = 2I_a$$

$$= 2(80.684 \text{ A})$$

$$= 161.4 \text{ W}$$

(c) Since the rotational loss is given as 650 W, the total power loss is

$$\sum P_{loss} = P_f + P_a + P_{brush} + P_{rot}$$

$$= 579 + 651 + 161.4 + 650$$

$$= 2,041.4 \text{ W}$$

(d) The percent efficiency of the motor is

$$\% \eta = \left(1 - \frac{\sum P_{loss}}{P_{in}}\right) \times 100$$

$$= \left(1 - \frac{2,041.4 \text{ W}}{20,750 \text{ W}}\right) \times 100$$

$$= 90.16\%$$

Example 8.6

Redo the Example 8.5 by using MATLAB®.

(a) Write the MATLAB program script.
(b) Give the MATLAB program output.

Solution

(a) Here is the MATLAB program script:

```
clear
clc
%System Parameters
```

```
Vt = 250;
IL = 83;
Ra = 0.1;
Rf = 108;
Prot = 650;
%Solution for part (a)
Pin = Vt*IL
%Solution for part (b)
Vf = Vt;
Pf = Vf^2/Rf
If = Pf/Vf
Ia = IL-If
Pa = Ia^2*Ra
Pbrush = 2*Ia
%Solution for part (c)
Ploss = Pf+Pa+Pbrush+Prot
%Solution for part (d)
percent _ n = (1-Ploss/Pin)*100
```

(b)   Here is the MATLAB program output:
```
Pin =
    20750
Pf =
    578.7037
If =
    2.3148
Ia =
    80.6852
Pa =
    651.0099
Pbrush =
    161.3704
Ploss =
    2.0411e+003
percent _ n =
    90.1635
>>
```

8.20   DC MOTOR CHARACTERISTICS

As previously stated, a dc machine can be used both as a generator and as a motor. In fact, in certain applications, dc machines operate alternately as a motor and as a generator.

However, in general there are some design differences, depending on whether the dc machine is intended for operation as a motor or as a generator. Unlike dc generators, dc motors are still very much in use in many industrial applications because of their attractive performance characteristics.

8.20.1   Speed Regulation

The dc motors excel in speed control applications, where they are compared by their speed regulations. The *speed regulation (Speed Reg)* of any motor is determined from

$$\text{Speed Reg} = \frac{\omega_{nl} - \omega_{fl}}{\omega_{fl}} \tag{8.62}$$

Thus, the percent speed regulation is

$$\%\text{Speed Reg} = \left(\frac{\omega_{nl} - \omega_{fl}}{\omega_{fl}} \right) \times 100 \qquad (8.63)$$

where
  ω_{nl} is the no-load angular speed in rad/s
  ω_{fl} is the full-load angular speed in rad/s

Alternatively, if the speeds are given in rpm then the percent speed regulation of a motor is found from

$$\%\text{Speed Reg} = \left(\frac{n_{nl} - n_{fl}}{n_{fl}} \right) \times 100 \qquad (8.64)$$

The magnitude of the speed regulation indicates the steepness of the slope of the torque–speed characteristic.

8.20.2 SPEED–CURRENT CHARACTERISTIC

In general, motors are designed to provide a rated horsepower at a rated speed. Since the internal generated voltage is a function of the angular velocity, then

$$E_a = K_a \times \Phi_d \times \omega_m \qquad (8.65)$$

from which the angular velocity ω_m can be expressed as

$$\omega_m = \frac{E_a}{K_a \times \Phi_d} \qquad (8.66)$$

By substituting Equation 8.65 into Equation 8.66, the angular velocity or shaft speed can be found as

$$\omega_m = \frac{V_t - I_a R_a}{K_a \times \Phi_d} \qquad (8.67)$$

This equation is called the *motor speed equation*. Notice that the speed of a dc motor depends on the applied terminal voltage V_t, the armature current I_a, the resistance R_a, and the field flux per pole Φ_d. The K_a is a design constant and cannot be changed to control the speed. Equation 8.67 can be expressed as

$$\omega_m = \frac{V_t}{K_a \times \Phi_d} - \frac{I_a R_a}{K_a \times \Phi_d} \qquad (8.68)$$

Since at no load the second term becomes zero,* the speed is

---

* In reality, however, the armature current I_a at no load is not zero, but is about 5% of the full-load current due to rotational losses.

$$\omega_{nl} = \frac{V_t}{K_a \times \Phi_d} \tag{8.69}$$

For a given armature current, the speed of a dc motor will decrease as the armature-circuit resistance increases. As can be observed in Equation 8.67, such a speed adjustment can be made by weakening the field flux by inserting resistance in the field circuit using a field rheostat. Such speed control is a smooth and efficient means of changing the motor speed from basic speed to maximum speed.

However, *if the field circuit is opened accidentally, the field flux will suddenly decrease to its relatively small residual value. If the armature circuit is not opened immediately, the motor speed will increase to dangerously high values and the motor will destroy itself in a few seconds* either by the windings being forced from the slots or the commutator segments being thrown out by centrifugal force.

Since the sudden decrease in the field flux reduces the counter voltage to a very small amount, the armature current of the motor will increase to a very high value. This will take place before the motor starts to rotate at a high speed. A properly sized circuit breaker inserted into the armature circuit can prevent such a disaster.

Succinctly put, the *field circuit of a shunt motor must never be opened if the motor is running. Otherwise, the motor will "run away" and will destroy itself in a few seconds!*

Similarly, as the load is removed from a series motor, its field flux will decrease. If all the mechanical load is removed from its shaft, the field flux will decrease to almost zero, and the motor speed will increase to a dangerously high level until it destroys itself.

Thus, a series motor must never be run without a load! Nor should it be connected to a mechanical load by a belt since it can break or slip. Figure 8.31 shows the speed–current characteristics of shunt, series, and cumulative compound motors.

8.20.3　Speed–Torque Characteristic

Figure 8.32 shows the speed–torque characteristics of shunt, series, and cumulative compound motors. Notice that the speed of a shunt motor changes very little (in fact, less than 5% in large motors or less than 8% in small motors). Because of this, shunt motors are classified as *constant-speed motors*.

The speed of a series motor changes drastically as the load or the load torque changes. However, the cumulative compound motor combines the operating characteristics of the shunt and series motors.

Unlike a series motor, a compound motor has a definite no-load speed and can be safely operated at no load. As the load or load torque is increased, the growth in the field flux decreases the speed more in a series motor than in a shunt motor.

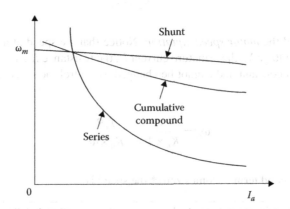

FIGURE 8.31 Speed–current characteristics.

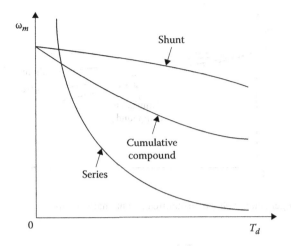

FIGURE 8.32   Speed–torque characteristics.

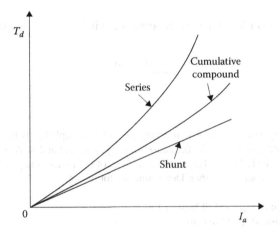

FIGURE 8.33   Torque–load current characteristics.

8.20.4 TORQUE–CURRENT CHARACTERISTIC

Figure 8.33 shows the torque–load current characteristics of shunt, series, and cumulative compound motors. Notice that except for lighter loads, the series motor has a much greater torque than the shunt or the cumulative compound motor for a given armature current.

Since a series motor has a much greater starting torque, it is exceptionally well suited for staring heavy loads at a reduced speed.

8.20.5 INTERNAL GENERATED VOLTAGE–CURRENT CHARACTERISTIC

Figure 8.34 shows the internal generated voltage–armature current characteristics of shunt, series, and cumulative compound motors. In a motor, the internal generated voltage is called the *counter-voltage* (i.e., the *counter emf* or *back emf*).

Notice that the generated voltage decreases as the load increases because of the increased voltage drop $I_a \sum R_a$ in the armature circuit. In series and compound motors, the armature-circuit resistance $\sum R_a$ is the sum of the armature winding resistance R_a and the series-field winding resistance R_{se}. Thus,

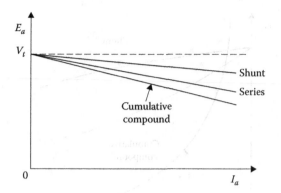

FIGURE 8.34 Internal generated voltage–armature current characteristics.

$$\sum R_a = R_a + R_{se} \tag{8.70}$$

However, in the shunt motor it is equal to the armature winding resistance R_a. That is,

$$\sum R_a = R_a \tag{8.71}$$

Example 8.7

Assume that the separately excited dc machine given in Example 8.3 is being used as a shunt motor to drive a mechanical load. Its field current is kept constant at 2 A. As before, the full-load current of the machine is 100 A. The terminal voltage of the motor is kept variable at 150, 200, and 250 V by using a *control rectifier*. Determine the following:

(a) The developed torque at full load of 100 A
(b) The ideal no-load speed in rpm
(c) The full-load speed in rpm
(d) A sketch of the torque–current characteristic based on the results found in Part (a)
(e) A sketch of the speed–current characteristics based on the results found in Parts (b) and (c)

Solution

(a) Since the magnetization curve is drawn at the rated and constant speed of $n_{mc} = 1200$ rpm at $I_f = 2$ A, the corresponding voltage from the magnetization curve given in Table 8.1 is found as $E_{a(mc)} = 235$ V. However, at a different speed n_m, the corresponding voltage would be E_a. Therefore, from Equation 8.14,

$$\frac{E_a}{E_{a(mc)}} = \frac{\omega_m}{\omega_{mc}}$$

or

$$\frac{E_a}{\omega_m} = \frac{E_{a(mc)}}{\omega_{mc}}$$

The right side of this equation represents *constant excitation on the d-axis*. Since

$$E_a = K_a \times \Phi_d \times \omega_m$$

from which

$$K_a \times \Phi_d = \frac{E_a}{\omega_m}$$

Therefore, the developed torque is

$$T_d = K_a \times \Phi_d \times I_a$$

$$= \left(\frac{E_a}{\omega_m}\right) I_a$$

$$= \left(\frac{E_{a(mc)}}{\omega_{mc}}\right) I_a$$

Hence, at full load

$$T_d = \frac{(235\ \text{V})(100\ \text{A})}{(1200\ \text{rpm})\left(\frac{2\pi}{60}\right)}$$

$$= 187\ \text{N} \cdot \text{m}$$

(b) Since

$$\frac{E_a}{E_{a(mc)}} = \frac{\omega_m}{\omega_{mc}}$$

or

$$\frac{E_a}{E_{a(mc)}} = \frac{n_m}{n_{mc}}$$

then

$$n_m = \left(\frac{E_a}{E_{a(mc)}}\right) n_{mc}$$

where $n_{mc} = 1200$ rpm and $E_{a(mc)} = 235$ V from the me curve as long as $I_f = 2$ A is kept constant. Since at no load the terminal voltage V_t and the internal generated voltage E_a are the same at $E_a = V_t = 150$ V, the ideal no-load speed is

$$n_{nl} = \left(\frac{150\ \text{V}}{235\ \text{V}}\right)(1200\ \text{rpm}) = 766\ \text{rpm}$$

The other ideal no-load speeds that correspond to the terminal voltages of 200 and 250 V are given in Table 8.3.

TABLE 8.3

Other Full-Load Voltages for V_t = 200 and 250 V

V_t	At I_a = 100 A $\sum R_a = 0.10\,\Omega$ $\sum I_a R_a$	At I_a = 0 A $E_{a,\,nl}$	At I_a = 100 A $E_{a,\,fl}$	At I_f = 2.0 A $E_{a,\,(mc)}$	At I_a = 0 A n_{nl}	At I_a = 100 A n_{fl}
150 V	10 V	150 V	138 V	235 V	766 rpm	705 rpm
200	10	200	188	235	1021 rpm	960
250	10	250	238	235	1277	1215

(c) Similarly, the speed at full load is

$$n_{fl} = \left(\frac{E_{a,\,fl}}{E_{a(mc)}} \right) n_{mc}$$

where the internal generated voltage is

$$E_{a,\,fl} = V_t - \sum I_a R_a - 2.0$$

If the applied terminal voltage V_t is 150 V, then the internal generated voltage is

$$E_{a,\,fl} = (150\ \text{V}) - (100\ \text{A})(0.10\ \Omega) - 2.0$$

$$= 138\ \text{V}$$

Therefore, the corresponding speed at full load is

$$n_{fl} = \left(\frac{138\ \text{V}}{235\ \text{V}} \right)(1200\ \text{rpm})$$

$$= 705\ \text{rpm}$$

The other full-load speeds that correspond to the terminal voltages of 200 and 250 V are given in Table 8.3.

(d) Figure 8.35 shows the torque–current characteristic based on the results found in Part (a). Note that if the excitation current is increased, the slope of the characteristic increases and the motor provides more torque. Similarly, if the excitation current is decreased, the slope of the characteristic decreases and the motor provides less torque.

(e) Figure 8.36 shows the speed–current characteristics based on the results found in Parts (b) and (c).

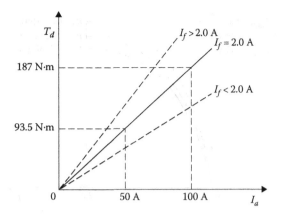

FIGURE 8.35   Torque–current characteristics.

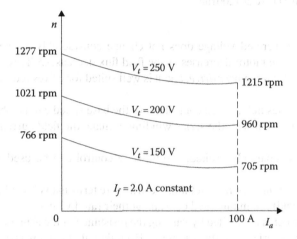

FIGURE 8.36   Speed–current characteristics.

For the sake of comparison, Figure 8.37 shows the magnetization curves corresponding to the no-load speed of 1277 rpm, the full-load speed of 1215 rpm, and also the mc curve drawn at 1200 rpm.

8.21   CONTROL OF DC MOTORS

The speed of a dc motor can be controlled with relative ease over a wide range above and below the base (rated) speed.* Speed control methods for dc motors are simpler and less expensive than those for ac motors. As can be observed from Equation 8.67, the speed of a dc motor can be changed by using the following methods:

(a) Field control method
(b) Armature resistance control method
(c) Armature voltage control method

The *field control method* is the simplest, cheapest, and is most applicable to shunt motors. In this method, the armature-circuit resistance R_a and the terminal voltage V_t are kept constant, and the speed is controlled by changing the field current I_f. As the value of the rheostat resistance in the shunt-field circuit is increased, the flux Φ_d decreases and the speed increases.

---

* The base speed is the speed obtained with rated armature voltage, normal field flux, and normal armature resistance.

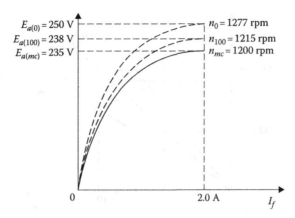

FIGURE 8.37 Magnetization curves corresponding to the no-load speed of 1277 rpm, the full-load speed of 1215 rpm, and also the mc curve at 1200 rpm.

Thus, the internal generated voltage does not change considerably as the speed is increased. However, the torque of the motor decreases as the field flux decreases.* This speed control method is also called a *constant-horsepower drive*, and it is well suited for drives requiring increased torque at low speeds.

If the motor has a series field, speed control above the base speed can be obtained by inserting a diverter-resistance in parallel with the series winding to make the field current less than the armature current.

If the shunt field is separately excited, a solid-state control can be used without a significant change in motor losses.

In the *armature resistance control* method, the armature terminal voltage V_t and the field current I_f (and therefore, the field) are maintained constant at their rated values.

The speed of the motor is controlled by varying the resistance of the armature circuit by inserting an external resistance in series with the armature.[†] Even though it can also be applied to compound and series motors, it is more easily applied to shunt motors.[‡]

The armature resistance control method is simple to perform and requires a small initial investment, but has the disadvantage of considerable power loss and low overall efficiency with the full armature current passing through the external resistance connected in series.

Today, this speed control method is still used in various transit system vehicles. The same armature rheostat control can be used for both starting and speed control. The speed range begins at zero speed.

In the *armature voltage control method*, the armature-circuit resistance R_a and the field current if are kept constant,[§] and the speed is controlled by varying the armature terminal voltage V_t. This is the most flexible method of speed control and avoids the disadvantages of poor speed regulation and low efficiency that are characteristic of the armature resistance method. It can be applied to shunt, series, and compound motors.

The speed is easily controlled from zero to a maximum safe speed in either forward or backward directions. The controlled-voltage source may be a dc machine or a solid state–controlled rectifier. If a dc machine is used, the speed control system is called a *Ward Leonard system*.

---

* The process of decreasing the field current is also known as the field weakening. By inserting external resistance in series with the motor field, the speed of a motor can only be increased from a minimum speed, that is, the base speed.

[†] By inserting external resistance in series with the armature, the speed of a motor can only be decreased from a maximum speed, that is, the base speed.

[‡] In shunt and compound motors, the external resistor must be connected between the shunt-field winding and the armature, not between the line and the motor.

[§] The field current is usually kept constant at its rated value.

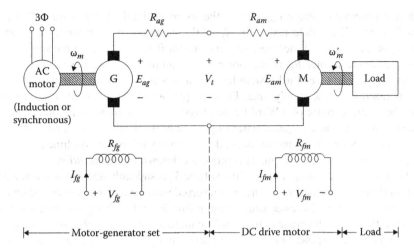

FIGURE 8.38   Ward Leonard system for dc motor speed control.

Figure 8.38 shows a Ward Leonard system for dc motor speed control. In such a system, a three-phase induction motor or a three-phase synchronous motor drives a separately excited dc generator, with its armature connected directly to the armature of the separately excited dc motor that drives a mechanical load. Note that the motor of the motor-generator set operates at a constant speed. The armature voltage of the dc drive motor can be controlled by changing the field current of the dc generator.

Thus, such control of the armature voltage allows for smooth control of the motor's speed from a very small value up to the base speed.* The motor speed for speeds above the base speed can be controlled by reducing the field current I_{fm} in the motor.

Figure 8.39 shows the torque and power limits as a function of speed for a shunt motor that has a combined armature voltage and field resistance control. Notice that the range below base speed is a constant-torque drive since the flux and permissible armature current are almost constant. The range above base speed represents a constant-horsepower drive. In the event that the field current I_{fg} of the generator is reversed, the polarity of the generator armature voltage is also reversed and so is the direction of rotation.

A Ward Leonard system is capable of providing a wide variation of speed in both forward and reverse directions. Increasing the field resistance R_{fg} of the generator decreases its field current I_{fg}

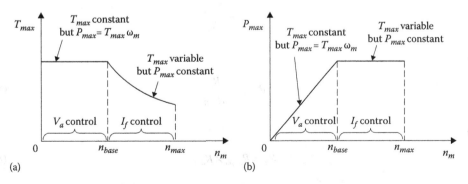

FIGURE 8.39   (a) Torque and (b) power limits as a function of speed for a shunt motor having combined armature voltage and field resistance control.

---

* Here, the base speed is the rated speed of the motor. The rated speed is the speed that can be obtained when the motor is operating at its rated terminal voltage, power, and field current.

and its internal generated voltage E_{ag}. Hence, the speed of the dc drive motor will decrease. The opposite will be true if the field resistance R_{fg} of the generator is decreased. Increasing the field resistance R_{fm} of the motor will decrease its field current *if M* and increase the speed of the motor. Decreasing the field resistance R_{fm} of the motor will result in a decrease in speed.

Furthermore, the Ward Leonard system has the ability to "regenerate," that is, to return stored energy in the machine to the supply lines. For example, when a heavy load is first lifted and then lowered by the dc drive motor of a Ward Leonard system, the dc motor starts to act as a generator as the load is coming down, using its countertorque as a brake. Under these conditions, the "generator" itself starts to operate in the motor mode, driving the synchronous machine as a generator and supplying power back to the ac system. This process is known as *regeneration*.

While the dc drive motor is operating, if the voltage V_t is suddenly decreased to a value below the counter emf of the motor, the armature current is reversed with the motor acting as a generator, driving the dc generator as a motor. This establishes dynamic *braking*, which brings the motor to a quick stop.

In summary, the main advantage of the Ward Leonard system is that the speed is adjustable over a wide range without large power losses which results in high efficiency at all speeds. This system is satisfactory for a maximum-to-minimum speed range of 40–1, but must be modified for greater speed ranges.

The disadvantage is that a special motor-generator set is needed for each dc drive motor. Instead of one machine, three machines of basically equal ratios must be purchased and therefore a greater initial investment is required. If there are long periods when the motor is operating under light load, the losses in the motor-generator set are high. It is relatively inefficient since several energy transformations are involved.

In recent years, the application of silicon-controlled rectifiers (SCR) has resulted in solid-state dc motor drives with precise speed control. However, they will not be presented here.

8.22 DC MOTOR STARTING

Only small dc motors of 1 hp or less can be connected directly to a line of rated voltage safely, but they must have very small moments of inertia. Any dc motor larger than 1 hp requires a starting device to protect the armature from excessive current during the starting operation.

Typically, the armature resistance of a dc motor is about 0.05 per unit. If such a motor is connected directly to a line of rated voltage $V_{a,B}$ the armature current on starting is

$$I_{a,start} = \frac{V_a}{R_a} \tag{8.72}$$

$$= \frac{V_{a,B}}{R_a}$$

Since

$$R_{a,pu} = \frac{R_a}{R_{a,B}} \tag{8.73}$$

then

$$R_a = R_{a,pu} \times R_{a,B} \tag{8.74}$$

Therefore,

$$I_{a,start} = \frac{V_{a,B}}{R_{a,pu} \times R_{a,B}} \tag{8.75}$$

or

$$I_{a,\,start} = \frac{I_{a,\,B}}{R_{a,\,pu}} \tag{8.76}$$

where

$I_{a,\,B}$ is the base or rated armature current
$R_{a,\,pu}$ is the armature resistance per unit

Since the typical $R_{a,\,pu}$ is about 0.05 per unit, then

$$I_{a,\,start} = \frac{I_{a,\,B}}{0.05} = 20 \times I_{a,\,B} \tag{8.77}$$

In other words, *the armature current of a dc motor on starting is about 20 times its base* or *rated armature current*! This is due to the fact that the motor is at a standstill on starting and the counter emf is zero. Therefore, the armature starting is limited only by the resistance of the armature circuit.

All except very small dc motors are started with variable external resistance in series with their armatures to limit the starting current to the value (about 1.5–2 times rated value) that the motor can commutate without any damage. Such starting resistance is taken out of the circuit either manually or automatically as the motor comes to speed.

To develop maximum starting torque, the shunt and compound motors are normally started with full field excitation (i.e., full line voltages are applied across the field circuits with their field rheostat resistances set at zero). Of course, the series motor is always started under load. Figure 8.40a illustrates how to start a shunt motor with a dc motor starter. Figure 8.40b illustrates how to start a dc motor with starting resistors and accelerating contactors.

Note that the starting resistor in each case consists of a series of pieces, each of which is cut out from the circuit* in succession as the speed of the motor increases. This limits the armature current of the motor to a safe value, and for rapid acceleration it does not allow the current to decrease to a value that is too low, as shown in Figure 8.41.

As the armature resistance is reduced at each step the motor accelerates, but requires less time to reach its asymptotic speed (zero acceleration) and reduces its current to approximately the rated load. The number of acceleration steps is almost a function of the horsepower capacity of the motor. Accordingly, larger motors with more inertia require more steps and a longer time interval to attain a given asymptotic speed.

Figure 8.42 shows a simplified diagram of an automatic starter[†] using "counter-emf" relays for a dc motor. It is based on the principle that as the motor speeds up, its countervoltage E_a starts to increase from an initial value of zero and causes a reduction in the armature current. The segments of the starting resistor are then cut out in steps as the counter emf E_a increases. When the E_a has increased to an adequately high value the starting resistor is short-circuited entirely, and the motor is then connected directly to the line.

---

* If the motor starting is achieved by using a manual dc motor starter, the handle is moved to position 1 at start up so that all the resistances, R_1, R_2, R_3, R_4, and R_5, are in series with the armature to reduce the starting current. As the motor speed increases, the handle is moved to positions 2, 3, 4, 5, and finally the "run" position. At the "run" position, all the resistances in the starter are cut out of the armature circuit. If the motor starting is achieved by using starting resistors and accelerating contactors, the individual segments of the resistor are cut out of the circuit by closing the 1A, 2A, and 3A contactors.

† Such starters are also called controllers. An electric controller is defined as a device, or group of devices, which governs in some predetermined fashion the electric power delivered to the apparatus to which it is connected. Therefore, an electric starter is defined as a controller whose main task is to start and accelerate a motor.

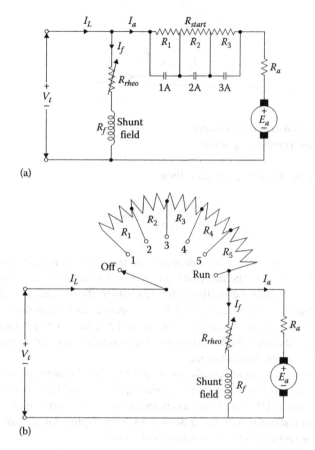

(a)

(b)

FIGURE 8.40 Starting a shunt motor with (a) a manual dc motor starter and (b) starting resistors and accelerating contactors.

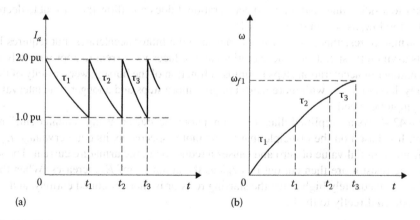

(a)   (b)

FIGURE 8.41 (a) Armature current versus time and (b) speed versus time during the starting of a dc motor.

Figure 8.43 shows some typical symbols used in automatic starter circuits. An automatic starter such as the one shown in Figure 8.42 is operated simply by pushing a button. Note that the field circuit has a relay labeled FL known as the *field loss relay*. In the event that the field current is lost, this field loss relay is de-energized and cuts off the power to the M relay by deactivating the normally open FL contacts even after the start button is pushed. It also causes the normally open M contacts in the armature circuit to open as well. The three relays that are located across the armature are

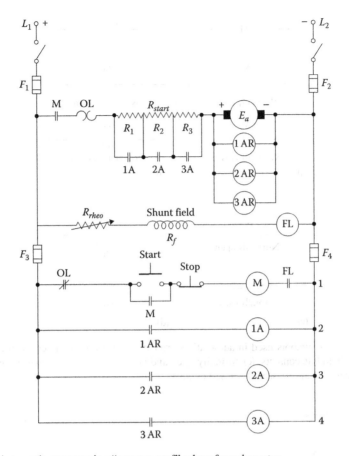

FIGURE 8.42 Automatic starter using "counter-emf" relays for a dc motor.

called the *accelerating relays* (AR). They are fast-acting relays with their contacts located in control lines 2, 3, and 4, respectively, as shown in Figure 8.43.

The control relays are labeled as 1A, 2A, and 3A and are located on control lines 2, 3, and 4. They short out the segments of the starting resistance as the counter emf of the motor increases. If the overload device OL that is located in the armature circuit heats up excessively due to excessive power demands on the motor, its normally closed contact located on the control line 1 will open and de-energize the main relay M. When the relay M is de-energized its contacts will open and disconnect the motor from the line.

Figure 8.44 shows a simplified diagram of an automatic starter for a dc motor using time delay relays. The operation is based on series time delay relays adjusted in a predetermined manner to close their individual contacts and short out each section of the starting resistor at the proper time intervals.

Example 8.8

Assume that a 100 hp, 240 V dc motor has a full-load current of 343 A and an armature resistance of 0.05 per unit. Determine the following:

(a) The base value of the armature resistance in ohms
(b) The value of the armature resistance in ohms
(c) The value of the armature current on starting in amps
(d) The value of the external resistance required if twice the full-load current is permitted to flow through the armature at the time of starting

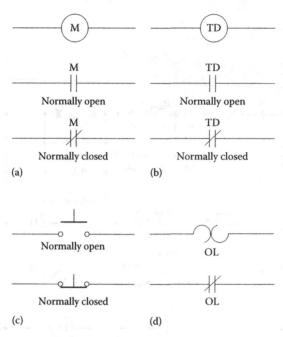

FIGURE 8.43 Typical symbols used in automatic starter diagrams: (a) a relay coil and its contacts, (b) a time delay relay coil and its contacts, (c) normally open and normally closed push-button switches, and (d) a thermal overload device and its normally closed contacts.

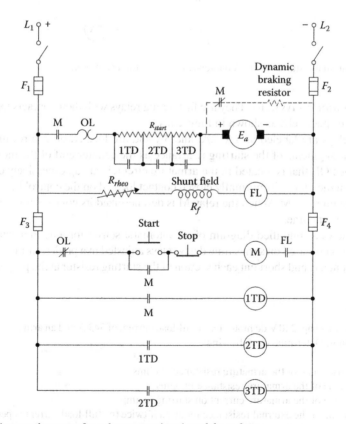

FIGURE 8.44 Automatic starter for a dc motor using time delay relays.

Solution

(a) The base value of the armature resistance is

$$R_{a,B} = \frac{V_{a,B}}{I_{a,B}}$$

$$= \frac{240 \text{ V}}{343 \text{ A}}$$

$$= 0.6997 \ \Omega$$

(b) Since

$$R_{a,pu} = \frac{R_a}{R_{a,B}}$$

then the value of the armature resistance in ohms is

$$R_a = R_{a,pu} \times R_{a,B}$$

$$= (0.05 \text{ pu})(0.6997 \ \Omega)$$

$$= 0.035 \text{ A}\Omega$$

(c) The value of the armature current on starting is

$$I_{a,start} = \frac{V_t - E_a}{R_a}$$

$$= \frac{240 - 0 \text{ V}}{0.035 \ \Omega}$$

$$= 6{,}857.1 \ \Omega$$

or by using Equation 8.75 directly,

$$I_{a,start} = 20 \times I_{a,B}$$

$$= 20(343 \text{ A})$$

$$= 6{,}860 \text{ A}$$

(d) The value of the external resistance required is

$$R_{ext} = \frac{V_t}{2I_{a,fl}}$$

$$= \frac{240 \text{ V}}{2(343 \text{ A})} - 0.035$$

$$= 0.3149 \ \Omega$$

8.23   DC MOTOR BRAKING

By using automatic motor starters, a number of additional control actions can be accomplished. These actions may include dynamic braking, reversing, jogging, plugging, and regenerative braking.*

When the stop button is pushed, the power supply of the motor is cut off and the motor coasts to a stop. Since the only braking effect is mechanical friction, it will take some time for the motor to come to rest. The time that it takes for the motor to stop completely is a function of the kinetic energy that is stored in the motor armature and the attached mechanical load.

As briefly discussed in Section 8.22, a motor can be stopped quickly by the use of the *dynamic braking* technique. In dynamic braking, the shunt field of the motor is left connected to the supply after the armature is disconnected by the opening of the main (M) contactor.

When the M contactor opens, a resistor called a *dynamic braking resistor* is connected across the armature terminals. With its shunt field energized, the dc machine behaves like a generator and produces a countertorque that quickly slows the armature by releasing the stored kinetic energy in the resistor as heat. During this braking operation, a current flows in the armature winding in a direction opposite to that of the motor mode of operation.

A shunt motor operation can be smoothly converted from a motor mode to a generator mode without changing the field winding connections. In a series machine, either the series winding connections have to be reversed or it has to be connected to a separate voltage source to achieve good braking.

In a compound machine, the series winding is left disconnected during the braking operation. In dynamic braking, the amount of braking effort is a function of the motor speed, the motor-field strength, and the value of the resistance. It is used extensively in the control of elevators and hoists and in other applications in which motors have to be started, stopped, and reversed frequently.

In some motor applications, it may be necessary to quickly reverse the direction of rotation. This can be achieved by using dynamic braking to stop the motor quickly and then reversing the voltage applied to the armature. The operation of running a motor for only a fraction of a revolution or a few revolutions without going through the starting sequence is called *jogging*. It is often used for positioning applications.

Plugging is used when a motor has to be brought to a stop quickly or when a fast reversal of the direction of rotation is needed. It can be used in some motor applications where there is a sudden reversal in direction at full speed. This is done by reversing the armature connections by leaving the field winding connections undisturbed to maintain the magnetic field direction the same. Because the armature winding resistance is very small, the counter emf is almost equal and opposite to the applied voltage.

However, the counter emf and the applied voltage are in the same direction at the time of plugging. Therefore, the total voltage in the armature circuit is almost twice that of the applied voltage.

To protect the motor from the sudden increase in armature current, an external resistance (known as the *plugging resistance*) must be inserted in the armature circuit. The armature current reverses its direction and develops a force that tends to rotate the armature in a direction opposite to that of its initial rotation and brings the motor to a stop. The kinetic energy of the armature and mechanical load is hence being dissipated as heat in the plugging resistor.

In general, *regenerative braking* is used in motor applications where the motor speed is likely to increase from its normal speed. Such applications include electric railway locomotives, elevators, cranes, and hoists.

The speed of motors driving such loads can be reduced significantly without mechanical braking by using regenerative braking to feed electrical energy back into the electrical system. Note that as the speed increases, so does the counter emf in the motor.

---

* In addition, eddy-current braking can be used. An eddy-current brake is a disc of conducting material affected by the magnetic field of a coil. This disc rotates with the shaft of the motor. When the motor is turned off, the coil is energized. As the shaft continues to rotate, the eddy currents produced in the disc develop torque in the opposite direction of the rotation and stop the motor.

When the counter emf becomes greater than the supply voltage, the current in the armature winding reverses its direction, causing the motor to operate as a generator. Regenerative braking can be used to maintain safe speeds but cannot be used to stop a mechanical load. For this action, dynamic braking, plugging, or mechanical braking* are required.

Example 8.9

Assume that a 240 V, self-excited shunt motor is supplied by a line current of 102.4 A when it is loaded with a full load at a speed of 1000 rev/min. The armature-circuit resistance and the shunt-field circuit resistance of the motor are 0.1 and 100 Ω, respectively. Assume that a braking resistor of 1.05 Ω is used for *dynamic braking* and determine the following:

(a) The value of the counter emf
(b) The value of the armature winding current at the time of initial braking
(c) The full-load torque
(d) The value of the initial dynamic braking torque

Solution

(a) The field current of the motor is

$$I_f = \frac{V_f}{R_f}$$

$$= \frac{240 \text{ V}}{100 \text{ Ω}}$$

$$= 2.4 \text{ A}$$

Therefore, the armature winding current is

$$I_a = I_L - I_f$$

$$= 102.4 \text{ A} - 2.4 \text{ A}$$

$$= 100 \text{ A}$$

Hence, the value of the counter emf generated is

$$E_a = V_t - I_a R_a$$

$$= 240 \text{ V} - (100 \text{ A})(0.1 \text{ Ω})$$

$$= 230 \text{ V}$$

(b) At the time of the initial braking, since E_a, speed, and flux have not changed, then the value of the armature winding current is

$$I_{a,brake} = \frac{E_a}{R_a + R_{brake}}$$

$$= \frac{230 \text{ V}}{0.1 \text{ Ω} + 1.05 \text{ Ω}}$$

$$= 200 \text{ A}$$

---

* This mechanical braking may be operated by a magnetic solenoid.

Note that during the braking, the value of the armature winding current has increased about twice.

(c) The full-load torque of the motor is

$$T_d = \frac{E_a I_a}{\omega_m}$$

$$= \frac{E_a I_a}{2\pi n_m / 60}$$

$$= \frac{(230 \text{ V})(100 \text{ A})}{2\pi(1000 \text{ rev/min})/60}$$

$$= 219.6 \text{ N} \cdot \text{m}$$

(d) Therefore, the value of the initial dynamic braking torque is

$$T_d = 2(219.6 \text{ N} \cdot \text{m})$$

$$= 439.2 \text{ N} \cdot \text{m}$$

PROBLEMS

8.1 An eight-pole 500 V, 500 kW dc generator has a lap winding with 640 armature conductors. If the generator has six commutating poles, determine the following:
(a) The number of turns in the commutating winding if the mmf of the commutating poles is 1.8 times that of the armature
(b) The number of conductors for the compensating winding in each pole face if the generator has a compensating winding and the pole face covers 80% of the pole span
(c) The number of turns per pole in the commutating winding when the compensating winding is in the circuit

8.2 Resolve Example 8.1 assuming that the winding of the armature is lap-wound.

8.3 Assume that the armature of a dc machine operating at 1800 rpm is lap-wound with 720 conductors and that the machine has four poles. If the flux per pole is 0.05 Wb, determine the following:
(a) The induced armature voltage
(b) The induced armature voltage if the armature is wave-wound

8.4 Suppose that a separately excited shunt dc machine has a rated terminal voltage of 250 V and a rated armature current of 100 A. Its armature winding, commutating winding, and compensating winding resistances are 0.1, 0.02, and 0.009 Ω, respectively. Determine the following:
(a) The induced armature voltage if the machine is operating as a generator at full load
(b) The induced armature voltage if the machine is operating as a motor at full load

8.5 Suppose that a four-pole wave-wound dc machine is operating at 1050 rpm at a terminal voltage of 250 V and that the resistance of the winding between terminals is 0.15 Ω. The armature winding has 100 coils of three turns each. If the cross-sectional area of each pole face is 150 cm² and the average flux density in the air gap under the pole faces is 0.75 T, determine the following:
(a) The total number of conductors in the armature winding.
(b) The flux per pole.
(c) The armature constant K_a.
(d) The induced armature voltage.
(e) Is the machine operating as a motor or a generator?
(f) The armature current.

(g) The developed power.

(h) The developed torque and its direction with respect to the direction of rotation.

8.6 A shunt motor operating at 1200 rev/min has an armature current of 38 A from a 240 V source when providing 8398 W of mechanical power. If the armature winding resistance is 0.2 Ω, determine the following:

(a) The loss torque of the motor at the given speed

(b) The required armature current to provide half the mechanical (shaft) power at the same speed

8.7 Consider the shunt motor in Example 8.4 and determine the following:

(a) The developed torque if the load current is 125 A

(b) The developed torque if a 25% increase in full-load current results in a 12% increase in the flux due to the demagnetizing effect of the armature reaction

8.8 A series motor operating at a full-load speed of 1200 rev/min has a terminal voltage of 240 V and a rated load current of 74 A. The speed at which its magnetization curve has been developed is 1200 rev/min. From its magnetization curve, its no-load voltages (i.e., internal generated voltages) are given as 230 and 263.5 V at the series-field currents of 74 and 100 A, respectively. The armature winding and the series-field winding resistances of the motor are 0.085 and 0.05 Ω, respectively. Ignore the effect of armature reaction and determine the torque, speed, and power output of the motor when the line current is 100 A. The developed torque is 125.66 N m at the rated load current of 74 A.

8.9 Consider the shunt generator of Example 8.3. Assume that it is used as a separately excited cumulative compound generator by the addition of a series winding of five turns per pole. The resistance of the series winding is 0.03 Ω. The applied voltage at the terminals of the shunt field is 250 V and the resistance of the shunt-field rheostat setting is 25 Ω. The generator is driven at a speed of 1200 rpm at no load and at 1150 rpm at a full load of 100 A. As before, its armature-circuit resistance is 0.1 Ω, and its shunt-field winding has a 100 Ω resistance and is made up of 1000 turns per pole. The total brush-contact voltage drop is 2 V and the demagnetization of the d-axis by the armature mmf is neglected. Use the magnetization curve data given in Table 8.1 and the magnetization curve plotted in Figure 8.26. Determine the following:

(a) The excitation current, the terminal voltage, and the developed torque at no load

(b) The excitation current, the terminal voltage, and the developed torque at a full load of 100 A

8.10 Consider the speed–current characteristics developed in Example 8.6 and describe how and why they would be altered if the demagnetization of the d-axis (i.e., the armature reaction) by the armature mmf (due to the load current) were accounted for.

8.11 Assume that a 100 kW, 250 V long-shunt compound generator is driven at its rated speed of 1800 rpm. Its armature winding resistance, the series winding resistance, and the interpole winding resistance are given as 0.018, 0.006, and 0.006 Ω, respectively. Its shunt-field current is 3 A. Its no-load rotational loss is 4500 W. Assume that its brush-contact voltage drop is 2 V and that its stray-load loss is 1% of the machine output. Determine the following:

(a) The total armature-circuit resistance excluding the brush-contact resistance

(b) The armature current

(c) The total losses

(d) The efficiency at the rated load

8.12 Assume that a 125 kW, 250 V long-shunt compound generator is driven at its rated speed of 1000 rpm. Its armature winding resistance, the series winding resistance, and the shunt winding resistance are given as 0.03, 0.01, and 35 Ω, respectively. Its stray-load loss at the rated voltage and speed is 1250 W. Its rated field current is 4 A. If its rotational losses are 1250 W, determine the following:

(a) The shunt-field copper loss

(b) The series-field copper loss

(c) The total losses

(d) The percent efficiency of the machine

(e) The maximum percent efficiency at its rated speed

8.13 Consider the results of Example 8.5 and determine the following for the shunt motor:

(a) The armature terminal power

(b) The developed power

(c) The shaft power

8.14 A 20 kW, 250 V, self-excited dc shunt generator is driven at its rated speed of 1200 rpm. Its armature-circuit resistance is 0.15 Ω, and the field current is 2 A when the terminal voltage is 250 V at rated load. If its rotational loss is given as 1000 W, determine the following:

(a) The internal generated voltage

(b) The developed torque

(c) The percent efficiency of the generator

8.15 Assume that the separately excited dc machine given in Example 8.3 is being used as a separately excited shunt motor to drive a mechanical load. Its terminal voltage is kept constant at 200 V. The full-load current of the machine is 100 A as before. Its field current is kept variable at 1.0, 1.5, and 2.5 A by using the field rheostat. For each value of the field current, determine the following:

(a) The developed torque at a full load of 100 A.

(b) The developed power at a full load of 100 A.

(c) The ideal no-load speed in rpm.

(d) The full-load speed in rpm.

(e) Sketch the torque–current characteristic based on the results found in Part (a).

(f) Sketch the speed–current characteristic based on the results found in Parts (c) and (d).

8.16 Assume that the dc machine given in Example 8.3 is being used as a self-excited cumulative compound motor to drive a mechanical load. Its series winding has 5 turns per pole and a resistance of 0.03 Ω. Its shunt-field current is kept constant at 2 A by using a 25 Ω shunt-field rheostat setting. If the applied terminal voltage is 250 V, determine the following:

(a) The short-shunt connection diagram of the motor.

(b) The ideal no-load speed in rpm.

(c) The full-load speed in rpm.

(d) The developed torque at a full load of 100 A.

(e) Sketch the speed–current characteristic of the machine and compare it with the one given in Example 8.3. Explain the difference in performance.

(f) Sketch the torque–current characteristic of the machine and compare it with the one given in Example 8.3. Explain the difference in performance.

8.17 Consider the solution of Problem 8.16 and account approximately for the demagnetization of the d-axis by the armature mmf on the q-axis. This demagnetization is due to the nonlinearity of the magnetization curve. Assume that the armature reaction constant K_d is 2.0 ampere-turns per pole and that the machine in Problem 8.16 has no compensating winding. Determine the *modified value* of the number of turns of the series winding N_{se} (i.e., $N_{se} = 5$ turns) that will give approximately the same performance found in Problem 8.16 when the demagnetization due to the armature mmf was ignored.

8.18 Explain the effect that demagnetization due to armature mmf has on ideal (i.e., with no demagnetization) dc generator external characteristics.

8.19 Explain the effect of demagnetization due to armature mmf on the developed torque and the speed of the motor found in Problem 8.17.

8.20 Assume that the dc machine given in Example 8.3 is being used as a series motor to drive a mechanical load. Its series winding has 25 turns per pole and a resistance of 0.08 Ω. Its shunt field is totally disconnected. If the applied terminal voltage is 250 V, determine the following:

(a) The speed in rpm when the load current is 20 A

(b) The developed torque when the load current is 20 A

8.21 Consider the dc machine of Problem 8.20 and determine the following:
   (a) The speed in rpm when the load current is 100 A
   (b) The developed torque when the load current is 100 A
8.22 Consider the solutions of Problems 8.20 and 8.21 and sketch the following characteristics:
   (a) The torque–speed characteristic
   (b) The speed–current characteristic
   (c) The torque–current characteristic
8.23 Assume that the dc machine given in Example 8.3 is being used as a self-excited shunt motor. The machine is being considered for an application that requires the motor to have a developed torque of 375 N · m at the start. The armature current at starting is desired to be as small as possible but not to be greater than 200% of the rated full-load armature current. The motor starter to be designed will have a connector to short-circuit the field rheostat at starting. The supply line voltage is maintained at 250 V. Determine the following:
   (a) The armature current at starting if there is no starting resistance connected
   (b) The value of the field current at starting
   (c) The value of $K_a \Phi_d$ of the motor at starting
   (d) The armature current at starting if the 375 N · m starting torque is to be developed
   (e) The value of the starting resistance
8.24 Suppose that the dc machine given in Example 8.3 is being used as a self-excited shunt motor and that its speed control is achieved by inserting an external armature-circuit resistance of 1.10 Ω into its circuit. Assume that the supply line voltage is maintained at 250 V and that the field current is 2.0 A. Also assume that the core loss and the friction and windage losses of the machine are 1% of the machine's rating. Determine the following:
   (a) The developed torque at full load
   (b) The ideal no-load speed in rpm
   (c) The full-load speed in rpm
   (d) The full-load (overall) efficiency of the motor
   (e) The conversion efficiency of the motor
8.25 Assume that the dc machine given in Example 8.3 is being used as a self-excited shunt motor and that its speed control is achieved by using the variable voltage method. The thyristor equipment used in such an application is made up of two solid state–controlled rectifiers. Both rectifiers are continuously adjustable from 250 V terminal dc voltage: one of them has 3 A continuous rating, the other 100 A. The current and the resistance of the shunt field are 2 A and 100 Ω, respectively. Assume that the core loss and the friction and windage losses of the machine are 1% of the machine's rating. To achieve the maximum efficiency and minimum losses, the field rheostat resistance is set at zero. Determine the following:
   (a) The voltages V_f and V_t to be set to duplicate the performance of Problem 8.24 at full load of 100 A armature current and also to achieve the maximum possible efficiency
   (b) The full-load (overall) efficiency of the motor
   (c) The conversion efficiency of the motor
8.26 A 200 hp, 240 V dc motor has a full-load current of 675 A and an armature resistance of 0.05 per unit. Determine the following:
   (a) The base value of the armature resistance in ohms
   (b) The value of the armature resistance in ohms
   (c) The value of the armature current on starting in amps
8.27 A 50 hp, 240 V dc motor has a full-load current of 173 A and an armature resistance of 0.05 per unit. Determine the following:
   (a) The base value of the armature resistance in ohms
   (b) The value of the armature resistance in ohms
   (c) The value of the armature current on starting in amps

(d) The value of the external resistance needed if twice the full-load current is permitted to flow through the armature at the time of starting.

8.28 A 250 V, self-excited dc shunt motor draws 102 A when it is loaded with a full load at a speed of 15 rev/s. Its armature-circuit resistance is 0.15 Ω and the shunt-field current is 2 A when the terminal voltage is 250 V at the rated load. An external plugging resistance is inserted into the armature circuit so that the armature current does not exceed 150% of its rated load value when the motor is plugged. Determine the following:
   (a) The value of the plugging resistance
   (b) The braking torque at the instant of plugging
   (c) The braking torque when the motor reaches zero speed

8.29 A 250 V, self-excited shunt motor is supplied by a line current of 102.5 A when it is loaded with a full load at a speed of 1200 rev/min. The armature-circuit resistance and the shunt-field circuit resistance of the motor are 0.1 and 100 Ω, respectively. If a braking resistor of 1.1 Ω is used for dynamic braking, determine the following:
   (a) The value of the counter emf
   (b) The value of the armature winding current at the time of initial braking
   (c) The full-load torque of the motor
   (d) The value of the initial dynamic braking torque

8.30 A 20 kW, 250 V, self-excited generator supplying the rated load has an armature-circuit voltage drop of 4% of the terminal voltage and a shunt-field current equal to 4% of rated load current. Determine the resistance of the armature circuit and that of the field circuit.

8.31 Redo Example 8.5 by using MATLAB assuming that the new terminal voltage is 260 V and the new rotational losses are 700 W. Use the other given values and determine the following:
   (a) Write the MATLAB program script.
   (b) Give the MATLAB program output.

9 Single-Phase and Special-Purpose Motors

The conscious mind allows itself to be trained like a parrot, but the unconscious does not-which is why St. Augustine thanked God for not making him responsible for his dreams.

Carl G. Jung, Psychology and Alchemy, 1952

Mind moves matter.

Virgil

What is mind? No matter. What is matter? Never mind.

Thomas H. Key

9.1 INTRODUCTION

Today it can be said without exaggeration that about 90% of all motors manufactured are the single-phase type. They are used extensively in homes, businesses, farms, and small industries. Most of them are built as fractional-horsepower (hp) or subfractional-horsepower motors (1 hp is equal to 746 W). Standard ratings for *fractional-horsepower motors* range from 1/20 to 1 hp. The small motors rated for less than 1/20 hp, called *subfractional-horsepower motors*, are rated in millihorse-power (mhp) and range from 1 to 35 mhp.

The single-phase motors manufactured in standard integral horsepower sizes are in the 1.5, 2, 3, 5, 7.5–10 hp range. However, special integral horsepower sizes can range from several hundreds up to a few thousands, for example, in locomotive service using single-phase ac series motors.

They can also be designed for very rugged use in cranes and hoists. Unlike *integral horsepower motors*, small single-phase motors are manufactured in many different types of designs with different characteristics. This is especially true of *subfractional-horsepower motors*. There are three basic types of single-phase ac motors: single-phase induction motors, universal motors, and single-phase synchronous motors.

9.2 SINGLE-PHASE INDUCTION MOTORS

Single-phase induction motors generally have a distributed stator winding and a squirrel-cage rotor. Figure 9.1 shows a schematic diagram of a single-phase induction motor. The ac supply voltage is applied to the stator winding, which in turn creates a nonrotating (i.e., stationary in position and pulsating with time) magnetic field.*

As shown in Figure 9.1a, currents are induced in the squirrel-cage rotor windings by transformer action. These currents produce an mmf opposing the stator mmf. Since the axis of the rotor-mmf wave coincides with that of the stator field, the torque angle is zero and no starting torque develops. At standstill, therefore, the motor behaves like a single-phase nonrotating transformer with a short-circuited secondary.

---

* This type of mmf field is sometimes referred to as a breathing field because it expands and contracts in the same place on the stator. In other words, the stator winding does not provide a rotating mmf field for the rotor mmf to chase.

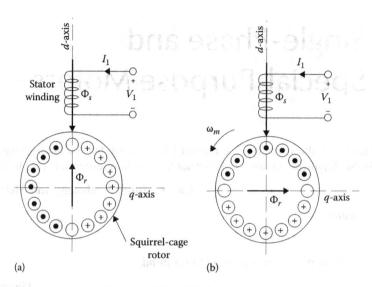

FIGURE 9.1   Single-phase motor: (a) at standstill and (b) during rotation.

Hence, a single-phase induction motor is not self-starting. However, if the rotor of a single-phase induction motor is given a spin or started by auxiliary means, it will continue to run and develop power.* As shown in Figure 9.1a, the single-phase induction motor can develop torque when it is running. This phenomenon can be explained by the *double-revolving field theory*.[†]

According to the double-revolving field theory, a pulsating mmf (or flux) field can be replaced by two rotating fields half the magnitude but rotating at the same speed in opposite directions. For a sinusoidally distributed stator winding, the mmf along a position θ can be expressed as

$$\mathcal{F}(\theta) = Ni\cos\theta \tag{9.1}$$

where
N is the effective number of turns of the stator winding
i is the instantaneous value of the current in the stator winding

Thus,

$$i = I_{max}\cos\omega t \tag{9.2}$$

Hence, the mmf can be written as a function of space and time as

$$\mathcal{F}(\theta,t) = NI_{max}\cos\theta\cos\omega t \tag{9.3a}$$

$$= \frac{NI_{max}}{2}\cos(\omega t - \theta) + \frac{NI_{max}}{2}\cos(\omega t + \theta) \tag{9.3b}$$

$$\mathcal{F}(\theta,t) = \mathcal{F}_f + \mathcal{F}_b \tag{9.3c}$$

---

* Historically, the first single-phase induction motors were started by wrapping a rope or a strap around the shaft and pulling to spin to rotor. Fortunately, today the necessity for manual starting can be overcome by various relatively simple methods.

† It can also be explained by the *cross-field theory*. For a discussion of the cross-field theory, see Chapman (1985) and Veinott (1959).

In other words, the stator mmf is the sum of a positive- and a negative-traveling mmfs in the direction of 9. ($\mathcal{F}_f$ *is the rotating mmf in the direction of* θ only it represents the *forward-rotating field*; $\mathcal{F}_b$ *is the rotating mmf in the* opposite direction and represents the *backward-rotating field*.) Here, it is assumed that the rotational direction of the forward-rotating field is the same as the rotational direction of the rotor.[*]

As shown in Figure 9.2, the forward-rotating mmfs and backward-rotating mmfs both produce induction motor action, that is, they both produce a torque on the rotor, though in opposite directions. Notice that at standstill, the torques that are caused by the fields are equal in magnitude and their resultant starting torque is zero. At any other speed, however, the torques are not equal and therefore the resultant torque causes the motor to rotate in the rotational direction of the motor.

Assume that the rotor is made to rotate at a speed of n_m rpm in the forward direction and that the synchronous speed is n_s rpm. The slip with respect to the forward-rotating field is

$$s_f = s = \frac{n_s - n_m}{n_s} = 1 - \frac{n_m}{n_s} \tag{9.4}$$

However, because the direction of rotation is opposite that of the backward-rotating field, the slip with respect to the backward field is

$$s_b = \frac{n_s - (-n_m)}{n_s} = \frac{n_s + n_m}{n_s} = 1 + \frac{n_m}{n_s} = 2 - s \tag{9.5}$$

or

$$s_b = 2 - s_f \tag{9.6}$$

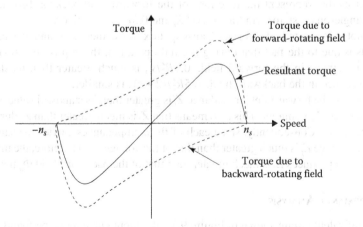

FIGURE 9.2 Torque–speed characteristics of a single-phase induction motor based on the revolving field theory.

---

[*] Since by definition the forward direction is that direction in which the motor is initially started, it is also referred to as the *positive sequence*. Similarly, the backward direction is also referred to as the *negative sequence*.

9.2.1 EQUIVALENT CIRCUIT

At standstill, a single-phase induction motor behaves like a transformer with its secondary short-circuited. Figure 9.3a shows the corresponding equivalent circuit where R_1 and X_1 are the resistance and reactance of the stator winding, respectively. Here, X_m is the magnetizing reactance, and R_2 and X_2 are the standstill values of the rotor resistance and reactance referred to the stator winding by the use of the appropriate turns ratio. The core losses of the motor are not shown but are included in the rotational losses along with the mechanical and stray losses.

Based on the double-revolving field theory, the equivalent circuit can be modified to include the effects of the two counterrotating fields of constant magnitude. At standstill, the magnitudes of the forward and backward resultant mmf fields are both equal to half the magnitude of the pulsating field. Therefore, the rotor-equivalent circuit can be split into equal sections. The equivalent circuit of a single-phase induction motor, then, consists of the series connection of a forward- and a backward-rotating field equivalent circuits, as shown in Figure 9.3b.

After the motor has been brought up to speed by the use of an auxiliary winding (which is switched out again after obtaining the proper speed) and is running in the direction of the forward-rotating field at a slip s, its equivalent circuit has to be modified, as shown in Figure 9.3c. Therefore, the rotor resistance in the forward equivalent circuit is $0.5R_2'/s$.

Also, since the rotor is rotating at a speed that is s less than the forward-rotating field, the difference* in speed between the rotor and the backward-rotating field is $2 - s$. Hence, the rotor resistance in the equivalent backward circuit is represented by $0.5R_2'/(2-s)$.

To simplify the calculations, the impedances shown in Figure 9.3c corresponding to the forward and backward fields are defined, respectively, as

$$\mathbf{Z}_f = R_f + jX_f = \frac{jX_m(R_2'/s + jX_2')}{jX_m + (R_2'/s + jX_2')} \tag{9.7}$$

and

$$\mathbf{Z}_b = R_b + jX_b = \frac{jX_m[R_2'/(2-s) + jX_2']}{jX_m + [R_2'/(2-s) + jX_2']} \tag{9.8}$$

These impedances that represent the reactions of the forward- and backward-rotating fields with respect to the single-phase stator winding are $0.5\mathbf{Z}_f$ and $0.5\mathbf{Z}_b$, respectively.

After the motor is started, the forward air-gap flux wave increases and the backward wave decreases. This is due to the fact that during normal operation, the slip is very small. Because of this, the rotor resistance in the forward field, $0.5R_2'/s$, is much greater than its standstill value, whereas the resistance in the backward field, $0.5R_2'/(2-s)$, is smaller.

As a result of this, the forward-field impedance $\mathbf{Z}_f$ is greater than its standstill value, while the backward-field impedance $\mathbf{Z}_b$ is smaller. This also means that $\mathbf{Z}_f$ is much greater than $\mathbf{Z}_b$ during the normal operation of the motor. Consequently, since each of these impedances carries the same current, the magnitude of the voltage E_f is much greater than that of the voltage E_b. Therefore, the magnitude of the forward field Φ_f that produces E_f is much greater than that of the backward field Φ_b that produces E_b.

9.2.2 PERFORMANCE ANALYSIS

Based on the equivalent circuit shown in Figure 9.3c, the input current can be found from

$$I_1 = \frac{V_1}{R_1 + jX_1 + 0.5\mathbf{Z}_f + 0.5\mathbf{Z}_b} \tag{9.9}$$

---

* The total difference in speed between the forward- and backward-rotating fields is 2.

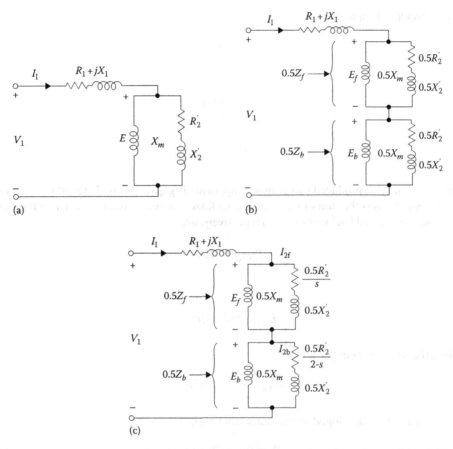

FIGURE 9.3 Equivalent circuit of a single-phase induction motor based on the revolving-field theory: (a) conventional configuration at standstill, (b) modified configuration at standstill, and (c) typical torque–speed characteristic.

Therefore, the air-gap powers developed by the forward and backward fields, respectively, are

$$P_{g,f} = I_1^2(0.5R_f) \tag{9.10}$$

and

$$P_{g,f} = I_1^2(0.5R_b) \tag{9.11}$$

Hence, the total air-gap power is

$$P_g = P_{g,f} - P_{g,b} \tag{9.12}$$

Thus, the developed torques due to the forward and backward fields, respectively, are

$$T_{d,f} = \frac{P_{g,f}}{\omega_s} \tag{9.13}$$

$$T_{d,b} = \frac{P_{g,b}}{\omega_s} \tag{9.14}$$

The total developed torque is

$$T_d = \frac{P_g}{\omega_s} \tag{9.15a}$$

$$= \frac{P_{g,f} - P_{g,b}}{\omega_s} \tag{9.15b}$$

$$= T_{d,f} - T_{d,b} \tag{9.15c}$$

Since the rotor currents produced by the two component air-gap fields are different frequencies, the total rotor copper loss is the sum of the rotor copper losses caused by each field. These rotor copper losses of the forward and backward fields, respectively, are

$$P_{2,Cu,f} = sP_{g,f} \tag{9.16}$$

and

$$P_{2,Cu,b} = (2 - s)P_{g,b} \tag{9.17}$$

Therefore, the total rotor copper loss is

$$P_{2,Cu} = P_{2,Cu,f} + P_{2,Cu,b} \tag{9.18}$$

The mechanical power developed in the motor can be found from

$$P_d = P_{mech} = T_d \times \omega_m \tag{9.19a}$$

$$= T_d \times \omega_g (1 - s) \tag{9.19b}$$

$$= (1 - s)P_g \tag{9.19c}$$

$$= (1 - s)(P_{g,f} - P_{g,b}) \tag{9.19d}$$

$$= 0.5I_1^2 (R_f - R_b)(1 - s) \tag{9.19e}$$

Hence, the output power is

$$P_{out} = P_d - P_{rot} \tag{9.20a}$$

$$= P_d - (P_{core} + P_{FW} + P_{stray}) \tag{9.20b}$$

Example 9.1

A 1/4 hp, single-phase, 120 V, 60 Hz, two-pole induction motor has the following constants in ohms referred to the stator:

$$R_1 = 2.0 \, \Omega \quad R_2' = 4.1 \, \Omega$$

$$X_1 = 2.5\,\Omega \quad X_2' = 2.2\,\Omega$$

$$X_m = 51\,\Omega$$

The core losses of the motor are 30 W; and the friction, windage, and stray losses are 15 W. The motor is operating at the rated voltage and frequency with its starting winding open. For a slip of 5%, determine the following:

(a) The shaft speed in rpm
(b) The forward and backward impedances of the motor
(c) The input current
(d) The power factor
(e) The input power
(f) The total air-gap power
(g) The developed power
(h) The output power
(i) The developed torque
(j) The output torque
(k) The efficiency of the motor

Solution

(a) The synchronous speed is

$$n_s = \frac{120 f_1}{p}$$

$$= \frac{120(60\,\text{Hz})}{2}$$

$$= 3600\ \text{rev/min}$$

Thus, the rotor's mechanical shaft speed is

$$n_m = (1 - s)n_s$$

$$= (1 - 0.05)(3600\,\text{rev/min})$$

$$= 3420\ \text{rev/min}$$

(b) The forward impedance of the motor is

$$Z_f = R_f + jX_f$$

$$= \frac{jX_m(R_2'/s + jX_2')}{jX_m + (R_2'/s + jX_2')}$$

$$= \frac{j51(4.1/0.05 + j2.2)}{j51 + 4.1/0.05 + j2.2}$$

$$= 42.8\angle 58.56°$$

$$= 22.32 + j36.52\,\Omega$$

Similarly, the backward impedance of the motor is

$$Z_b = R_b + jXb$$

$$= \frac{jXm[R_2'(2-s) + jX_2']}{jXm + [R_2'(2-s) + jX_2']}$$

$$= \frac{j51[4.1/(2-0.05) + j2.2]}{j51 + [4.1/(2-0.05) + j2.2]}$$

$$= 2.915\angle 48.56°$$

$$= 1.929 + j2.185 \ \Omega$$

(c) The stator input current of the motor is

$$I_1 = \frac{V_1}{R_1 + jX_1 + 0.5Z_f + 0.5Z_b}$$

$$= \frac{120\angle 0°}{2.0 + j2.5 + 0.5(22.32 + j36.51) + 0.5(1.929 + j2.185)}$$

$$= \frac{120\angle 0°}{14.1245 + j21.8475}$$

$$= 4.61\angle -57.12° \ A$$

(d) The stator power factor of the motor is

$$PF = \cos 57.12°$$

$$= 0.543 \ \text{lagging}$$

(e) The input power of the motor is

$$P_{in} = VI \cos\theta$$

$$= (120)(4.61)\cos 57.12°$$

$$= 300.4 \ W$$

(f) The air-gap power due to the forward field is

$$P_{g.f} = I_1^2(0.5R_f)$$

$$= (4.61)^2(0.5)(22.32)$$

$$= 237.17 \ W$$

and the air-gap power due to the backward field is

$$P_{g.b} = I_1^2(0.5R_b)$$

$$= (4.61)^2(0.5)(1.929)$$

$$= 20.5 \ W$$

Therefore, the total air-gap power of the motor is

$$P_g = P_{g,f} - P_{g,b}$$

$$= 237.17 - 20.5$$

$$= 216.67 \text{ W}$$

(g) The developed mechanical power is

$$P_d = P_{mech} = (1 - s)P_g$$

$$= (-0.05)216.67$$

$$= 205.84 \text{ W}$$

(h) The output (shaft) power is

$$P_{out} = P_d - (P_{core} + P_{FW} + P_{stray})$$

$$= 205.84 - (30 + 15)$$

$$= 160.84 \text{ W}$$

(i) The developed torque is

$$T_d = \frac{P_g}{\omega_s}$$

$$= \frac{216.67 \text{ W}}{(3600 \text{ rev/min})(1 \text{ min}/60 \text{ s})(2\pi \text{ rad/rev})}$$

$$= 0.575 \text{ N} \cdot \text{m}$$

(j) The output torque is

$$T_{out} = T_{load} = \frac{P_{out}}{\omega_m}$$

$$= \frac{160.84 \text{ W}}{(3420 \text{ rev/min})(1 \text{ min}/60 \text{ s})(2\pi \text{ rad/rev})}$$

$$= 0.45 \text{ N} \cdot \text{m}$$

(k) The efficiency of the motor is

$$\eta = \frac{P_{out}}{P_{in}} \times 100$$

$$= \frac{160.84 \text{ W}}{300.4 \text{ W}} \times 100$$

$$= 53.5\%$$

9.3 STARTING OF SINGLE-PHASE INDUCTION MOTORS

As stated previously, a single-phase induction motor cannot be started by its main winding alone, but must be started by an auxiliary (starting) winding or some other means. The auxiliary winding may be disconnected automatically by the operation of a centrifugal switch at about 75% of synchronous speed. Once the motor is started, it continues to run in the same direction.

A single-phase motor is designed so that the current in its auxiliary winding leads that of the main winding by 90 electrical degrees.* Accordingly, the field of its auxiliary winding builds up first. The direction of rotation of the motor can be reversed by reversing the connections of the main or the auxiliary winding. However, reversing the connections of both the main and auxiliary windings will not reverse the direction of rotation.

Consider the phasor diagram of a motor at starting shown in Figure 9.4b. The phase angle α between the two currents I_m and I_a is about 30°–45°. Therefore, the starting torque can be expressed as

$$T_{start} \propto I_m I_a \sin \alpha \tag{9.21}$$

$$T_{start} = K I_m I_a \sin \alpha \tag{9.22}$$

where K is a constant. Thus, the starting torque is a function of the magnitudes of the currents in the main and auxiliary windings and the phase difference between these two currents.

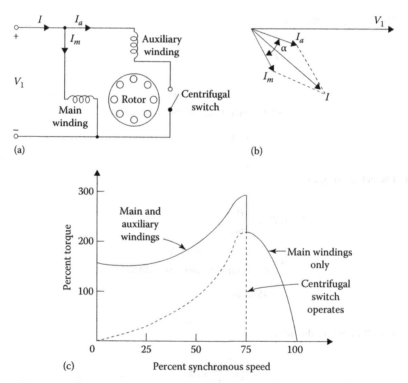

FIGURE 9.4 Split-phase induction motor: (a) schematic diagram, (b) phasor diagram at starting, and (c) typical torque–speed characteristic.

---

* Therefore, the operation of a single-phase induction motor is very similar to that of two-phase motors. The two windings of a single-phase induction motor are placed in the stator with their axes displaced 90 electrical degrees in space.

9.4 CLASSIFICATION OF SINGLE-PHASE INDUCTION MOTORS

Single-phase induction motors are categorized based on the methods used to start them. Each starting method differs in cost and in the amount of starting torque it produces.

9.4.1 SPLIT-PHASE MOTORS

A split-phase motor is a single-phase induction motor with two stator windings: a main (stator) winding, m, and an auxiliary (starting) winding, a, as shown in Figure 9.4a. The axes of these two windings are displaced 90 electrical degrees in space and somewhat less than 90° in time.* As shown in Figure 9.4b, the auxiliary winding has a higher resistance-to-reactance ratio than the main winding, so that its current *leads* the current in the main winding.

The most common way to obtain this higher R/X ratio is to use smaller wire for the auxiliary winding. This is acceptable since the auxiliary winding is in the circuit only during the starting period. The auxiliary winding is disconnected by a centrifugal switch or relay when the speed of the motor reaches about 75% of the synchronous speed.

The rotational direction of the motor can be reversed by switching the connections of the auxiliary winding, while the connections of the main winding remain the same.[†]

A typical torque–speed characteristic of the split-phase motor is shown in Figure 9.4c. A higher starting torque can be obtained by inserting a series resistance in the auxiliary winding. Alternatively, a series inductive reactance can be inserted into the main winding to achieve the same result. Split-phase motors that are rated up to 1/2 hp are relatively less costly than other motors and are used to drive easily started loads, such as fans, blowers, saws, pumps, and grinders.

When the motor is at standstill, the impedances of the main and the auxiliary windings, respectively, are

$$Z_m = R_m + jX_m \tag{9.23}$$

$$Z_a = R_a + jX_a \tag{9.24}$$

Thus, the magnitude of the auxiliary (starting) winding current can be determined from

$$I_a = \frac{V_1}{\sqrt{(R_a^2 + X_a^2)}} \tag{9.25}$$

where

$$R_a = \frac{X_a}{X_m}(R_m + Z_m) \tag{9.26}$$

or

$$R_a = \left(\frac{N_a}{N_m}\right)^2 (R_m + Z_m) \tag{9.27}$$

---

* Note that when two identical motor stator windings spaced 90 electrical degrees apart are connected in parallel to a single-phase source, the currents through the two windings lag the applied voltage by the same angle. Connecting a resistance in series with one winding causes the current in that winding to be more nearly in phase with the applied voltage. Since the current in the first winding is not affected by the added resistance, the currents in the two windings are displaced in time phase. This is the required condition to produce a revolving field. A motor using this method of phase splitting is called a resistance-start motor, a *resistance split-phase* motor, or simply a *split-phase motor.*

† However, such reversal (*plugging*) can never be done under running conditions even though it is sometimes done with polyphase induction motors.

and

$$X_a = \left(\frac{N_a}{N_m}\right)^2 X_m \tag{9.28}$$

Hence, for *design purposes*, it is easier to assume a number of turns for the auxiliary winding N_a to determine the value of R_a for maximum starting torque and the current of the auxiliary winding. If the optimum values of starting torque and current are not achieved, the process can be repeated until the proper design is found.

9.4.2 CAPACITOR-START MOTORS

A capacitor-start motor is also a split-phase motor. As shown in Figure 9.5a, a capacitor is connected in series with the auxiliary winding of the motor. By selecting the proper capacitor size, the current in the auxiliary winding can be made to *lead* the voltage V_1 and to bring about a 90° time displacement between the phasors of currents I_m and I_a, as shown in Figure 9.5b.

This produces a much greater starting torque than resistance split-phase starting, as shown in Figure 9.5c. The auxiliary winding is disconnected by a centrifugal switch when the speed of the motor reaches about 75% of the synchronous speed. In contrast to the split-phase motor, the capacitor-start motor is a *reversible* motor.

To reverse direction of the motor, it is temporarily disconnected and its speed is allowed to drop to a slip of 20% (about four times the rated slip of 5%). At the same time, its centrifugal switch is closed over a reversely connected (with respect to the main winding) auxiliary winding. These two simultaneous actions reverse the rotational direction of the motor.

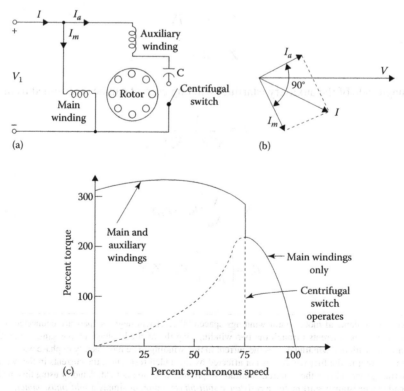

FIGURE 9.5 Capacitor-start induction motor: (a) schematic diagram, (b) phasor diagram at starting, and (c) typical torque–speed characteristic.

The cost of the capacitor is an added cost and makes these motors more expensive* than split-phase motors. They are used in applications that require high-starting torques, such as compressors, pumps, air conditioners, conveyors, larger washing machines, and other hard-to-start loads.

For design purposes, the value of the capacitive reactance which is connected in series with the auxiliary winding and provides the maximum starting torque can be expressed as

$$X_c = X_a + \frac{R_a R_m}{X_m + Z_m}$$
(9.29)

The value of this capacitance can be found from

$$C = \frac{1}{\omega X_c}$$
(9.30)

$$C = \frac{1}{\omega \left(X_a + R_a R_m / (X_m + Z_m) \right)}$$
(9.31)

However, as suggested by Sen (1989), the best design for the motor may be found by maximizing the starting torque per ampere of starting current rather than by maximizing the starting torque alone. The value of this capacitive reactance can be determined from

$$X_c = \frac{X_a + \{-X_m R_a + [R_a (R_a + R_m)]^{1/2} Z_m\}}{R_m}$$
(9.32)

The value of the capacitance is determined from

$$C = \frac{1}{\omega X_c}$$
(9.30)

9.4.3 CAPACITOR-RUN MOTORS

The *capacitor-split-capacitor motor* is also called the *permanent split-capacitor motor* or simply the *capacitor motor*, because it is designed to operate with its auxiliary winding and its series capacitor permanently connected, as shown in Figure 9.6a. It is simpler than the capacitor-start motor since there is no need for any centrifugal switch. Its torque,[†] efficiency, and power factor are also better since the motor runs effectively as a two-phase motor.

In this motor, the value of the capacitor is based on its optimum *running* rather than its starting characteristic. Since at starting the current in the capacitive branch is very low, the capacitor motor has a very low starting torque, as shown in Figure 9.6b. The *reversible* operation is not only possible, but also more easily done than in other motors. Its speed can be controlled by varying its stator voltage using various methods.

Capacitor-run motors are used for fans, air conditioners, and refrigerators. Since at starting, slip s is unity and R_f is equal to R_b, the starting torque of a capacitor-run induction motor is determined from

---

* However, since the capacitor is in the circuit only during the relatively short starting period, it can be an inexpensive ac electrolytic type.

[†] It produces a constant torque, not a pulsating torque as in other single-phase motors. Therefore, its operation is smooth and quiet.

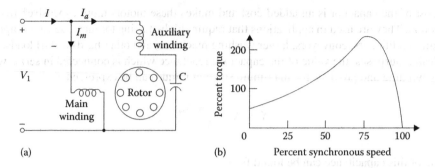

FIGURE 9.6 Capacitor-run induction motor: (a) schematic diagram and (b) typical torque–speed characteristic.

$$T_{start} = \frac{2aI_aI_m(R_f + R_b)}{\omega s}\sin(\theta_a + \theta_m) \qquad (9.33)$$

$$T_{start} = KI_mI_a \sin\alpha \qquad (9.22)$$

as before. Here, a is the turns ratio of the auxiliary and main windings, and θ_a and θ_m are the impedance angles of the auxiliary and main windings, respectively.

9.4.4 CAPACITOR-START CAPACITOR-RUN MOTORS

The *capacitor-start capacitor-run motor* is also called the *two-value capacitor motor*. In this motor, the high-starting torque of the capacitor-start motor is combined with the good running performance of the capacitor-run motor, as shown in Figure 9.7b. This is achieved by using two capacitors, as shown in Figure 9.7a. Both the auxiliary winding capacitor and the capacitor C_{run} are usually the electrolytic type and are connected in parallel at starting. Since the running capacitor C_{run} must have a continuous rating, this motor is expensive but provides the best performance.

9.4.5 SHADED-POLE MOTORS

The shaded-pole induction motor is used widely in applications that require 1/20 hp or less. As shown in Figure 9.8a, the motor has a salient-pole construction, with one-coil-per-pole main windings, and a squirrel-cage rotor. One portion of each pole has a *shading band* or *coil*. The shading

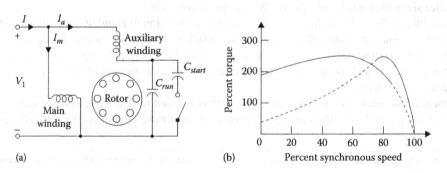

FIGURE 9.7 Capacitor-start capacitor-run induction motor: (a) schematic diagram and (b) typical torque–speed characteristic.

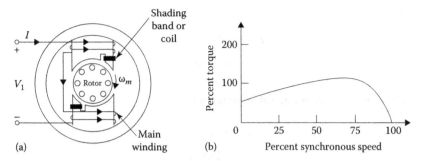

FIGURE 9.8 Shaded-pole induction motor: (a) schematic diagram and (b) typical torque–speed characteristic.

band is simply a short-circuited copper strap (or single-turn solid copper ring) wound around the smaller segment of the pole piece.

The purpose of the shading band is to retard, in time, the portion of the flux passing through it in relation to the flux coming out of the rest of the pole face. In other words, the current induced in the shading band causes the flux in the shaded portion of the pole to lag the flux in the unshaded portion of the pole.

Therefore, the flux in the unshaded portion reaches its maximum before the flux in the shaded portion. The result is like a rotating field moving from the unshaded to the shaded portion of the pole, and causing the motor to produce a slow starting torque. The shaded-pole motor is rugged, cheap, small in size, and needs minimum maintenance. It has very low starting torque, efficiency, and power factor, and is used in turntables, motion-picture projectors, small fans, and vending machines.

Example 9.2

Assume that a single-phase, 120 V, 60 Hz, two-pole induction motor has the following *standstill* impedances when tested at rated frequency:
   Main winding: $Z_m = 1.6 + j4.2\,\Omega$
   Auxiliary winding: $Z_a = 3.2 + j6.5\,\Omega$
   Determine the following:

(a) The value of external resistance that needs to be connected in series with the auxiliary winding to have maximum starting torque, if the motor is operated as a resistance split-phase motor
(b) The value of the capacitor to be connected in series with the auxiliary winding to have maximum starting torque, if the motor is to be operated as a capacitor-start motor
(c) The value of the capacitor that needs to be connected in series with the auxiliary winding to have maximum starting torque per ampere of the starting current as a capacitor-start motor

Solution

(a) The value of the external resistance that needs to be connected in series with the auxiliary winding is found from Equation 9.26 as

$$R_a = \frac{X_a}{X_m}(R_m + Z_m)$$

$$= \frac{6.5}{4.2}(1.6 + 4.495)$$

$$= 9.43\,\Omega$$

where

$$Z_m = Z_m \angle \theta_m$$

$$= 4.495 \angle 69.15° \ \Omega$$

Therefore, the value of external resistance is found as

$$R_{ext} = 9.43 - 6.5$$

$$= 2.93 \ \Omega$$

(b) The value of the capacitive reactance that needs to be connected in series with the auxiliary winding is found from Equation 9.29 as

$$X_c = X_a + \frac{R_a R_m}{X_m + Z_m}$$

$$= 6.5 + \frac{(3.2)(1.6)}{4.2 + 4.495}$$

$$= 7.089 \ \Omega$$

Thus, the value of the capacitance is found as

$$C = \frac{1}{\omega X_c}$$

$$= \frac{10^6}{2\pi 60(7.089)}$$

$$= 374.19 \ \mu F$$

(c) To have maximum starting torque per ampere of the starting current, the value of the capacitive reactance is found from Equation 9.32 as

$$X_c = \frac{X_a + \{-X_m R_a + \left[R_a(R_a + R_m)\right]^{1/2} Z_m\}}{R_m}$$

$$= \frac{6.5 + \{-4.2(3.2) + [3.2(3.2 + 1.6)]^{1/2} 4.4951\}}{1.6}$$

$$= 9.11 \ \Omega$$

Hence, the value of the capacitance is found as

$$C = \frac{1}{\omega X_c}$$

$$= \frac{10^6}{2\pi 60(9.11)}$$

$$= 291.17 \ \mu F$$

9.5 UNIVERSAL MOTORS

A universal motor is a single-phase **series** motor that can operate on either alternating or direct current with similar characteristics as long as both the stator and the rotor cores are completely laminated. It is basically a series dc motor with laminated stator and rotor cores without laminated cores, the core losses would be tremendous if the motor were supplied by an ac power source. Since such a motor can run from either an ac (at any frequency up to design frequency) or a dc (zero frequency) power source, it is often called a *universal motor*. Here, the main field and armature field are in phase, because the same current flows through the field and armature.[*]

When it is supplied by an ac power source, both the main field and armature field will reverse at the same time, but the torque and the rotational direction will always be in the same direction. Like all series motors, the *no-load speed* of the universal motor is usually high, often in the range of 1,500–20,000 rpm, and is limited by windage and friction.

It is typically used in fractional horsepower ratings (1/20 hp or less) in many commercial appliances,[†] such as electric shavers, portable tools, sewing machines, mixers, vacuum cleaners, small hand-held hair dryers, drills, routers, and hand-held grinders. In such applications, it is always directly loaded with little danger of motor runaway.

The best way to control the speed and torque of the universal motor is to vary its input voltage by using a solid-state device (an SCR or a TRIAC).

There are also large (in the range of 500 hp) single-phase series ac motors that are still extensively used for traction applications such as electric locomotives.

Under dc excitation, the developed torque and induced voltage of a universal motor can be expressed, respectively, as

$$T_d = K_a \times \Phi_{d(dc)} \times I_a \tag{9.34}$$

$$E_a = K_a \times \Phi_{d(dc)} \times \omega_{m(dc)} \tag{9.35}$$

If magnetic linearity can be assumed, then the developed torque and induced voltage can be expressed, respectively, as

$$T_d = K \times I_a^2 \tag{9.36}$$

$$E_a = K \times I_a \times \omega_{m(dc)} \tag{9.37}$$

Under ac excitation, the average developed torque and the rms value of the induced voltage of a universal motor can be expressed, respectively, as

$$T_d = K_a \times \Phi_{d(ac)} \times I_a \tag{9.38}$$

$$E_a = K_a \times \Phi_{d(ac)} \times \omega_{m(ac)} \tag{9.39}$$

---

[*] Therefore, a shunt dc motor cannot operate on an ac power source due to the fact that the shunt field is highly inductive, and the armature is basically highly resistive. Hence, the armature and the field are not in phase. The high inductance of the field winding causes the field current to lag the armature current by such a large angle that a very low net torque is produced.

[†] Such applications require a motor with relatively high starting torques and speeds that exceed the maximum synchronous speed of 3600 rpm at 60 Hz. They are built for voltages from 1.5 to 250 V. Therefore, universal motors are ideal for such applications.

where

$\Phi_{d(ac)}$ is the rms value of the d-axis flux
I_a is the rms value of the motor current

If magnetic linearity can be assumed, then the average developed torque and the rms value of the induced voltage can be expressed, respectively, as

$$T_d = K \times I_a^2 \tag{9.40}$$

$$E_a = K \times I_a \times \omega_{m(ac)} \tag{9.41}$$

Since the developed mechanical power is

$$P_d = P_{mech(ac)} = E_a \times I_a \tag{9.42}$$

then

$$T_d = \frac{P_d}{\omega_{m(ac)}} = \frac{E_a \times I_a}{\omega_{m(ac)}} \tag{9.43}$$

Notice that the terminal voltage (under ac excitation) is

$$V_1 = E_a + I_a Z_a + I_a Z_{se} \tag{9.44}$$

$$Z_a = R_a + jX_a \tag{9.45}$$

$$Z_{se} = R_{se} + jX_{se} \tag{9.46}$$

Therefore, the input voltage is

$$V_1 = E_a + I_a(R_a + jX_a) + I_a(R_{se} + X_{se}) \tag{9.47}$$

$$V_1 = E_a + I_a(R_a + R_{se}) + jI_a(X_a + X_{se}) \tag{9.48}$$

Hence, the induced voltage can be expressed as

$$E_a = V_1 - I_a(R_a + R_{se}) - jI_a(X_a + X_{se}) \tag{9.49}$$

Based on the assumption that the armature current under dc excitation and the rms value of the armature current under ac excitation are the same, it can be shown that

$$\frac{E_{a(dc)}}{E_{a(ac)}} = \frac{K_a \times_a \Phi_{d(dc)} \times \omega_{m(dc)}}{K_a \times_a \Phi_{d(ac)} \times \omega_{m(ac)}} \cong \frac{\omega_{m(dc)}}{\omega_{m(ac)}} \tag{9.50}$$

Furthermore, if saturation takes place while the motor is under ac excitation, then the flux under $\Phi_{d(ac)}$ is a little less than the flux under dc excitation $\Phi_{d(dc)}$. Thus, the ratio of the induced voltage becomes

$$\frac{E_{a(dc)}}{E_{a(ac)}} \cong \frac{1}{\cos\theta} \tag{9.51}$$

which is greater than unity. When the terminal voltage, armature current, and torque are constant, the speed of a universal motor is lower under ac excitation than under dc excitation. In summary, under ac excitation, the universal motor produces a lower speed, a poorer power factor, and a pulsating torque.

Example 9.3

A single-phase, 120 V, 60 Hz universal motor is operating at 1800 rpm and its armature current is 0.5 A when it is supplied by a 120 V dc source. Its resistance and reactance are 22 and 100 Ω, respectively. If the motor is supplied by ac power, determine the following:

(a) The speed of the motor when it is connected to an ac source
(b) The power factor of the motor when it is connected to an ac voltage source
(c) The developed torque of the motor when it is connected to an ac voltage source

Solution

(a) When the motor is supplied by the dc source:

$$E_{a(dc)} = V_1 - I_a R_a$$

$$= (120\,\text{V}) - (0.5\,\text{A})(22\,\text{W})$$

$$= 109\ \text{V}$$

When the motor is supplied by the ac source:

$$E_{a(ac)} + I_a R_a = [V_1^2 - (I_a X)^2]^{1/2}$$

or

$$E_{a(ac)} = [V_1^2 - (I_a X)^2]^{1/2} - I_a R$$

$$= \{(120\,\text{V}) - [(0.5\,\text{A})(100\,\Omega)]^2\}^{1/2} - (0.5\,\text{A})(22\,\Omega)$$

$$= 98.09\ \text{V}$$

By assuming the same flux for the same current under the dc and ac operation from Equation 9.50,

$$\frac{E_{a(dc)}}{E_{a(ac)}} = \frac{n_{dc}}{n_{ac}}$$

Thus, the speed of the motor when it is connected to an ac source is

$$n_{ac} = \frac{n_{dc} E_{a(ac)}}{E_{a(dc)}}$$

$$= \frac{(1800\,\text{rpm})(98.09\,\text{V})}{109\,\text{V}}$$

$$= 1619.83\ \text{rpm}$$

(b) The power factor of the rotor is found as

$$\cos\theta = \frac{E_a + I_a R_a}{V_1}$$

$$= \frac{(98.09\,\text{V}) + (0.5\,\text{A})(22\,\Omega)}{120\,\text{V}}$$

$$= 0.91\,\text{lagging}$$

(c) The developed (mechanical) power of the motor is

$$P_d = P_{mech} = E_a I_a$$

$$= (98.09\,\text{V})(0.5\,\text{A})$$

$$= 49\,\text{W}$$

Therefore, the developed torque of the motor is

$$T_d = \frac{P_d}{\omega_m}$$

$$= \frac{P_d}{n_m(2\pi/60)}$$

$$= \frac{49\,\text{W}}{(1619.83\,\text{rpm})(2\pi/60)}$$

$$= 0.289\,\text{N} \cdot \text{m}$$

9.6 SINGLE-PHASE SYNCHRONOUS MOTORS

Single-phase synchronous motors are used for applications that require precise speed. They include the reluctance motor, the hysteresis motor, and the stepper motor. Reluctance and hysteresis motors are used in electrical clocks, timers, and turntables. Stepper motors are used in electrical typewriters, printers, computer disk drives, VCRs, and other electronic equipment.

9.6.1 RELUCTANCE MOTORS

A *reluctance motor** is a salient-pole synchronous machine with no field excitation. The operation of this type of motor depends on reluctance torque that tends to align the rotor under the nearest pole of the stator and defines the direction of rotation. The torque applied to the rotor of the motor is proportional to sin 2δ, where δ is defined as the electrical angle between the rotor and stator magnetic fields. Hence, the reluctance torque of the motor becomes maximum when the angle between the rotor and stator magnetic fields is 45°.

In general, any induction motor can be modified into a self-starting reluctance type synchronous motor. This can be done by altering the rotor so that the laminations have salient rotor poles, as shown in Figure 9.9a. Notice that the saliency is introduced by removing some rotor teeth from the proper sections to make a four-pole rotor structure.

This rotor structure can then be used for a four-pole reluctance motor. The reluctance of the air-gap flux path will be far greater at the places where there are no rotor teeth. Thus, the reluctance motor can start as an induction motor as long as the squirrel-cage bars and end rings are left in place.

---

* It is also referred to as a *single-phase salient-pole synchronous-induction motor* or simply *as a synchronous motor*.

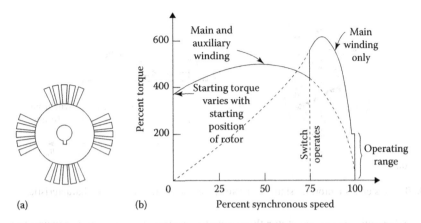

FIGURE 9.9   Reluctance motor: (a) rotor design and (b) typical torque–speed characteristic.

This motor, coming up to speed as an induction motor, will be pulled into synchronism by the pulsating ac single-phase field due to a reluctance torque produced by the salient iron poles with lower-reluctance air gaps.

In summary, the torque develops because of the tendency of the rotor to align itself with the rotating field so that a reluctance motor starts as an induction motor, but continues to operate as a synchronous motor.

There are two stator windings, namely, a main winding and an auxiliary winding. When the motor starts as an induction motor, it has both windings energized. At a speed of approximately 75% of the synchronous speed, a centrifugal switch disconnects the auxiliary winding so that the speed of the motor increases to almost the synchronous speed. At that time, as a result of the reluctance torque, the rotor snaps into synchronism and continues to rotate at synchronous speed.*

Figure 9.9b shows the torque–speed characteristic of a typical single-phase reluctance motor. Note that the value of the starting torque depends on the position of the unsymmetrical rotor with respect to the field winding.

Also, since there is no dc excitation in the rotor of a reluctance motor, it develops less torque than an excited synchronous motor of the same size. Since the volume of a machine is approximately proportional to the torque, the reluctance motor is about three times larger than a synchronous motor with the same torque and speed.

9.6.2   HYSTERESIS MOTORS

These motors use the phenomenon of hysteresis to develop a mechanical torque. The rotor of a hysteresis motor is a smooth cylinder made up of a special magnetic material such as hard steel, chrome, or cobalt, and has no teeth, laminations, or windings.

The stator windings are made up of distributed windings in order to have a sinusoidal space distribution of flux. The stator windings can be either single or three phase. In single-phase motors, the stator windings are customarily permanent-split-capacitor type, as shown in Figure 9.6a. If the stator windings are energized, a revolving magnetic field is developed, rotating at synchronous speed. This rotating field magnetizes the metal of the rotor and induces **eddy currents**. Due to the hysteresis, the magnetization of the rotor lags with respect to the inducing revolving field, as shown in Figure 9.10a.

The lag angle δ exists because the metal of the rotor has a large hysteresis loss. The angle by which the rotor magnetic field lags the stator magnetic field depends on the hysteresis loss of the

---

* If the load of this type of reluctance motor increases significantly, the motor will slip out of synchronism. However, it will continue to run with some slip just like an induction motor

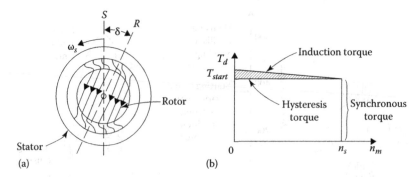

FIGURE 9.10   Hysteresis motor: (a) stator and rotor field and (b) torque–speed characteristic.

rotor. At synchronous speed, the stator flux stops to sweep across the rotor, causing the eddy currents to disappear, and the rotor behaves like a permanent magnet. At that time, the developed torque in the motor is proportional to the angle δ between the rotor and stator magnetic fields, which is dictated by the hysteresis of the motor. Consequently, a constant torque (indicated as the hysteresis torque in Figure 9.10b) exists from zero up to and including synchronous speed.

As indicated in the figure, a hysteresis motor, whose rotor is round and not laminated, has an induction torque which is added to the hysteresis torque until synchronous speed is reached. Hysteresis motors are self-starting and are manufactured up to about 200 W for use in precise-speed drives. The applications include clocks, record players, compact disk players, and servomechanisms.

9.6.3   STEPPER MOTORS

These motors are also referred to as *stepping* or *step motors*. Basically, a stepper motor is a type of ac motor that is built to rotate a specific number of degrees in response to a digital input in the form of a pulse. Step sizes typically vary from 1°, 2°, 2.5°, 5°, 7.5°, 15°, or more for each electrical pulse.

Stepper motors are often used in digital control systems, where the motor is given open-loop commands in the form of a train of pulses and the controller directs pulses sequentially to the motor windings to turn a shaft or move an object a specified distance.

They are excellent devices for accurate speed control or precise position control without any feedback. In such usage, the axis of the motor's magnetic field *steps* around the air gap at a speed that is based on the frequency of pulses. The rotor inclines to align itself with the axis of the magnetic field. Therefore, the rotor *steps* in synchronism with the motion of the magnetic field. Because of this, the motor is referred to as a *stepper motor*.

These motors are relatively simple in construction and can be controlled to step in equal increments in either direction. They are increasingly used in digital electronic systems because they do not need a position sensor or a feedback system to make the output response follow the input command.

Figure 9.11 illustrates a primitive form of control implementation in a stepper motor. Notice that a train of f pulses per second is furnished to the digital driver circuit and that the input of the controller is divided so that the output is sent in sequence to one phase winding at a time. In the event that 2p is the number of phases and k is the number of teeth, then the rotor angular motion per pulse is *a step* of π/kp radians. Accordingly, the rotor moves n steps per second. Hence, the angular speed is exactly $\pi n/kp$ rad/s.

Stepper motors are classified according to the type of motor used. If a permanent-magnet motor is used, it is called a *permanent-magnet stepper motor*.

If a variable-reluctance motor is used, it is called a *variable-reluctance stepper motor*. Permanent-magnet stepper motors have a higher inertia and thus a slower acceleration than variable-reluctance stepper motors. For example, the maximum step rate for permanent-magnet

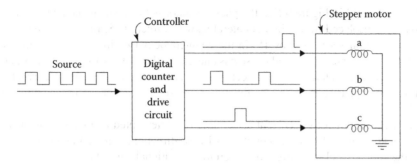

FIGURE 9.11 Driver for a stepper motor.

stepper motors is 300 pulses per second, but it can be as high as 1200 pulses per second for variable-reluctance stepper motors.

The permanent-magnet stepper motor develops more torque per ampere stator current than the variable-reluctance stepper motor. There is also a *hybrid stepper motor* that has a rotor with an axial permanent magnet in the middle and ferromagnetic teeth at the outer sections.

The hybrid stepper motor combines the characteristics of the variable-reluctance and permanent-magnet stepper motors. A variable-reluctance stepper motor can be the single-stack type or the multiple-stack type. The latter one is used to provide smaller step sizes. Its motor is segmented along its axial length into magnetically isolated sections which are called *stacks* which are excited by a separate winding called a **phase**. Even though variable-reluctance stepper motors with up to seven stacks and phases are used, three-phase arrangements are more often used.

Figure 9.12b shows a variable-reluctance stepper motor that has a rotor with eight poles and three separate eight-pole stators arranged along the rotor. If phase-a poles of a stator are energized by a set of series-connected coils with current i_a, the rotor poles align with the stator poles of phase a.

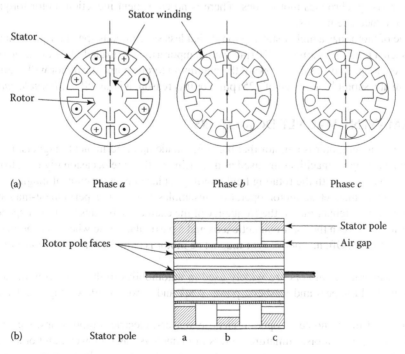

FIGURE 9.12 A variable-reluctance stepper motor having a rotor with eight poles and three separate eight-pole stators arranged along the rotor: (a) cross section of a stepper motor operation and (b) cut off view of the motor.

As can be observed in Figure 9.12a, the phase-*b* stator is the same as the phase-*a* stator except that its poles are displaced by 15° in a counterclockwise direction. Similarly, the phase-*c* stator is displaced from the phase-*b* stator by 15° in the counterclockwise direction. When the flow of the current i_s in phase *a* is interrupted and phase *b* is energized, the motor will develop a torque, rotating its rotor by 15° in the counterclockwise direction. Similarly, when the flow of the current i_b in phase *b* is interrupted and phase *c* is energized, the motor will rotate another 15° in the counterclockwise direction.

Finally, when the flow of the current i_c in phase *c* is interrupted and phase *a* is energized, the motor will rotate another 15° in the counterclockwise direction, completing a one-step (i.e., 45°) rotation in the counterclockwise direction. Therefore, additional current pulses in the *abc* sequence will develop additional counterclockwise stepping motions. Reversing the current-pulse sequence to *abc* will develop reversed rotation.

For an *n*-stack motor, the rotor or stator (but not both) on each stack is displaced by 1/*n* times the pole-pitch angle. Permanent-magnet stepper motors require two phases and current polarity is important. The hybrid stepper motor varies significantly from a multistack variable-reluctance stepper motor in that the stator pole structure is continuous along the length of the rotor.

9.7   SUBSYNCHRONOUS MOTORS

A subsynchronous motor has a rotor with an overall cylindrical outline and yet it is as toothed as a many-pole salient-pole rotor. For example, a typical motor may have 16 teeth or poles, and in combination with a 16-pole stator will normally rotate at a synchronous speed of 450 rpm when operated at 60 Hz.

The motor starts as a hysteresis motor. At synchronous speed, the rotor poles induced in a hysteresis rotor stay at fixed spots on the rotor surface as the rotor rotates into synchronism with the rotating magnetic field of the stator. The hysteresis torque is in effect when the rotor rotates at less than synchronous speed.

Subsynchronous motors, which are self-starters, start and accelerate with hysteresis torque just as the hysteresis synchronous motor does. There is no equivalent induction-motor torque like the one found in reluctance motors.

This type of motor has a higher starting torque but less synchronous speed torque than reluctance torque. If such a motor operating at 450 rpm were temporarily overloaded, it would drop out of synchronism. As the speed drops down toward the maximum torque point, the motor will again lock into synchronism at a submultiple speed of 225 rpm. For this reason, it is called a *subsynchronous motor*.

9.8   PERMANENT-MAGNET DC MOTORS

A permanent-magnet motor is a motor that has poles made up of permanent magnets. Even though most permanent-magnet machines are used as dc machines, they are occasionally built to operate as synchronous machines with the rotating field winding replaced by a permanent magnet.

The permanent-magnet ac-motor operation resembles that of the permanent-magnet stepper motor. Just as in the stepper motor, the frequency of the excitation dictates the motor speed, and the angular position between the rotor magnetic axis and a particular phase when it is energized affects the developed torque. Often, a permanent-magnet ac motor is called a *brushless motor* or *brushless dc motor*.

Permanent-magnet dc motors are widely used in automobiles to drive air conditioners, heater blowers, windshield wipers and washers, power seats and power windows, tape decks, and radio antennas.

They are used in the home to operate electric shavers, electric toothbrushes, carving knives, vacuum cleaners, power tools, miniature motors in many toys, lawn mowers, and other equipment that uses batteries. They are used as starter motors for outdoor motors. In computers, they are used for capstan and tape drives.

They can also be used in control systems such as dc servomotors and tape drives. In these applications, they are often used as fractional-horsepower motors for economic reasons. However, they can also be built in sizes greater than 200 hp.

Since there is no field winding in a permanent-magnet dc motor, it has a smooth stator structure on which a cylindrical shell made up of a permanent magnet is mounted.

Hence, the magnetic field is produced by the permanent magnet. The rotor of this permanent-magnet motor is a wound armature. The dc power supply is connected directly to the armature conductors through a brush/commutator assembly.

In these motors, there are basically three types of permanent magnets, namely, alnico magnets, ceramic (or ferrite) magnets, and rare-earth magnets (samarium-cobalt magnets). Ceramic magnets are usually used for low-horsepower slow-speed motors. They are most economical in fractional horsepower motors and are also less expensive than alnico in motors up to 10 hp.

The rare-earth magnets are very expensive; however, they have proven to be the most cost effective in very small motors. In general, alnico magnets are used in very large motors up to 200 hp. It is also possible to use special combinations of magnets and ferromagnetic materials to achieve high performance (i.e., high torque, high efficiency, and low volume) at a low cost.

Figure 9.13 shows a cutaway view of a permanent- magnet dc motor. A permanent-magnet dc motor is basically a shunt dc motor with its field circuit replaced by permanent magnets. Since the flux of the permanent magnet cannot be changed, its speed can only be controlled by varying its armature voltage and armature circuit resistance.

Therefore, the equivalent circuit of a permanent-magnet dc motor is made up of an armature connected in series with the armature-circuit resistance R_a. Hence, the internal generated voltage can be determined from

$$E_a = K_a \times \Phi_d \times \omega_m \tag{9.52}$$

where
   K_a is the armature constant
   Φ_d is the net flux per pole

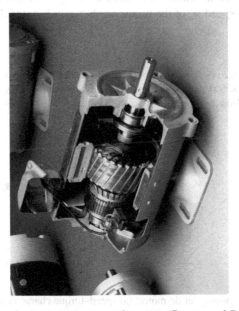

FIGURE 9.13 Cutaway view of a permanent-magnet dc motor. (Courtesy of General Electric, Fairfield, CT.)

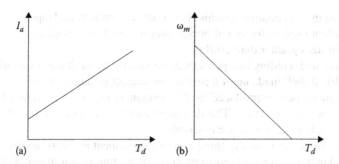

FIGURE 9.14 For a permanent-magnet dc motor: (a) typical current–torque characteristic and (b) typical speed–torque characteristic.

In a permanent-magnet dc machine Φ_d is constant; thus,

$$E_a = K \times \omega_m \tag{9.53}$$

where

$$K = K_a \times \Phi_d \tag{9.54}$$

and is called the *torque constant of the motor*. It is determined by the armature geometry and the properties of the permanent magnet used. The developed torque of the motor is found from

$$T_d = \frac{E_a \times I_a}{\omega_m} = K \times I_a \tag{9.55}$$

Figure 9.14a and b shows typical current–torque and speed–torque characteristics of a permanent-magnet dc motor, respectively. Varying terminal voltage V_t of the motor changes the no-load speed of the motor, but the slope of the curves remains constant, as shown in Figure 9.15a.

However, varying the armature-circuit resistance R_a changes the speed–torque characteristic, but does not affect the no-load speed ω_o of the motor, as shown in Figure 9.15b.

Example 9.4

Assume that the armature resistance of a permanent-magnet dc motor is 1.2 Ω. When it is operated from a dc source of 60 V, it has a no-load speed of 1950 rpm and is supplied by 1.5 A at no load. Determine the following:

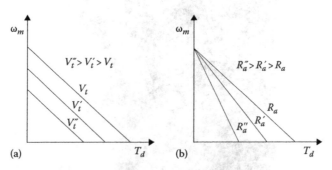

FIGURE 9.15 For a permanent-magnet dc motor: (a) speed–torque characteristics for different supply voltages and (b) speed–torque characteristics for different armature circuit resistances.

(a) The torque constant
(b) The no-load rotational losses
(c) The output in horsepower if it is operating at 1500 rpm from a 50 V source

Solution

(a) The internal generated voltage of the motor is

$$E_a = V_t - I_a R_a$$

$$= (60\,\text{V}) - (1.5\,\text{A})(1.2\,\Omega)$$

$$= 58.2\ \text{V}$$

At speeds of 1950 rpm, its speed is

$$K = \frac{E_a}{\omega_m} = \frac{58.2\,\text{V}}{204.2\,\text{rad/s}} = 0.285\,\text{V/(rad/s)}$$

(b) Since all the power supplied at no load is used for rotational losses of the motor

$$P_{rot} = E_a I_a$$

$$= (58.2\,\text{V})(1.5\,\text{A})$$

$$= 87.3\ \text{W}$$

(c) At 1500 rpm,

$$\omega_m = 1500\left(\frac{2\pi}{60}\right) = 157.01\ \text{rad/s}$$

hence,

$$E_a = K\omega_m$$

$$= (0.285\,\text{V/(rad/s)})(157.01\,\text{rad/s})$$

$$= 44.75\ \text{V}$$

Therefore, the input power is

$$P_{shaft} = P_d = E_a I_a$$

$$= (44.75\,\text{V})(4.38\,\text{A})$$

$$= 196\ \text{W}$$

Since the rotational losses are approximately constant, the output power of the motor is

$$P_{out} = P_{shaft} - P_{rot}$$

$$= 196 - 87.3$$

$$= 108.7\ \text{W}$$

or in horsepower,

$$P_{out} = \frac{108.7 \text{ W}}{746 \text{ W/hp}} = 0.1457 \text{ hp}$$

PROBLEMS

9.1 A 1/2 hp, single-phase, 120 V, four-pole induction motor has the following constants in ohms referred to the stator:

$$R_1 = 1.95 \ \Omega \quad R_2 = 3.5 \ \Omega$$

$$X_1 = 2.5 \ \Omega \quad X_2 = 2.5 \ \Omega$$

$$X_m = 62 \ \Omega$$

The core losses of the motor are 35 W, and the friction, windage, and stray losses are 14 W. The motor is operating at the rated voltage and frequency with its starting winding open. For a slip of 4%, determine the following:
(a) The shaft speed in rpm
(b) The forward and backward impedances
(c) The input current
(d) The power factor
(e) The input power
(f) The total air-gap power
(g) The developed power
(h) The output power
(i) The developed torque
(j) The output torque
(k) The input power
(l) The efficiency of the motor

9.2 A 1/4 hp, single-phase, 120 V, 60 Hz, four-pole induction motor has the following constants in ohms referred to the stator:

$$R_1 = 2.5 \ \Omega \quad R_2 = 3.8 \ \Omega$$

$$X_1 = 2.2 \ \Omega \quad X_2 = 1.9 \ \Omega$$

$$X_m = 59 \ \Omega$$

The core losses of the motor are 30 W, the friction and windage losses are 10 W, and the stray losses are 4 W. The motor is operating at the rated voltage and frequency with its starting windings open. For a slip of 5%, determine the following:
(a) The forward and backward impedances
(b) The input current
(c) The input power
(d) The total air-gap power
(e) The developed torque
(f) The developed power
(g) The output power in watts and horsepower
(h) The output torque
(i) The efficiency of the motor

9.3 Use the data given in Problem 9.1 and determine the following for the single-phase induction motor:
(a) The air-gap power due to the forward field
(b) The air-gap power due to the backward field
(c) The rotor copper loss due to the forward field
(d) The rotor copper loss due to the backward field
(e) The total rotor copper loss

9.4 Use the data given in Problem 9.2 and determine the following:
(a) The air-gap power due to the forward field
(b) The air-gap power due to the backward field
(c) The rotor copper loss due to the forward field
(d) The rotor copper loss due to the backward field
(e) The total rotor copper loss

9.5 Determine the developed torque given in Problem 9.1, if it is operating at 4% slip and its terminal voltage is
(a) 208 V
(b) 240 V

9.6 Determine the developed torque in the motor given in Problem 9.2, if it is operating at 5% slip and its terminal voltage is
(a) 208 V
(b) 240 V

9.7 Assume that the currents in the main and the auxiliary windings of a single-phase induction motor are given, respectively, as

$$i_m = \sqrt{2}I_m\cos\omega t \quad \text{and} \quad i_a = \sqrt{2}I_a\cos(\omega t + \theta_a)$$

and that the windings are located in quadrature with respect to each other. If the effective number of turns for the main and auxiliary windings are N_m and N_a, determine the following:
(a) A mathematical expression for the rotating mmf wave of the stator
(b) The magnitude and the phase angle of the auxiliary winding current needed to produce a balanced, two-phase system

9.8 Consider the solution of Example 9.2 and develop a table to compare the starting torques and starting currents in part (a), (b), and (c) expressed as per unit of the starting torque without any external element in the auxiliary circuit, when connected to a supply of 120 V at 60 Hz.

9.9 Assume that the impedances of the main and auxiliary windings of a single-phase 120 V, 60 Hz, capacitor-start induction motor are

$$Z_m = 4.2 + j3.4\,\Omega \quad \text{and} \quad Z_a = 9 + j3\,\Omega$$

Determine the value of the starting capacitance that will cause the main and auxiliary winding currents to be in quadrature at starting.

9.10 Assume that the armature resistance of a permanent-magnet dc motor is 1.4 Ω. When it is operating from a dc source of 75 V, it has a no-load speed of 2200 rpm and is supplied by 1.7 A at no load. Determine the following:
(a) The torque constant
(b) The no-load rotational losses
(c) The output in horsepower, if it is operating at 1800 rpm from a 70 V source

9.11 Determine the best motor selection for the following applications and explain the reasoning:
(a) electric drill, (b) electric clock, (c) refrigerator, (d) vacuum cleaner, (e) air conditioner fan, (f) air conditioner compressor, (g) electric sewing machine, (h) electric shaver, and (i) electric toothbrush.

9.12 Assume that a permanent-magnet dc motor is operating with a magnetic flux of 5 mWb, that its armature resistance is 0.7 Ω and the supply voltage is 30 V. If the motor load is 2 N · m and its armature constant is 110, determine the following:

(a) The operating speed of the motor

(b) The developed torque under a blocked-rotor condition

10 Transients and Dynamics of Electric Machines

Our understanding of a phenomenon becomes complete when, and only when, we can measure the quantities and factors involved and can assign numbers to them and can tie these numbers together with mathematical equations.

William Thomson (Lord Kelvin)

10.1 INTRODUCTION

Steady-state operation and behavior of ac and dc electromechanical machines have been reviewed in previous chapters. However, disturbances or sudden changes (e.g., faults, sudden load changes or shifts, or sudden changes in network configurations or in supply voltages) can cause these machines to behave quite differently. During such transient periods, it often becomes very important to have a knowledge of machine behavior.

Here, the *transient period* is defined as the time period between the beginning of the disturbance and the following steady-state operating conditions. In this chapter, a brief review of both the *electrical transient behavior* and the *mechanical transient behavior* (*dynamic response*) of dc and ac machines is presented. However, an in-depth study of this topic is outside the scope of this book.*

10.2 DC MACHINES

In this type of study, it is customary to simplify the problem by making various assumptions. For example, *it is assumed that the field mmf acts only along the d-axis and the armature mmf acts only along the q-axis.*

In other words, there is no mutual inductance between the field circuit and the armature circuit. Therefore, the armature reaction has no demagnetizing effect. However, the effects of the armature reaction may be added later as an additional field excitation requirement. It can also be assumed that magnetic saturation does I_f.

A given dc machine can be represented by two coupled electrical circuits both with resistances and inductances, as shown in Figure 10.1. These circuits, representing the field and the armature of the dc machine, are coupled through the electromagnetic field.

Similarly, the electrical system is coupled to the mechanical system through the developed electromagnetic torque T_d and external mechanical torque, which can be either an input torque T_s from a prime mover or a load torque T_L.

10.3 SEPARATELY EXCITED DC GENERATOR

A schematic representation of a separately excited dc generator is shown in Figure 10.1a. Notice that the inductances of the armature winding and field winding are represented by L_a and L_f, respectively. The developed torque is given by

---

* For an excellent reference, see *Power System Control and Stability*, P. M. Anderson and A. A. Fouad, IEEE Press, New York, 1994.

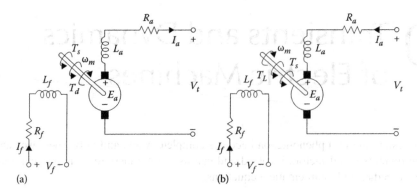

FIGURE 10.1 Schematic representation of a separately excited dc machine: (a) generator and (b) motor.

$$T_d = K_a \times \Phi_d \times I_a \qquad (10.1)$$

and the internal generated voltage is given by

$$E_a = K_a \times \Phi_d \times \omega_m \qquad (10.2)$$

Since *magnetic linearity* (i.e., the air-gap flux is directly proportional to the field current I_f) is assumed,

$$T_d = K_f \times I_f \times I_a \qquad (10.3)$$

$$E_a = K_f \times I_f \times \omega_m \qquad (10.4)$$

where K_f is a constant.

Assume that the generator is driven at a constant speed of ω_m by the prime mover. The voltage equation for the field circuit can then be expressed as

$$V_f = I_f R_f + L_f \frac{dI_f}{dt} \qquad (10.5)$$

where I_f, R_f, and L_f are the current, resistance, and self-inductance of the field circuit, respectively. Therefore,

$$\frac{V_f}{R_f} = I_f + \tau_f \left(\frac{dI_f}{dt} \right) \qquad (10.6)$$

where $\tau_f = L_f / R_f$ is the time constant of the field circuit. Assuming that the effect of saturation is negligible, the internal generated voltage is

$$E_a = K_f \times I_f \times \omega_m = K_g \times I_f \qquad (10.7)$$

where $K_g = K_f \times \omega_m$ is the slope of the linear section of the magnetization curve. For the armature circuit, the voltage equation can be written in terms of the generator terminal voltage as

$$E_a - V_t = R_a \times I_a + L_a \left(\frac{dI_a}{dt} \right) \tag{10.8}$$

or

$$\frac{E_a - V_t}{R_a} = I_a \times \tau_a \left(\frac{dI_a}{dt} \right) \tag{10.9}$$

where $\tau_a = L_a/R_a$ is the time constant of the armature circuit. When the separately excited dc generator is providing an armature current I_a to an electrical load, the developed electromagnetic torque is

$$T_d = \frac{P_d}{\omega_m} = \frac{E_a \times I_a}{\omega_m} \tag{10.10}$$

The *dynamic equation* of the dc machine is a function of the mechanical torque applied to its shaft by the prime mover. Therefore,

$$T_{shaft} = T_d + J \left(\frac{d\omega_m}{dt} \right) \tag{10.11}$$

or

$$T_{shaft} - T_d = J \left(\frac{d\omega_m}{dt} \right) \tag{10.12}$$

where J is the moment of inertia of the rotor and the prime mover.

The electrical *transient* behavior of a dc generator involves both the transient behavior of its field circuit and the transient behavior of its armature circuit. It is easier to study this behavior of the machine by first finding the appropriate transfer function and then applying the techniques of Laplace transform theory.

10.3.1 Field-Circuit Transient

Assume that the generator is being run by a prime mover at a constant speed of ω_m, with its armature circuit open and its field circuit having just been closed, as shown in Figure 10.1a. The Laplace transform of the voltage equation (10.5) for the field circuit with zero initial conditions can be given as

$$V_f(s) = I_f(s)R_f + L_f s I_f(s) \tag{10.13}$$

$$V_f(s) = I_f(s) (R_f + sL_f) \tag{10.14}$$

where
$I_f(s)$ is the Laplace transform of the time function I_f
$V_f(s)$ is the Laplace transform of V_f

Thus, the transfer function relating the field current to the field voltage is given by

$$\frac{I_f(s)}{V_f(s)} = \frac{1}{R_f + sL_f} \tag{10.15}$$

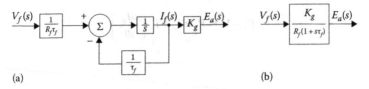

(a)   (b)

FIGURE 10.2 Block diagram of a separately excited dc generator: (a) with the armature circuit closed and (b) with simplified diagram.

$$\frac{I_f(s)}{V_f(s)} = \frac{1}{R_f(1+s\tau_f)} \tag{10.16}$$

where $\tau_f = I_f/R_f$ as before. From Equation 10.7, the Laplace transform of the internal generated voltage is

$$E_a(s) = K_g I_f(s) \tag{10.17}$$

Therefore, the total transfer function relating the internal generated voltage to the field circuit voltage can be expressed as

$$\frac{E_a(s)}{V_f(s)} = \frac{E_a(s)}{I_f(s)} \times \frac{I_f(s)}{V_f(s)} = \frac{K_g}{R_f(1+s\tau_f)} \tag{10.18}$$

The block diagram representation of this equation is shown in Figure 10.2. The time domain response associated with this total transfer function can be expressed as

$$e_a(t) = \frac{K_g V_f}{R_f}(1-e^{t/\tau_f}) \tag{10.19}$$

$$E_a(t) = E_a(1-e^{t/\tau_f}) \tag{10.20}$$

Therefore, in the steady-state, the internal generated voltage becomes

$$E_a = \frac{K_g V_f}{R_f} = K_g I_f \tag{10.21}$$

where the steady-state field current is

$$I_f = \frac{V_f}{I_f} \tag{10.22}$$

10.3.2 ARMATURE-CIRCUIT TRANSIENT

Assume that the armature circuit of the generator has just been closed over an electrical load and that the armature speed and the field current are kept constant. Therefore, the internal generated voltage of the generator can be expressed as

$$E_a = R_a I_a + L_a\left(\frac{dI_a}{dt}\right) + R_L I_a + L_L\left(\frac{dI_a}{dt}\right) \tag{10.23}$$

where R_L and L_L are the resistance and the inductance of the electrical load, respectively. Thus,

$$E_a = (R_a + R_L) + (L_a + L_L)\left(\frac{dI_a}{dt}\right) \tag{10.24}$$

Its Laplace transform is

$$E_a(s) = I_a(s)(R_a + R_L)(1 + s\tau_{at}) \tag{10.25}$$

where τ_{at} is the armature-circuit time constant determined by

$$\tau_{at} = \frac{L_{at}}{R_{at}} = \frac{L_a + L_L}{R_a + R_L} \tag{10.26}$$

Therefore, the transfer function is

$$\frac{I_a(s)}{E_a(s)} = \frac{1}{R_{at}(1 + s\tau_{at})} \tag{10.27}$$

Hence, its time domain response is

$$i_a(t) = \frac{E_a}{R_{at}}(1 - e^{t/\tau_{at}}) \tag{10.28}$$

Similarly, the total transfer function relating the armature current to the field voltage can be expressed as

$$\frac{I_a(s)}{V_f(s)} = \frac{I_a(s)}{E_a(s)} \times \frac{E_a(s)}{V_f(s)} \times \frac{K_g}{R_f R_{at}(1 + s\tau_f)(1 + s\tau_{at})} \tag{10.29}$$

Figure 10.3 shows the corresponding block diagram. Since the Laplace transform of the field voltage for a step change of voltage is

$$V_f(s) = \frac{V_f}{s} \tag{10.30}$$

$$I_a(s) = \frac{K_g V_f}{R_f R_{at} s(1 + s\tau_f)(1 + s\tau_{at})} \tag{10.31}$$

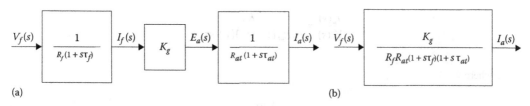

(a)   (b)

FIGURE 10.3   (a) Block diagram of a separately excited dc generator with the armature circuit closed and (b) simplified diagram.

Example 10.1

Assume that a separately excited dc generator has an armature-circuit resistance and inductance of $0.2\,\Omega$ and $0.2\,H$, respectively. Its field winding resistance and inductance are $120\,\Omega$ and $30\,H$, respectively. The generator constant K_g is $120\,V$ per field ampere at rated speed. Assume that the generator is driven by the prime mover at rated speed and that a $240\,V$ dc supply is suddenly connected to the field winding. Determine the following:

(a) The internal generated voltage of the generator
(b) The internal generated voltage in the steady state
(c) The time required for the internal generated voltage to rise to 99% of its steady-state value

Solution

(a) The time constant of the field circuit is

$$\tau_f = \frac{L_f}{R_f} = \frac{30}{120} = 0.25\ \text{s}$$

From Equation 10.19,

$$E_a(t) = \frac{K_g V_f}{R_f}(1 - e^{-t/\tau_f}) = \frac{(120)(240)}{120}(1 - e^{-t/0.25}) = 240(1 - e^{-4t})$$

(b) The internal generated voltage in the steady state is

$$E_a(\infty) = 240\ \text{V}$$

(c) The time required for the internal generated voltage to rise to 99% of its steady-state value is found from

$$0.99(240) = 240(1 - e^{-4t}) \quad t = 1.15\ \text{s}$$

Example 10.2

Assume that the generator given in Example 10.1 is driven by the prime mover at rated speed and is connected to a load made up of a resistance of $3\,\Omega$ and inductance of $1.4\,H$ that are connected in series with respect to each other. If a $240\,V$ dc supply is suddenly connected to the field winding, determine the armature current as a function of time.

Solution

From Equation 10.29, the transfer function relating the armature current to the field voltage is

$$\frac{I_a(s)}{V_f(s)} = \frac{K_g}{R_f R_{at}(1 + s\tau_f)(1 + s\tau_{at})}$$

where

$$\tau_f = \frac{L_f}{R_f} = \frac{30}{120} = 0.25\ \text{s}$$

and

$$\tau_{at} = \frac{L_{at}}{R_{at}} = \frac{L_a + L_L}{R_a + R_L} = \frac{0.2 + 1.4}{0.2 + 3} = 0.5 \text{ s}$$

Therefore,

$$\frac{I_a(s)}{V_f(s)} = \frac{120}{(120)(3.2)(1 + 0.25\, s)(1 + 0.5\, s)}$$

The Laplace transform of the field voltage for a step change of 240 V is

$$V_f(s) = \frac{240}{s}$$

Thus,

$$I_a(s) = \frac{240}{s} \times \frac{120}{384(1 + 0.25\, s)(1 + 0.5\, s)} = \frac{600}{s(s+4)(s+2)} = \frac{A_0}{s} \times \frac{A_1}{(s+4)} \times \frac{A_2}{(s+2)}$$

where

$$A_0 = \frac{600}{s(s+4)(s+2)}\bigg|_{s=-4} = 75$$

$$A_1 = \frac{600}{s(s+2)}\bigg|_{s=-4} = 75$$

$$A_2 = \frac{600}{s(s+4)}\bigg|_{s=-2} = -150$$

Hence,

$$I_a(s) = \frac{75}{s} + \frac{75}{s+4} - \frac{150}{s+2}$$

Therefore, taking the inverse Laplace transform, the armature as a function of time can be found as

$$I_a(s) = 75 + 75e^{-4t} - 150e^{-2t}$$

Example 10.3

Redo the Example 10.2 by using MATLAB®.

    (a) Write the MATLAB program script.
    (b) Give the MATLAB program output.

Solution

    (a) Here is the MATLAB program script:

```
clear
clc
syms s
```

```
%System Parameters
 Lf = 30;
 Rf = 120;
 Kg = 120;
 Vf = 240; % in V
 Rl = 3;
 Ll = 1.4;
 Ra = 0.2;
 La = 0.2;
 Vstep = 240;
%Solution
 Lat = La+Ll;
 Rat = Ra+Rl;
 tf = Lf/Rf
 tat = Lat/(Rat)
 Rf
 Rat
 F = (Vstep/s)*Kg/(Rf*Rat*(1+tf*s)*(1+tat*s))
 disp('* or *')
 F = simplify(F)
%Taking the inverse Laplace Transform
 Ia = ilaplace(F)
```

(b) Here is the MATLAB program output:

```
tf =
0.2500
tat =
0.5000
Rf =
120
Rat =
3.2000
F =
28800/(s*(s/2+1)*(96*s+384))
* or *
F =
600/(s*(s+2)*(s+4))
Ia =
75/exp(4*t) - 150/exp(2*t) + 75
»
```

10.4   SEPARATELY EXCITED DC MOTOR

Assume that a separately excited dc motor is operating at a constant field current I_f, as shown in Figure 10.1b, and that its speed is controlled by changing its terminal voltage V_t. Ignoring the effects of any saturation, the developed torque T_d and the internal generated voltage E_a can be expressed as

$$T_d = K_f \times I_f \times I_a = K_m \times I_a \tag{10.32}$$

$$E_a = K_f \times I_f \times \omega_m = K_m \times \omega_m \tag{10.33}$$

where $K_m = K_f I_f$ is a constant. This motor constant K_m can also be found from the magnetization curve as

$$K_m = \frac{E_m}{\omega_m} \tag{10.34}$$

The Laplace transforms of Equations 10.32 and 10.33 are

$$T_d(s) = K_m \times I_a(s) \tag{10.35}$$

$$E_a(s) = K_m \times \omega_m(s) \tag{10.36}$$

When the armature circuit is just energized at $t=0$, the terminal voltage of the motor is

$$V_t = E_a + R_a I_a + L_a \frac{dI_a}{dt} \tag{10.37}$$

By substituting Equation 10.33 into this equation,

$$V_t = K_m \omega_m + R_a I_a + L_a \frac{dI_a}{dt} \tag{10.38}$$

Its Laplace transform then is

$$V_t(s) = K_m \omega_m(s) + R_a I_a(s) + L_a s I_a(s) \tag{10.39}$$

or

$$V_t(s) = K_m \omega_m(s) + I_a(s) R_a (1 + s\tau_a) \tag{10.40}$$

where $\tau_a = L_a/R_a$ is the electrical time constant of the armature.

The developed torque of the motor must be equal to the sum of all opposing torques. Thus, the *dynamic equation* of the motor can be expressed as

$$T_d = K_m \times I_a = J \frac{d\omega_m}{dt} \tag{10.41}$$

where J is the moment of inertia including the load. If B is the equivalent viscous friction constant of the motor including the load, and T_L is the mechanical load torque. Here, $B\omega_m$ is the rotational loss torque of the system. The Laplace transform of Equation 10.41 is

$$T_d(s) = K \times J_a(s) = Js\omega_m(s) + B\omega_m(s) + T_L(s) \tag{10.42}$$

so that

$$\omega_m(s) = \frac{T_d(s) - T_L(s)}{B(1 + sJ/B)} \tag{10.43}$$

or

$$\omega_m(s) = \frac{K_m I_a(s) - T_L(s)}{B(1 + s\tau_m)} \tag{10.44}$$

where $\tau_m = J/B$ is the mechanical time constant of the system. Therefore, the Laplace transform of the armature current can be found from Equations 10.36 and 10.40 as

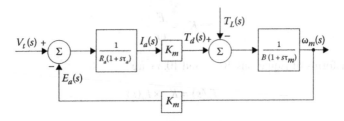

FIGURE 10.4 Block diagram of a separately excited dc motor with the armature circuit closed.

$$I_a(s) = \frac{V_t(s) - E_a(s)}{R_a(1 + s\tau_a)}$$ (10.45)

or

$$I_a(s) = \frac{V_t(s) - K_m\omega_m(s)}{R_a(1 + s\tau_a)}$$ (10.46)

The corresponding block diagram is shown in Figure 10.4.

If mechanical damping B is ignored, the overall transfer function can be found as

$$\frac{\omega_m(s)}{V_t(s)/K_m} = \frac{1}{\tau_i s(1 + s\tau_a) + 1}$$ (10.47)

$$\frac{\omega_m(s)}{V_t(s)/K_m} = \frac{1/(\tau_a\tau_i)}{s(1/\tau_a + s) + 1/(\tau_a\tau_i)}$$ (10.48)

where $\tau_i = JR_a/K_m^2$ is the inertial time constant. The corresponding block diagram is shown in Figure 10.5. From the overall transfer function, the characteristic equation of the speed response to the voltage input is determined as

$$s\left(\frac{1}{\tau_a} + s\right) + \frac{1}{\tau_a\tau_i} = s^2 + s\left(\frac{1}{\tau_a} + s\right) + \frac{1}{\tau_a\tau_i} = 0$$ (10.49)

The standard form of the characteristic equation of a second-order system is

$$s^2 + 2\alpha s + \omega_n^2 = 0$$ (10.50)

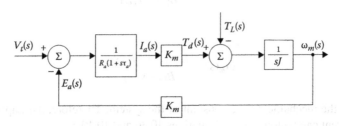

FIGURE 10.5 Block diagram of a separately excited dc motor with the mechanical damping neglected.

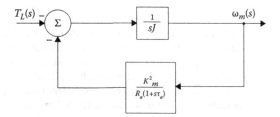

FIGURE 10.6 Simplified block diagram of a dc motor.

By comparing Equations 10.49 and 10.50, the undamped natural frequency ω_n can be expressed as

$$\omega_n = \left(\frac{1}{\tau_a \tau_i} \right)^{1/2} \tag{10.51}$$

and the damping factor α is

$$\alpha = \frac{1}{2\tau_a} \tag{10.52}$$

Therefore, the damping ratio is

$$\zeta = \frac{\alpha}{\omega_m} = \left(\frac{\tau_i}{\tau_a} \right)^{1/2} \tag{10.53}$$

Note that to study the response of the motor to load changes, the block diagram shown in Figure 10.5 can be modified to the one shown in Figure 10.6.

The overall transfer function relating the speed response to a change in load torque with $V_i = 0$ can be determined from Figure 10.5 by eliminating the feedback path as

$$\frac{\omega_m(s)}{T_L(s)} = \frac{1 + s\tau_a}{Js(1 + s\tau_a) + \left(K_m^2/R_a \right)} \tag{10.54}$$

or

$$\frac{\omega_m(s)}{T_L(s)} = \frac{1/\tau_a + s}{s(1/\tau_a + s) + 1/(\tau_a \tau_i)} \tag{10.55}$$

Also notice that its characteristic equation has the same undamped natural frequency and damping factor given in Equations 10.51 and 10.52.

For a given mechanical load torque T_L, the required armature current can be found from the dynamic equation of the motor. Therefore, dividing Equation 10.41 by K_m and substituting $\omega_m = E_a/K_m$, the armature current is found as

$$I_a = \frac{J}{K_m^2} \times \frac{dE_a}{dt} + \frac{BE_a}{K_m^2} + \frac{T_L}{K_m} \tag{10.56}$$

Example 10.4

Assume that a 250 V separately excited dc motor has an armature-circuit resistance and inductance of 0.10 Ω and 14 mH, respectively. Its moment of inertia J is 20 kg/m^2 and its motor constant K_m is 2.5 N m/A. The motor is supplied by a 250 V constant-voltage source. Assume that the motor is initially operating at steady-state without a load. Its no-load armature current is 5 A. Ignore the effects of saturation and armature reaction. If a constant load torque T_L of 1000 N m is suddenly connected to the shaft of the motor, determine the following:

    (a) The undamped natural frequency of the speed response
    (b) The damping factor and damping ratio
    (c) The initial speed in rpm
    (d) The initial acceleration
    (e) The ultimate speed drop

Solution

    (a) The armature time constant of the motor is

$$t_a = \frac{L_a}{R_a} = \frac{0.014}{0.1} = 0.14 \text{ s}$$

and its inertial time constant is

$$\tau_i = \frac{JR_a}{K_m^2} = \frac{(20)(0.1)}{2.5^2} = 0.32 \text{ s}$$

Thus, the undamped natural frequency of the motor is

$$\omega_n = \left(\frac{1}{\tau_a \tau_i}\right)^{1/2} = \left(\frac{1}{(0.14)(0.32)}\right)^{1/2} = 4.72 \text{ rad/s}$$

    (b) The damping factor of the system is

$$\alpha = \frac{1}{2\tau_a} = \frac{1}{2(0.14)} = 3.571$$

and its damping ratio is

$$\zeta = \frac{\alpha}{\omega_n} = \frac{0.3571}{4.72} = 0.7567$$

    (c) When the mechanical load is suddenly connected at $t=0$, the internal generated voltage is

$$E_a = V_t - I_a R_a = 250 - (5)(0.1) = 249.5 \text{ V}$$

Hence, the corresponding initial speed of the motor is

$$\omega_m = \frac{E_a}{K_m} = \frac{249.5}{2.5} = 99.8 \text{ rad}$$

    (d) Assuming that the losses are small enough to ignore, the initial acceleration of the motor can be found from Equation 10.41 as

$$\alpha = \frac{d\omega_m}{dt} = \frac{K_m I_a - T_L}{J} = \frac{(2.5)5 - 1000}{20} = 49.38 \text{ rad/s}^2$$

(e) By applying the final-value theorem of Laplace transforms to Equation 10.55, the ultimate drop in speed of the motor can be found as

$$\Delta\omega_m = \lim_{s\to\infty}\left\{\frac{s\left[-(1/J)(1/\tau_a + s)\right]}{s(1/\tau_a + s) + 1/(\tau_a\tau_i)}\right\} = -\frac{\tau_i\Delta T_L}{J} = -\frac{(0.32)1000}{20} = -152.8 \text{ rpm}$$

10.5 SYNCHRONOUS GENERATOR TRANSIENTS

Whenever there is a sudden change in the shaft torque applied to a synchronous generator or in its output load, there is always a transient that lasts for a very short time before the generator resumes its steady-state operation. Such operation of a generator is defined as *transient operation* and can be electrical or mechanical in nature. For example, a sudden three-phase short circuit at the stator terminals is an *electrical transient* and a sudden load change may result in a *mechanical transient*.

10.6 SHORT-CIRCUIT TRANSIENTS

The most severe transient in a synchronous generator takes place when its three stator terminals are suddenly shorted out while the generator is operating at synchronous speed with constant excitation under no-load conditions. Such a short on a power system is called a *fault*.

Figure 10.7 represents a typical short-circuit oscillogram* that shows the three-phase armature current waves as well as the field current. Notice that the traces of the armature-phase currents are not symmetrical at the zero-current axis, and clearly exhibit the dc components responsible for the offset waves. In other words, each phase current can be represented by a dc transient component of current added on top of a symmetrical ac component. The symmetrical ac component of current by itself is shown in Figure 10.8. As shown in Figure 10.7, the dc component of the armature current is different in each phase and depends on the point of the voltage wave at which the fault takes place.

The initial value of the dc component[†] of the fault current can be as large as the magnitude of the steady-state current. The decay rate of this dc component is found from the resistance and equivalent inductance seen at the stator terminals. The dc component plays a role here because the synchronous generator is essentially inductive and in an inductive circuit a current cannot vary instantaneously.

As previously stated, Figure 10.8 shows the ac symmetrical component of the short-circuit armature current. This symmetrical trace can be found oscillographically if the short circuit takes place at the instant when the prefault flux linkage of the phase is zero. The envelope of the short-circuit current represents three different periods: the *subtransient period*, the *transient period*, and the *steady-state period*. For the purpose of short-circuit current calculations, the variable reactance of a synchronous machine can be represented by the following reactance values:

X_s'' = *subtransient reactance* determines the short-circuit current during the first cycle or so after the short circuit occurs. In about 0.05–0.1 s, this reactance increases to X_s'.

X_s' = *transient reactance* determines the short-circuit current after several cycles at 60 Hz. In about 0.2–2 s, this reactance increases to X_s.

X_s = *synchronous reactance* determines the short-circuit current after a steady-state condition is reached.

---

* These oscillograms of short-circuit currents can be used to determine the values of some of the reactances and time constants.

† Also note that the presence of the dc component in the stator phase coil induces an additional current component in the field winding that assumes a damped sinusoidal character.

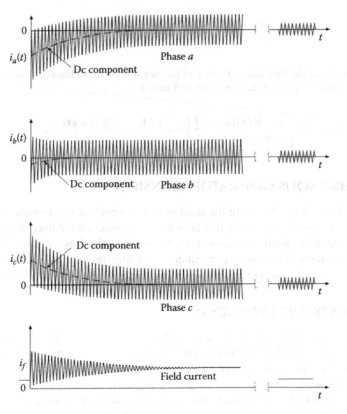

FIGURE 10.7 Three-phase short-circuit armature currents and field current waves during a three-phase fault at the terminals of a synchronous generator as a function of time.

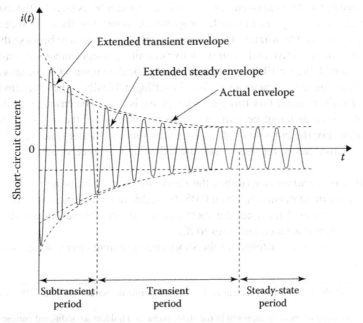

FIGURE 10.8 The symmetrical ac component of a short-circuit armature current of a synchronous generator.

This representation of the synchronous machine reactance by three different reactances is due to the fact that the flux across the air gap of the machine is much greater at the instant the short circuit takes place than it is a few cycles later. When a short circuit occurs at the terminals of a synchronous machine, it takes time for the flux to decrease across the air gap. As the flux lessens, the armature current decreases since the voltage generated by the air gap flux regulates the current. The subtransient reactance X_s includes the leakage reactances of the stator and rotor windings of the generator and the influence of the damper windings.* Therefore, the subtransient reactance can be expressed as

$$X_s'' = X_a + \frac{X_{ar}X_fX_d}{X_fX_d + X_{ar}X_d + X_{ar}X_f} \tag{10.57}$$

where
X_a is the leakage reactance of the stator winding per phase
$\equiv \omega L_a$
Xf is the leakage reactance of the field winding
$\equiv \omega L_f$
X_{ar} is the reactance associated with the armature mmf acting in the mutual flux path
$\equiv \omega L_{ar}$
X_d is the leakage reactance of the damper winding
$\equiv \omega L_d$
L_d is the leakage inductance of the damper winding

The transient reactance X_d'' includes the leakage reactances of the stator and excitation windings of the generator. It is usually larger than the subtransient reactance. The transient reactance can be expressed as

$$X_s'' = X_a + \frac{X_fX_{ar}}{X_f + X_{ar}} \tag{10.58}$$

Alternatively, the three reactance values can be determined from

$$X_s'' = \frac{E_a}{I_a''} \tag{10.59}$$

$$X_s' = \frac{E_a}{I_a'} \tag{10.60}$$

$$X_s = \frac{E_a}{I_a} \tag{10.61}$$

where
E_a is the internal generated voltage
I_a'' is the subtransient current
I_a' is the transient current
I_a is the steady-state current

It is interesting to observe in Figure 10.8 that the ac component of the short-circuit current consists of the steady-state value and the two components that decay with time constants T_s' and T_s''. Thus,

---

* They are located in the pole faces of generator and are used to reduce the effects of hunting.

the rms magnitude of the ac current* at any time after a three-phase short circuit at the generator terminals can be determined from

$$I_{ac}(t) = I_a + \left(I_a' - I_a\right)\exp\left(-\frac{t}{T_s'}\right) + \left(I_a'' - I_a'\right)\exp\left(-\frac{t}{T_s''}\right) \tag{10.62}$$

where all quantities are in rms values and are equal but displaced 120 electrical degrees in the three phases. However, the instantaneous value of the ac short-circuit current is

$$i_{ac}(t) = \sqrt{2}\left[I_a + \left(I_a' - I_a\right)\exp\left(-\frac{t}{T_s'}\right) + \left(I_a'' - I_a'\right)\exp\left(-\frac{t}{T_s''}\right)\right]\sin\omega t \tag{10.63}$$

The dc (i.e., unidirectional) component of the short-circuit current is different in each phase, and its maximum value is

$$I_{dc,max} = \sqrt{2}I_a'' \tag{10.64}$$

Since it decays with the armature time constant T_a,[†] it can be expressed as

$$I_{dc} = \sqrt{2}I_a''\exp\left(-\frac{t}{T_a}\right) \tag{10.65}$$

Also, the maximum value of the dc component depends upon the point in the voltage cycle where the short circuit takes place. Thus, the dc component of the short-circuit current can be determined from

$$I_{dc} = \sqrt{2}I_a''(\cos\alpha)\exp\left(-\frac{t}{T_a}\right) \tag{10.66}$$

where a is the switching angle. Therefore, the total short-circuit current with the dc offset can be expressed as

$$i_{tot} = \sqrt{2}\left[I_a + \left(I_a' - I_a\right)\exp\left(-\frac{t}{T_s'}\right) + \left(I_a'' - I_a'\right)\exp\left(-\frac{t}{T_s''}\right)\right]\sin\omega t + \sqrt{2}I_a''(\cos\alpha)\exp\left(-\frac{1}{T_a}\right) \tag{10.67}$$

Since dc components of current decay very fast, as a rule of thumb, the value of the ac component of current should be multiplied by 1.6 to find the total initial current.

Now assume that the synchronous generator is supplying power to a bus before such a short circuit takes place. Since the reactance of the generator changes from X to X_s'', the corresponding internal generated voltages must also vary to maintain the initial condition of flux linkage constancy. Therefore, the internal generated voltages can be expressed as

$$E_a'' = V_t + jI_aX_a'' \tag{10.68}$$

---

* If such a short circuit does not include all three phases, then the fault currents are determined by using symmetrical *components* methods, which are beyond the scope of this book. For further information, see Gönen (1988).
† Typically, it varies between 0.1 and 0.2 s.

$$E'_a = V_t + jI_aX'_s \tag{10.69}$$

$$E_a = V_t + jI_aX_sI_a \tag{10.70}$$

where I_a is the prefault load current. Thus, the total short-circuit current with the dc offset can be determined from

$$i_{tot} = \sqrt{2}\left[I_a + (I'_a - I_a)\exp\left(-\frac{t}{T'_d}\right) + (I''_a - I'_a)\exp\left(-\frac{t}{T''_d}\right)\right]\sin\omega t + \sqrt{2}I''_a(\cos\alpha)\exp\left(-\frac{1}{T_d}\right) \tag{10.71}$$

where

$$T'_d \cong \frac{X'_s}{X_s}T'_s \tag{10.72}$$

$$T''_d \cong \frac{X''_s}{X_s}T''_s \tag{10.73}$$

Example 10.5

Assume that a three-phase, 200 MVA, 13.2 kV, 60 Hz, wye-connected synchronous generator has the following parameters:

$$X_s = 1.0 \text{ pu} \quad X'_s = 0.23 \text{ pu} \quad X''_s = 0.12 \text{ pu}$$

$$T'_s = 1.1 \text{ s} \quad T''_s = 0.035 \text{ s} \quad T_a = 0.16 \text{ s}$$

Assume that the machine is operating at no load when a three-phase short circuit takes place at its terminals. If the initial dc component of the short-circuit current is about 60% of the initial ac component of the short-circuit current, determine the following:

(a) The base current of the generator
(b) The subtransient current in per unit and amps
(c) The transient current in per unit and amps
(d) The steady-state current in per unit and amps
(e) The initial value of the ac short-circuit current in amps
(f) The initial value of the total short-circuit current in amps
(g) The value of the ac short-circuit current after two cycles
(h) The values of the ac short-circuit current after 4, 6, and 8 s, respectively

Solution

(a) From Equation B.60, the base current of the generator is

$$I_B = \frac{S_{3\phi,base}}{\sqrt{3}V_{L,base}} = \frac{200 \times 10^6}{\sqrt{3}(13,200)} = 8,748 \text{ A}$$

(b) The subtransient current is

$$I''_a = (8.333 \text{ pu})(8,748 \text{ A}) = 72,897 \text{ A}$$

(c) The transient current is

$$I_a' = \frac{E_a}{X_s'} = \frac{1.0}{0.23} = 0.438 \text{ pu}$$

or

$$I_a' = (4.348 \text{ pu})(8,748 \text{ A}) = 38,036 \text{ A}$$

(d) The steady-state current is

$$I_a = \frac{E_a}{X_s} = \frac{1.0}{1.0} = 1.0 \text{ pu}$$

or

$$I_a = (1.0 \text{ pu})(8,748 \text{ A}) = 8,748 \text{ A}$$

(e) The initial value of the ac short-circuit current is

$$I_a'' = 72,897 \text{ A}$$

(f) The initial value of the total short-circuit current is

$$I_{tot} = 1.6 I_a'' = 1.6(72,897 \text{ A}) = 116,635 \text{ A}$$

(g) From Equation 10.62, the rms value of the ac short-circuit current as a function of time is

$$I_{ac}(t) = I_a + (I_a' - I_a)\exp\left(-\frac{t}{T_s'}\right) + \left(I_a'' - I_a'\right)\exp\left(-\frac{t}{T_s''}\right)$$

$$= 8,748 + 29,288\exp\left(-\frac{t}{1.1}\right) + 34,861\exp\left(-\frac{t}{0.035}\right) = 50,611.91 \text{ A}$$

Therefore, the value of the ac current after two cycles (i.e., $t = 1/30$ s) is

$$I_{ac}(t) = 8,748 + 29,288\exp\left(-\frac{1/30}{1.1}\right) + 34,861\exp\left(-\frac{1/30}{0.035}\right) = 50,611.91 \text{ A}$$

(h) The value of the ac current after 4 s is

$$I_{ac}(t) = 8,748 + 772 + 0 = 9,520 \text{ A}$$

After 6 s, it is

$$I_{ac}(t) = 8,748 + 125 + 0 = 8,873 \text{ A}$$

After 8 s, it is

$$I_{ac}(t) = 8,748 + 20 + 0 = 8,768 \text{ A}$$

10.7 TRANSIENT STABILITY

Power system stability can be defined as *the ability of the numerous synchronous machines of a given power system to remain in synchronism* (i.e., in step) *with each other following a disturbance. In a stable power system*, if synchronous machines are disturbed they will return to their original operating state if there is no change of power, or they will attain a new operating state without a loss of synchronism.

The disturbance often causes a transient that is oscillatory in nature, but if the system is stable the oscillations will be damped. In a synchronous generator, the most severe disturbance is caused by a short circuit across its terminals. In a synchronous motor, a disturbance may be caused by a sudden application of load torque to the shaft.

Stability can be classified as either *transient* or *dynamic stability.** The definition of *transient stability* includes stability after a sudden large disturbance such as a fault, loss of a generator, a sudden load change, or a switching operation. Transient stability is a short-term problem.

Today, *dynamic stability* is defined as *the ability of various machines to regain and maintain synchronism after a small and slow disturbance*, such as a gradual change in load. Dynamic stability is a long-term problem with time constants running into minutes. In this study, the effects of regulators, governors, and modern exciters, as well as other factors that affect stability may be included.

Assume that the power system is made up of two synchronous machines and that one of them is operating as a generator and the other as a motor. Also assume that the two-machine system is operating at steady state at point 1 on the power-angle curve, given in Figure 10.9a. Suppose that the generator is supplying electrical power P_1 at an angle $\delta 1$ to the motor and that the motor is driving a mechanical load connected to its shaft. Consider the following specific cases of the operation of this system:

Case 1. Assume that the shaft load of the motor is slowly increasing. The resulting net torque tends to slow down the motor and decreases its speed, causing an increase in the power angle δ. This, in turn, causes the input power to increase until an equilibrium is achieved at a new operating point 2, which is higher than 1.

Case 2. Suppose that the load of the motor is increased suddenly by connecting a large load. The resultant shortage in input power will be temporarily brought about by the decrease in kinetic energy. Consequently, the speed of the motor will decrease, causing the power angle δ and the input power to increase. As long as the new load is less than P_{max}, the power angle δ will increase to a new

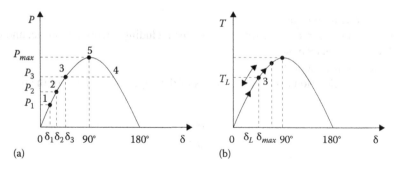

(a)   (b)

FIGURE 10.9   (a) Electric power input to a motor as a function of torque angle δ and (b) torque versus the torque angle δ characteristic.

---

* Today, the IEEE does not recognize steady-state stability as a separate class of stability. In present practice, it is included in the definition of dynamic stability. However, some authors still refer to the dynamic stability by using the ambiguous name of "steady-state stability."

value so that the input power of the motor is equal to its load. At this point the motor may still be running slowly, causing the power angle to increase beyond its proper value. Also an accelerating torque that increases the speed of the motor develops. However, it is possible that when the motor regains its normal speed, the power angle δ may have gone beyond point 4 and the motor input will be less than the load. This causes the motor to pull out.

Case 3. Assume that the load of the motor is increased slowly until point 5 of maximum power is reached. Any additional load will increase the power angle δ beyond 90°, causing the input power to decrease further and the net retarding torque to increase again. This torque slows the motor down even more until it pulls out of step.

Case 4. Suppose that the load of the motor is increased suddenly, but the additional load is not too large. The motor will regain its normal speed before the power angle δ becomes too large. The situation is illustrated in Figure 10.9b in terms of the resultant torque and δ. When the torque angle δ reaches the value $δ_L$, the load torque is the same as the torque developed by the motor. However, due to inertia, δ will increase beyond $δ_L$ and the motor will develop more torque than the load needs. The deceleration will decrease, causing the angle δ to reach a maximum value $δ_{max}$ and then swing back. The angle δ will oscillate around $δ_L$. Such oscillations will later die out because of damping in the system, and the motor will settle down to stable operating conditions at point 3.

The *transient stability limit* can be defined as the *upper limit to the sudden increment in load that the rotor can have without pulling out of step.* This limit is always less than the steady-state limit discussed in Section 7.8. Furthermore, the transient stability limit may have different values depending on the nature and magnitude of the disturbance involved.

10.8   SWING EQUATION

Without ignoring the torque that is due to friction, windage, and core losses, the net accelerating torque of a synchronous machine can be expressed, based on Newton's law of rotation, as

$$T_a = T_m - T_d = J\alpha \tag{10.74}$$

or in terms of the angular position as

$$J\frac{d^2\theta_m}{dt^2} = T_a = T_m - T_d \tag{10.75}$$

where
   T_a is the net accelerating torque
   T_m is the shaft torque corrected for rotational losses including friction, windage, and core losses
   T_d is the developed electromagnetic torque
   J is the moment of inertia of the rotor
   α is the angular acceleration expressed in terms of the angular position θ of the rotor

$$= \frac{d^2\theta}{dt^2}$$

It is customary to use the values of T_m and T_d as positive for generator action and negative for motor action. It is convenient to measure angular position and angular velocity with respect to a reference axis rotating at synchronous speed. Hence, the rotor position can be expressed as

$$\theta_m = \omega_{sm}t + \delta_m \tag{10.76}$$

Taking the derivatives of θ_m with respect to t,

$$\frac{d\theta_m}{dt} = \omega_{sm} + \frac{d\delta_m}{dt} \tag{10.77}$$

$$\frac{d^2\theta_m}{dt^2} = \frac{d^2\delta_m}{dt^2} \tag{10.78}$$

By substituting Equation 10.78 into Equation 10.75,

$$J\left(\frac{d^2\delta_m}{dt^2}\right) = T_a = T_m - T_d \tag{10.79}$$

and by multiplying both sides of this equation by the angular velocity

$$J\omega_m\left(\frac{d^2\delta_m}{dt^2}\right) = \omega_m T_a = T_m - T_d \tag{10.80}$$

Thus, the swing equation can be obtained as

$$M\left(\frac{d^2\delta_m}{dt^2}\right) = P_a = P_d - P_d \tag{10.81}$$

where
 $M = J\omega_m$ is the inertia constant
 $P_a = P_m - P_d$ is the net accelerating power
 $P_m = \omega_m T_m$ is the shaft power input corrected for rotational losses
 $P_d = \omega_m T_d$ is the electrical power output corrected for electrical losses

The swing equation describes how the machine rotor moves (swings) with respect to the synchronously rotating reference frame in a given disturbance (i.e., when the net accelerating power is not zero). The inertia constant for the synchronous machine is expressed as

$$H = \frac{\text{Kinetic energy of all rotating parts at synchronous speed}}{S_{rated}} \tag{10.82a}$$

$$H = \frac{J\omega_{sm}^2/2}{S_{rated}} \tag{10.82b}$$

or

$$H = \frac{1}{2}\frac{M\omega_{sm}}{S_{rated}} \tag{10.82c}$$

Equation 10.82 can be expressed in terms of per-unit quantities with respect to the rated (3ϕ) power of the synchronous generator as

$$\frac{2H}{\omega_{sm}} \frac{d^2\delta_m}{dt^2} = \frac{P_a}{S_{rated}} = \frac{P_m - P_d}{S_{rated}} \tag{10.83}$$

where the angle δ_m and angular velocity ω_m are in mechanical radians and mechanical radians per second, respectively. For a synchronous generator with p poles, the electrical power angle and radian frequency are associated with the corresponding mechanical variables as

$$\delta(t) = \frac{p}{2} \delta_m(t) \tag{10.84}$$

$$\omega(t) = \frac{p}{2} \omega_m(t) \tag{10.85}$$

Also the synchronous electrical radian frequency is related to the synchronous angular velocity as

$$\omega_s = \frac{p}{2} \omega_m \tag{10.86}$$

Thus, the rated per-unit swing equation (10.83) can be expressed in electrical units as

$$\frac{2H}{\omega_s} \times \frac{d^2\delta_m}{dt^2} = P_a = P_m - P_d \tag{10.87}$$

when δ is in electrical radians

$$\frac{2H}{\omega_s} \times \frac{d^2\delta}{dt^2} = P_a = P_m - P_d \tag{10.88}$$

when δ is in electrical degrees

$$\frac{H}{180f} \times \frac{d^2\delta}{dt^2} = P_a = P_m - P_d \tag{10.89}$$

The solution of this swing equation is called the *swing curve* $\delta(t)$.

If the synchronous machine is connected to an infinite bus through an external reactance, the electrical power output of the synchronous generator can be expressed as

$$P_d = P_{max}\sin\delta \tag{10.90}$$

By substituting this equation into Equation 10.87, the swing equation can be found as

$$P_m = \frac{2H}{\omega_s} \times \frac{d^2\delta}{dt^2} P_{max}\sin\delta \tag{10.91}$$

Since the resulting equation is nonlinear, it is often necessary to use a numerical technique to solve it.

Example 10.6

Assume that a three-phase, 250 MVA, 15 kV, 60 Hz, six-pole generator is connected to an infinite bus through a purely reactive network. Also assume that the inertia constant of the generator is 6 MJ/MVA and that it is supplying power of 1.0 per unit to the infinite bus at the steady state. The maximum power that can be supplied is 2.4 per unit. If the output power of the generator becomes zero at a three-phase fault, determine the following:

(a) The angular acceleration of the generator
(b) The shaft speed of the generator at the end of 12 cycles
(c) The change in the power angle S at the end of 12 cycles

Solution

(a) Since the generator is operating at steady state before the fault $P_m = P_d = 1.0$ pu
    Hence, from Equation 10.89,

$$\frac{H}{180f} \times \frac{d^2\delta}{dt^2} = P_m - P_d$$

from which the accelerating torque can be found as

$$\alpha = \frac{d^2\delta}{dt^2} = \frac{180f}{H}(P_m - P_d) = \frac{(180)(60)}{6}(1.0 - 0) = 1800 \text{ electrical degrees/s}^2$$

Since the machine has six poles,

$$\alpha = 300\left(\frac{60 \text{ s/min}}{360°/\text{rev}}\right) = 50 \text{ rpm/s}$$

(b) The synchronous speed of the machine is

$$\omega_{sm} = \frac{120f}{p} = \frac{120(60)}{6} = 1200 \text{ rpm}$$

and a 12-cycle interval is

$$t = \frac{60}{p} = 0.2 \text{ s}$$

$$\omega_m = \omega_{sm} + \alpha t = 1200 + (50)(0.2) = 1210 \text{ rpm}$$

(c) Since the generator is initially operating at the power angle δ from

$$P_0 = P_{max} \sin\delta_0$$

the initial power angle can be found as

$$\delta_0 = \sin^{-1}\left(\frac{P_0}{P_{max}}\right) = 24.62°$$

Therefore,

$$\delta = \delta_0 + \frac{1}{2}at^2 = 24.62° + 2(1800)(0.2)^2 = 60.62°$$

PROBLEMS

10.1 Assume that a 250 V separately excited dc generator has an armature-circuit resistance and inductance of 0.10 Ω and 1.2 mH, respectively. Its field winding resistance and inductance are 10 Ω and 52 H, respectively. The generator constant K_g is 104 V per field ampere at 1200 rpm. Assume that the field and armature circuits are initially open and that the prime mover is driving the machine at a constant speed of 1200 rpm. Derive a mathematical expression for the armature terminal voltage as a function of time in terms of the unit step function, if the field circuit is connected to a constant-voltage source of 260 V at time $t = 0$.

10.2 Consider Problem 10.1 and assume that after the field circuit has reached a steady state, the armature circuit of the generator is suddenly connected to a load made up of a resistance and inductance that are connected in series with respect to each other. If the values of the resistance and inductance are 1.2 Ω and 1.4 mH, respectively, determine the following in terms of unit step functions:
(a) The armature current
(b) The terminal voltage
(c) The developed electromagnetic torque

10.3 Assume that a 250 V separately excited dc motor has an armature-circuit resistance and inductance of 0.15 Ω and 15 mH, respectively. Its moment of inertia J is 18 kg/m² and its motor constant K_m is 2.0 N m/A. The motor is supplied by a constant-voltage source of 250 V. Assume that the motor is initially operating at steady state without load. Its no-load armature current is 10 A. Ignore the effects of saturation and armature reaction. If a constant load torque of 500 N m is suddenly connected to the shaft of the motor, determine the following:
(a) The undamped natural frequency of the speed response
(b) The damping factor and damping ratio
(c) The initial speed in rpm
(d) The initial acceleration
(e) The ultimate speed drop

10.4 A separately excited dc generator has an armature-circuit resistance and inductance of 1 Ω and 3 H, respectively. Its field winding resistance and inductance are 100 Ω and 20 H, respectively. The generator constant K_g is 100 V per field ampere at rated speed. Assume that the generator is driven by the prime mover at rated speed and that a 250 V dc supply is suddenly connected to the field winding. Determine the following:
(a) The internal generated voltage of the generator
(b) The internal generated voltage in the steady state
(c) The time required for the internal generated voltage to rise to 99% of its steady-state value

10.5 Assume that the generator given in this problem is driven by the prime mover at rated speed and is connected to a load made up of a resistance of 15 Ω and inductance of 5 H that are connected in series with respect to each other. If a 250 V dc supply is suddenly connected to the field winding, determine the armature current as a function of time.

10.6 Consider the synchronous generator given in Example 10.5 and determine the following:
(a) The instantaneous values of the ac short-circuit current for any given time t
(b) The maximum value of the dc component of the short-circuit current
(c) The dc component of the short-circuit current at $t = 0.1$ s, if the switching angle α is given as 45°

(d) The total short-circuit current with the dc offset at $t=0.1$ s, if the switching angle α is zero.

(e) The maximum rms value of the total short-circuit current

10.7 Redo Example 10.2 by using MATLAB, assuming that the new V_f is 150 V and the new V_{step} is 150 V. Use the other given values and determine the following:

(a) Write the MATLAB program script.

(b) Give the MATLAB program output.

11 Renewable Energy

11.1 INTRODUCTION

The part of the electric utility system which is between the distribution substation and the distribution transformers is called the *primary* system. It is made of circuits known as *primary* (or *main*) *feeders* or *primary distribution feeders*.

Figure 1.2 shows a one-line diagram of a typical primary distribution feeder. A feeder includes a "main" or main feeder, which usually is a three-phase four-wire circuit, and branches or laterals, which usually are single-phase or three-phase circuits tapped off the main. Also sublaterals may be tapped off the laterals as necessary. In general, laterals and sublaterals located in residential and rural areas are single phase and consist of one phase conductor and the neutral. The majority of the distribution transformers are single phase and connected between the phase and the neutral through fuse cutouts.

Typically, a residential area served by such feeder, illustrated in Figure 1.2, serves approximately 1000 homes per square mile. The feeder area is 1–4 square miles, depending on load density of the area. Usually, there are 15–30 single-phase laterals per feeder. Also typically, 150–500 short-circuit MVA is available at the substation bus.

A given feeder is sectionalized by reclosing devices at various locations in such a manner as to remove as little as possible of the faulted circuit so as to hinder service to as few consumers as possible. This can be achieved through the coordination of the operation of all the fuses and reclosers.

It appears that, due to growing emphasis on the service reliability, the protection schemes in the future will be more sophisticated and complex, ranging from manually operated devices to remotely controlled automatic devices based on supervisory-controlled or computer-controlled systems.

The congested and heavy-load locations in metropolitan areas are served by using underground primary feeders. They are usually radial three-conductor cables. The improved appearance and less-frequent trouble expectancy are among the advantages of this method. However, it is more expensive, and the repair time is longer than the overhead systems. In some cases, the cable can be employed as suspended on poles. The cost involved is greater than that of open wire but much less than that of underground installation.

There are various and yet interrelated factors affecting the selection of a primary-feeder rating. Examples are as follows:

1. The nature of the load connected
2. The load density of the area served
3. The growth rate of the load
4. The need for providing spare capacity for emergency operations
5. The type and cost of circuit construction employed
6. The design and capacity of the substation involved
7. The type of regulating equipment used
8. The quality of service required
9. The continuity of service required

The voltage conditions on distribution systems can be improved by using shunt capacitors, which are connected as near the loads as possible to derive the greatest benefit. The use of shunt capacitors also improves the power factor involved, which in turn lessens the voltage drops and currents, and therefore losses, in the portions of a distribution system between the capacitors and the bulk power buses. The capacitor ratings should be selected carefully to prevent the occurrence of excessive overvoltages at times of light loads due to the voltage rise produced by the capacitor currents.

The voltage conditions on distribution systems can also be improved by using series capacitors. But the application of series capacitors does not reduce the currents, and therefore losses, in the system.

11.2 RENEWABLE ENERGY

Renewable energy is of many types, including wind, solar, hydro, geothermal (earth heat), and biomass (waste material). All renewable energy with the exception of tidal and geothermal power, and even the energy of fossil fuels, ultimately comes from the sun. About 1%–2% of energy coming from the sun is converted into wind energy. Figure 11.1 shows the flow paths of renewable energy. Today, the most prevalent renewable energy resources are wind and solar.

Renewable resources can generate both electricity and heat; the term *green power* is used in a narrow sense to mean electric products that generated from renewable sources that are environmentally and socially acceptable. Most renewable energy forms are readily converted to electricity. Solar energy, geothermal energy, and biomass can also be used to supply heat.

Renewable energy is also a naturally distributed resource, that is, it can provide energy to remote areas without the requirement for elaborate energy transportation systems. However, it needs to be pointed out that it is not always a requirement that the renewable energy has to be converted into electricity. Solar water heating and wind-powered water pumping are good examples for it.

Presently, the largest renewable energy technology application (with the exception of hydro) has taken place is in wind power, with 95 GW, worldwide by the end of 2007. In 2003, renewable energy contributed 13.5% of world total primary energy (2.2% hydro, 10.8% combustible renewables and waste, and 0.5% geothermal, solar, and wind). Even though the combustible renewables are used for heat, the contributions for electric generation were somewhat different: hydro contributed 15.9%, and geothermal, solar, wind, and combustibles contributed 1.9%.

The capacity of world hydro plant is over 800 GW and the capacity of fast developing wind energy has sustained a 25% compound growth for well over a decade and was about 60 GW by the end of 2005. Also, in 2003, world electricity production from hydro was about 2654 TWh and all other renewables provided 310 TWh.

It is known that world primary energy demand almost doubled between 1971 and 2003. Furthermore, it is projected to increase by another 40% by 2020 [2]. It is also known that in the last 30 years, there has been a considerable shift way from oil and toward the natural gas. The natural

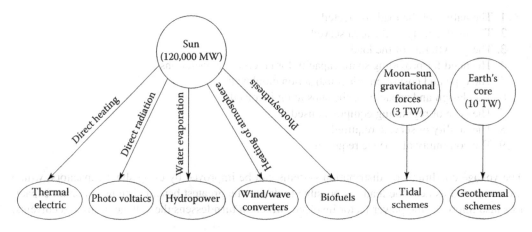

FIGURE 11.1   Renewable energy flow paths.

gas accounted for 21% of primary energy and 19% of electricity generation worldwide, in 2003. The impetus for this change has primarily been the increasing concerns over global warming due to carbon dioxide emissions. The Kyoto protocol, under the auspices of the United Nations, finally became legally binding on February 16, 2005, after Russia's ratification in November 2004.

11.3 IMPACT OF DISPERSED STORAGE AND GENERATION

Following the oil embargo and the rising prices of oil, the efforts toward the development of alternative energy sources (*preferably renewable resources*) for generating electric energy have been increased. Furthermore, opportunities for small power producers and cogenerators have been enhanced by recent legislative initiatives, for example, the *Public Utility Regulatory Policies Act* (PURPA) of 1978, and by the subsequent interpretations by the *Federal Energy Regulatory Commission* (FERC) in 1980.

The following definitions of the criteria affecting facilities under PURPA are given in Section 201 of PURPA:

- A *small power production facility* is one which produces electric energy solely by the use of primary fuels of biomass, waste, renewable resources, or any combination thereof. Furthermore, the capacity of such production sources together with other facilities located at the same site must not exceed 80 MW.
- A *cogeneration facility* is one which produces electricity and steam or forms of useful energy for industrial, commercial, heating, or cooling applications.
- A *qualified facility* is any small power production or cogeneration facility, which conforms to the previous definitions and is owned by an entity not primarily engaged in generation or sale of electric power.

In general, these generators are small (typically ranging in size from 100 kW to 10 MW and connectable to either side of the meter) and can be economically connected only to the distribution system. They are defined as *dispersed-storage-and-generation* (DSG) devices. If properly planned and operated, DSG may provide benefits to distribution systems by reducing capacity requirements, improving reliability, and reducing losses. Examples of DSG technologies include hydroelectric, diesel generators, wind electric systems, solar electric systems, batteries, storage space and water heaters, storage air conditioners, hydroelectric pumped storage, photovoltaics, and fuel cells.

11.4 INTEGRATING RENEWABLES INTO POWER SYSTEMS

Before going any further, it might be appropriate to define some terminology including the terms *grid*, *grid-connected*, and *national grid*. The term *grid* is usually used to describe the totality of the electric power network. The term *grid-connected* means connected to any part of the power network. On the other hand, the term *national grid* usually means the EHV transmission network.

The physical connection of a generator to the network is defined as *integrated*. But, it is required that the necessary attention must be given to the secure and safe operation of the system, and the control of the generator to achieve optimality in terms of the energy resource usage.

In general, the integration of the generators powered by the renewable energy sources essentially is the same as fossil-fuel powered generators and is based on the same methodology. However, renewable energy sources are very often variable and geographically dispersed.

A renewable energy generator can be defined as stand-alone or grid-connected. A *stand-alone renewable energy generator* provides for the greater part of the demand with or without other generators or storage. On the other hand, in *a grid-connected system*, the renewable energy generator supplies power to a large interconnected network which is also supplied power by other generators. Here, the power supplied by the renewable energy generator is only a small portion of the power supplied to the grid with respect to power supplied by other connected generators. The connection point is called the *point of common coupling** (PCC).

11.5 DISTRIBUTED GENERATION

It appears that there is a general consensus that by the end of this century, the most of our electric energy will be provided by renewable energy sources. As said previously, small generators cannot be connected to the transmission system due to high cost of high-voltage transformers and switchgear.

Thus, small generators must be connected to the distribution system network. Such generation is known as *distributed generation* (DG) or *dispersed generation*. It is also called *embedded generation* since it is embedded in the distribution network.

Power in such power systems may flow from point to point within the distribution network. As a result, such unusual flow pattern may create additional challenges in the effective operation and protection of the distribution network.

Due to decreasing fossil-fuel resources, poor energy efficiency, and environmental pollution concerns, the new approach for generating power locally at distribution voltage level by employing nonconventional/renewable energy sources such as natural gas, wind power, solar photovoltaic cells, biogas, cogeneration systems (which are combined heat and power [CHP] systems, Stirling engines, and microturbines.)

These new energy sources are connected to (or integrated into) the utility distribution network. As aforementioned, such power generation is called DG and its energy resources are known as distributed energy resources (DERs). Furthermore, the distribution network becomes active with the integration of DG, and thus is known as active distribution network. The properties of DG include the following:

1. It is normally less than 50 MW.
2. It is neither centrally dispatched nor centrally planned by the power utility.
3. The distributed generators or power sources are generally connected to the distribution systems, which typically have voltages of 240 V up to 34.5 kV.

The development and integration of the DG were based on the technical, economic, and environmental benefits, which include the following:

1. Reduction of environmental pollution and global warming concerns, as dictated by the Kyoto protocol, use of the nonconventional/renewable energy resources as a viable solution.
2. As a result of rapid load growth, the fossil-fuel reserves are increasingly depleted. Therefore, the use of nonconventional/renewable energy resources is increasingly becoming a requirement. Also, the use of DERs is to produce clean power without the associated pollution of the environment.
3. DERs are usually modular units of small capacity because of their lower energy density and dependence on geographically conditions of a region.
4. The overall power quality and reliability improves due to contributions of the stand-alone and grid-connected operations of DERs in generation augmentation. Such DG integration further increases due to deregulated environment and open-access to the distribution network.
5. The overall plant energy efficiency increases and also associated thermal pollution of the environment decreases because of the use of the DG such as cogeneration or CHP plants.

Furthermore, it is possible to connect a DER separately to the utility distribution network or it may be connected as a microgrid due to the fact that the power is produced at low voltage. Thus, the microgrid can be connected to the utility's network as a separate semiautonomous entity.

---

* See Gönen [4], Chapter 12.

11.6 RENEWABLE ENERGY PENETRATION

The proportion of electric energy or power being supplied from wind turbines or from other renewable energy sources is usually referred to as the penetration. It is usually given in percentage. The average penetration [3] is defined as

$$\text{Average penetration} = \frac{\begin{pmatrix} \text{Annual energy from renewable} \\ \text{energy powered generators (kWh)} \end{pmatrix}}{\begin{pmatrix} \text{Total annual energy} \\ \text{delivered to loads (kWh)} \end{pmatrix}} \quad (11.1)$$

The term *average penetration* is used when fuel or CO_2-emission savings are being considered.

However, for other purposes, including system control, use the following definition:

$$\text{Instantaneous penetration} = \frac{\begin{pmatrix} \text{Power from renewable energy} \\ \text{powered generators (kW)} \end{pmatrix}}{\begin{pmatrix} \text{(Power from renewable energy} \\ \text{powered generators [kW])/(Total power delivered} \\ \text{to loads [kW])} \end{pmatrix}} \quad (11.2)$$

In general, the maximum instantaneous penetration is much greater than the average penetration.

11.7 ACTIVE DISTRIBUTION NETWORK

It is also called *generation embedded distribution network*. In the past, distribution networks had a unidirectional electric power transportation. That is, distribution networks were stable passive networks.

Today, the distribution networks are becoming active by the addition of DG, which causes bidirectional power flows in the networks. Today's distribution networks started to involve not only demand-side management but also integration of DG.

In order to have good active distribution networks that have flexible and intelligent operation and control, the following should be provided:

1. Adaptive protection and control
2. Wide-area active control
3. Advanced sensors and measurements
4. Network management apparatus
5. Real-time network simulation
6. Distributive penetrating communication network
7. Knowledge and data extraction by intelligent methods
8. New and modern design of transmission and distribution systems

11.8 CONCEPT OF MICROGRID

A microgrid is basically an active distribution network and is made up of collection of DG systems and various loads at distribution voltage level. They are generally small low-voltage combined heat loads of a small community. The examples of such small community include university or school campuses, a commercial area, an industrial site, a municipal region or a trade center, or a housing estate, or a suburban locality.

The generators or micro sources used in a microgrid are generally based on renewable/nonconventional distribution energy resources. They are integrated together to provide power at distribution voltage level. In order to introduce the microgrid to the utility power system as a single controlled unit that meets local energy demand for reliability and security, the micro sources must have power electronic interfaces (PEIs) and controls to provide the necessary flexibility to the semiautonomous entity so that it can maintain the dictated power quality and energy output.

A *microgrid* is different than a conventional power plant. The differences include the following:

1. Power generated at distribution voltage level and can thus be directly provided to the utility's distribution system.
2. They are of much smaller capacity with respect to the large generators in conventional power plants.
3. They are usually installed closer to the customers' locations so that the electrical/heat loads can be efficiently served with proper voltage level and frequency, and ignorable line losses.
4. They are ideal for providing electric power to remote locations.
5. The fundamental advantage of microgrids to a power grid is that they can be treated as a controlled entity within the power system.
6. The fundamental advantage of microgrids to customers is that they meet the electrical/heat requirements locally. This means that they can receive uninterruptible power, reduced feeder losses, improved local reliability, and local voltage support.
7. The fundamental advantage to the environment is that they reduce environmental pollution and global warming by utilizing low-carbon technology.

However, before microgrids can be extensively established to provide a stable and secure operation, there are a number of technical, regulatory, and environmental issues that need to be addressed which include the establishment of standards and regulations for operating the microgrids in synchronism with the power utility, low energy content of the fuels involved, and the climate-dependent nature of production of the DERs.

Figure 11.2 shows a microgrid connection scheme. Microgrid is connected to the medium voltage (MV) utility "main grid" through the *point of common coupling* (PCC) circuit breaker. Micro source and storage devices are connected to the feeders *B* and *C* through *micro source controllers* (MCs). Some loads on feeders *B* and *C* are considered to be priority loads (i.e., needing uninterruptible power supply), while the rest are non-priority loads. On the other hand, feeder *B* had only non-priority electrical loads.

The microgrid has two modes of operations: (1) grid-connected and (2) stand-alone. In the first mode, the microgrid imports or exports power from or to the main grid. In the event of any disturbance in the main grid, the microgrid switches aver to stand-alone mode, but still supplying power to the priority loads. This is achieved by opening the necessary circuit breakers. But, feeder *A* will be left alone so that it can ride through the disturbance.

The main functions of central controller (CC) include *energy management module* (EMM), and *protection coordination module* (PCM). The EMM supplies the set points for active and reactive power output, voltage, and frequency to each microgrid controller (MC). This is done by advanced communication and artificial intelligent techniques. Whereas the PCM answers to microgrid and main grid faults and loss of grid situations so that proper protection coordination of the microgrid is achieved.

Chowdhury et al. [1] define the functions of the CC in the grid-connected mode and in the stand-alone mode. The functions of the CC in the grid-connected mode include the following:

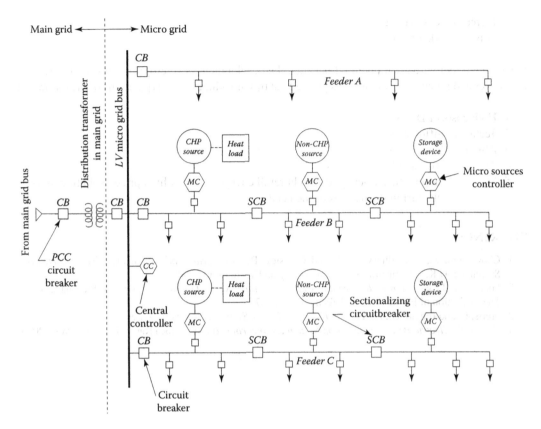

FIGURE 11.2 A typical microgrid connection scheme.

1. Monitoring system diagnostics by gathering information from the micro sources and loads.
2. Performing state estimation and security assessment evaluation, economic generation scheduling and active and reactive power control of the micro sources and demand-side management functions by employing collected information.
3. Ensuring synchronized operation with the main grid maintaining the power exchange at priori contract points.

The functions of the CC in the stand-alone mode are as follows:

1. Performing active and reactive power control of the micro sources to keep stable voltage and frequency at load ends.
2. Adapting load interruption/load-shedding strategies using demand side management with storage device support for maintaining power balance and bus voltage.
3. Beginning a local "cold start" to ensure improved reliability and continuity of service.
4. Switching over the microgrid to grid-connected mode after main grid supply is restored without hindering the stability of either grid.

Chowdhury et al. [1] list the following technical and economic advantages of microgrid for the electric power industry:

1. Reducing environmental problems and issues
2. Reducing some operational and investment issues
3. Improving power utility and reliability

4. Increasing cost savings
5. Solving market issues

Even though there are great potential benefits, the development of microgrids suffers from several challenges and potential drawbacks, as pointed out by Chowdhury et al. [1] in terms of the following:

1. High costs of DERs
2. Technical difficulties
3. Absence of standards
4. Administrative and legal barriers
5. Market monopoly (i.e., microgrids might retail energy at a very high price exploiting market monopoly during the periods of the need)

REFERENCES

1. Chowdhury, S., Chowdhury, S. P., and Crossley, P., *Microgrids and Active Distribution Networks*, Stevenage, U.K.: The Institution of Engineering and Technology, 2009.
2. Fox, B. et al., *Wind Power Integration, Connection and System Operational Aspects*, Stevenage, U.K.: The Institution of Engineering and Technology, 2007.
3. Freris, L. and Infield, D., *Renewable Energy in Power Systems*, New York: Wiley, 2008.
4. Gönen, T., *Electric Power Distribution System Engineering*, 2nd Ed., Boca Raton, FL: CRC Press, 2009.

12 Wind Energy and Wind Energy Conversion System (WECS)

12.1 INTRODUCTION

The wind results from the large-scale movements of air masses in the air in motion. Succinctly put, wind is air in motion. These movements of air created on a global scale primarily by differential solar heating of the earth's atmosphere. Thus, wind energy, like hydro, is also an indirect form of solar energy. Air in the equatorial regions is heated more strongly than at other latitudes, causing it to become lighter and less dense. This warm air rises to altitudes and then starts to flow northward and southward toward the poles where the air near the surface is cooler. This movement stops at about 30°N and 30°S, where the air starts to cool down and sink and a return flow of this cooler air takes place in the lowest layers of the atmosphere. The areas where air is descending are called the *low-pressure zones*. Hence, this horizontal pressure gradient forces the flow of air from high to low pressure, which dictates the speed and initial direction of wind motion. Naturally, the greater the pressure gradient, the greater is the wind speed. But, as soon as wind motion is created, a deflective force takes place due to the rotation of the earth, which changes the direction of motion. This force is called the *Coriolis force*.

Like the sun, wind energy depends on geographic location at the time of the year. In addition, there are also a number of local effects. For example, differential heating of the sea and the land also affects the general flow. The type of the terrain, such as mountains and valleys or more local obstacles in terms of buildings and trees, also has an important effect on the resulting wind.

The lower region of the atmosphere where the wind speed is slowed down by frictional forces on the earth's surface is known as the *boundary layer*. Because of this, wind speed increases with the height from ground. This is true up to the heights (approximately 1000 m) of the boundary layer. However, it also depends on the atmospheric conditions. The fact that wind speed increase as a function of height is called the *wind shear*. As a result, the hub height of the modern wind turbines of the multimegawatt wind machines now is over 100 m. A wind generator can provide power day and night.

Besides the home wind-electric generation, a number of electric utilities around the world have built larger wind turbines to supply electric power to their customers. In the United States, the largest wind turbine built before the late 1970s was a 1250 kW machine built on Grandpa's Knob, near Rutland, Vermont, in 1941. This wind turbine, called the Smith–Putnam machine, had a tower that was 34 m height and a rotor 53 m in diameter. The rotor turned an ac synchronous generator that produced 1250 kW of electrical power at wind speeds above 13 m/s. However, after the World War II, the interest in wind power died down due to cheap oil imported from the Middle East. But, after 1973 oil embargo, the interest in wind power started to increase again worldwide. This is especially true for the United States, Denmark, and Germany. In 2009, more than 1,000,000 windmills of about 120 GW installed power generation capacity are in operation worldwide, as given in Table 12.1. This was based on the understanding that ultimately, additional energy sources causing less pollution are necessary [5]. Due to favorable tax regulations in the 1980s, about 12,000 wind turbines providing power ranging from 20 kW to about 200 kW were installed in California. In Europe, a lot of tax money was spent on the development of bigger wind turbines and on marketing them. Vestas, a Danish company, has more installed wind turbine capacity worldwide than any other manufacturer [7].

However, Germany had the leadership until 2007. Then, the United States has taken over the leadership. The average commercial size of WESC was 300 kW until the mid-1990s. Today, there

TABLE 12.1

Installed Wind Power Capacity Worldwide, As of 2009

	Rated Capacity (MW)	Share Worldwide (%)
United States	25,200	21
Germany	23,900	20
Spain	16,800	14
China	12,200	10
India	9,600	8
Italy	3,700	3
France	3,400	3
United Kingdom	3,200	3
Denmark	3,100	2
Portugal	2,800	2
The rest of the world	16,700	14
Total	120,000	100

are wind turbines with a capacity up to 6 MW developed and installed. Since 1973, prices have dropped as performance has improved. Today, the cost of a wind turbine is below $2 per watt of installed capacity, large wind farms with several hundred megawatt capacities are being developed over several months. For example, it is now quite common for wind power plants (wind farms) with collections of utility-scale turbines to be able to sell electricity for fewer than 4 cents/kWh. Early developments in California were basically in the form of wind farms, with tens of wind turbines, even up to 100 or more in some cases. The reasons for this development include the economies of scale that can be achieved by building wind farms, especially in construction and grid connection costs, and even possibility by getting quantity discounts from the turbine manufacturers.

It is interesting to point out that the market introduction of wind energy is being done in industrialized countries as well as in developing countries like, for example, India. As given in Table 12.1, India ranked fifth in the world with wind power–installed capacity of 9600 MW at the end of 2008. Even with its technology level, the onshore potential of India for utilization of wind energy for electricity generation is estimated to be about 48,000 MW.

The European accessible onshore wind resource has been estimated at* 4800 TWh/year, taking into account typical wind turbine efficiencies, with the European offshore resource in the region of 3000 TWh/year although this is very dependent on the assumed allowable distance from the shore. According to a recent report [13], by 2030 the EU could be generating 965 TWh from onshore and offshore wind, amounting to 22.6% of electricity requirements. The world onshore resource is approximately 53,000 TWh/year, considering siting constraints. Note that the annual electricity demand for the United Kingdom and the United States are 350 and 3500 TWh, respectively.

12.2 ADVANTAGES AND DISADVANTAGES OF WIND ENERGY CONVERSION SYSTEMS

The wind energy is the fastest growing energy source in the world due to many advantages that it offers. Continuous research efforts are being made even further to increase the use of wind energy.

---

* Note that a terawatt is denoted as TW so that
  $1\,TWh = 1 \times 10^{12}\,Wh = 1000\,GWh$
  $T = Tera = 10^{12}$
  $1\,MWh = 1000\,kWh$
  $1\,MW$ wind power produces $2\,GWh$/year on land and $3\,GWh$/year offshore.

12.2.1 Advantages of a Wind Energy Conversion System

1. It is one of the lowest-cost renewable energy technologies that exist today.
2. It is available as a domestic source of energy in many countries worldwide and not restricted to only few countries, as in the case of oil.
3. It is energized by naturally flowing wind; thus, it is a clean source of energy. It does not pollute the air and cause acid rain or greenhouse gases.
4. It can also be built on farms or ranches and, hence, can provide the economy in rural areas using only a small fraction of the land. Thus, it still provides opportunity to the landowners to use their land. Also, it provides rent income to the landowners for the use of the land.

12.2.2 Disadvantages of a Wind Energy Conversion System

1. The main challenge to using wind as a source of power is that the wind is intermittent and it does not always blow when electricity is needed. It cannot be stored; not all winds can be harnessed to meet the timing of electricity demands. At the present time, the use of energy storage in battery banks is not economical for large wind turbines.
2. Despite the fact that the cost of wind power has come down substantially in the past 10 years, the technology requires a higher initial investment than the solutions using fossil fuels. Hence, depending on the wind profile at the site, the wind farm may or may not be as cost competitive as a fossil fuel–based power plant.
3. It may have to compete with other uses for the land and those alternative uses may be more highly valued than electricity generation.
4. It is often that good sites are located in remote locations, far from cities where the electricity is needed. Thus, the cost of connecting remote wind farms to the supply grid* may be prohibitive.
5. There may be some concerns over the noise generated by the rotor blades, and aesthetic problems that can be minimized through technological developments or by correctly siting wind plants [12].

12.3 CATEGORIES OF WIND TURBINES

Wind turbines turn the kinetic energy of the moving air into electric power or mechanical work. There are various WECSs. They can be classified as (1) horizontal-axis converters, (2) vertical-axis converters, and (3) upstream power stations.

The most common type is the horizontal-axis converter. It has only a few rotor blades. Another conventional (older) type of horizontal-axis rotor is the multiblade wind converter. The horizontal-axis converters are of two types: fast rotation and slow rotation.

The vertical-axis converters are of two types: (1) Darrieus and (2) Savonius. The Darrieus converter has a vertical-axis construction. They do not depend on the direction of the wind. But, they have a low starting torque. Because of this, they need the help of a generator working as a motor or the help of Savonius rotor installed on top of the vertical axis.

The wind velocity increases substantially with height, as a result, the horizontal-axis wheels on towers are more economical. In the 1980s, a large number of Darrieus converters were installed in California, but a further expansion into a higher power range and their application worldwide has not happened. The Savonius rotor is used as a measurement device especially for wind velocity. However, it is used for power production for very small capacities under 100 W.

---

* The term "grid" is often used loosely to describe the totality of the network. For example, grid connected means connected to any part of the electrical network. The term "national grid" usually means the EHV transmission network. Similarly, integration means the physical connection of the generator to the network for secure, safe, and optimal operation of the electrical system.

The last technique mentioned above is also known as "upstream power station" or "thermal tower." It is a mix between a wind converter and a solar collector. The top of a narrow and high tower has a wind wheel on a vertical axis driven by the rising warm air. A solar collector is installed around the foot of the tower driven by the rising warm air. Its design is very simple: a transparent plastic foil is fixed several meters above the ground in a circle around the tower. Thus, the station requires a lot of space, and the tower has to be very high. Its advantage is simplicity in design. But, it has a very poor efficiency, only about 1%. It has been built only in Spain and in India so far.

Figure 12.1 shows the most popular type of horizontal-axis three-blade wind energy converter for generating electricity worldwide. It shows the front and side views of a three-blade horizontal-axis wind energy converter.

Note that the terms "wind energy converters," "windmills," or "wind turbines" represent the same thing. The first one is the technical name of the system, whereas the other two are popularly used terms. Today, there are various types of wind energy converters that are in operation, as shown in Figure 12.2. Figure 12.3 shows eight different classes of wind turbines used in the Altamont pass in California [8].

Over the last 25 years, the size of the largest commercial wind turbines, as shown in Figure 12.4, has increased from approximately 50 kW to 2 MW, with machines up to 6 MW under design [11]. Figure 12.5 shows the main subsystems of a typical horizontal design. Figure 12.5 shows the main subsystems of a typical horizontal-axis wind turbine. These include the rotor, including the blades

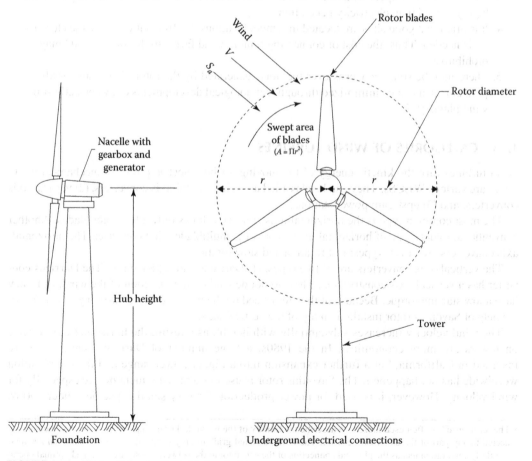

FIGURE 12.1 Horizontal-axis, three-blade wind energy.

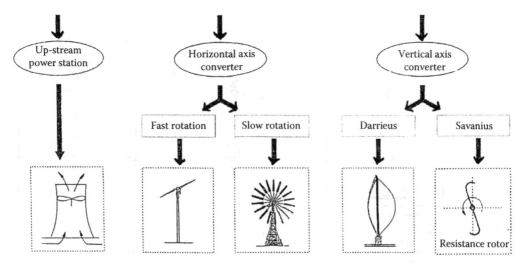

FIGURE 12.2   Overview of differential types of wind energy converters.

and supporting hub, the drive train, which includes the rotating parts of the wind turbine (except the rotor), including shafts gearbox, coupling, a mechanical brake, and the generator; the nacelle and main frame, including wind turbine housing, bedplate, and the yaw system; the tower and the foundation; and the machine controls, the switchgear, transformers, and possibly electronic power converters.

There are a number of options in wind machine design and construction. These options include the number of blades (normally two or three); the blade material, construction method, and profile; the rotor orientation: downward or upward of tower; hub design: rigid, teetering, or hinged; fixed or variable rotor speed; orientation by self-aligning action (free yaw) or direct control (active yaw); power control via aerodynamic control (stall control) or variable pitch blades (pitch control); synchronous or induction generator; and gearbox or direct-drive generator.

Almost all wind turbines use either induction or synchronous generators. Both of these designs entail a constant or nonconstant rotational speed of the generator when the generator is directly connected to a utility network. The majority of wind turbines installed in grid-connected applications use induction generators. An induction generator operates within a narrow range of speeds slightly higher than its synchronous speed. The main advantage of induction generators is that they are rugged, inexpensive, and easy to connect to an electric network. An induction generator is much simpler to connect to the grid than is a synchronous generator.

The nacelle of horizontal-axis turbine contains a bedplate on which the components are mounted. There is a main shaft with main bearings, a generator, and a yaw motor that turns the nacelle and rotor into the wind. The nacelle cover protects the contents from the weather. Nacelle and yaw system include the wind turbine housing, the machine bedplate or main frame, and the yaw orientation system. The mainframe provides for the mounting and proper alignment of the drive train components.

A yaw orientation system is needed to keep the rotor shat properly aligned with the wind. The main component is a large bearing that connects the mainframe to the tower. An active yaw drive, generally used with an upwind turbine, has one or more yaw motors, each of which drives a pinion gear against a bull gear attached to the yaw bearing. This mechanism is controlled by an automatic yaw control system with its wind direction sensor usually mounted on the nacelle of the wind turbine. Sometimes, yaw brakes are used with this type design to hold the nacelle of the wind turbine. Free yaw systems are normally used on downwind wind machines. They can self-align with the wind. The control system of a wind turbine includes sensors, controllers, power amplifiers, and actuators.

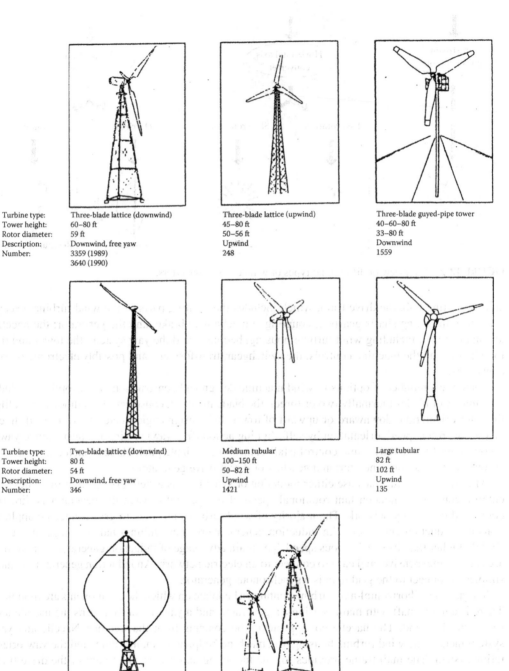

Turbine type:	Three-blade lattice (downwind)	Three-blade lattice (upwind)	Three-blade guyed-pipe tower
Tower height:	60–80 ft	45–80 ft	40–60–80 ft
Rotor diameter:	59 ft	50–56 ft	33–80 ft
Description:	Downwind, free yaw	Upwind	Downwind
Number:	3359 (1989)	248	1559
	3640 (1990)		

Turbine type:	Two-blade lattice (downwind)	Medium tubular	Large tubular
Tower height:	80 ft	100–150 ft	82 ft
Rotor diameter:	54 ft	50–82 ft	102 ft
Description:	Downwind, free yaw	Upwind	Upwind
Number:	346	1421	135

Turbine type:	Vertical axis	Windwall
Tower height:	90–106 ft	140 ft
Rotor diameter:	56–62 ft	59 ft
Description:	-	Downwind, free yaw
Number:	169	103

FIGURE 12.3 Eight categories of wind turbines used in the Altamont Pass in California. (From Orloff, S. and Flannery, A., Wind turbine effects on avian activity, habitat use, and mortality in Altamont Pass and Solano County wind resource areas: 1989–1991, *California Energy Commission Report*, No. P700-92,002, 1992.)

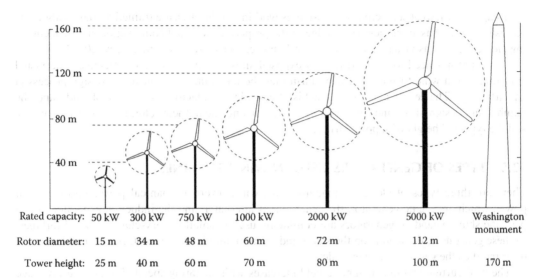

Rated capacity: 50 kW   300 kW   750 kW   1000 kW   2000 kW   5000 kW   Washington monument

Rotor diameter: 15 m   34 m   48 m   60 m   72 m   112 m

Tower height: 25 m   40 m   60 m   70 m   80 m   100 m   170 m

FIGURE 12.4 Representative size, height, and diameter of wind turbines.

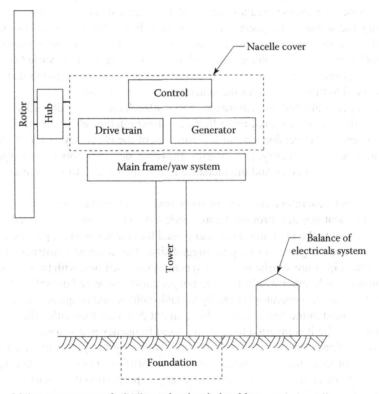

FIGURE 12.5 Major components of a horizontal-axis wind turbine.

12.4 VISUAL IMPACT OF WIND TURBINES

One of wind power is perceived adverse environmental impact factors, and a major concern of the public is its visibility. At the same time, the visual impact is one of the least quantifiable environmental impacts. For example, the public's perceptions may change with knowledge of the technology involved, location of wind turbines, and many other factors. Even though the assessment of a landscape is somewhat subjective, professionals working in this area are trained to make judgments on visual impact based on their knowledge of the properties of visual composition and by identifying elements, such as visual clarity, harmony, balance, focus, order, and hierarchy [10,11].

Wind turbines need to be sited in well-exposed sites in order to be cost-effective. The visual appearance of a wind turbine or a wind farm must be taken into account in the design process at an early stage. The degree of visual impact is affected by such factors as the type of landscape, the number and design of turbines, the pattern of their arrangement, their color, the number of blades, and the type and height of support structures.

12.5 TYPES OF GENERATORS USED IN WIND TURBINES

There are three types of electrical machines that can convert mechanical power into electrical power, which are the dc generator, the synchronous alternator, and the induction generator. In the past, the shunt-wound dc generators are commonly used in small battery-charging wind turbines. In these generators, the field is on the stator and the armature is on the rotor. A commutator on the rotor rectifies the generated power to dc.

The field current and thus magnetic field increases with operating speed. The armature voltage and electrical torque also increase with speed. The actual speed of the turbine is determined by a balance between the torque from the turbine rotor and the electrical torque. Since the wind speed is variable over a wide range, some regulation method must be used, as shown in Figure 12.7.

By regulating the speed of the generator (i.e., wind turbine) and/or its field, the dc voltage can be maintained with a specified range. Speed regulation is usually performed by changing the pitch of the propeller blades. If the dc voltage is sensed, the field strength can be varied according to the control of the generated voltage. As illustrated in Figure 12.6a, a transmission that increases the rotating blade speed to that required for the generator has to be included.

A wind machine typically rotates in the range of 50–100 rev/min (i.e., about 5–10 rad/s). Depending on the generator, this has to be geared up to 1000–2000 rev/min (i.e., about 100–200 rad/s). The net efficiency of the energy conversion system is a function of the efficiency of the blades, transmission, generator, regulating circuitry, and inverter. However, dc generators of this type are seldom used today because of high costs and maintenance requirements (due to the commutators and the brushes).

Permanent-magnet generators are used in most small wind turbine generators, up to at least 10 kW. Here, permanent magnets provide the magnetic field. Hence, there is no need for field windings, or supply current to the field, nor there is any need for commutators, slip rings, or brushes.

The permanent-magnet generator is quite rugged since the machine construction is so simple. Their operating principles are similar to that of synchronous machines, with the exception that they are run asynchronously. In other words, they are not generally connected directly to the ac network. The power produced by the generator is initially variable voltage and frequency ac. This ac variable voltage is often rectified immediately to dc. The resultant dc power then either directed to dc loads or battery storage, or else it is inverted to ac with a fixed frequency and voltage.

Synchronous machines operate at constant speed, with only the power angle changing as the torque varies. Synchronous machines hence have a very "stiff" response to fluctuating conditions. An alternator produces an ac voltage whose frequency is proportional to shaft speed. Even with speed regulation, there will still be enough of a variation in frequency and phase to prevent connection of the alternator directly to the utility grid.

Therefore, the alternator is permitted to turn at different speeds, producing a variable-frequency output. The alternator output is then rectified, converting it to dc, as shown in Figure 12.5. The magnitude will be constant since the alternator field is constant. It is usually a permanent-magnet alternator. The dc is now fed to a synchronous inverter, whose line-frequency output can be connected directly to the utility grid. Here, the need for transmission is eliminated and the alternator can be connected directly to the wind wheel.

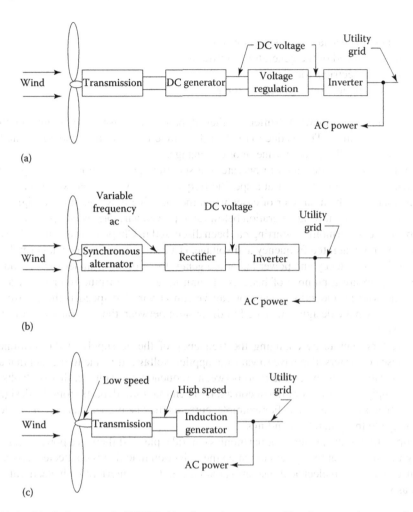

FIGURE 12.6   Block diagram of a WESC: (a) using a dc generator, (b) using a synchronous alternator, and (c) using induction generator.

The induction generator is well suited for a wind energy system provided that utility power is available. But, in order for the induction machine to operate as a generator, a separate source of reactive power is necessary to excite the machine. Also, the induction generator must be driven slightly faster than synchronous speed. However, it is not necessary for the speed to be constant, merely to maintain a negative slip. Rated power and peak efficiency are generally achieved at about −3% slip, not the speed of its rotor.

The only components required for this WESC are a transmission to gear the speed of the blades up to that necessary for the negative slip and the induction generator, as represented in Figure 12.6c. However, a loss of utility power automatically disables the WESC since the field excitation no longer exists. The net system efficiency depends on the efficiency of the blades, transmission, and generator. But, some means of speed regulation is required to maintain the required slip.

Note that when a constant torque is applied to the rotor of an induction machine, it will operate at a constant slip. If the applied torque is varying, then the speed of the rotor will vary as well. This relationship can be described by the following equation:

$$J\frac{d\omega_r}{dt} = Q_e - Q_r \tag{12.1}$$

where

    J is the moment of inertia of the generator rotor

    ω_r is the angular speed of the generator rotor (rad/s)

    Q_e is the applied electrical torque

    Q_r is the torque applied to the generator rotor

Induction machines are somewhat "softer" in their dynamic response to changing conditions than are synchronous machines. This is due to the fact that induction machines undergo a small but significant speed change (slip) as the torque in or out changes.

Induction machines are designed to operate at a specific operating point. This operating point is usually defined as the rated power at a specific frequency and voltage. However, in wind turbine applications, there may be a number of cases when the machine may run at off-design conditions. These conditions include starting, operation below rated power, variable speed operation, and operation in the presence of harmonics. Starting has been discussed in Chapter 6 previously. The operation below rated power, but at rated frequency and voltage, is a common occurrence. It normally presents few problems. But, efficiency and power factor are generally both lower under such conditions.

In general, there are a number of benefits of running a wind turbine rotor at variable speed. A wind turbine with an induction generator can be run at variable speed if the electronic power converter of approximate design is included in the system between the generator and the rest of the electrical network.

Such converters operate by changing the frequency of the ac supply at the terminals of the generator. These converters also have to vary the applied voltage. It is due to the fact that an induction machine performs best when the ratio between frequency and voltage, that is, "volts to Hertz ratio," of the supply is constant or almost constant. When that ratio departs from the design value, a number of problems can take place. For example, currents may be higher, causing higher losses and possible damage to the generator windings.

Finally, operation in the presence of harmonics can take place, if there is a power electronic converter of significant size on the system to which the induction machine is connected. Also, harmonics may cause bearing and electrical insulation damage and may interfere with electrical control or data signal as well.

12.6 WIND TURBINE OPERATING SYSTEMS

Depending on controllability, wind turbine operating systems are categorized as (1) constant-speed wind turbines and (2) variable-speed wind turbines.

12.6.1 CONSTANT-SPEED WIND TURBINES

They operate at almost constant speed as predetermined by the generator design of gearbox ratio. The control schemes are always aimed at maximizing either energy capture by controlling the rotor torque or the power output at high winds by regulating the pitch angle. Based on the control strategy, constant-speed wind turbines are again subdivided into (a) stall-regulated turbines and (b) pitch-regulated turbines.

Constant-speed stall-regulated turbines have no options for any control input. Its turbine blades are designed with a fixed pitch to operate near the original tip speed ratio (TSR) for a given wind speed. When wind speed increases, it causes a reduced rotor efficiency and limitation of the power output. The same result can be achieved by operating the wind turbine at two distinct constant operating speeds by either changing the number of poles of the induction generator or changing the gear ratio.

The stall regulation has the advantage of simplicity. But it has the disadvantage of not being able to capture wind energy in an efficient manner at wind speeds other than the design speed. They use pitch regulation for starting up. They have the following advantages:

1. They have a simple, robust construction and electrically efficient design.
2. They are highly reliable since they have fewer parts.
3. No current harmonics are produced since there is no frequency conversion.
4. They have a lower capital cost in comparison to variable-speed wind turbines.

On the other hand, their disadvantages include the following:

1. They are aerodynamically less efficient.
2. They are prone to mechanical stress and are noisier.

12.6.2 Variable-Speed Wind Turbine System

Figure 12.7 shows a typical variable-speed pitch-regulated wind turbine system. It has two methods for controlling the turbine operation in terms of speed changes and blade pitch changes. The control strategies that are usually used are power optimization strategy and power limitation strategy.

Power optimization strategy is used when the wind speed is below the rated value. It optimizes the energy capture by keeping the speed constant based on the optimum TSR. However, if speed is changed because of load variation, the generator may be overloaded for wind speeds above nominal value. In order to prevent this, methods like generator torque control are employed to control the speed.

On the other hand, the power limitation strategy is used for wind speeds above the rated value by changing the blade pitch to reduce the aerodynamic efficiency. The advantages of the variable-speed wind turbine systems include the following:

1. They are subjected to less mechanical stress and they have high energy capture capacity.
2. They are aerodynamically efficient and have low transient torque.
3. They require no mechanical damping systems since the electrical system can effectively provide the damping.
4. They do not suffer from synchronization problems or voltage sags because they have stiff electrical controls.

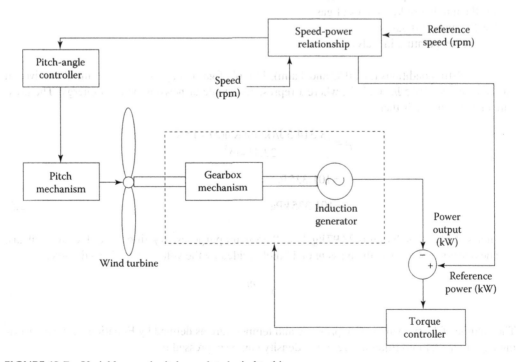

FIGURE 12.7   Variable-speed, pitch-regulated wind turbine.

The disadvantages of the variable-speed wind turbine systems include the following:

1. They are more expensive.
2. They may require complex control strategies.
3. They have lower electrical efficiency.

12.7 METEOROLOGY OF WIND

The fundamental driving force of air is a difference in air pressure between two regions. This air pressure is governed by various physical laws. One of them is known as *Boyle's law*. It states that *the product of pressure and volume of a gas at a constant temperature must be constant*. Thus,

$$p_1 v_1 = p_2 v_2 \tag{12.2}$$

Another is *Charles' law*. It states that *for a constant pressure, the volume of a gas varies directly with absolute temperature*. Hence,

$$\frac{v_1}{T_1} = \frac{v_2}{T_2} \tag{12.3}$$

Therefore, at −273.15°C or 0 K, the volume of a gas becomes zero.

Charles' law and Boyle's law can be combined into the ideal gas law. That is,

$$pv = nRT \tag{12.4}$$

where
p is the pressure in pascal (N/m²)
v is the volume of gas in cubic meters
n is the number of kilo moles of gas
R is the universal gas constant
T is the temperature in Kelvin

At standstill conditions (i.e., 0°C and 1 atm), 1 kmol of gas occupies 22.414 m³ and the universal gas constant is 8314.5 J/(kmol K), where J represents a joule or newton meter of energy. The pressure of 1 atm at 0°C is then

$$p = \frac{[8,314.5\ \text{J/(kmol K)}](273.15\ \text{K})}{22.414\ \text{m}^3}$$

$$= 101,325\ \text{Pa}$$

$$= 101.325\ \text{kPa} \tag{12.5}$$

The mass of 1 kmol of dry air is 28.97 kg. For all ordinary purposes, dry air behaves like an ideal gas. The density ρ of a gas is the mass m of 1 kmol divided by the volume v of that kilomole:

$$\rho = \frac{m}{v} \tag{12.6}$$

The volume of 1 kmol varies with pressure and temperature as defined by Equation 12.4. By inserting Equation 12.4 into Equation 12.6, the density can be expressed by

$$\rho = \frac{mp}{RT}$$

$$= \frac{3.484p}{RT} \text{ kg/m}^3 \tag{12.7}$$

where
  p is in kilopascal
  T is in Kelvin (K)

This expression yields a density for dry air at standard conditions of 1.293 kg/m³.

The common unit of pressure used in the past for meteorological work has been the bar (i.e., 100 kPa) and the millibar (100 Pa). A standard atmosphere is 1.01325 bar or 1013.25 millibar.

Atmospheric pressure has also been given by the height of mercury in an evacuated tube. This height is 29.92 in. or 760 mm of mercury for a standard atmosphere. Also note that the chemist uses 0°C as standard temperature, whereas engineers have often used 68°F (20°C) or 77°F (25°C) as standard temperature. Therefore, here standard conditions are always defied to be 0°C and 101.3 kPa pressure.

Most wind-speed measurements are made about 10 m above the ground. Typically, small wind turbines are mounted 20–30 m above ground level, while the propeller tip may read a height of more than 100 m on the large turbines. Thus, an estimate of wind-speed variation with height is needed. Here, let us examine a property that is known as *atmospheric stability* in the atmosphere.

Pressure decreases quickly with height at low attitudes, where density is high, and slowly at high altitudes, where density is low. At sea level and a temperature of 273 K, the average pressure is 101.3 kPa. A pressure of half this value is reached at about 5500 m.

A temperature decrease of 30°C will often be related to a pressure increase of 2–3 kPa. The atmospheric pressure tends to be a little higher in the early morning than in the middle of the afternoon. Winter pressure tends to be higher than summer pressures.

The power output of a wind turbine is proportional to air density, which in turn is proportional to air pressure. Hence, a wind speed produces less of power from a given wind turbine at higher elevations, due to the fact that the air pressure is less. A wind turbine located at an elevation of 1000 m above sea level will produce only about 90% of the power it would produce at sea level, for the same wind speed and air temperature.

However, there are many good wind sites in the United States at elevations above 1000 m. The air density at a proposed wind turbine site is estimated by determining the average pressure at that elevation from Figure 12.8 and then using Equation 12.7 to find density. The ambient temperature must be used in the equation.

Example 12.1

Consider a wind turbine that is rated at 100 kW in a 10 m/s wind speed in air at standard conditions. If power output is directly proportional to air density, determine the power output of the wind turbine in a 10 m/s wind speed at a temperature of 20°C at a site that has the elevation of

  (a) One thousand meters above sea level
  (b) Two thousand meters above sea level

Solution

  (a) From Figure 12.8, the average pressure at the 1000 m elevation is 90 kPa and from Equation 12.7, the density at 20°C = 293 K is

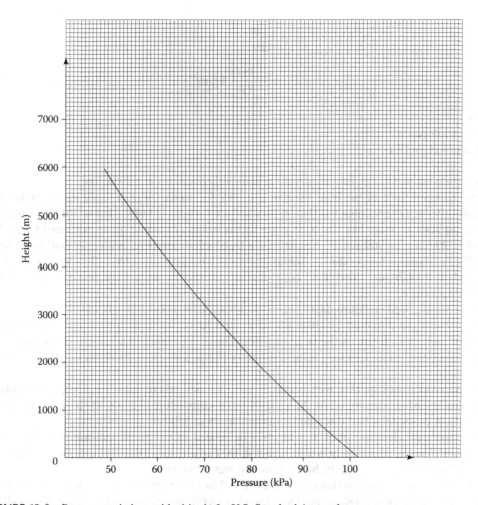

FIGURE 12.8 Pressure variations with altitude for U.S. Standard Atmosphere.

$$\rho = \frac{3.484p}{T}$$

$$= \frac{3.484(90)}{293}$$

$$= 1.070$$

Thus, the power output at the conditions is just the ratio of this density to the density at standard conditions times the power at standard conditions:

$$P_{new} = P_{old}\left(\frac{\rho_{new}}{\rho_{old}}\right)$$

$$= 100\left(\frac{1.070}{1.293}\right)$$

$$= 82.75\,kW$$

(b) From Figure 12.8, the average pressure of the 2000 m elevation is 80 kPa and since the temperature is still 20°C = 293 K, the density is

$$\rho = \frac{3.484p}{T}$$

$$= \frac{3.484(80)}{293}$$

$$= 0.951$$

Hence, the power at the 2000 m elevation is

$$P_{new} = P_{old}\left(\frac{\rho_{new}}{\rho_{old}}\right)$$

$$= 100\left(\frac{0.951}{1.293}\right)$$

$$= 73.55\,kW$$

Note that the power output has dropped 100–82.75 kW at the same wind speed at the 1000 m elevation and to 73.55 kW at the 2000 m elevation due to the fact that there are lesser air densities at the higher elevations.

12.8 POWER IN THE WIND

The wind speed is always fluctuating, and thus, the energy content of the wind is always changing. The variation depends on the weather and on local surface conditions and obstacles to the wind flow. Power output from a wind turbine will vary as the wind varies, even though the most rapid variations will to some extent be compensated for by the inertia of the wind turbine rotor.

It is common knowledge around the globe that it is windier during the day time than at night. This variation is mostly as a result of temperature differences that tend to be larger during the day than at night.

Furthermore, the wind is also more turbulent and tends to change direction more frequently during the day than at night. Therefore, forecasting the amount of electrical energy that can be harnessed over a period of time is extremely difficult.

Consider the wind turbine shown in Figure 12.1 and assume that the wind blows perpendicularly through a circular cross-sectional area, as illustrated in Figure 12.1. A wind generator will capture only the wind power caught by the given swept area A can be expressed in watts in SI system as

$$P = \frac{1}{2}\rho_a A v^3 \tag{12.8}$$

where
  ρ_a is the mass density of air (and is relatively constant)
  A is the circular cross-sectional area in m^2 (i.e., $A = \pi r^2$)
  r is the radius of the circular cross-sectional area in m
  v is the wind velocity in m/s

For the average mass density of air, $\rho_a = 1.24\,kg/m^3$.
Or in British system in ft lb/s as

$$P = \frac{1}{2}\rho_a A v^3\left(\frac{746}{550}\right) \tag{12.9a}$$

or

$$P = 0.678\rho_a A v^3 \tag{12.9b}$$

where $\rho_a = 0.0024\,\mathrm{lb}\ \mathrm{s^2/ft^4}$, A is in $\mathrm{ft^2}$, v is in ft/s. Note that since the wind speed is usually given in miles per hour (mph), it needs to be converted into ft/s by using

$$v_{\mathrm{ft/s}} = 1.47 v_{\mathrm{mph}} \tag{12.10}$$

The following equation gives an improved version of the above equation to determine the power in the wind in watts:

$$P = \frac{1}{2}\rho_a A v^3 C_p \tag{12.11}$$

where C_p is the *turbine power coefficient*, which represents the power conversion efficiency of wind turbine. It gives a measure of the amount of power extracted by the turbine rotor. Its value varies with *rotor design* and the TSR.

TSR is the *relative speed of the rotor* and the wind and has a maximum practical value of about 0.4. The ratio of the tip speed of the machine turbine blades to wind speed is found from

$$\lambda = \frac{r \times \Omega}{v} \tag{12.12}$$

where
  r is the radius of the circular cross-sectional area (i.e., turbine radius)
  Ω is the tip speed of the machine turbine blades
  v is the wind speed

Figure 12.9 shows various tip speed diagrams for various types of wind energy converters.

Here, the TPR is the relation between the speed v_{tip} and the undisturbed wind speed v_0 and is signified by λ. Thus,

$$\lambda = \frac{\text{Tangential velocity of blade tip}}{\text{Wind speed}} \tag{12.13a}$$

or

$$\lambda = \frac{v_{tip}}{v_0} \tag{12.13b}$$

Previously, the power present in a wind for a given velocity and swept area was given by Equation 12.8 or 12.9b. However, all of this power cannot be collected by wind turbine. The theoretical maximum fraction of available wind power that can be collected by a wind turbine is given by the *Betz coefficient*.

The energy in the wind is kinetic energy. In order to capture this energy, the blades of a wind turbine have to slow down as it passes through them. Hence, after the wind has passed through the wind turbine, its velocity (thus, its kinetic energy) is less than it originally had.

Here, the energy it lost has been converted to the kinetic energy of the rotating blades. If after passing through the blades, the wind speed has decreased to one-third of its initial value, the blades will have theoretically captured a maximum fraction of the available wind energy. This maximum energy is given by

Beta coefficient = 0.5926

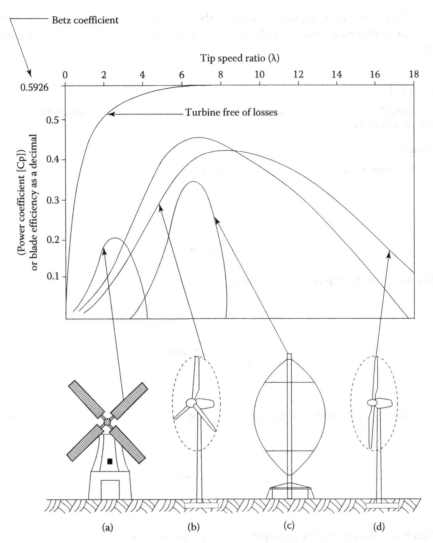

FIGURE 12.9 TSR diagrams for various types of wind energy converters. (The power coefficient gives a measure of how large a share of the wind's power a turbine can utilize.) The theoretical maximum of the value is 16/27 = 0.5926. The diagram shows the relation between TSR and power coefficient for different types of wind turbines: (a) windmill, (b) modern turbine with three blades, (c) vertical-axis Darrieus turbine, and (d) modern turbine with two blades.

This means that the actual power input for a wind turbine will be (at best) 59% of the power provided by Equation 12.8 or 12.9b. The actual blade efficiency is somewhat less than the Betz coefficient. It is a function of a quantity called the TPR λ, as explained in Equation 12.13.

The power coefficient C_p gives a measure of how large portion of the wind's power a turbine can utilize. The theoretical maximum value of C_p is 16/27 = 0.5926. The curves in Figure 12.9 show the relation between TPR and power coefficient for different types of wind turbines: (a) windmill, (b) modern turbine with three blades, (c) vertical-axis Darrieus turbines, and (d) modern turbine with two blades.

Note that the turbine power coefficient C_p is maximum at the $\lambda_{optimal}$. Also note that the wind turbine system uses induction generators that are independent of torque variation while speed varies between 1% and 2%.

In general, there is a great amount of power in the wind. However, this mechanical power when it is converted to electrical power is reduced substantially. A typical WECS has an efficiency of 20%–30%.

Example 12.2

Determine the amount of power that is present in a 10 m/s wind striking a windmill whose blades have a radius of 5 m.

Solution

The area swept by the blades of the wind turbine is

$$A = \pi r^2$$

$$= \pi (5 \text{ m})^2$$

$$\cong 78.54 \text{ m}^2$$

Thus, the power that is present in the wind is

$$P = \frac{1}{2} \rho_a A v^3$$

$$= \frac{1}{2} (1.24)(78.54)(10)^3$$

$$\cong 48,695 \text{ W}$$

If the turbine power coefficient (C_p) is 0.20, then the amount that will be converted to usable electrical power is

$$P = 48,695 C_p$$

$$= 48,695(0.20)$$

$$\cong 9.739 \text{ W}$$

which is considerably lesser than the power that is present in the wind.

12.9   EFFECTS OF A WIND FORCE

In any WECS, the support of the tower on which the wind generator is mounted must be considered. When a wind blows on a wind turbine, it applies a force on the blades. This wind force applied to the blades determined in SI or British system from

$$F_w = 0.44 \rho_a A v^2 \tag{12.14}$$

Additionally, the wind force applied on the tower (F_t) carrying the wind turbine has to be considered. The resultant effect of these forces is to develop a moment about the tower base in the clockwise direction. This overturning moment is a function of the wind speed, size of the blades, and the height of the wind turbine.

Because of this, large wind turbines mounted on high towers must be properly supported. Also, many wind turbines have an automatic high-wind shutdown feature. This future automatically turns the blades so that they become parallel to the wind and it can escape any damage to the WECS system.

12.10 IMPACT OF TOWER HEIGHT ON WIND POWER

As general rule, a taller tower is expected to result in higher-speed winds to the wind turbine. However, surface winds can also be affected by the irregularities or roughness of the earth's surface or by the existing forest and/or buildings in the vicinity. The relationship between the wind speed and the height of the wind turbine can be expressed as

$$\frac{v}{v_0} = \left(\frac{H}{H_0}\right)^{\alpha} \tag{12.15}$$

where
   v is the wind speed at height H
   v_0 is the reference (or known) wind speed at reference height of H_0
   α is the roughness (friction) sufficient

In Europe, the relationship in Equation 12.14 is modified as

$$\frac{v}{v_0} = \frac{\ln(H/2)}{\ln(H_0/2)} \tag{12.16}$$

There are many factors that affect wind. For example, elevation, contour of the ground in the surrounding areas, tall buildings, and trees. The average wind speed will be probably different at different tower height. In the event that the average wind speed at different heights is the same, the location with shorter height should be considered since such application results in with less-expensive tower.

Furthermore, at a higher elevation having greater wind, it is possible to use smaller wind turbine with shorter blade diameter rather than using a large turbine with larger blade diameter at a lower elevation for obtaining the same amount of power.

The value of the exponent α in Equation 12.14 depends on the roughness of the terrain given in Table 12.2.

Example 12.3

If the average wind speeds on an open plain (roughness class 1) is known to be 6 m/s at 10 m height, determine the wind speed at 50 m height.

Solution

From Table 12.1, $\alpha = 0.15$ and using Equation 12.15,

$$\frac{v}{v_0} = \left(\frac{H}{H_0}\right)^{\alpha} = \left(\frac{50}{10}\right)^{0.15}$$

TABLE 12.2
Roughness Coefficient for Various Class Types of Terrain

Roughness Class	Terrain Description	Roughness Coefficient (α)
Class 0	Open water	$\alpha = 0.1$
Class 1	Open plain	$\alpha = 0.15$
Class 2	Countryside with farms	$\alpha = 0.2$
Class 3	Villages and low forest	$\alpha = 0.3$

or

$$\frac{v_{50}}{6} = \left(\frac{50}{10}\right)^{0.15}$$

Thus, at 50 m height,

$$v_{50} = 6\left(\frac{50}{10}\right)^{0.15} = 7.6 \text{ m/s}$$

Example 12.4

Assume that the average wind speed at a point A is 6 m/s, while at point B is 7 m/s. In order to capture 2 kW, determine the blade diameter d for a wind turbine operating

(a) At point A
(b) At point B

Solution

(a) Using Equation 12.10, the given wind speed needs to be converted to ft/s as

$$v = 1.47 \times v_{(mph)}$$

$$= 1.47 \times 6$$

$$= 8.82 \text{ ft/s} \quad \text{(at point } A\text{)}$$

and

$$v = 1.47 \times v_{(mph)}$$

$$= 1.47 \times 7$$

$$= 10.29 \text{ ft/s} \quad \text{(at point } B\text{)}$$

From Equation 12.9b, at point A,

$$A = \frac{P}{0.678 \rho_a v^3}$$

$$= \frac{2000 \text{ W}}{0.678 \times 0.0024 \times 8.82^3}$$

$$\cong 1791.4 \text{ ft}^3$$

Since

$$A = \pi \frac{d^2}{4}$$

then

$$d = \sqrt{\frac{4A}{\pi}}$$

$$= \sqrt{\frac{4(1791.4)}{\pi}}$$

$$\cong 47.76 \text{ ft}$$

(b) At point B,

$$A = \frac{P}{0.678\rho_a v^3}$$

$$= \frac{2000 \text{ W}}{0.678 \times 0.0024 \times 10.29^3}$$

$$\cong 1128.09 \text{ ft}^3$$

Thus,

$$d = \sqrt{\frac{4A}{\pi}}$$

$$= \sqrt{\frac{4(1128.09)}{\pi}}$$

$$\cong 37.9 \text{ ft}$$

Therefore, a smaller (cheaper) wind turbine could be employed at point A and provide the same power as a larger wind turbine at point B.

12.11   WIND MEASUREMENTS

Wind measurement equipment usually consists of an anemometer, which measure wind speed, and a wind vane, which measures wind direction. In most countries, a national meteorological institute has measured and collected data on the winds since the nineteenth century. They register wind speed, wind direction, temperature, and other kinds of meteorological data several times a day (every 4 h, day and night) all year around. These data are reported daily to a central institution.

Nowadays, wind data are registered automatically. These observations make up the basis for the *wind statistics* that are used to describe wind climate in different regions and to create so-called *wind atlas data* that are used to calculate how much wind turbines can be expected to produce at different sites.

However, in the past, weather observers read the anemometer every 4 h, day and night. They observed the anemometer for a couple of minutes and recorded the average wind speed for that period. However, wind-speed data are affected by the height of the anemometer, the human factor in reading the wind speed, and the quality and maintenance of the anemometer.

A typical wind-cup anemometer works with a diametric flow of air. As the wind blows, the anemometer rotates at a speed proportional to the wind speed. Typically, a permanent-magnet dc generator is connected to the rotating shaft. A voltage is thus produced, which is proportional to the wind speed at every instant of time. The second instrument that is required is a *wind data compilator*. It is an electronic instrument that is connected to the anemometer and records the wind speed continuously.

Example 12.5

Consider a wind turbine that has blades with 8 ft radius. At its location where it is mounted, data was taken and it was discovered that the wind speed was 3 mph for 3 h and 12 mph for another 3 h time period. Determine the amount of energy that can be intercepted by the wind turbine.

Solution

The energy needs to be determined independently for each 3 h period.

During the first 3 h time period:
The average wind speed is

$$v_{avg} = \left(1.47\frac{ft/s}{mph}\right)(3\,mph) = 4.41\,ft/s$$

By using Equation 12.9b,

$$P = 0.678 \times 0.0024 \times 314.16 \times 4.41^3$$

$$= 43.84\,W$$

where

$$A = \pi(10\,ft)^2$$

$$\cong 314.16\,ft^2$$

$$Energy = (43.84\,W)(3\,h)$$

$$= 131.52\,Wh$$

$$= 0.13152\,kWh$$

During the second 3 h time period:
The average wind speed is

$$v_{avg} = \left(1.47\frac{ft/s}{mph}\right)(12\,mph) = 17.64\,ft/s$$

By using Equation 12.9b,

$$P = 0.678 \times 0.0024 \times 314.16 \times 17.64^3$$

$$= 2806\,W$$

Therefore, the total energy generated during the total period:

$$Energy = (2.806\,kW)(3\,h)$$

$$= 8.418\,kWh$$

Hence, total energy is

$$Total\ energy = 0.13152 + 8.418$$

$$\cong 8.56\,kWh$$

12.12 CHARACTERISTICS OF A WIND GENERATOR

The most important characteristic of a wind generator is its power curve. Normally, it is a graph provided by the manufacturer of a particular wind turbine. It shows the approximate power output as a function of wind speed. Figure 12.10 shows a typical power curve for a wind turbine rated 3 kW/25 mph. The power curve of a wind generator provides several important information.

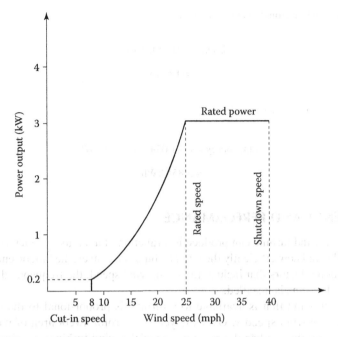

FIGURE 12.10 A typical power curve for a wind turbine.

In addition, to provide information for the obtainable power output at any given wind speed, it provides information about the cut in speed, the rated power, the rated speed, and the shutdown speed.

Here, the minimum wind speed required to start the blades turning and producing a useful output is defined as the *cut-in speed*. The maximum power output that the wind turbine will produce is called the *rated power*.

The minimum wind speed needed for the wind turbine to produce rated power is known as the *rated speed*. The *shutdown speed* is also called the *furling speed*. It is the maximum operational speed of the wind turbine. Beyond this speed, in order to prevent damage to the system from high winds, the blades are either folded back or turned to a high-pitch position.

Example 12.6

Consider the wind turbine whose power curve of its generator is shown in Figure 12.10. It is rated 3 kW/25 mph, as indicated in the figure. Assume that during an 8 h period, the wind had the following average speeds: 6 mph for 2 h duration, 10 mph for 3 h duration, 15 mph for 2 h duration, and 20 mph for 1 h duration. Determine the resultant electrical output for the 8 h period.

Solution

The energy calculated for each of the four wind speeds and time intervals.
At 6 mph: It is below the cut-in speed in Figure 12.10; thus, the output is zero.
At 10 mph: The output from the curve is 0.35 kW

$$\text{Energy} = (0.35\,\text{kW})(3\,\text{h})$$

$$= 1.05\,\text{kWh}$$

At 15 mph: The output from the curve is 0.85 kW

$$\text{Energy} = (0.85\,\text{kW})(2\,\text{h})$$

$$= 1.7\,\text{kWh}$$

At 20 mph: The output from the curve is 1.65 kW

$$\text{Energy} = (1.65 \text{ kW})(1 \text{ h})$$

$$= 1.65 \text{ kWh}$$

Thus, the total energy for 8 h duration is

$$\text{Total energy} = 0 + 1.05 + 1.7 + 1.65 \text{ kWh}$$

$$= 3.855 \text{ kWh}$$

12.13   EFFICIENCY AND PERFORMANCE

How much energy a wind turbine can produce is a function of a number of factors: the rotor swept area, the hub height, and how efficiently the wind turbine can convert the kinetic energy of the wind. Also are the additional factors that include the mean wind speed, the frequency distribution at the site where the wind turbine is installed.

The power of the wind that is available to a turbine is proportional to the rotor swept area *A* and the cube of the wind speed *v*. Over the years, the rotor swept area of wind turbines has increased steadily and thus so has the rated power of the wind turbines, as given in Table 12.2. Note that the production figures given in the table are based on a site with average wind resources. It appears that since 1980s, the power of wind turbines has doubled every 4–5 years on the average (Table 12.3).

Example 12.7

Assume that a wind generator, whose power curve is shown in Figure 12.10, has a blade diameter of 16 ft. If its power output is at 120 V at 60 Hz, determine the net efficiency of this WECS at a wind speed of 20 mph.

Solution

First, it is necessary to convert the wind speed from mph to ft/s by using Equation 12.10,

$$v_{(\text{ft/s})} = 1.47 \times v_{(\text{mph})}$$

$$= 1.47 \times 20$$

$$= 29.4 \text{ ft/s}$$

TABLE 12.3
Development of Wind Turbine Size, 1980–2005

Year	1980	1985	1990	1995	2000	2005
Power (kW)	50	100	250	600	1000	2500
Diameter (m)	15	20	30	40	55	80
Swept area (m²)	177	314	706	1256	2375	5024
Production (MWh/year)	90	150	450	1200	2000	5000

Source: Wizelius, T., *Developing Wind Power Projects: Theory and Practice*, Earthscan, London, U.K., 2007.

The input power is found from Equation 12.9b as

$$P_{in} = 0.678\rho_a A v^3$$

$$= 0.678\rho_a \times \pi r^2 \times v^3$$

$$= 0.678 \times 0.0024 \times \pi \times 8^2 \times 29.4^3$$

$$= 8314 \text{ W}$$

From Figure 12.9, the output power at 20 mph is

$$P_{out} = 1.65 \text{ kW} = 1650 \text{ W}$$

Thus, the efficiency of the system is

$$\eta = \frac{P_{out}}{P_{in}} \times 100$$

$$= \frac{1650 \text{ W}}{8314 \text{ W}} \times 100$$

$$= 19.84\%$$

Example 12.8

Assume that the wind turbine with three blades, as in Example 12.7, is rotating at 100 rpm. Find the blade efficiency at a wind speed of 20 mph.

Solution

$$V_0 = \text{wind speed}$$

$$= 20 \times 1.47$$

$$= 29.4 \text{ ft/s}$$

The circumference that the blade tip traces out is

$$2\pi r = 2\pi \times (8 \text{ ft})$$

$$= 50.27 \text{ ft}$$

The blade-tip speed is

$$v_{tip} = (50.27 \text{ ft/rev}) \, (100 \text{ rev/min}) \left(\frac{1}{60 \text{ s/min}} \right)$$

$$= 83.78 \text{ ft/s}$$

From Equation 12.13b

$$\lambda = \frac{v_{tip}}{v_0}$$

$$= \frac{83.78 \text{ ft/s}}{29.4 \text{ ft/s}}$$

$$\cong 2.85$$

From Figure 12.10 for $\lambda = 2.85$, the blade efficiency is about 13%. Note that the share of power in the wind that can be utilized by the rotor is called the *power coefficient*, C_p.

Example 12.9

Assume that a WECS shown in Figure 12.5 uses a three-phase, six-pole induction machine. The line frequency is 60 Hz and the average wind speed is 12 mph. The blades have a 30 mph diameter and peak efficiency when the TSR is 8.3. If the generator efficiency is a maximum at a negative slip of 3.3%, determine the transmission gear ratio for the peak system efficiency.

Solution

At first, the speeds required for the blades and generator has to be found. Then the transmission will be selected to match the two speeds. The required generator speed is

$$n_g = [1 - (-s)]\frac{120f}{p}$$

$$= [1 - (-0.033)]\frac{120 \times 60}{6}$$

$$= 1239.6 \text{ rev/min}$$

From Equation 12.10, the average wind speed is

$$v_0 = 1.47(12 \text{ mph})$$

$$= 17.64 \text{ ft/s}$$

From Equation 12.13b, the blade-tip speed is

$$v_{tip} = \lambda \times v_0$$

$$= 8.3 \times 17.64$$

$$= 146.412 \text{ ft/s}$$

The circumference traced out by the blade tip is

$$2\pi(15 \text{ ft}) = 94.248 \text{ ft/rev}$$

Hence, the blade-tip speed must be

$$\frac{146.412 \text{ ft/s}}{94.248 \text{ ft/rev}} = 1.5535 \text{ rev/s}$$

Converting this to rev/min

$$(1.5535 \text{ rev/s})(60 \text{ s/min}) = 93.21 \text{ rev/min}$$

Therefore, the transmission must gear up from 93.21 to 1239.6 rev/min. Hence, the required gear ratio is

$$\text{Gear ratio} = \frac{1239.6 \text{ rev/min}}{93.21 \text{ rev/min}}$$

$$\cong 13.3$$

12.14 EFFICIENCY OF A WIND TURBINE

In order to calculate the efficiency of a wind turbine, the efficiency of its components has to be calculated at first.

12.14.1 GENERATOR EFFICIENCY

A wind turbine can never utilize all the power in the wind. The amount of power that can be utilized by a wind turbine is given by power coefficient, C_p. It is known that (*based on Betz' law*) the maximum value of this coefficient is 0.59. It varies with the wind speed. For the most wind turbines, the maximum value varies between 0.45 and 0.50 at a wind speed of 8–10 m/s.

In order to convert the power in the wind from the revolving rotor to electric power, it is passed through a gearbox and a generator, or, for direct-drive turbines, through a generator and an inverter. In this conversion process, some power will be lost. Also, the efficiency of the individual components will vary with the wind speed.

It is known that a generator is most efficient when it is running at its nominal power. On a wind turbine, most of the time the generator is operating on *partial load*, that is, it runs on lower power when the wind speed is lower than the nominal wind speed. As a result, the standard generator efficiency will then be reduced, as given in Table 12.4.

There is also a relationship between the physical size of a generator and efficiency. That is, efficiency increases with the size of the generator, since losses to heat are reduced, as given in Table 12.5.

For example, a 1 MW wind turbine running at 20% of its nominal power (200 kW) has an efficiency of $0.95 \times 0.90 = 85\%$. Note that the relationship between efficiency, size, and partial load can also differ between different models and manufacturers.

12.14.2 GEARBOX

Typically, on a large modern wind turbine, the rotor has a rotational speed of 20–30 rpm, while the generator will need to rotate at 1520 rpm. In order to increase the speed, a gearbox is used. If the turbine rotor runs at 30 rpm, a gear change of 30:1520 = 1:50.7 is required. That is, one revolution of the main shaft has to be increased to 50.7 revolutions on the secondary shaft that is connected to the generator.

Generally, a gearbox has several steps, thus the rotational speed is increased stepwise. Losses can be estimated at 1% per step. In wind turbines, three-step gearboxes are usually used and the efficiency of the gearbox will then be about 97%.

However, wind turbines with a direct-drive generators and variable speed do not need any gearbox. Instead, the frequency and voltage of the electrical current will vary with the rotational speed.

TABLE 12.4
Generator Efficiency

Full load (%)	5	10	20	50	100
Efficiency	0.4	0.8	0.90	0.97	1.00

TABLE 12.5
Relationship between Size and Efficiency

Nominal power (kW)	5	50	500	1000
Efficiency	0.84	0.89	0.94	0.95

Thus, the current has to be rectified to dc and then converted by an inverter to ac with the same frequency and voltage as the grid. The efficiency of such an inverter is also about 97%.

12.14.3 OVERALL EFFICIENCY

In summary, the overall efficiency η_{total} of a wind turbine is the product of the turbine rotor's power coefficient C_p and the efficiency of the gearbox (or inverter) and generator

$$\eta_{total} = C_p \times \mu_{gear} \times \mu_{generator} \qquad (12.17)$$

Often, C_p is set to 0.59 and μ_{rotor} (or μ_r) is used to show how large a share of the theoretically available power the rotor can utilize. For example, if the power coefficient $C_p = 0.49$, the rotor turbine charges is then

$$\mu_r = \frac{0.49}{0.59} = 0.83$$

The efficiency of a wind turbine changes with the wind speed. When the wind speed is below the nominal wind speed, the efficiency of the generator will decrease and if the turbine has a fixed rotational speed, the TSR will change, that is, the an ever smaller share of the power in the wind will be utilized and C_p will decrease successfully.

Since the wind turbines are used to convert wind power to electric power, another coefficient is used, C_p, which indicates how large a share of the wind power is converted to electric power at different wind speeds. For example, a Siemens 1300 turbine is most efficient at wind speeds between 6 and 8 m/s, operating with $C_e \cong 0.46$.

12.15   OTHER FACTORS TO DEFINE THE EFFICIENCY

In order to estimate efficiency the following factors are also often used:

$$\text{Power/swept area} = \frac{\text{Production per year}}{\text{Rotor swept area}} \text{ kWh/m}^2 \qquad (12.18)$$

$$\text{Power production/nominal power} = \frac{\text{Production per year}}{\text{Rotor swept area}} \text{ kWh/kW} \qquad (12.19)$$

$$\text{Capacity factor} = \left(\frac{\text{Production per year}}{(\text{Nominal power}) \times 8760} \right) \times 100\% \qquad (12.20)$$

$$\text{Full load hours} = \left(\frac{\text{Production per year}}{\text{Nominal power}} \right) \times 100\% \qquad (12.21)$$

$$\text{Cost efficiency} = \frac{\text{Investment cost}}{\text{Production per year}} \text{ \$/kWh/year} \qquad (12.22)$$

$$\text{Availability} = \left(\frac{8760\,\text{h} - \text{stop hours}}{8760\,\text{h}} \right) \times 100\% \qquad (12.23)$$

Availability is the technical reliability of a wind turbine. If the wind turbine is out of operation due to faults or scheduled service and maintenance for 5 days a year, the technical availability is 98.6%. (A year is normally taken as 360 days for such calculations.)

It means that the turbine could produce power for 98.6% of the time, if there was always enough wind to make the run. The technical lifetime for a turbine is estimated at 20–25 years. However, its economic life time can be shorter due to increased maintenance costs as the turbine gets old.

There is another factor that is used to indicate the capacity factor. It is called annual load factor [4] and defined as

$$\text{Annual load duration factor} = (\text{Capacity factor}) \times 8760\% \tag{12.24}$$

Here, the significance of load duration is that it expresses that number of hours for which the wind turbine can be considered to be virtually operating at its rated capacity in 1 year.

In general, in order to indicate how much wind power is there in a country, the total installed capacity is used as a measure. Every wind turbine has a rated power (maximum power) that can vary from a few hundred watts to 5000 kW (5 MW). The number of turbines does not give any information on how much wind power they can produce.

How much wind a wind turbine can produce depends not only on its rated power, but also on the wind conditions. In order to get an indication of how much a certain amount of installed (rated) power will produce per year, use the following rule of thumb: "1 MW wind power produces 2 GWh* per year on land and 3 GWh per year offshore."

Example 12.10

Consider a 4 MW wind turbine that is under maintenance for 400 h in 1 year out of a total of 8760 h of 1 year. If it actually produced 8000 MWh due to fluctuations in wind availability, determine the following:

(a) The availability factor of the wind turbine
(b) The capacity factor of the wind turbine
(c) The annual load duration of the wind turbine

Solution

(a) The availability factor of the wind turbine is

$$\text{Availability factor} = \left(\frac{8760\,\text{h} - \text{stop hours}}{8760\,\text{h}} \right) \times 100\%$$

$$= \frac{8760 - 400}{8760} \times 100 = 0.9543 \text{ or } 95.43\%$$

(b) The capacity factor of the wind turbine is

$$\text{Capacity factor} = \left(\frac{\text{Production per year}}{(\text{Nominal power}) \times 8760} \right) \times 100\%$$

$$= \frac{8000\,\text{MWh}}{(2\,\text{MW}) \times 8760} \times 100$$

$$= 0.4566 \text{ or } 45.66\%$$

---

* Note the following relations:
  1 TWh (terawatt hour) = 1000 GWh (gigawatt hour)
  1 GWh = 1000 MWh (megawatt hour)
  1 MWh = 1000 kWh (kilowatt hour)
  1 kWh = 1000 Wh (watt hours)

(c) The annual load duration for the wind turbine is

$$\text{Annual load duration factor} = (\text{Capacity factor}) \times 8760\%$$

$$= (0.4566) \times 8760$$

$$= 3999.8\,\text{h}$$

However, the capacity factor of 0.4566 or 45.66% does not mean that the wind turbine is only running less than half of the time. Rather, a wind turbine at a typical location would normally run for about 65%–90% of the time. But much of the time it will be generating at less than full capacity, causing its capacity factor lower.

12.16   GRID CONNECTION

The term "grid" is often used loosely to describe the totality of the network. Specifically, *grid-connected* means connected to any part of the network. The term "national grid" usually means the EHV transmission network.

Integration particularly means the physical connection of the generator to the network with due regard to the secure and safe operation of the system and the control of the generator so that the energy resource is utilized optimally. The integration of generators power from wind turbine (or any other renewable energy sources) is basically similar to that of fossil fuel powered generator and is based on the same principles. However, renewable energy sources are often variable and geographically dispersed. The point on the connected to the network is referred to as the *point of common coupling* (PCC).

Wind power can be classified as small and non-grid-connected, small and grid-connected, large and non-grid-connected, and large and grid-connected. The small and non-grid-connected type of wind turbine can be used in a location that is not served by a utility. It can be improved by adding batteries to level out supply and demand. The cost will be high about $0.50 per kWh. The small and grid-connected wind turbine is usually not economically feasible.

The economic feasibility can be improved, if the local utility is willing to provide an arrangement that is called *net metering*. In such system, the meter runs backward when the turbine is generating more than the owner is consuming at the moment. The owner pays a monthly charge for the wires to his home.

In general, utilities want to buy at wholesale and sell at retail. It is often that the owner might pay $0.08–$0.15/kWh, and get paid $0.02/kWh for the wind-generated electricity which is far from enough to economically justify a wind turbine.

Wind speed is the main factor in determining electricity cost, in terms of influencing the energy yield, and approximately, at the locations with wind speeds of 8 m/s will yield electricity at one-third of the cost for a 5 m/s site. Wind speeds of approximately 5 m/s can typically be found at the locations away from the coastal areas. However, wind energy developers usually intend to find higher wind speeds. Levels at about 7 m/s can be found in many coastal regions.

The large and non-grid-connected wind turbines are installed on islands or in some native villages where it is virtually impossible to connect to a large grid. In such places, one or more wind turbines can be installed in parallel with the diesel generators, and so that the wind turbines can act as fuel savers when the wind is blowing. This system can operate easily. In general, the justification for having the small or the large wind turbines must be based on whether or not it will result in *a lower net cost to society*, including the environmental benefits of wind generation. Today, wind turbines with ratings near 1 MW or more are now common.

However, this is still small compared to the needs of a utility, so clusters of turbines are placed together to form *wind farms* or *wind plants* with total ratings of 10–100 MW, or even more. Presently,

Southern California Edison (SCE) Company is working on Tehachapi Renewable Transmission Project (TRTP) of 500 kV [6].

The purpose of the proposed TRTP project is to provide the electrical facilities necessary to integrate levels of new wind generation in excess of 700 MW and up to approximately 4500 MW in the future in the Tehachapi Wind Resource Area (TWRA) in Southern California.

The voltage level of large wind turbines, in general, is 600 V, also called *industrial voltage*. Therefore, they can be connected to a factory without a transformer. Smaller wind turbines, up to 300 kW, which were common in the near past, have a voltage of 480 V and can be connected directly via a feeder cable to a farm or a house. However, usually wind turbines are connected to the power grid through a transformer that increases the voltage level from 480 or 600 V to the higher voltage, normally 10 or 20 kV, in the distribution grid.

A suitable transformer is installed on the ground next to the tower for small- and medium-sized wind turbines. But in large wind turbines, the transformer is often a component of the turbine itself.

In modern wind turbines, the power that is supplied into the power grid can be converted by power electronics to achieve the phase angle and reactive power that the grid need at the point where the wind turbine is connected to improve power quality in the grid.

However, the power electronic equipment can cause a main problem, namely, *harmonics*, that is, currents with frequencies that are multiples of 60 Hz, and has a negative effect on power quality. Such "dirt" can, to some extent, be "cleaned of" by different kinds of filters. Unfortunately, such equipment is expensive and seldom takes care of all the "dirt."

12.17 SOME FURTHER ISSUES RELATED TO WIND ENERGY

In general, integration of wind power plants into the electric power system presents challenges to power system planners and operators. Wind plants naturally operate when the wind blows, and their power levels vary with the strength of the wind. Thus, they are not dispatchable in the traditional sense. Wind is primarily an energy source. Its main function is displacement of fossil fuel combustion in existing generating units.

These units maintain system balance and reliability, so no new conventional generation is required as *"backup"* for wind plants. Wind also provides some effective load-carrying capability and therefore contributes to planning reserves but not day-to-day operating reserves. Wind's variability and uncertainty do increase the operating costs of the non-wind portion of the power system, but generally by modest amounts.

Nowadays, wind studies in the United States employ sophisticated atmospheric (*mesoscale numerical weather prediction*) models to develop credible wind power time series for use in the integration analysis. Today, it is in general accepted that integration studies should use this type of data, synchronized with load data, when actual wind data not available [9].

According to Smith et al. [9], wind integration studies performed in recent years have provided important new insights into the *impact of wind's variability*, and uncertainty will have on system operation and operating costs. Their conclusions include the following:

1. Several studies of very high penetrations of wind (up to 25% energy and 35% capacity) have concluded that the power system can handle these high penetrations without compromising system operation.
2. The importance of detailed wind resource modeling has been clearly demonstrated.
3. The importance of increased flexibility in the non-wind portion of the generating mix has been clearly demonstrated.
4. The value of good wind forecasting has been clearly demonstrated to reduce unit commitment costs in the day-ahead time frame.
5. The difficulties of maintaining system balance under light-load conditions with significant wind variability constitute a serious problem.

6. Even though wind is mainly an energy resource, it does provide modest amounts of additional installed capacity for planning-reserve purposes.
7. There is a great value sharing balancing functions over large regions with a diversity of loads, generators, and wind resources.

12.18 DEVELOPMENT OF TRANSMISSION SYSTEM FOR WIND ENERGY IN THE UNITED STATES

In the United States, existing wind farms are in remote areas with respect to load centers. Transmission system owners have been unable to build new high-voltage transmission lines to remote areas where there may be a high-potential wind energy source but little existing generation or load.

Also, it is uneconomic to build transmission capacity to the peak power capacity of wind farms. But if transmission capacity is built to a number lower than the peak, it can lead to congestion when wind production is greater than the transmission capacity. That is, wind developers may find it economical to build wind capacity even though they know that congestion may develop and remain for a period of time [14].

When it comes to building new transmission lines, it appears that, due to limited funds, the emphasis is on the *eliminating bottlenecks in high-load corridors*. Also, in the past, new transmission lines have been approved only if there is *a proven need for improved system reliability*.

Because of these concerns, the utility companies that are interested in building wind farms have not been able to build new power plants in remote but wind-rich areas if there is no transmission line that has the capacity to transfer the plant output to major load centers. As a result, this chicken-and-egg dilemma delays the development if new wind plants and transmission lines to deliver the wind energy to load centers.

However, there has been some progress in California, Texas, and Colorado. For example, in California, the Tehachapi region has the potential for more than 7000 MW (7 GW) of new wind generation, but the opportunity to develop it was stalled because there was no way to fund the necessary expansion of the bulk 500 kV transmission system. SCE received the California Independent System Operator's (CALISO) approval for the $1.4 billion Tehachapi Transmission project in 2007. Some transmission segments are now under construction, and a few more are in the proposal stage. The project completion date is given as 2013.

12.19 ENERGY STORAGE

When wind production exceeds the transmission system capacity or congestion takes place on the system, storage can capture the "lost" energy and can then discharge back to the grid when the congestion eases. Here, the main idea is to use storage to increase the effective capacity of the wind farm.

The unconstrained wind farm output is known as *potential capacity factor*. It is the total capacity if transmission capacity is built to wind peak. Whereas, the *actual capacity factor* defines the real capacity in case the transmission constraint restricts the output. The *effective capacity factor* is defined as the capacity that can be achieved through the use of storage.

If there is unlimited storage capacity, the effective capacity factor would equal the potential full capacity factor of the wind farm. But, other factors such as cost and size force wind developers to establish a balance between the maximum (*unconstrained*) and minimum (take no action) capacity factors. The economics of such application is a function of the wind farm power duration, the transmission congestion duration, and the ratio of the storage capacity to wind farm capacity (both in terms of power and duration) [3].

Succinctly put, wind is *not a constant resource*. Wind velocities follow *regular diurnal patterns,* that is, wind does not blow consistently throughout the day, but rather reaches peaks at specific and typical times and declines in the same manner. In the plains, wind velocities might be greatest at nighttime.

In mountain regions, on the other hand, wind velocity might be greatest in the early morning and late afternoon as well as early evening and lowest during the daytime or in the middle of the night. Offshore wind is typically more reliable; still, the pattern throughout the day varies.

According to Fioravanti et al. [3], although emerging storage technologies are making great progress, the megawatt capacity needed to shift the potential generation of wind farms tends to outsize the capabilities of the storage technologies.

There are basically two main technologies that have the capacity to perform in this application: *pumped hydro* and *compressed-air energy storage* (CAES). There are other storage technologies that include batteries, flywheels, ultra capacitors, and, to some extent, photovoltaics.

Most of these technologies are best suited for power quality and reliability enhancement applications, due to their relative energy storage capabilities and power density characteristics, even though some large battery installations could be used for peak shaving.

All of the storage technologies have a power electronic converter interface and can be used together with other distributed utilities (DU) technologies to provide "seamless" transitions when power quality is a requirement.

The earliest known use of pumped hydro technology was in Zurich, Switzerland, in 1982. The relative, low-efficiency and low-cost pumped hydro plants are more often compensated by the ability to avoid expensive peaking power. When available, pumped hydro plants are excellent solutions to solve the diurnal problems. But, there are presently limited siting possibilities for new pumped hydro resources.

An alternative is the *CAES*. In such system, the off-peak electrical energy is used to run meters and compressors to pump air into a limestone cavern. At times of the need that is during peak, the pressured air is let through a recuperator-turbine-generator system. The produced electric energy is returned to the power system at the time of peak. This system of CAES has been developed by the Electric Power Research Institute (EPRI) over the years.

However, CAES is not a pure storage system since natural gas is added to the compressed air going through the turbine to boost power production and overall efficiency. Without this injection, the overall efficiencies of the cycle would be low about 70%–80%.

CAES is a peaking gas turbine power plant that consumes less than 40% of the gas used in a combined-cycle gas turbine to produce the same amount of electric output power.

This is accomplished by blending compressed air to the input fuel to the turbine. By compressing air during off-peak periods when energy prices are very low, the plant's output can produce electricity during peak periods at lower costs than conventional stand-alone gas turbines can achieve.

Today, the EPRI has an advanced CAES system designed around a simpler system using advanced turbine technology. It is developed for plants in the 150–400 MW range with underground storage reservoirs of up to 10 h of compressed air at 1500 lb/in.^{2}

Depending on the reservoir size, multiple units can be deployed. The largest plant under construction in the United States would have an initial rating of 800 MW. EPRI is also studying an aboveground CAES alternative with high-pressure air stored in a series of large pipes. These smaller systems are targeted at ratings of up to 15 MW for 2 h. CAES has the potential to be very large scale when underground storage is used. The first commercial CAES was a 290 MW unit built in Huntorf, Germany, in 1978 [3].

12.20 WIND POWER FORECASTING

Wind power forecasting plays a key role in dealing with the challenge of balancing the system supply and demand, given the uncertainty involved with the wind plant output. According to Smith et al. [9], wind forecasting is a prerequisite for the integration of a large share of wind power in an electricity system, as it links the weather-dependent production with the scheduled production of conventional power plants and the forecast of the electricity demand, the latter being predictable with reasonable accuracy.

The essential application of wind power forecasting is to reduce the need for balancing energy and reserve power, which are needed to integrate wind power into the balancing of supply system and demand in the electricity supply system (i.e., to optimize the power plant scheduling). This leads to lower integration costs for wind power, lower emissions from the power plants used for balancing and subsequently to a higher value of wind power.

A second application is to provide forecasts of wind power feed-in for grid operation and grid security evaluation, as wind farms are often connected to remote areas of the transmission grid. In order to forecast congestion as well as losses due to high physical glows, the grid operator required to know the current and future wind power feed-in at each grid connection point [9].

Therefore, as wind power capacity rapidly increases, forecast accuracy becomes increasingly important. This is especially true for large onshore or offshore wind farms, where an accurate forecast is crucial due to the high concentration of capacity in a small area.

Luckily, in recent years, the forecasting accuracy has improved steadily, and will be more likely even better in the future. It has been discovered that if many wind farms are forecasted together, the forecast error decreases. The larger the involved regions the more accurate the resultant forecast since the forecast errors of different regions will partially cancel each other out.

Today, utility companies study *not only wind-integration costs* but also *operational savings due to disputed fuel and emissions*. Since 2006, wind-integration studies have evolved to consider higher wind penetration and larger regions, which leads to a greater focus in new transmission needs. According to Corbus et al. [2], such a regional study approach dictates that *additional questions to be answered* including

1. How do local wind resources compare with higher capacity-factor wind that requires more transmission?
2. How does the geographic diversity of wind power reduce wind-integration costs (i.e., by spreading the wind over a larger region and thereby "smoothing out" some of the variability)?
3. How does offshore wind compare with onshore wind?
4. How does balancing area consolidation or cooperation affect wind power integration costs?
5. How much new transmission is needed to facilitate high penetration of wind power?
6. What is role and value of wind forecasting?
7. What role do shorter scheduling intervals have to play?
8. What are the wind power integration costs spread over large market footprints and regions?
9. What additional operating reserves are needed for large wind power developments?

According to Ackermann et al. [1], experience with the integration of high amounts of wind generation into power systems around the world has shown no incidents in which wind generation has directly or indirectly caused unmanageable operational problems. The key elements *for the successful integration of high-penetration* levels of wind power are as follows:

1. There must be well-functioning markets over large geographic areas—combining a number of balancing areas—that enable an economical way of sharing balancing resources. This situation also enables aggregation of a more diverse portfolio of wind plants, which reduces the output variability. *Well-functioning markets* must also offer a range of scheduling periods (i.e., day ahead, hour ahead, and real time) to accommodate the uncertainty in wind-plant forecasts. The basic requirement for such a well-functioning market overlarge geographic areas is an appropriately designed transmission system to interconnect the different network areas.
2. Advanced wind-forecasting systems based on a variety of weather input and their active integration into power-system operation are needed.

3. New simulation tools are necessary to evaluate the impact of wind power on the security of supply and load balancing in near real time.
4. The corresponding "right to curtail" wind power, when necessary from the system security point of view.

PROBLEMS

12.1 Consider a wind turbine that is rated at 100 kW in a 10 m/s wind speed in air at standard conditions. If power output is directly proportional to air density, what is the power output of the wind turbine in a 10 m/s wind speed at a elevation 2000 m above sea level at a temperature of 20°C?

12.2 Consider Example 12.4 and assume that the average wind speed at the point a is 10 mph, while at the point b is 8 mph. In order to capture 2 kW, determine the blade diameter (d) for a wind turbine operating
(a) At point a
(b) At point b

12.3 Consider a wind turbine with blades of 10 ft radius. At the location, wind speed was 5 mph for 3 h and 15 mph for another 3 h time period. Determine the amount of energy that can be intercepted by the wind turbine.

12.4 Consider the wind turbine given in Example 12.10 and assume that the wind turbine is 2 MW and is under maintenance for 200 h in 1 year out of a total of 8760 h of 1 year. If it is actually produced 4000 MWh due to fluctuations in wind availability, determine the following:
(a) The availability factor of the wind turbine
(b) The capacity factor of the wind turbine
(c) The annual load duration of the wind turbine

12.5 Consider Example 12.6 and assume that during an 8 h period, the wind had the following average speeds: 4 mph for 2 h duration, 12 mph for 2 h duration, 17 mph for 1 h duration, and 23 mph for 3 h duration. Determine the resultant electric output for the 8 h period.

12.6 Assume that a wind generator whose power curve is shown in Figure 12.9 has a blade diameter of 18 ft. If its power output is at 120 V and 60 Hz, determine the net efficiency of the WECS at a wind speed of 15 mph.

12.7 Assume that the wind turbine with three blades in Problem 12.6 has been replaced by one with two blades that is rotating at 90 rpm. Find the blade efficiency at a wind speed of 16 mph.

12.8 A WECS that has a eight-pole, 60 Hz, three-phase synchronous alternator driven at synchronous speed. The blades have an 8 m diameter and a peak efficiency when the TSR = 5. Determine the transmission gear ratio for peak system efficiency at a wind speed of 6 m/s?

12.9 A WECS uses a six-pole, 60 Hz, three-phase induction generator. It is excited by a three-phase, 60 Hz power line. The blades have an 11 m diameter and peak efficiency when the TSR = 6. If the generator efficiency is a maximum at a slip of −3.3%, what should the transmission gear ratio be for peak system efficiency at a wind speed of 5 m/s?

12.10 Assume that the WECS shown in Figure 12.5 uses a four-pole, three-phase induction machine. The line frequency is 60 Hz and the average wind speed is 15 mph. The blades have a 32 ft diameter and peak efficiency when the TSR is 6. If the generator efficiency is a maximum at a negative slip of 3.3%, determine the transmission gear ratio that is necessary for the peak system efficiency.

REFERENCES

1. Ackermann, T. et al., Where the Wind Blows, *IEEE Power & Energy Magazine,* 7(6), November/December 2009, 65–75.

2. Corbus, D. et al., Up with Wind, *IEEE Power & Energy Magazine*, 7(6) , November/December 2009, 34–46.

3. Fioravanti, R., Khoi, V., and Stadlin, W., Large-scale solutions, *IEEE Power & Energy Magazine*, 7(4), July/August 2009, 48–57.

4. Gönen, T., *Electric Power Distribution System Engineering*, 2nd Ed., Boca Raton, FL: CRC Press, 2008.

5. Grant, W. et al., Change in the air, *IEEE Power & Energy Magazine*, 7(6), November/December 2009, 36–46.

6. Grant, W. et al., Change in the air, *IEEE Power & Energy Magazine*, 7(6), November/December 2009, 47–64.

7. Manwell, J. F., McGowan, J. G., and Rogers, A. L., *Wind Energy Explained: Theory, Design, and Application*, West Sussex, U.K.: Wiley, 2003.

8. Orloff, S. and Flannery, A., Wind turbine effects on avian activity, habitat use, and mortality in Altamont Pass and Solano County wind resource areas: 1989–1991, *California Energy Commission Report*, No. P700-92,002. 1992.

9. Smith, C. et al., A mighty wind, *IEEE Power & Energy Magazine*, 7(2), March/April 2009, 41–57.

10. Stanon, C., Wind farm visual impact and design of wind farms in the landscape, *Wind Energy Conversion*, BWEA, 1994, pp. 249–255.

11. Stanon, C., Wind farm visual impact and its assessment, *Wind Directions*, BWEA, August 1995, pp. 8–9.

12. Wagner, H. J. and Mathur, J., *Introduction to Wind Energy Systems*, Berlin, Germany: Springer, 2009.

13. Wind Advisory Council, Wind energy: A vision for Europe in 2030, Report from TP Wind Advisory Council, European Wind Energy Technology Platform, 2007.

14. Wizelius, T., *Developing Wind Power Projects: Theory and Practice*, London, U.K.: Earthscan, 2007.

13 Solar Energy Systems

13.1 INTRODUCTION

Even though solar energy is a very small portion of the energy system today, the size of the resource is enormous. The average intensity of light outside the atmosphere (known as the solar constant) is about $1353\,\text{W/m}^2$. In order to produce a gigawatt of power, an area of about $5\,\text{km}^2$ would be needed, assuming a conversion of 20%. The earth receives more energy from the sun in 1 h than the global population uses in an entire year.

Furthermore, the *solar photovoltaic* (PV) industry is growing very fast, sustaining an annual growth rate of more than 40% for the last decade. Because of this fast growth, decreasing costs, and a vast technical potential, solar energy is becoming an important alternative for the future energy needs. With increasing applications of distributed and utility-scale PV as well as *concentrating solar power* (CSP), solar technologies have started to play an important role in meeting world's energy demand.

The term "photovoltaic" describes the conversion process to convert light energy directly to electrical energy. The developments in semiconductor technology cause the invention of the PV cell (also known as solar cell) in the early 1950s.

PV and CSP technologies both use the sun to generate electricity. However, they do it in different ways. The earth's surface receives sunlight in either a direct or diffuse form. Direct sunlight is solar radiation whose path is directly from the sun's disk and shines perpendicular to the plane of a solar device. This is the form used by CSP systems and concentrating PV systems, in which the reflection or focusing of the sun's light is essential to the electricity-generating process. Flat-plate or nonconcentrating PV systems can also use direct sunlight.

PV (or *solar electric*) systems use semiconductor solar cells to convert sunlight directly into electricity. In contrast, CSP (or *solar thermal electric*) systems use mirrors to concentrate sunlight and exploit the sun's thermal energy. This energy heats a fluid that can be used to drive a turbine or piston, hence producing electricity. Succinctly put, PV uses the sun's light to produce electricity directly, whereas CSP uses the sun's heat to produce electricity indirectly. Again, PV and CSP both use the sun to produce electricity. However, they use different forms of the sun's radiation [2].

Other solar radiation is diffuse, meaning the sunlight reaches the earth's surface after passing through thin cloud cover or reflecting off of particles or surfaces. *Global radiation* is the sum of the direct and diffuse components of sunlight. This global radiation, as well as direct or diffuse radiation alone, can be used by *flat-plate* PV systems to generate electricity [3].

As said before, PV and CSP are different forms of solar technologies. Similarly, there are different PV materials and designs for generating electricity, which include crystalline silicon, thin films, concentrating PV, and future-generation PV, in addition to associated balance of systems components.

The new technologies, such as depositing solar modules onto a flexible plastic substrate or using solar "inks" (e.g., copper indium gallium selenide) and a "printing" process to produce film solar panels, are ready to drastically reduce the cost of solar power plants to less than $1 per watt.

Solar power plants can be built where they are most needed in the grid because siting PV arrays is usually much easier than siting a conventional power plant. Also, unlike conventional power plants, modular PV plants can be expanded incrementally as demand increases. It is expected that municipal solar power plants with few megawatt capacity built close to load centers will become common during the next decade.

13.2 CRYSTALLINE SILICON

Silicon was one of the very first materials that was used in early PV devices, and it still has more than 90% of market share in today's commercial solar cell market. The silicon-based cells are known as *first-generation PV*. Pure silicon is mixed with very small amounts of other elements such as boron and phosphorous, which become positive- and negative-type semiconductor materials, respectively. Putting the two materials in contact with one another creates a built-in potential field [6].

Thus, when this semiconductor device is subject to the sunlight, the energy of the sunlight frees electrons that then move out of the cell, because of the potential field, into wires that form an electric circuit. Such "PV" effect needs no moving parts and does not use up any of the material in the process of generating electricity.

A typical solar cell has a glass or plastic cover or other encapsulated cover, an antireflective surface layer, a front contact to permit electrons to enter a circuit, and a black contact to permit the semiconductor layers where the electrons start and finish their flow. The thickness of a crystalline silicon (c-Si) cell may be 170–200 μm (10^{-6}).

Figure 13.1 shows a typical 3 in diameter cell. It will produce a voltage of 0.57 V when sunlight shines upon it under open-circuit conditions. This voltage would be the same regardless of how big the size of cell is. But, its current supply is directly proportional to its surface area. In that sense, a solar cell can be considered a constant current source as well as a voltage source. The solar cell is a nonlinear device and its performance is subject to its characteristic curve.

Figure 13.2 shows a typical I–V characteristic for a PV cell. Note that when the current is zero (i.e., no load), the voltage (i.e., the *open-circuit voltage* V_{oc}) is about 0.6 V. As the load resistance increases, causing the voltage output of the cell to increase, the current remains relatively constant until the "knee" of the curve is reached. The current then drops off quickly, with only a small increase in voltage, until the open circuit condition is reached. At this point, the open-circuit voltage is obtained and no current is drawn from the device [1].

The power output of any electrical device, including a solar cell, is the output voltage times the output current under the same conditions. The open-circuit voltage is a point of no power, that is, the current is zero. Similarly, the short-circuit condition produces no power because the voltage is zero. The maximum power point is the best combination of voltage and current, as shown in Figure 13.2. This is the point at which the load resistance matches the solar cell internal resistance.

The power into the cell is a function of the cell area and the power density of light. Once these are fixed, the peak efficiency takes place when the power output is maximum. Thus, the maximum power point should be selected as the operating point of the cell. As indicated in Figure 13.2, the maximum power output takes place somewhere around the center of the knee of the curve.

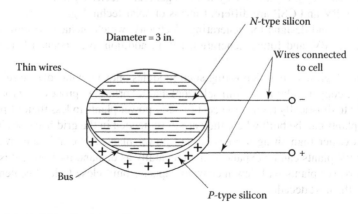

FIGURE 13.1   A typical 3 in. diameter cell.

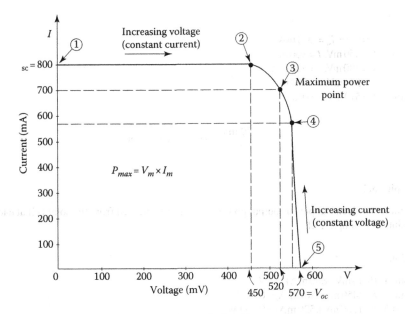

FIGURE 13.2   Typical I-V characteristic for a photovoltaic cell.

The power into the solar cell is a function of the cell area and the power density of the light. Hence, for a given cell, the peak efficiency takes place when the power output is a maximum. In general, the solar cell responds well to all forms of visible light. Thus, they can operate indoors from incandescent or florescent lamps [4].

The peak power current changes proportionally to the amount of sunlight, but the voltage drops only slightly with large changes in the light intensity. Hence, a solar cell system can be designed to extract enough usable power to trickle-charge a storage battery even on a cloudy day.

It is estimated that the sun is constantly emitting 1.7×10^{23} kW of power. A very small portion of this (about 8.5×10^{23} kW) reaches the earth. About 30% of this is lost and 70% (about 6×10^{13} kW) penetrates our atmosphere. The amount of power per unit area that is received from the sum is defined as *power density* [5]. With the sun directly overhead on a clear day, the power density of sunlight is about $100\,mW/cm^2$. The power density of sunlight is also defined with a unit called the SUN. Hence,

$$1\,SUN = 100\,mW/cm^2 = 1\,kW/m^2 \tag{13.1}$$

On a cloudy day, the *power density* of sunlight might be

$$30\,mW/cm^2 = 0.3\,SUN$$

Energy density is another quantity that is also used to measure sunlight. Its unit is Langley

$$1\,Langley = 11.62\,Wh/m^2 \tag{13.2}$$

Example 13.1

Assume that the solar cell, whose characteristic is shown in Figure 13.2, is connected a resistive load. Determine the required value of load resistance R_L to obtain at each of the operating points of 1, 2, 3, 4, and 5 on the characteristics:

Solution

At point 1: $V=0$, $I=I_{sc}=800\,mA$
At point 2: $V=450\,mV$, $I=800\,mA$
At point 3: $V=700\,mV$, $I=520\,mA$
At point 4: $V=570\,mV$, $I=550\,mA$
At point 5: $V=570\,mV$, $I=0\,mA$

$$R_L = \frac{570\,mV}{0} = \infty \Rightarrow \text{open circuit}$$

Example 13.2

Determine the electrical power (output power) that can be obtained from the solar cell at each of the points in Example 13.1.

Solution

At point 1: $P=(0\,mV)(800\,mA)=0\,W$
At point 2: $P=(450\,mV)(800\,mA)=0.36\,W$
At point 3: $P=(700\,mV)(520\,mA)=0.364\,W$
At point 4: $P=(570\,mV)(550\,mA)=0.313\,W$
At point 5: $P=(570\,mV)(0\,mA)=0\,W$

Notice that a good power output is obtained between points 2 and 3. In fact, maximum power output is obtained at the point 3, that is, the *maximum power point*.

Example 13.3

Determine the energy density in Langley, if the strength of sunshine is 1 SUN for a period of 2 min.

Solution

$$\text{Power density} = 1\,SUN = 1\,kW/m^2 = 1000\ W/m^2$$

$$\text{Energy} = \text{power} \times \text{time}$$

$$= (1000\,W/m^2)\left(\frac{1\,min}{60\,min/h}\right)$$

$$= 16.67\,Wh/m^2$$

To determine the energy density, use Equation 13.2,

$$\text{Energy density} = \frac{16.7\,Wh/m^2}{11.62\,Wh/m^2}$$

$$= 1.424\,\text{Langleys}$$

Example 13.4

Determine the energy density in Langley, if the strength of sunlight is 1/2 SUN for a period of 2 min.

Solution

$$\text{Power density } = \frac{1}{2}\,\text{SUN}$$

$$= 0.5 \text{ kW/m}^2$$

$$= 500 \text{ W/m}^2$$

$$\text{Energy} = \text{power} \times \text{time}$$

By using Equation 13.2, the energy density in Langley is

$$\text{Energy density } = \frac{16.67 \text{ Wh/m}^2}{11.62 \text{ Wh/m}^2}$$

$$= 1.434 \text{ Langleys}$$

It can be seen that still the same amount of energy density is obtained but it would take twice as long as 1/2 SUN.

Example 13.5

Consider the rooftop of a home measuring 12×15 m, which is all covered with PV cells to provide the electrical energy requirement by that home. Assume that the sun is at its peak (i.e., having strength of 1 SUN) for 3 h every day and the efficiency of a solar cell is 10% (i.e., 10% of the sunlight power that falls on the cell is converted to electric power). Determine the average daily electrical energy converted by the rooftop.

Solution

The area of the roof is

$$A = 12\,\text{m} \times 15\,\text{m} = 180\,\text{m}^2$$

The amount of power collected by the rooftop is

$$P = (1\,\text{kW/m}^2)(180\,\text{m}^2) = 180\,\text{kW}$$

Since the cells are 10% efficient, the electric power converted by the cells will be

$$P_{elec} = 0.1 \times 180\,\text{kW} = 18\,\text{kW}$$

Assuming that a peak sun exists for 2 h every day, the average daily energy is

$$18\,\text{kW} \times 3\,\text{h} = 54\,\text{kWh}$$

Example 13.6

Consider the results of Example 14.3 and assume that the cost of electrical energy to the home owner is $0.10/kWh charged by the utility company. If the cost of solar roof is about $5000, determine the following:

(a) The amount of electrical energy produced by solar cells per year
(b) The amount of savings on electrical energy to the home owner per year
(c) The break-even period for the solar panels to pay for themselves in months

Solution

(a) The amount of electrical energy produced by the solar panels per year is

$$\text{Annual electric energy produced} = (54\,\text{kWh/day})(365\,\text{days/year})$$

$$= 19,710\,\text{kWh/year}$$

(b) The amount of savings to the home owner due to electrical energy produced by the solar panels per year is

$$\text{Annual savings} = (19,710\,\text{kWh/year})(\$0.10/\text{kWh})$$

$$= \$1,871/\text{year}$$

(c) The break-even time for the solar panels to pay for themselves is

$$n = \frac{\$5,000}{\$1,871/\text{year}}$$

$$= 2.67\,\text{years}$$

13.3 EFFECT OF SUNLIGHT ON SOLAR CELL'S PERFORMANCE

The I–V characteristic given in Figure 13.2 is based on the assumption that the solar cell is operating under a bright noontime sun. In the event that the power density of the sunlight decreases, the output of the cell decreases accordingly. The reasons for such decrease may include the following: The sun is not shining directly on the cell due to the fact that it is just rising or setting; the sun is not shining on the cell because it is winter. (In the northern hemisphere, the sun has a southern exposure. On the other hand, in the southern hemisphere, the opposite is true, that is, in the winter the sun has a northern exposure.) There may exist tall trees or structures that cast a shadow on the cell, during certain times of the day. It is a cloudy or overcast day [7].

The angular position on the earth's surface north or south of the equator is defined as the *latitude*. The equator itself has 0° latitude and divides the earth into two equal hemispheres. For every 69 miles north or south of the equator, the latitude increases by 1°. For instance, the north pole has a latitude of 90° north and the south pole has a latitude of 90° south. Table 13.1 gives the latitudes of selected cities around the world.

Sun's position varies at a given time. For example, at noontime, the sun's position varies about 40° from its highest position in June to its lowest position in December. In March and September (equinox), the sun's center is directly over the equator. Its apparent position in the sky then is approximately equal to your latitude. Thus, in New York City, the noontime sun will appear to be 40° south of vertical during the equinox. In June, it will be 20° (40° − 20°) south of vertical, and in December, 60° (40° + 20°) south of vertical.

For maximum energy absorption from the sun, a solar cell should be tilted south (in the Northern Hemisphere) by the angle of the latitude of the location on the earth.

Example 13.7

At what angle, a solar cell is tilted to get the most energy from the sun if it is located in

(a) Quito, Ecuador?
(b) Athens, Greece?
(c) Valparaiso, Chile?

TABLE 13.1
The Latitudes of Selected Cities around the World

Location	Latitude	Location	Latitude
Athens, Greece	38° N	Madrid, Spain	40° N
Berlin, Germany	53° N	Miami, Florida	26° N
Bogota, Columbia	2° N	Montreal, Quebec	46° N
Bombay, India	20° N	Moscow, Russia	55° N
Buenos Aries, Argentina	20° N	Munich, Germany	48° N
Cairo, Egypt	30° N	Oslo, Norway	60° N
Edinburgh, Scotland	56° N	Paris, France	49° N
Entebbe, Uganda	0°	Quito, Ecuador	0°
Honolulu, Hawaii	20° N	Rio de Janeiro, Brazil	23° S
Houston, Texas	30° N	Rome, Italy	42° N
Kansas City, Missouri	39° N	Seattle, Washington	47° N
Las Vegas, Nevada	36° N	Sidney, Australia	35° S
Lima, Peru	12° S	Thule, Greenland	77° N
London, England	52° N	Tokyo, Japan	36° N
Los Angeles, California	34° N	Valparaiso, Chile	36° N

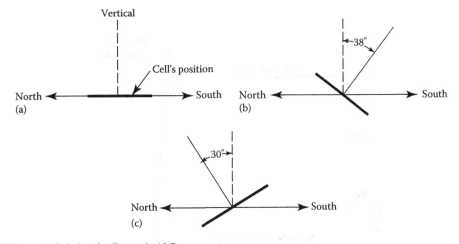

FIGURE 13.3 Solution for Example 13.7.

Solution

(a) From Table 13.1, the latitude of Quito, Ecuador, is 0°. Hence, the position of the solar cell is as shown in Figure 13.3a.
(b) From Table 13.1, the latitude of Athens, Greece, is 38° N. Hence, the position of the solar cell is as shown in Figure 13.3b.
(c) From Table 13.1, the latitude of Valparaiso, Chile, is 33° S. Hence, the position of the solar cell is as shown in Figure 13.3c.

13.4 EFFECTS OF CHANGING STRENGTH OF THE SUN ON A SOLAR CELL

It is often that characteristics of a solar cell include the effects of a variation in power density. Figure 13.4 shows the effects of changing sunlight power density on the I–V curve of a solar cell. The current output of a cell is directly proportional to the sunlight [8,9].

Thus, under a very weak sun, the cell puts out very little current. However, its voltage is still quite high. Thus, one can conclude that under open-circuit conditions, the voltage is relatively independent of sunlight. But, the effect of the load on the operating point as the power density varies is more important.

Example 13.8

Assume that a fixed resistive load selected to force the solar cell shown in Figure 13.4 operate at the point where maximum power conversion takes place.

(a) If that point has full sunlight (1 SUN), determine the value of the load resistance R_L at the maximum power point a in Figure 13.4.
(b) The value of the power at such point a.
(c) Let us say due to passing by cloud now, the power density drops to 0.5 SUN and as a result, the operating point of the cell changes. At the corresponding lower sun curve (with the current output of 300 mA) and the fixed load resistance, determine the value of the voltage.
(d) Correspondingly, the new operating point is at b in Figure 13.4. Determine the power at this point.
(e) Find the new load resistance that is necessary to achieve this.

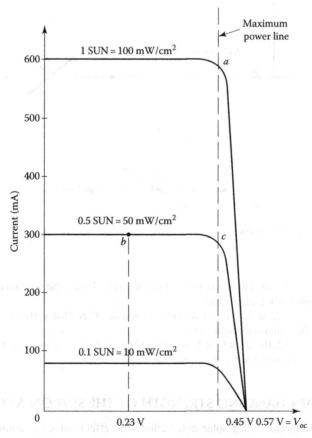

FIGURE 13.4   Variation of I–V characteristic of a solar cell due to changing power density.

(f) As it can be observed that the solar cell is no longer operating at the knee of the curve. Consequently, the maximum power is not being converted. To achieve the maximum power conversion, the operating point has to be moved to the point c in Figure 13.4. As a result, there has to be a new load resistance. Determine its value.

(g) Determine the corresponding power at the point c. (Note that the power at the point c is one-half the power at the point a.)

Solution

(a) The value of the load resistance is

$$R_L = \frac{0.45\,V}{0.58\,A} \cong 0.78\,\Omega$$

(b) The value of the power at such point a is

$$P_a = (0.45\,V)(0.58\,A) = 0.26\,W$$

(c) The value of the voltage at point a is

$$V = R_L I$$

$$= (0.78\,\Omega)(0.3\,A)$$

$$= 0.23\,V$$

(d) Thus, the power at the new point b is

$$P_b = (0.23\,V)(0.3\,A)$$

$$= 0.069\,W$$

Consequently, the maximum power is not being converted.

(e) The new load resistance that is necessary to achieve the maximum power, the operating point has to be moved to the point c in Figure 12.4.

(f) As a result there has to be a new load resistance. Hence,

$$new\,R_L = \frac{0.45\,V}{0.29\,A} \cong 1.55\,\Omega$$

(g) Hence, the corresponding power at point c becomes

$$P_c = (0.45\,V)(0.29\,A)$$

$$\cong 0.13\,W$$

Note that the power at the point c is one-half the power at the point a.

Therefore, it can be concluded that the current output is directly proportional to the sunlight (i.e., power density). However, the power will not be proportional unless the load is changed as the intensity of the light varies. Accordingly, the maximum power output of solar cell is a function of not only sunlight but also the load. Because of this, inverters employed for solar energy conversion systems have special tracking circuitry that continuously adjusts the loading on the solar cells while monitoring power output.

The predictability of how much power can be obtained from the sun from hour to hour or day to day is a very serious problem from the scheduling point of view. This problem is also caused by the variation in sunlight. Because of this, it is often that the average data on sunlight are used.

Example 13.9

In order to determine how much energy can be received from the sun by one solar cell in a city in California, data were collected on a sunny day and the data are given in Table 13.2. During the day, the sun rises at 7 AM and sets at 7 PM.

Solution

Since from Table 13.2, the total energy density for the day is 8.4 SUN-hours, it is equivalent to having a full or peak sun (i.e., 1 SUN) for 8.4 h. Keeping this in mind and using the power calculated in part (c) for 1 SUN power density in Example 13.8,

$$P = 0.26\,\text{W} \quad (\text{for 1 SUN})$$

Thus, the total energy* that can be produced by one solar cell can be found as

$$\text{Total energy} = (0.26\,\text{W/SUN-h})(8.4\,\text{SUN-h})$$

$$= 2.184\,\text{W}$$

13.5 TEMPERATURE'S EFFECT ON CELL CHARACTERISTICS

The ratings of solar cells are based on the minimum current they supply at 0.45 V under a full sun at 25°C (77°F). Its output is a function of its cell temperature. As temperature increases, the current will increase while the voltage will decrease by about 2.1mV/°C. As a result, its power output and consequently its cell efficiency decrease [10,11].

The opposite takes place as the temperature decreases, that is, cell operates more efficiently when they are cooler. Because of this, commercially manufactured solar panels, that is, groups of

TABLE 13.2
Data for Example 13.9

Power Density (SUNS)	Time (h)	Energy Density (SUN-Hours)
0.2	1	0.2
1.0	1	1.0
0.9	2	1.8
1.0	3	3.0
0.9	2	1.8
0.3	1	0.3
0.2	1	0.2
0.1	1	0.1
		Total = 8.4 SUN-hours

* An alternative method would be to use a set of curves to the ones in Figure 13.5. Hence, for each power density of the sunlight in Table 13.2, calculate the power at the knee of the curve, assuming that the load is adjusted for maximum power. To determine the energy for each time interval, the calculated maximum powers are multiplied by the time of duration. The total energy is found by adding energies for the day.

interconnected cells, have a metal (usually, aluminum) that plays the role of a heat sink. Otherwise, in areas of low latitude, the temperatures of a solar cell can reach 80°C (i.e., 176°F) without a heat sink. At other temperatures, the voltage and the current of the cell can be determined from

$$E_0 = E_R - 0.0021(T - 25) \tag{13.3}$$

and

$$I_0 = I_R + 0.025A(T - 25) \tag{13.4}$$

where E_R and I_R are the cell ratings in volts and mA at 25°C. E_0 and I_0 will be the cell voltage and current at the new temperature T in degrees Celsius. A is the cell area in square centimeter.

Example 13.10

Assume that a solar cell is rated 600 mA, 0.45 V, at 25°C, and that the cell area is 30 cm². If as the cell is under a full sun and providing maximum power, the temperature increases to about 50°C, determine the following:

   (a) Its power output at 25°C
   (b) Its voltage, current, and power output at 50°C
   (c) The amount of percentage drop in power output due to the increased temperature

Solution

   (a) Its power output at 25°C can be found from its rated voltage and current values as

$$P = (0.45\,\text{V})(600\,\text{mA})$$

$$= 270\,\text{mW}$$

   (b) From Equation 13.3, its new voltage is

$$E_0 = E_R - 0.0021(T - 25)$$

$$= 0.45 - 0.0021(50 - 25)$$

$$\cong 0.40\,\text{V}$$

And from Equation 13.4, the new current is

$$I_0 = I_R + 0.0025A(T - 25)$$

$$= 600\,\text{mA} + \left(0.025\,\frac{\text{mA}}{\text{deg-cm}^2}\right)(30\,\text{cm}^2)\big[(50 - 25)\text{deg}\big]$$

$$= 618.75\,\text{mA} \cong 619\,\text{mA}$$

Hence, the new power output of the cell is

$$P = (0.40\,\text{V})(618.75\,\text{mA})$$

$$= 247.5\,\text{mW}$$

(c) The amount of drop power output in percentage is

$$\%P_{drop} = \frac{270 - 247.5}{270} \times 100$$

$$\cong 8.33\%$$

Temperature effects have to be considered when a PV system is designed. In general, the power estimate is increased by about 10% to take into account the loss due to increased cell temperature.

13.6 EFFICIENCY OF SOLAR CELLS

In general, the solar cell efficiency is defined as the ratio of the electrical power output to the sunlight power it receives. The maximum theoretical efficiency of a silicon solar cell is about 25%. Today's cells have rated efficiencies of 10%–16% [12].

The efficiency of a cell is a function of number of things, including the number and thickness of the wires connected to the top of the cell and light reflected from the surface of the cell. Also, when cells are connected together to form panels, the panel efficiency will be based on the cell's shape.

Example 13.11

Consider a circular cell that has a diameter of 2.5 in. If it has ratings at 25°C of 1200 mA and 0.45 V in a full sun, determine the cell's efficiency.

Solution

In order to determine the cell's efficiency, first let us find the cell area. Since the radius of the cell is

$$r = \frac{d}{2} = \frac{2.5 \text{in.}}{2} = 1.25 \text{in.}$$

or

$$r = (1.25 \text{in.})(2.54 \text{cm/in.}) = 3.175 \text{cm}$$

and

$$A = \pi r^2 = \pi (3.175 \text{cm})^2 = 31.67 \text{cm}^2$$

Since 1 SUN = 100 mW/cm²

$$P_{in} = (100 \text{mW/cm}^2)(31.67 \text{cm}^2) = 3167 \text{mW}$$

$$P_{out} = (0.45 \text{V})(1200 \text{mA})$$

$$= 540 \text{mW}$$

Thus, the efficiency is

$$\eta = \frac{P_{out}}{P_{in}} \times 100$$

$$= \frac{540\,\text{mW}}{3{,}167\,\text{mW}} \times 100$$

$$= 17\%$$

13.7   INTERCONNECTION OF SOLAR CELLS

Typical solar cell output is 800 mA at 0.45 V. But, most everyday applications dictate more than 800 mA at 0.45 V. However, solar cells can be treated just like batteries. The net voltage can be increased by connecting them in series, and the net current can be increased by connecting them in parallel. For instance, if 12 identical cells each rated 1 A and 0.45 V in a full sun are connected as shown in Figure 13.5, the net output of the system will be 3 A at 1.8 V. For each parallel path, the current increases by 1 A, and for each cell in series, the voltage increases by 0.45 V [13].

It may be of an interest that solar cells are more flexible than batteries in that they can be broken into pieces to get odd ratings. The voltage output from a piece of a cell will still be the rated voltage of the whole cell. But, the current will be proportional to the area of the piece of cell. It is a common practice to cut the circular cells in halves and quadrant cells. If many solar cells are connected in series and parallel to form a permanent unit, it is called a *solar panel.*

However, the current produced by a series connection of multiple cells will be the minimum of all the cells. For example, if three cells each rated 1 A is connected in series with a half of the same cell (rated 0.5 A, the current will be 0.5 A). In practice, solar panels are manufactured in different sizes and ratings. Such panels can be interconnected to form *solar arrays* or *modules.*

Example 13.12

Consider a solar panel that is rated 20 W at 1.0 V. Determine the following:

(a) How many of these panels are needed to supply 1 A of current at 120 V?
(b) How should they be connected?

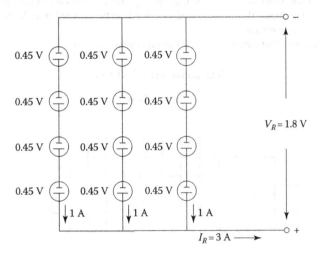

FIGURE 13.5   Connection of 12 identical cells.

(c) The value of the necessary load resistance.

(d) The total power that can be obtained from this array of solar cells.

Solution

(a) Since each panel produces 12 V, to get 120 V the number of panels required in series is

$$\text{\# of panels in series} = \frac{\text{Total voltage}}{\text{volts/panel}}$$

$$= \frac{120\,\text{V}}{10\,\text{V/panel}}$$

$$= 12\,\text{panels}$$

(b) The current rating of each panel can be found from panel's power and voltage ratings. Hence,

$$I = \frac{20\,\text{W}}{10\,\text{V}} = 2\,\text{A}$$

Thus, each path, of seriously connected panels, will provide 2 A. Therefore,

$$\text{\# of paths} = \frac{\text{Total current}}{\text{Current/path}} = \frac{10\,\text{A}}{2\,\text{A/path}} = 5\,\text{paths}$$

Hence, 5 parallel paths each having 12 panels in series are required to form an array to meet the power requirement. The resultant panel arrangement is shown in Figure 13.6.

(c) The value of the required load resistance is found as

$$R_L = \frac{120\,\text{V}}{10\,\text{A}} = 12\,\Omega$$

This input resistance facilitates the maximum power conversion in a full sun. However, as the lighting changes, this input impedance needs to be changed in order to get the maximum power conversion.

(d) The total power that can be obtained from this solar array is

$$\text{Total power} = (120\,\text{V})(10\,\text{A})$$

$$= 1.2\,\text{kW}$$

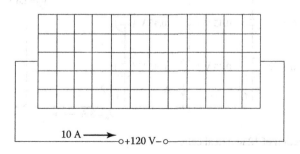

FIGURE 13.6 Panel arrangement for Example 13.12.

Alternatively,

$$\text{Total power} = (60 \text{ panels})\left(\frac{20\,\text{W}}{\text{panel}}\right)$$

$$= 1.2\,\text{kW}$$

13.8 OVERALL SYSTEM CONFIGURATION

PV cells and modules are configurable from 1 to 5 MW. A solar generator has basically two possible fundamental configurations. The first one is a stand-alone system, as shown in Figure 13.7a. It can be used in the location where there is no utility power available. During peak-sun hours, the solar array supplies all the ac power needs and keeps the storage batteries fully charged. The storage batteries have to have the capability of storing enough energy to supply the power required when the sun goes down.

In that second system, shown in Figure 13.7b, the solar generator is connected to the utility grid. Here, there is no need for the storage batteries. The dc power from the solar array gets inverted to ac power. The inverter output provides the ac power needs, during peak-sun hours. In the event that these needs are low, power will be fed back into the utility grid for credit. On the other hand, during nighttime hours, the ac power requirements are met by power supplied by the utility grid. Figure 13.7 only shows the overall system configuration, without the necessary circuit breakers and/or meters.

The overall system frequency must take into account the effectiveness of all other components (such as, inverter, batteries, and any additional circuitry) in addition to the efficiency of the solar array.

A given solar energy system is designed based on the size of the solar array requirement (for a given specific location and electrical energy needs for that location) and the amount of electrical energy that the array can supply. After the location and solar array size and its rating are known, the energy output from the system can be easily determined.

Example 13.13

Mr. Smith lives in Miami, Florida (with yearly average of 4.7 peak SUN-hours per day), and is considering placing a PV array on his south-facing roof. The roof is unshaded and has a size of $17 \times 36\,\text{ft}$ (approximately $6 \times 12\,\text{m}$). The electrical power produced is to be converted to ac and used in his home. Any excess power produced will be back into the local utility company's grids system. Analyze the system configuration and determine how much energy Mr. Smith can expect to get from the solar installation.

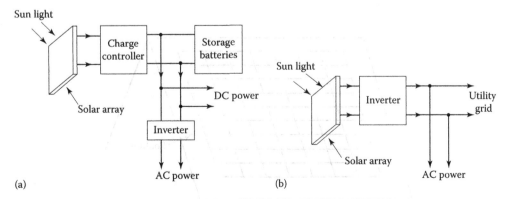

FIGURE 13.7 Two fundamental solar generator configurations: (a) stand-alone system and (b) supplemental or cogeneration system.

Solution

The system configuration will be like the one shown in Figure 13.7b. But, in addition, there will be a circuit breaker and an ac watt-hour meter will be placed between the inverter and the utility grid. The amount of energy that will be sold to the utility company through the grid connection will be measured by the watt-hour meter. The local utility company by law (PURPA Act of 1078) is obliged to pay for this energy.

The solar panel chosen for this application is rated 16.2 V, 2.4 A, 39 W, and has 10% efficiency at 25°C in full sun. Its dimensions are 1×4 ft. At 50°C, the ratings of the solar panel become 14.7 V, 2.27 A, and 55 W. If about 4 ft is permitted at the edges of the roof for a work area, 63 solar panels can easily be placed on the roof. They could be placed end to end with seven in a row ($7 \times 4 = 28$ ft) and nine rows total ($9 \times 1 = 9$ ft), as illustrated in Figure 13.8. Hence, the maximum power output will be

$$(63 \text{ panels})\left(\frac{39 \text{ W}}{\text{panel}}\right) = 2457 \text{ W}$$

A 2.5 kW, single-phase inverter will be employed. Its maximum input current is specified as 25 A and the input voltage can vary from 60 to 120 V dc. The output voltage is 120 V ac. The seven panels in each row will be connected in series. This will provide a range of voltage (from 25°C to 50°C) of

$$7 \times 16.2 = 113.4 \text{ V} \quad (25°C)$$

to

$$7 \times 14.7 = 102.9 \text{ V} \quad (50°C)$$

which is within the range of the inverter. The nine rows will be connected in parallel. Hence, the peak current from the solar array (at 50°C) will be

$$9 \times 2.27 = 20.43 \text{ A}$$

which is within the inverter limit. The panel should be tilted 26° toward south. Since in Miami, Florida, one can expect an average of 4.7 peak SUN-hours per day, multiplying it by the peak power output at 50°C will provide the average daily supplied by the solar array. Hence,

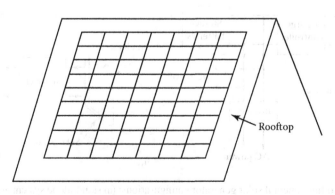

FIGURE 13.8 Installation of solar panels in Example 13.13.

$$\text{At } 50°C: \quad P = (102.9\,\text{V})(20.43\,\text{A}) \cong 2102 = 2.102\,\text{kW}$$

$$\text{Average daily energy} = (2.1\,\text{kW})(4.7\,\text{h}) \cong 9.87\,\text{kWh}$$

If the inverter efficiency is 95%, the average energy produced by the system per day will be

$$(9.87\,\text{kWh/day})(0.95) = 9.4\,\text{kWh/day}$$

By multiplying this by 30 days,

$$(9.4\,\text{kWh/day})(30\,\text{days}) = 282\,\text{kWh/month}$$

or

$$(9.4\,\text{kWh/day})(365\,\text{days/year}) = 3431\,\text{kWh/year}$$

At 10 cents/kWh, Mr. Smith will save about \$343/year in electric bills. The solar array would cost

$$(2457\,\text{W})(\$1/\text{W}) = \$2457$$

Thus, the break-even point for this investment is

$$n = \frac{\$2457}{\$343} = 7.2\,\text{years}$$

Mr. Smart will recover his initial investment in 7.2 years. But, if the cost of solar array drops down to \$0.50 per peak watt (as predicted), the investment cost would be

$$(2457\,\text{W})(\$0.50/\text{W}) = \$1228.50$$

Thus,

$$n = \frac{\$1228.50}{\$343/\text{year}} \cong 3.6\,\text{years}$$

At this price, Mr. Smart will recover his investment in less than 4 years.

13.9 THIN-FILM PV

A more recent development is called second-generation PV devices. They are made of layers of semiconductor materials that are much thinner than those in silicon cells. The thickness of a thin-film cell is on the order of only 2–3 μm thick.

In the event that silicon is used, it is typically in the form of *amorphous* (i.e., not crystallized) *silicon* (a-Si), which has no discernible crystal structure. Also, microcrystalline silicon thin-film devices are being developed. In addition, other thin-film materials have also been developed and commercialized, including *cadmium telluride* (CdTe) and *copper indium gallium diselenide* (CIGS). These PV devices need much less material than to traditional c-Si devices [14].

According to Kroposki et al. [10], thin films generally have lower solar conversion efficiency than the c-Si cells. Here, the conversion efficiency is defined as the percentage of the sun's power shining on the cell. For instance, if 1000 W of solar power illuminate a cell, and 150 W of electricity are generated, then the cell has a solar conversion efficiency of 15%.

A commercial silicon cell may have an efficiency of about 20%, whereas a commercial CdTe cell's efficiency is about 11%. The thin-film cell uses less material and can be deposited with a method that is much less energy-intensive than silicon. Less material also causes lighter weight.

Also, some thin-film technologies do not use rigid wafers; instead, they can be deposited on flexible layers of stainless steel or plastic. Depending on the application, such flexibility might be highly desirable. In general, thin-film PV is less expensive to manufacture and easier to implement [10].

13.10   CONCENTRATING PV

There is another type of second-generation PV device, which is based on high-efficiency multifunction cell that uses compounds from the group III and group V elements of the periodic table of elements. For example, such multifunction solar cell design has three layers, each of which absorbs a different portion of the solar spectrum to use in generating electricity.

The top layer may be made of gallium indium phosphide, and the bottom layer of germanium. This type design may have a high efficiency of about 40%. This is due to the fact that each layer in this design is designed to absorb and use a different portion of the solar spectrum.

But, such design is expensive since the group III and V materials are costly to produce. However, the cost can go down substantially, if a relatively inexpensive lens or mirror can be employed to focus sunlight on just a small area of cells [10]. For instance, if a 10 × 10 in. lens focuses this area of incident sun onto a 0.5 × 0.5 in. cell, the concentration factor becomes 400×, that is, 100 m²/0.25 in.² Thus, such cell with the lens can produce as much power as a 10 × 10 in. cell with lens, but at about 1/400 of the cell cost. Even though they are not suitable for small projects, concentration systems could be very effective in large power generation for several homes.

13.11   PV BALANCE OF SYSTEMS

Balance of systems compromises all of the components of a PV system beyond the actual PV module that produces the power. A frame structure is also required to hold the module, keep it oriented toward the sun, stabilizing it in the outer elements, including wind and snow.

PV systems produce direct current (dc) electricity. Hence, if alternating current (ac) is needed, the balance of systems has to include an inverter. But, the inverter decreases the overall system efficiency by an additional 5%–10%. However, the system efficiency can be improved by connecting a tracking system to the solar modules. The trackers can be of single axis or dual axis.

The single-axis trackers aligned with the axis in a north–south direction permit the module to trade the sun's progress across the sky from East to West during the day. Dual-axis trackers further improve the module's orientation, permitting the sun to always illuminate the cells perpendicular to the plane of the module. The result is the maximum energy output from the system.

In general, residential and commercial PV systems are directly connected to the grid without energy storage. Including battery energy storage would increase the reliability of the system. Hence, the batteries store excess power that is generated from the PV array to be used later.

13.12   TYPES OF CONVERSION TECHNOLOGIES

There are two primary technologies for the conversion of sunlight into electricity. PV cells depend on the use of semiconductor devices for the direct conversion of the solar radiation into electrical energy. The typical efficiencies of such commercial crystalline PV cells are in the range of 12%–18%, even though there have been experimental cells built that are capable of over 30%.

The second type of technology is based on *solar thermal systems*. It is known as *concentrating solar power*. It involves intermediate conversion of solar energy into thermal energy in the form of steam, which in turn is employed to drive a turbogenerator. To have high temperatures, thermal systems invariably use concentrators by the use of mirrors either in the form of parabolic troughs or thermal towers.

But, presently, generation of electricity by either technology is considerably more expensive than traditional means. The CSP systems are essentially categorized based on how the systems collect solar energy. The three basic systems are the linear, tower, and disk systems.

13.13   LINEAR CSP SYSTEMS

In such systems, CSP collectors capture the sun's energy with large mirrors that reflect and focus the sunlight onto a linear receiver tube. Inside the receiver, there is a fluid that is heated by the sunlight and then employed to create superheated steam that causes a turbine to rotate in order to drive a generator to produce electricity.

It is also possible to produce the steam directly in the solar field. Here, no heat exchanger is employed, but the system uses expensive high-pressure piping system in the entire solar field. It has a lower operating temperature.

Essentially, concentrating collector fields in such systems are made of a large number of collectors in parallel rows that are usually aligned in north–south orientation to increase both summertime and annual energy collection. Using its single-axis sun-tracking system, the system facilitates the mirrors to track the sun from East to West during the day, causing the sun to reflect continuously onto the receiver tubes.

Such trough designs can use thermal storage. In that case, the collector field is built oversized in order to heat a storage system during the day that in the evening can be used to produce additional steam to generate electricity.

It is also possible to design the parabolic trough plants as hybrid systems that consume fossil fuel to supplement the solar output during periods of low solar radiation. In such applications, usually a natural gas–fired heater or a gas-steam boiler/reheater is used [10].

13.14   POWER TOWER CSP SYSTEMS

In such system, several large, flat, sun-tracking mirrors, which are called heliostats, focus sunlight onto a receiver at the top of a tower. The receiver has a heat-transfer fluid that is heated to produce steam. The heated steam in turn is employed in a typical turbine generator to generate electricity. The heat-transfer fluid is usually water/steam although in advanced designs is replaced by molten nitrate salt due to its better heat-transfer and energy-storage capabilities. Presently, such systems have been developed to produce up to 200 MW of electricity [16].

13.15   DISH/ENGINE CSP SYSTEMS

According to Kroposki et al. [10], these systems generate relatively small amounts (3–25 kW) of electricity with respect to other CSP technologies. Here, a solar concentrator (or dish) collects the solar energy radiating directly from the sun. The resultant beam of concentrated sunlight is reflected onto a thermal receiver that collects the solar heat. The dish is attached on a structure that tracks the sun continuously throughout the day to reflect the highest percentage of sunlight that is possibly onto the thermal receiver.

The power conversion unit is made of the *thermal receiver* and the *engine/generator*. The thermal receiver is the interface between the dish and the engine/generator. Its function is to absorb the concentrated beams of solar energy and after converting them to heat, to transfer this heat to the engine/generator.

The *thermal receiver* can be made of a bank of tubes with a cooling fluid (hydrogen or helium), which is used as a transfer medium. Other thermal receivers are made of heat pipes, where the boiling and condensing of an intermediate fluid transfers the heat to the engine.

The *engine/generator system* is the subsystem of the dish/engine CSP system. It takes the heat from the thermal receiver and uses it to produce electricity. Most commonly, a *Stirling engine* is used as the heat engine. It uses a heated fluid to move pistons and create mechanical power to rotate the shaft of the generator to produce electric power.

The last subsystem of the CSP system is *thermal energy storage system*. It provides a solution for the curtailed energy production when the sun sets or is blocked by the clouds.

13.16 PV APPLICATIONS

PV cells were first used in power satellites. By the end of the 1990s, PV electrical generation was cost-competitive with the marginal cost of production with gas turbines. As a result, a number of utilities have introduced utility-interactive PV systems to supply portion of their total customer demand.

Some of these systems have been residential and commercial rooftop systems, and other systems have been larger ground-mounted systems. PV systems are classified as utility-interactive (grid-connected) or stand-alone systems.

13.16.1 UTILITY-INTERACTIVE PV SYSTEMS

Utility-interactive PV systems are categorized by IEEE Standard 929 as small, medium, or large PV systems. Small systems are less than 10 kW, medium systems range from 10 to 500 kW, and large systems are larger than 500 kW. Each size dictates different consideration for the utility interaction.

Since the output of PV modules is dc, it is, of course, necessary to convert this output to ac before connecting it to the grid; it is done by an inverter. It is also called a power conditioning unit (PCU). A typical small utility system of a few kilowatts consists of an array of modules selected either by a total cost or by an available roof area. The modules are connected to produce an output voltage ranging from 48 to 300 V, as a function of the dc input requirements of the PCU. One or two PCUs are used to interface the PV output to the utility at 120 or 120/240 V.

The point of utility connection is the load side of the circuit breaker in the distribution panel of the building of the PV system is connected on the customer side of the revenue meter. However, medium- and large-scale utility-interactive systems differ from small-scale systems only in the possibility that the utility may dictate different interfacing conditions with respect to disconnect and/or power quality.

13.16.2 STAND-ALONE PV SYSTEMS

They are used when it is not possible to connect to the utility grid. Examples include water-pumping systems, PV-powered fans, power systems for remote installations, portable highway signs, etc. Some of them include battery storage to operate the system under sun or no-sun situations [15].

PROBLEMS

13.1 Determine the energy density in Langley, if the strength of sunlight is 1 SUN for a period of 15 min.

13.2 Consider type rooftop of a home measuring 15 × 20 m, which is all covered with PV cells to provide the electrical energy requirement by that home. If the peak SUN per day is 4 h, efficiency of solar cell is 10%, find the average daily electrical energy converted by the rooftop.

13.3 Consider the results of Problem 13.2, and assume that the cost of electrical energy to the homeowner is $0.19/kWh which is changed by the utility company. If the cost of solar roof is about $5000, determine the following:

TABLE P13.1

The Necessary Energy Data for the Cottage

Appliance	Current (A)	Time (h)	Battery Drain (A·h)	Power (W)	Energy (W·h)
Refrigerator	2	22	44	25	550
Miscellaneous (radio, TV, lighting, toaster, etc.)	5	4	20	75	300
			—	—	—
Total			64	100	850

    (a) The amount of electrical energy produced by the solar cells per year

    (b) The amount of savings on electrical energy produced by the solar cells per year

    (c) The break-even period for the solar panels to pay for themselves in months

13.4 Consider a circular solar cell that has a 4 in. diameter which is rated 2.0 A at 0.45 V. If a certain application dictates 1.8 V and draws 0.5 A, how can the cell be modified to satisfy the requirements of such application?

13.5 A circular cell has a diameter of 3 in. If it has a rating at 25°C of 800 mA and 0.45 V in a full sun, determine its efficiency.

13.6 A circular cell has a diameter of 3 in. If it has a rating at 25°C of 1200 mA and 0.45 V in a full sun, determine its efficiency.

13.7 Assume that Mr. Smart owns a vacation cottage in Napa Valley. He is considering installing a solar system to meet the electricity needs at the cottage. Table P13.1 gives the energy demands at the cottage. Assume that there is 5 peak SUN-hours per day at the location of the cottage. All the electrical equipment requires 12 V dc. Design the system and determine the following:

    (a) Design the solar system.

    (b) The changing current of the battery for every hour of peak sun.

    (c) The size of the battery that is needed.

    (d) The peak power of the panels.

    (e) If the cost $1 per peak watt, find the cost of panels alone.

13.8 Consider a solar module rated 5.3 V, 38 W. If it is used to provide power to an application that needs 30 V and 25 A current, determine the following:

    (a) The total number of solar modules required

    (b) The type of connection that is needed in terms of series and parallel connected modules

    (c) The necessary load resistance value to get the rated power

13.9 Assume that a roof measures 12 m × 18 m and that the solar array under consideration is 10% efficient and that is tilted 30° toward the south. The yearly average peak SUN-hours per day in Dallas, Texas, and Sacramento, California, is 4.7 peak SUN-hours and 5 peak SUN-hours, respectively. Determine the average daily electrical energy converted by the rooftop array, if the house is located at

    (a) Dallas, Texas

    (b) Sacramento, California

REFERENCES

1. ANSI/IEEE P929, *IEEE recommended practice for utility interface of residential and intermediate photovoltaic (PV) systems*, IEEE Standards Coordinating Committee 21, Photovoltaics, Draft 10, February 1999.

2. Babic, J., Walling, R., O'Brien, K., and Kroposki, B., The Sun also rises, *IEEEE Power & Energy Magazine*, May/June 2009, 45–54.

3. Buresch, M., *Photovoltaic Energy Systems: Design & Installations*, New York: McGraw-Hill, 1953.
4. Denholm, P. and Margolis, R. M., Evaluating the limits of solar photovoltaics (PV) in electric power systems utilizing energy storage and other enabling technology, *Energy Policy*, 35(9), 2007, 4424–4433.
5. Denholm, P. and Margolis, R. M., Evaluating the limits of solar photovoltaics (PV) in traditional electric power systems, *Energy Policy*, 35(5), 2007, 3852–2861.
6. Jha, A. R., *Solar Cell Technology & Applications*, Boca Raton, FL: CRC Press, 2010.
7. Key, T., Finding a bright spot, *IEEE Power & Energy Magazine*, May/June 2009, 34–44.
8. Komp, R. J., *Practical Photovoltaics: Electricity from Solar Cells*, 2nd Edn., Ann Arbor, MI: Aatce Publications, 1984.
9. Komp, R., *Practical Photovoltaics: Electricity from Solar Cells*, 3rd Edn., Ann Arbor, MI: Aatec Publications, 2002.
10. Kroposki, B., Margolis, R., and Ton, D., Harnessing the Sun, *IEEE Power & Energy Magazine*, May/June 2009, 22–33.
11. Maycook, P. D. and Stirewalt, E. N., *Photovoltaics: Sunlight to Electricity in One Step*, Andover, MA: Brick House Publishing Company, 1981.
12. Mehos, M., Kabel, D., and Smithers, P., Planting the seed, *IEEE Power & Energy Magazine*, May/June 2009, 55–62.
13. Nelson, J., *The Physics of Solar Cells*, London, U.K.: Imperial College Press, 2003.
14. Pagliaro, M., Palmisano, G., and Criminna, R., *Flexible Solar Cells*, Weinheim, Germany: Wiley, Vch, 2008.
15. Sandia National Labs, *Stand-Alone Photovoltaic Systems: A Handbook of Recommended Design Practices*, Albuquerque, NM: Sandia National Laboratories, 1996.
16. Zweibel, K., *Harnessing Solar Power*, New York: Plenum Press, 1990.

14 Energy Storage Systems

14.1 INTRODUCTION

The use of intermittent or variable sources of energy, such as solar and wind energy, and some of the forms derived from moving water, often requires some means of energy storage.

Energy storage can not only potentially benefit solar energy systems as well as other renewable energy resources but also benefit the transmission and distribution systems because storage applications can be used to mitigate diurnal or other congestion patterns and, in effect, store energy until the transmission system is capable of delivering it where needed.

By storing energy from variable resources, such as wind and solar power, energy storage could provide firm generation from these units, permit the energy produced to be used more efficiently, and provide supplementary transmission benefits.

Therefore, the adverse impacts of large-scale photovoltaic (PV) power generation systems connected to the power grid and developing output control technologies with integrated battery storage are still under study. The sodium–sulfur (NaS) battery is designed to absorb fluctuations in the PV output within its limit of kW and kWh capacities. For more efficient and effective operation of the NaS battery, several control algorithms of a battery system for smoothing PV output are being developed by the industry [1,8].

14.2 STORAGE SYSTEMS

At the present time, there is a great interest in the possible applications of energy storage in power systems. The interested parties include the electric utilities, energy service companies, and automobile manufacturers (for electric vehicle applications). For example, the ability to store large amounts of energy would permit electric utility companies to have greater flexibility in their operation because with this option, the supply and demand do not have to be matched instantaneously. Hopefully, the availability of the proper battery at the proper price will finally make the electric vehicle a reality.

The battery technologies are diverse and at different stages of development. They include a variety of batteries, high-speed flywheels, supercapacitors, and regenerative fuel cells. Local energy storage would assist embedded generation from renewable energy by providing a buffer between the variability of supply and demand. Potential benefits include capacity reduction, frequency support, standing reserve provision, and cold start capability. Depending on technical requirements and geographical settings, a given utility may avail of one or more of these technologies.

Power applications, such as uninterruptible power supply (UPS) backup for data centers and automotive starting batteries, represent the largest market for lead-acid batteries, whereas laptop batteries and power tools have caused incredible growth for lithium ion. For bulk energy storage in utility grids, pumped hydropower plants dominate, with approximately 100 GW in service around the world.

Even though many utilities possess pumped storage plant, little focus has been placed on the potential roles that management of load blocks to fill demand troughs or shave-off demand peaks, and this way partially decouple energy production from energy consumption.

Energy storage can perform the same roles, but may also be used as a generation source, either replacing expensive, low-efficiency storage capability or load scheduling. The generation capacity would be required to meet only the average electrical demand rather than the peak demand. Expensive network upgrades can be deferred.

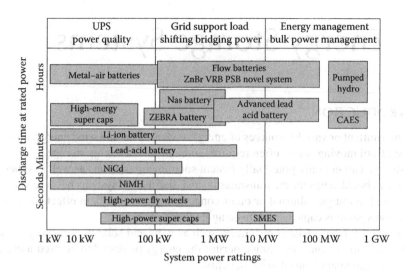

FIGURE 14.1 Comparison of storage technologies.

By enabling thermal generating units to operate closer to rated capacity, higher thermal efficiencies are achieved, and both system fuel costs and CO_2 emissions are reduced. Even further benefits also come from reducing demand variability and thus the requirement for load cycling of generating units and the requirements for additional regulating reserves.

As a result, the balancing costs that may be associated with wind variability can be reduced. Also, expensive standing reserve, in the form of open-cycle gas turbines, diesel engines, etc., can also be reduced, since both energy storage and load management can provide a similar role.

In general, power applications would be storage systems rated for 1 h or less, and energy applications would be for longer periods. Figure 14.1 presents a comparison of storage technologies in terms of power level applications and storage time.

Today, power applications for each of these technologies are being found in electric grid, for example, in the transmission system for bulk power storage as well as in the residential feeder circuits of smaller systems. The following abbreviations are used in Figure 14.1: Li-ion for lithium-ion battery, NiCd for nickel–cadmium battery, NiMH for nickel-metal hydride battery, CAES for compressed-air energy storage, SMES for superconducting magnetic energy storage, VRB for vanadium redox battery, ZnBr for zinc–bromine battery, NaS for sodium–sulfur battery, and ZEBRA battery for high-temperature battery (used at substations, and super caps for supercapacitors) [15–16].

14.3 STORAGE DEVICES

The list for conventional technologies includes the large hydro, compressed-air storage, and pumped hydro.

14.3.1 Large Hydro

It is the oldest renewable source for power/energy. Small hydro systems vary from 100 kW to 30 MW, while micro-hydropower plants are smaller than 100 kW. Small hydropower generators work at variable speed because the water upon which they depend flows at variable speeds. Induction generators are normally used with turbine system. The turbine converts the water's (kinetic) energy to mechanical rotational energy. The available power (P) from the water flow is expressed as

$$P_{avail} = Q \times H \tag{14.1}$$

where
   Q is the discharged water in m³/s
   H is the net head in m

Hydroelectric plants typically have fast ramp-up and ramp-down rates, proving strong regulating capabilities, and their marginal generation cost is close to zero. In many countries, a natural synergy exists between hydroelectric generation/pumped storage and wind power. Clearly, if hydro generation is being replaced by wind energy, then emission levels will not be affected, but the hydro energy can be transformed into potential energy stored for later use. Existing hydroelectric plant can reduce the output, using reservoirs as storage, to avoid wind energy curtailment.

14.3.2 COMPRESSED-AIR STORAGE

It involves the storage of compressed air in disused underground cavities, for example, exhausted salt mines. Alternatively, an underground storage complex can be created using a network of large-diameter pipes. Later, the compressed air can be released as part of the generation cycle, providing a cycle efficiency of approximately 75%. In an open-cycle gas turbine or combined-cycle gas turbine plant, incoming air is compressed by the gas turbine compressor before being ignited with the incoming fuel supply. The exhaust gases are then expanded within the turbine, driving both an electrical generator and the compressor.

A modern compressed-air energy storage is a peaking gas turbine power plant that consumes less than 40% of the gas used in a combined-cycle gas turbine (and 60% less gas than is used by a single-cycle gas turbine) to produce the same amount of electric output power. It is accomplished by blending compressed air to the input fuel to the turbine by compressing air during peak periods at lower costs than conventional stand-alone gas turbines.

It is required that plants are near proper underground geological formations, such as salt caverns, mines, or depleted gas wells. The first commercial CAES plant was a 290 MW unit built in Handorf, Germany, in 1978. The second one was a 110 MW unit built in McIntosh, Alabama, in 1991.

They are fast-acting units and typically can be put into service in 15 min when it is required. The Electric Power Research Institute (EPRI) has developed an advanced CAES system designed around a simpler way using advanced-turbine technology. The largest plant under consideration in the United States has a rating of 800 MW.

14.3.3 PUMPED HYDRO

The most widely established large-scale form of energy storage is hydroelectric pumped storage. It is an excellent energy storage technique, but unfortunately few attractive sites exist and initial investment costs are very high. Typically, such plant operates on a diurnal basis charging at night during periods of low demand (and low-priced energy) and discharging at during times of high or peak demand. A pumped storage plant may have the capacity for 4–8 h of peak generation with 1–2 h of reserve, although in some cases the discharge time can extend to a few days.

A typical pumped hydro plant consists of two interconnected reservoirs (lakes), tunnels that convey water from one reservoir to another, valves, hydro machinery (a water pump turbine), a motor generator, transformers, a transmission switchyard, and a transmission connection. The amount of stored electricity is proportional to the product of the total volume of water and the differential height between reservoirs. For example, storing 1000 MWh (deliverable in a system with an elevation change of 300 m) dictates a water volume of about 1.4 million cubic meters. The earliest application of pumped hydro technology was in Zurich, Switzerland, in 1882. It was realized early that a Francis turbine could also be used as a pump, such as the one used in the Hiwassee Dam Unit 2, in 1956, and has a rating of 59.5 MW.

Today, the global capacity of pumped hydro storage plants totals more than 95 GW, with approximately 20 GW operating in the United States. The original intent of these plants was to provide off-peak base loading for large coal and nuclear plants to optimize their overall performance and provide peaking energy each day. Since then, their duties have also included frequency regulation in the generation mode.

There are also less-conventional technologies, including hydrogen, flywheels, high-power fuel cells, high-power supercapacitors, SMES, heat or cold storage systems, and high-power batteries.

14.3.4 HYDROGEN

Hydrogen has been proposed as the energy store (carrier) for the future and the basis for a new transport economy. The reasons for this are simple: hydrogen is the lightest chemical element, thus offering the best energy/mass ratio of any fuel, and in a fuel cell can generate electricity efficiently and cleanly. Indeed, the waste product (water) can be electrolyzed to make more fuel (hydrogen).

Hydrogen can be transported conveniently over long distances using pipelines or tankers, so that generation and utilization take place in distinct locations, while a variety of storage forms are possible (gaseous, liquid, metal hydrating, etc.). It can be produced by the electrolysis of water using energy from a renewable resource. It can then be burned as a fuel to generate electricity [6].

Alternatively, it can be piped as a gas or liquid to consumers to be used locally providing both electricity and heating in a total energy scheme, or it can be used for transport. For transport needs, fuel cells in vehicles combine multi-fuel capability, high efficiency with zero (or low) exhaust emissions, and low noise.

The combustion of hydrogen provides energy plus water with no harmful emissions or by-products. If electricity is the final product, this process may not be attractive since the overall efficiency is usually below 50%. Because of this, the interest in hydrogen is usually for transportation purposes, which also depend on having proper storage systems.

In the future, hydrogen pipeline infrastructures are likely to be developed around the world. Excess hydrogen (i.e., energy) could be stored by temporarily increasing the gas pressure. Large wind farms could be used to power hydrogen-processing facilities and pipelines (in lieu of large electric transmission lines) could carry bulk hydrogen, as the energy source, to major population centers.

Thus, hydrogen (similar to transporting and storing natural gas) would be stored as necessary to match the demand for fuel cells for electricity and hydrogen-powered cars. This scheme has the further benefit of reducing wind power variability, since the wind energy is not directly used for electrical generation. For distances greater than 1000 km, energy transportation by hydrogen carrier should be more economical than high-voltage electrical transmission. However, there is a question on the overall efficiencies of creating large quantities of hydrogen to power fuel cells to create electricity [7,17].

14.3.5 HIGH-POWER FLOW BATTERIES

They operate similar to that of car batteries but without electrodes. Instead, when the flow cell is used as a "sink," the electric energy is converted into chemical energy by "charging" two liquid electrolyte solutions. The stored energy can be released on discharge. In common with all dc systems connected to the ac network, a bidirectional power electronic converter is needed.

Succinctly put, they use electrolyte liquids flowing through a cell stack with ion exchange through a microporous membrane to generate an electrical charge. Several different chemistries have been developed for use in utility power applications. Their advantage is their ability to scale systems independently in terms of power and energy.

More cell stacks means increased power rating. Also, a greater volume of electrolytes means an increased runtime. Furthermore, flow batteries operate at ambient temperature rather than high temperatures.

ZnBr flow batteries are being used for utility applications. The battery operates with a solution of zinc–bromide salt dissolved in water and stored in two tanks. The battery is charged or discharged by pumping the electrolytes through a reactor cell.

14.3.6 High-Power Flywheels

It is a kinetic-energy-storage device. In this method, energy is stored in very fast (approaching 75,000 rotations/min) rotating mass of flywheels. In the past, the flywheels had severe problems with maintenance, loses associated with bearings, material strength, and related severe failure management problems at high speeds.

Modern flywheels are made of fiber-reinforced composites. The flywheel motor/generator is interfaced to the main through a power electronic converter. At the present time, this technology is expensive and only used for select applications.

14.3.7 High-Power Supercapacitors

They are also called *ultra capacitors*. They consist of a pair of metal foil electrodes, each of which has an activated carbon material deposed on one side. These sides are separated by a proper membrane and then rolled into a package. Its operation is based on an electrostatic effect, whereby charging and discharging takes place with the totally physical (not chemical) reversible movement of ions.

Therefore, there are some fundamental differences between ultra capacitors and battery technologies including long shelf and operating life as well as large charge–discharge cycles of up to 500,000.

Supercapacitors are electrochemical capacitors. They look and perform similar to Li-ion batteries. They store energy in the two serious capacitors and the electric double layer which is formed between each of the electrodes and the electrolyte ions. The distance over which the charge separation takes place is just a few angstroms (a unit of length equal to 10^{-10} m). The extremely large surface area makes the capacitance and energy density of these devices thousands times larger than those of conventional electrolytic capacitors [5].

The electrodes are often made with porous carbon material. The electrolyte is either aqueous or organic. The aqueous capacitors have a lower energy density due to a lower cell voltage, but are less expensive and work in a wider temperature range. The asymmetrical capacitors that use metal for one of the electrodes have a significantly larger energy density than the symmetric ones do, and also have a lower leakage current.

In comparison to lead-acid batteries, electrochemical capacitors have lower energy density, but they can be cycled hundreds of thousands of times are much more powerful than batteries. They have fast chare and discharge capability.

They have been applied for blade-pitch control devices for individual wind turbine generators to control the rate at which power increases and decreases with changes in wind velocity. This is highly necessary if wind turbines are connected to weak utility grids [13].

In California, Palmdale Water District uses a 450 kW supercapacitor to regulate the output of a 950 kW wind turbine attached to the treatment plant microgrid. This arrangement helps to reduce network congestion in the area, while providing reliable supply to critical loads in the microgrid.

14.3.8 Superconducting Magnetic Energy Storage

As a result of recent developments in power electronics and superconductivity, the interest in using SMES units to store energy and/or damp power system oscillations has increased. It stores energy within a magnetic field created by the flow of direct current in a coil of superconducting material.

In a sense, SMES can be seen as a controllable current source whose magnitude and phase can be changed within one cycle. The upper limit of this source is imposed by the dc current in the

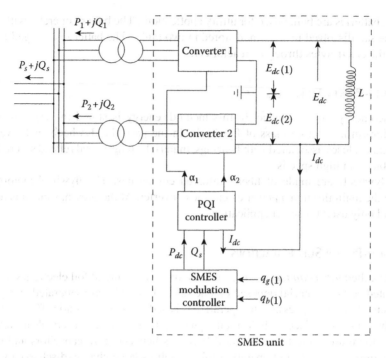

FIGURE 14.2   SMES unit with double GTO thyristor bridge. (From Gönen, T., *Electric Power Transmission System Engineering: Analysis and Design*, CRC Press, Boca Raton, FL, 2009. With permission.)

superconducting coil. Typically, the coil is maintained in its superconducting state through immersion in liquid helium at 4.2 K within a vacuum-insulated cryostat.

A power electronic converter interfaces the SMES to the grid and controls the energy flow bidirectionally. With the recent development of materials that exhibit superconductivity closer to room temperature, this technology may become economically viable [6].

Figure 14.2 shows a typical configuration of a SMES unit with a double gate turn-off (GTO) thyristor bridge. In the configuration, the superconducting coil L is coupled to the transmission system via two converters and transformers. The converter firing angles, α_1 and α_2, are determined by the PQI controller in order to control the real and reactive power outputs and the dc current I in the coil [5].

The control strategy is determined by the modulation controller of SMES to damp out power swings in the network. The active and reactive power available from SMES depends on the type of ac/dc tool for transient stability enhancement, and can be used to support primary frequency regulation [7].

14.3.9   HEAT OR COLD STORAGE

There has been a long tradition of using thermal storage to assist in power system operation, especially in the United Kingdom. This technology involves modulation of the energy absorbed by individual consumer electric heating elements and refrigeration systems for the benefit of overall system power balance.

An aggregation of a large number of dynamically controlled loads has the potential of providing added frequency stability and smoothing to power networks, both at times of sudden increase in demand (or less of generation) and during times of fluctuating wind or other renewable power.

Such devices could displace some reserve and may cause a substantial reduction in governor activity of remaining generators. The potential demand that could be operated under dynamic control is considerable. Deep-freeze units, industrial and commercial refrigeration, air conditioning, as well as water heating systems could provide dynamic demand control (DDC). The potential available in a developed country could be several GW. This concept is not limited for small applications. In Europe, a very large thermal storage system (up to 10,000 MWh) is being proposed [13].

14.4 BATTERY TYPES

Battery systems are quiet and nonpolluting. They can be installed near load centers and existing suburban substations. They have efficiencies in the range of 85%, and can respond to load changes within 20 ms. Lead-acid batteries as large as 10 MW with 4 h of storage have been used in several U.S., European, and Japanese utilities.

Although the input and output energy of a battery is electrical, the storage is in chemical form. Chemical batteries are individual cells filed with a conducting medium electrolyte that, when connected together, form a *battery*. Multiple batteries connected together form a *battery bank*. Essentially, there are two basic types of batteries: *primary batteries* (non-rechargeable) and *secondary batteries* (rechargeable).

14.4.1 SECONDARY BATTERIES

Secondary batteries are rechargeable batteries. They are further divided into two categories based on the operating temperature of the electrolytes. *Ambient operating temperature batteries* have either *aqueous* (flooded) or *nonaqueous* electrolytes. *High operating temperature batteries* (molten electrodes) have either solid or molten electrolytes. Rechargeable lead-acid and NiCd batteries have been used widely by utilities for small-scale backup, load leveling, etc.

The largest (NiCd) battery installation is a 45 MW, 10 MWh installations in Fairbanks, Alaska, built in 2003, and designed to provide a guaranteed 27 MW for at least 15 min following local power outages. For similar reasons, the largest (20 MW, 14 MWh) lead-acid system was installed by the Puerto Rico Electric Power Authority in 1994, and later repowered in 2004 [10,14].

But, given the fairly toxic nature of materials involved, low efficiency (70%–80%) and the limited life and energy density, secondary batteries based on other designs are being sought for utility applications. Batteries in electric vehicles are the secondary-rechargeable type and are in either of the two subcategories.

A battery for an electric vehicle has to satisfy certain performance goals that include quick discharge and charge capability, long-cycle life, low cost, recyclability, *high specific energy* (i.e., the amount of usable energy, measured in watt-hours per kilogram), *high energy density* (amount o energy stored per unit volume), *specific power* (defines the potential for acceleration), and the ability to work in extreme heat or cold.

However, at the present time, there is no battery that is available that meets all these criteria. Figure 14.3 shows exponential improvement in battery performance over the years. Today, a large variety of battery types are being used in electric power systems for grid support applications.

14.4.2 SODIUM–SULFUR BATTERIES

This battery is a high-performance battery, with the electrolyte operating at temperatures of 572°F (300°C). It consists of a liquid (molten) sulfur positive electrode and a molten sodium negative–positive electrode separated by a solid beta alumina ceramic electrode. The electrolyte permits only positive sodium ions to pass through it and combine with the sulfur to form sodium polysulfides [8–10].

The sodium component of this battery explodes on contact with water, which raises certain safety questions. The materials of the battery have to be capable of withstanding the high internal temperatures they create, as well as freezing and thawing cycles. This battery has a very high

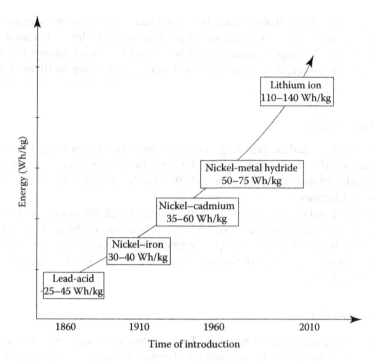

FIGURE 14.3 Exponential improvements in battery performance.

specific energy of 110 Wh/kg. During discharge, positive sodium ions flow through the electrolyte and electrons flow in the external circuit of the battery, providing about 2 V.

This technology for large-scale applications was perfected in Japan. Presently, there are 190 battery systems in service in Japan, totaling more than 270 MW of capacity with stored energy suitable for 6 h of daily peak shaving. The largest single NaS battery installation is a 34 MW, 245 MWh system for wind power stabilization in northern Japan. The battery will permit the output of the 51 MW wind farm to be 100% dispatchable during on-peak periods.

According to Roberts [13], in the United States, utilities have applied 9 MW of NS batteries for peak shaving, backup power, firming wind capacity, and other applications.

ZEBRA battery is another high-temperature battery and is based on sodium nickel chloride chemistry. It is used for electric transportation applications in Europe. Recently, it is being considered for utility applications as well [13].

14.4.3 Flow Battery Technology

The performance of flow batteries is similar to a hydrogen fuel cell. They use electrolyte liquids flowing through a microporous membrane to generate an electrical charge. They store and release electrical energy through a reversible electrochemical reaction between two liquid electrolytes.

The liquids are separated by an ion-exchange membrane, allowing the electrolytes to flow into and out from the cell through separate manifolds and to be transformed electrochemically within the cell. For their utility applications, various chemistries have been developed. In standby mode, the batteries have a response time of the order of milliseconds to seconds, making them suitable for frequency and voltage support. One of the advantages of such flow battery design is the ability to scale systems independently in terms of power and energy. For example, more cell stacks permit for an increase in power rating, a greater volume of electrolytes provides for more runtime. Plus, flow batteries operate at ambient (instead of high) temperature levels [2–4].

14.4.3.1 Zinc–Bromine Flow Battery

In utility applications, ZnBr flow batteries are being used. This battery operates with a solution of zinc–bromide salt dissolved in water and stored in two tanks. The battery is charged or discharged by pumping the electrolytes through a reactor cell.

During the charging cycle, metallic zinc from the electrolytes solution is plated onto the negative electrode surface of the reactor cell. The bromide is converted to bromine at the positive surface of the electrode in the reactor cell and then is stored in the other electrolyte tank as a safe chemically complex oily liquid.

During the discharge of the battery, the process is reversed, and the metallic zinc plated on the negative electrode is dissolved in the electrolyte solution and available for the next charge cycle.

In order to create different system ratings and duration times, flow battery manufacturers use modular construction. For example, a ZnBr flow battery package with a rating of 500 kW for 2 h. Other packages are being applied at utilities with ratings of up to 2.8 MWh packaged in a 53 ft trailer.

14.4.3.2 Vanadium Redox Flow Battery

Another type of flow battery is the VRB. During its charge and discharge cycles, positive hydrogen ions are exchanged between the two electrolyte tanks through a hydrogen-ion permeable polymer membrane. Similar to the ZnBr battery, the VRB system's power and energy ratings are independent of each other [13].

14.4.4 LITHIUM-ION BATTERIES

Among the available battery technologies today, the Li-ion battery has the greatest applications. It can be applicable in a large variety of shapes and sizes, permitting the battery to efficiently fill the available space, such as a cell phone or laptop computer.

They are also lighter in weight in comparison to other aqueous battery technologies, such as lead-acid batteries. They have the highest power density (110–140 Wh/kg) of all batteries on the commercial market on a per-unit-of-volume basis.

The leading Li-ion cell design is a combination of lithiated nickel, cobalt, and aluminum oxides, referred to as an NCA cell. There are two Li-ion designs that are starting to be employed in higher-power utility grid applications are lithium titanate and lithium iron phosphate [11,14,16].

14.4.4.1 Lithium Titanate Batteries

This battery uses manganese in the cathodes and titanate in the anodes. This chemistry provides for a very stable design with fast-charge capability and good performance at lower temperatures.

The batteries can be at lower temperatures. The batteries can be discharged to 0% and have a relatively long life. They are used in a utility power ancillary service application (e.g., frequency regulation).

14.4.4.2 Lithium Iron Phosphate Batteries

It is a newer and safer technology that is more difficult to release oxygen from the electrode, which reduces the risk of fire in the battery cells. It is more resistant to overcharge when operated in a range of up to 100% state of charge. They are also used in utility power ancillary service applications.

14.4.5 LEAD-ACID BATTERIES

They are the oldest and most mature among the all-battery technologies. Because of their large applications, lead-acid batteries have the lowest cost of all-battery technologies.

This battery operates at an ambient temperature and has an aqueous electrolyte. Even though the lead-acid battery is relatively inexpensive, it is very heavy, with a limited usable energy by weight (specific energy).

A cousin of this battery is the deep-cycle lead-acid battery, now widely used in golf carts and forklifts. The first electric cars built also employed this technology. Lead-acid batteries should not be discharged by more than 80% of their rated capacity or depth of discharge.

Exceeding the 80% of the depth discharge shortens the life of the battery. They are inexpensive, readily available, and are highly recyclable, using the elaborate recycling system already in place.

For utility application, a 40 MWh lead-acid battery was installed in the Southern California grid in 1988 to demonstrate the peak-shaving capabilities of batteries in a grid application.

The application of the battery demonstrated the value of stored energy in the grid; however, the limited cycling capability of lead acid made the overall economics of the system unacceptable.

However, for backup power sources in large power plants, lead-acid batteries are still used as "black start" sources in case of emergencies [13]. Their long life and lower costs make them ideal for applications with low-duty cycles.

Research continues to try to improve these batteries. For example, a lead-acid nonaqueous (gelled lead acid) battery uses an electrolyte paste instead of liquid.

These batteries do not have to be mounted in an upright position. There is no electrolyte to spill in the accident. But, nonaqueous lead-acid batteries typically do not have as high life cycle and are more expensive than flooded deep-cycle lead-acid batteries.

14.4.5.1 Advanced Lead-Acid Batteries

In order to significantly extend the life of lead-acid batteries, carbon is added to the negative electrode. As a result, their life is significantly extended in cycling applications. But, lead-acid batteries fail due to sulfation in the negative plate that increases as they are cycled more.

Adding as much as 40% of activated carbon to the negative electrode composition increases the battery's life up to 2000 cycles. This represents a three- to four-time improvement over current lead-acid designs. This extended life coupled with the lower costs will lead storage developers to revisit lead-acid technology for grid applications.

14.4.6 Nickel–Cadmium Batteries

Nickel iron (Edison cells) and NiCd pocket and sintered plate batteries have been in use for many years. Both of these batteries have a specific energy of approximately 25 Wh/lb (55 Wh/kg), which is higher than advanced lead-acid batteries. Both are nontoxic, while NiCds are toxic. They can be discharged to 100% of depth of discharge without damage.

The biggest obstacle to the utilization of these batteries is their cost. In the past, the NiCd batteries represented a substantial increase in battery power. They are rugged, durable with good cycling capability and a broad discharge range.

In power systems, NiCd batteries have been used in a variety of backup power applications and were chosen to provide "spinning reserve" for a transmission project in Alaska [13]. It involved a 26 MW NiCd battery rated for 15 min, which represents the largest battery in a utility application in North America. Today, they are still being used for utility applications. Examples include power ramp–rate control for smoothing with weak power grids (such as island power systems).

14.5 OPERATIONAL PROBLEMS IN BATTERY USAGE

The storage battery–integrated PV system recovers the energy that would have been lost when voltage is over the limitation value. Since the risk of overvoltages is higher when the reverse power flow is greater, the state of charge of the storage battery should not be full at around noon.

Thus, for efficient operation of the storage battery only part of the surplus power that is greater than the load demand should be charged into the storage battery.

According to Hara et al. [8], the following problems must be considered when operating the storage battery:

1. The storage battery must be at a discharge state in the morning to prepare for the charging around noon.
2. If the lead-acid battery is left in a discharge state, it may deteriorate and shorten the lifetime.
3. The frequency of use of the storage batteries may be varied by the impedance of the distribution line and by a power-flow condition.
4. There are round-trip energy losses of the storage battery and power conditioning systems (PCS) increases when charging and discharging larger amounts of energy.

14.6 FUEL CELLS

They were first developed in 1839 and put to practical use in the 1960s by NASA to generate fuel for electricity needed by the spacecrafts Apollo and Gemini. The stored hydrogen can be converted back to electricity using an open-cycle gas turbine. However, in that case electrical efficiency tends to be low, even ignoring transportation losses and those associated with converting the electricity to hydrogen in the first place [18].

Fuel cells are quiet, clean, and highly efficient onsite generators of electricity that use the electrochemical process to convert fuel into electricity. This is the reverse electrolysis. It has few moving parts, and produces very little waste heat or gas. In addition to generating electricity, fuel cells can also serve as a thermal energy source for water and space heating or for cooling absorption.

Fuel cells offer an alternative approach, and essentially consist of an electrolyte (liquid or solid) membrane sandwiched between two electrodes. A block diagram of a fuel cell is shown in Figure 14.4.

A single fuel cell produces output voltage less than 1 V. Thus, in order to produce higher voltages, fuel cells are stacked on top of each other and are serially connected, forming a full cell system. Electrical efficiencies of fuel cells lie between 36% and 60%, according to the type and system configuration. By using conventional heat recovery apparatus, the overall efficiency can be improved to about 85%.

Steam reforming of liquid hydrocarbons ($C_n H_m$) is a potential way of providing hydrogen-rich fuel for fuel cells. This is a preferred method since storage of hydrogen is quite hazardous and expensive. Reformers facilitate a continuous supply of hydrogen without having to use bulky pressurized hydrogen tanks or hydrogen vehicles for distribution. The endothermic reaction takes place in the reforming process in the presence of a catalyst is

$$C_n H_m + nH_2O \rightarrow nCO + \left(\frac{m}{2} + n\right)H_2 \tag{14.2}$$

and

$$CO + H_2O \rightarrow CO_2 + H_2 \tag{14.3}$$

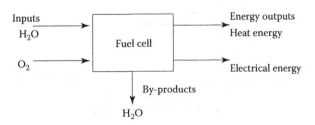

FIGURE 14.4 A block diagram of a fuel cell system.

Carbon monoxide combines steam to produce more hydrogen through the water gas shift reaction. Figure 14.5 shows the flows and reactions in a fuel cell.

Fuel cells are classified according to the nature of electrolyte used and the operating temperature, with each type requiring particular materials and fuels, as summarized in Table 14.1. The electrochemical efficiency tends to increase with fuel cell temperature. It is often the nature of the membrane that dictates the operating temperature, and expensive catalysts, such as platinum, may be required to step up the rate of electrochemical reactions. Fuel cells can run using hydrogen, natural gas, methanol, coal, or gasoline.

In addition to this raw fuel of hydrogen, more environmentally friendly fuels, such as biogas and biomass, may be used. For most fuel cells, such fuels must be transformed into hydrogen using a reformer or coal gasifier. However, high-temperature fuel cells can generally use a fossil fuel (natural gas, coal gas, etc.) directly. Polluting emissions are produced, but since hydrogen is passed over one electrode (anode), where hydrogen molecules separate to the cathode where they combine with oxygen to form water. The oxygen supply may be derived from air, or as a stored by-product from the water electrolysis (forming hydrogen).

For large-scale utility storage applications, the choice of technology will depend on the ability to use pure hydrogen (electrolyzed from water) as the fuel, the electrical efficiency of conversion, and the load-following capability of the fuel cell, thus providing a degree of regulation from fluctuating wind or other renewable sources. Of the various options available, SO and PEM seem most likely to succeed.

The efficiency for conversion of fuel to electricity can be as high as 65%, which is nearly twice as efficient as conventional power plants. Also, small-scale fuel cell plants are just as efficient as the large ones, whether they operate at the full load or not. Because of their modular nature, they can be placed at or near load centers, resulting in savings of transmission network expansion.

A fuel cell power plant is essentially made of three subsystems or sections. In the fuel-processing section, the natural gas or other hydrocarbon fuel is converted to hydrogen-rich fuel. This process is known as a steam catalytic reforming process. This fuel is then fed to the power section, where it reads with oxygen from the air in a large number of individual fuel cells to produce dc electricity, and by-product heat in the dorm of usable steam or hot water.

For a power plant, the number of fuel cells can vary from several hundreds (for a 40 kW plant) to several thousands (for a multimegawatt plant). In the final or third stage, the dc electricity is converted in the power-conditioning subsystem to electric utility-grade ac electricity.

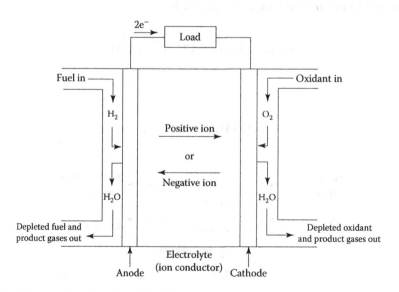

FIGURE 14.5 Flows and reactions in a fuel cell.

TABLE 14.1

A Brief Comparison of Five Fuel Cell Technologies

Type	Electrolyte	Operating Temperature (°C)	Applications	Advantages
PEM	Solid organic polymer	60–90	Electric utility, transportation, portable power	H_2
	Metal oxide (Y_2O_3/ZrO_2)	700–1000		H_2, CH_4, biogas, etc. Solid electrolyte reduces corrosion, low temperature, quick start-up. Efficiency is 35%–55%
Direct alcohol	Polymer membrane/ liquid alkaline	60–120	Transportation, portable power	H_2, CH_4, biogas, coal gas etc. Its efficiency is 35%–40%
Alkaline (AFC)	Aqueous solution of potassium hydroxide soaked in a matrix	50–90	Military, space	Cathode reaction faster in alkaline electrolyte; thus high performance. It uses H_2 as fuel. Efficiency is 50%–60%
Phosphoric acid (PAFC)	Liquid phosphoric acid soaked in a matrix	150–220	Electric utility, transportation and heat	Its efficiency is 45%–55%. Up to 85% efficiency in cogeneration of electricity
Molten carbonate (MCFC)	Liquid solution of lithium sodium, and/or potassium carbonates soaked in a matrix	600–750	Electric utility	Higher efficiency, fuel flexibility, inexpensive catalysts. It uses H_2, CH_4, biogas, coal gas, etc.
Solid oxide (SOFC)	Solid zirconium oxide to which a small amount of yttria is added	600–1000	Electric utility	Higher efficiency, fuel flexibility, inexpensive catalysts. Solid electrolyte advantage like PEM

In the power section of fuel cell, which has the electrodes and the electrolyte, two separate electrochemical reactions happen: an oxidation half-reaction, taking place at the anode, and a reduction half-reaction, occurring at the cathode. The anode and the cathode are separated from each other by the electrolyte. During the oxidation half-reaction at the anode, gaseous hydrogen produces hydrogen ions, which travel through the ionically conducting membrane to the cathode.

At the same time, electrons travel through an external circuit to the cathode. In the reduction half-reaction at the cathode, oxygen supplied from air combines with the hydrogen ions and electrons to form water and excess heat. Hence, the fuel products of the overall reaction are electricity, water, and excess heat.

14.6.1 TYPES OF FUEL CELLS

Since the electrolyte defines the key properties, specifically the operating temperature, of the fuel cell, fuel cells are categorized based on their electrolyte type, as described below:

1. Polymer electrolyte membrane (PEM)
2. Alkaline fuel cell (AFC)
3. Phosphoric acid fuel cell (PAFC)
4. Molten carbonate fuel cell (MCFC)
5. Solid oxide fuel cell (SOFC)

These fuel cells operate at different temperatures and each of them is best suited to specific applications. Table 14.1 gives a brief comparison of the five cell technologies introduced earlier.

14.6.1.1 Polymer Electrolyte Membrane

It is one of a family of fuel cells that are in various stages of development. The electrolyte in a PEM cell is a type of polymer and is usually referred to as a membrane, thus the name. PEMs are somewhat unusual electrolytes in that, in the presence of water, which the membrane readily absorbs, the negative ions are rigidly held within their structure. Only the positive (H) ions contained within the membrane are mobile and are free to carry positive charges through the membrane in one direction only, from anode to cathode.

At the same time, the organic nature of the PEM structure makes it an electron insulator, forcing it to travel through the outside circuit providing electric power to the load. Each of the two electrodes is made of porous carbon to which very small platinum particles are bonded. The electrodes are slightly porous so that the gases can diffuse through them to reach the catalyst. Also, as both platinum and carbon conduct electrons well, they are able to move freely through the electrodes [12]. Chemical reactions that take place inside a PEM fuel cell are as follows:
At anode:

$$2H_2 \rightarrow 4H^+ + 4e^- \qquad (14.4)$$

At cathode:

$$O_2 + 4H^+ + 4e^- \rightarrow H_2O \qquad (14.5)$$

Net reaction:

$$2H_2 + O_2 = 2H_2O \qquad (14.6)$$

Here, hydrogen gas diffuses through the polymer electrolyte until it meets a platinum particle in the anode. The platinum catalyzes dissociation of the hydrogen molecule into two hydrogen atoms (H) bonded to two neighboring platinum atoms. Only then can each H atom releases an electron to form a hydrogen ion (H$^+$) which travels to the same time, the free electron through the other circuit. At the cathode, the oxygen molecule interacts with the hydrogen ion and the electron from the outside circuit to form water. The performance of the PEM fuel cell is limited mainly by the slow rate of the oxygen reduction half-reaction at the cathode, which is 100 times slower than the hydrogen oxidation half-reaction at the anode [12].

14.6.1.2 Phosphoric Acid Fuel Cell

This technology has moved from the laboratory R&D to the first stages of the commercial application. Today, 200 kW plants are available and have been built at more than 70 sites in the United States, Japan, and Europe. Operating at approximately 200°C, the PAFC plant also produces heat for domestic hot water and space heating, and its electrical efficiency is close to 40%.

Its high cost is the only thing that stops it for its wide commercial acceptance. At the present time, capital costs of PAFC plant is about $2500–$4000/kW. According to Rahman [12], if it is reduced down to $1000 to $1500/kW, this technology may be accepted by the power industry. The chemical reactions that take place at two electrodes are as follows:
At anode:

$$2H_2 \rightarrow 4H^+ + 4e^- \qquad (14.7)$$

At cathode:

$$O_2 + 4H^+ + 4e^- \rightarrow 2H_2O \tag{14.8}$$

14.6.1.3 Molten Carbonate Fuel Cell

This technology is attractive because it offers several potential advantages over PAFC. Carbon monoxide, which positions the PAFC, is indirectly used as a fuel in the MCFC. The higher operating temperature of about 650°C makes the MCFC a better candidate for combined-cycle applications, whereby the fuel cell exhaust can be used as input to the intake of a gas turbine or the boiler of a steam turbine.

The total efficiency can approach 85%. It is just about to enter the commercial market. Capital costs involved are expected to be lower than PAFC. MCFCs are now being tested in full-scale demonstration plants [12]. The chemical reactions that take place inside the cell are as follows:
At anode:

$$2H_2 + 2CO_3^{2-} \rightarrow 2H_2O + 2CO_2 + 4e^- \tag{14.9}$$

and

$$2CO + 2CO_3^{2-} \rightarrow 4CO_2 + 4e^- \tag{14.10}$$

At cathode:

$$O_2 + 2CO_2 + 4e^- \rightarrow 2O_3^{2-} \tag{14.11}$$

14.6.1.4 Solid Oxide Fuel Cell

According to Rahman [12], a SOFC is currently being demonstrated at a 100 kW plant. This technology dictates very significant changes in the structure of the cell. It uses a solid electrolyte, a ceramic material, so the electrolyte does not need to be replenished during the operational life of the cell.

The results of this, is simplification in design, operation and maintenance, as well as having the potential to reduce costs. This offers the potential to reduce costs. This offers the stability, reliability of all solid-state construction and permits higher temperature operation.

The ceramic makeup of the cell lends itself to cost-effective fabrication techniques. Its tolerance to impure fuel streams make SOFC systems especially attractive for utilizing H_2 and CO from natural gas steam-reforming and coal gasification plants [12]. The chemical reactions that take place inside the cell are as follows:
At anode:

$$2H_2 + 2O^{2-} \rightarrow 2H_2O + 4e^- \tag{14.12}$$

and

$$2CO + 2O^{2-} \rightarrow 2CO_2 + 4e^- \tag{14.13}$$

At cathode:

$$O_2 + 4e^- \rightarrow 2O^{2-} \tag{14.14}$$

REFERENCES

1. Barak, M., *Electrochemical Power Sources*, Stevenage, U.K.: Peter Peregrinus Ltd., 1980.
2. Barclay, F., *Fuel Cells, Engines and Hydrogen*, New York: Wiley, 2006.
3. Béguin, F. and Frackowiak, E., *Carbons for Electrochemical Energy Storage and Conversion Systems*, Boca Raton, FL: CRC Press, 2010.
4. Gasik, M., *Materials for Fuel Cells*, Boca Raton, FL: CRC Press, 2008.
5. Gönen, T., *Electric Power Transmission System Engineering: Analysis and Design*, Boca Raton, FL: CRC Press, 2009.
6. Gönen, T., High-temperature superconductors, a technical article in *McGraw-Hill Encyclopedia of Science & Technology*, 7th Ed., McGraw-Hill, New York, Vol. 7, 1992, pp. 127–129.
7. Gönen, T., Anderson, P. M., and Bowen, D., Energy and the future, *Proceedings of the 1st World Hydrogen Energy Conference*, 3(2c), Miami, FL, 1977, pp. 55–78.
8. Hara, R. et al., Testing the technologies, *IEEE Power & Energy Magazine*, May/June 2009, 77–85.
9. Jasinski, R., *High-Energy Batteries*, New York: Plenium Press, 1967.
10. Kiehne, H. A., *Battery Technology Handbook*, New York: Marcel Dekker, Inc., 1987.
11. Ozawa, K., *Lithium Ion Rechargeable Batteries*, Weinheim, Germany: Wiley-VCH, 2009.
12. Rahman, S., Advanced energy technologies, in *Electric Power Generation, Transmission, and Distribution*, L. L. Grigsby, ed., Boca Raton, FL: CRC Press, 2007.
13. Roberts, B., Capital grid power, *IEEE Power & Energy Magazine*, July/August 2009, pp. 32–41.
14. Schalkwijk, W. A. and Scrosati, B., *Advances in Lithium-Ion Batteries*, New York: Kluwer Academic/ Plenum Publishers, 2002.
15. Soo, S. L., *Direct Energy Conversion*, Englewood Cliffs, NJ: Prentice-Hall, 1968.
16. Sutton, G. W., *Direct Energy Conversion*, New York: McGraw-Hill, 1966.
17. Zimmerman, A. H., *Nickel-Hydrogen Batteries*, Reston, VA: Aerospace Press, 2009.
18. Rajalakshmi, N. and Dhathathreyan, K. S., *Present Trends in Fuel Cell Technology Development*, New York: Nova Science Publishers, Inc., 2008.

15 The Smart Grid

15.1 INTRODUCTION

The electric power delivery system has often been referred to as the greatest and most complex machine ever built. It is made of wires, electrical machines, electric towers, transformers, and circuit breakers—all put together in some fashion. The existing electricity grid is unidirectional in nature. Its overall conversion efficiency is very low, about 33% or 34%.

That means it converts only one-third of fuel energy into electricity, without recovering even its waste heat. Approximately, 8% of its output is lost along its transmission lines, while 20% of its generation capacity exists just to meet its peak demand only (i.e., it is being used only 5% of the time). Furthermore, the existing electric system is vulnerable for domino-effect failures due to its hierarchical topology of its assets.

The electric power grid infrastructure of the United States was built more than 50 years ago. It has become a complex spider web of power lines and aging networks and systems with obsolescent technology and outdated communications. In addition, there is a growing demand for lower carbon emissions, renewable energy sources, improved system reliability, and security.

All these requirements dictate the modernization of the grid by installing intelligent electronic devices (IEDs) in terms of sensors, electronic switches, smart meters, and also advanced communication, and data acquisition and interactive software with real-time control that optimize the operation of the whole electrical system and make more efficient utilization of the grid assets. It is such grid that is called the *smart grid*.

It is envisioned that such smart grid would integrate the renewable energy sources, especially wind and solar, with conventional power plants in a coordinated and intelligent way that would not only improve reliability and service continuity, but also effectively reduce energy consumption and significantly reduce the carbon emissions. Based on a July 2009 Smart Grid Report of the United States Department of Energy [12], a smart grid has to have the following functions:

1. *Optimize asset utilization and operating efficiency*: The smart grid optimizes the utilization of the existing and new assets, improves load factors, and lowers system losses in order to maximize the operational efficiency and reduce the cost. Advanced sensing and robust communications will allow early problem detection, and prevent maintenance and corrective actions.
2. *Provide the power quality for the range of needs*: The smart grid will enable utilities to balance load sensitivities with power quality, and consumers will have the option of purchasing varying grades of power quality at different prices. Also, irregularities caused by certain consumer loads will be buffered to prevent propagation.
3. *Accommodate all generation and storage options*: The smart grid will integrate all types of electrical generation and storage systems, including small-scale power plants that serve their loads, known as *distributed generation*, with a simplified interconnection process analogous to "*plug and play.*"
4. *Enable informed participation by customers*: The smart grid will give consumers information, control, and options that enable them to become active participants in the grid. *Well-informed* customers will modify consumption based on balancing their demands and resources with the electric system's capability to meet those demands.

5. *Enable new products, services, and markets*: The smart grid will enable market participation, allowing buyers and sellers to bid on their energy resources through the supply and demand interactions of markets and real-time price quotes.

6. *Operate resiliently to disturbances, attacks, and natural disasters*: The smart grid operates resiliently, that is, it has the ability to withstand and recover from disturbances in a self-healing manner to prevent or mitigate power outages, and to maintain reliability, stability, and service continuity. The smart grid will operate resiliently against attack and natural disaster. It incorporates new technology and higher cyber security, covering the entire electric system, reducing physical and cyber vulnerabilities, and enabling a rapid recovery from disruptions.

Therefore, the next-generation electricity grid, known as the "smart grid" or "intelligent grid," is designed to address the major shortcomings of the existing grid. Basically, the smart grid is required to provide the electric power utility industry with full visibility and penetrative control and monitoring over its assets and services. It is required to be self-healing and resilient to system abnormalities.

Furthermore, the smart grid needs to provide an improved platform for utility companies to engage with each other and doing energy transactions across the system. It is expected to provide tremendous operational benefits to power utilities around the world, because it provides a platform for enterprise-wide solutions that deliver far-reaching benefits to both utilities and their end customers.

The development of new technologies and applications in distribution management can help drive optimization of smart grid and assets. Hence, the smart grid is the result of convergence of communication technology and communication technology with power system engineering, as depicted in Figure 15.1.

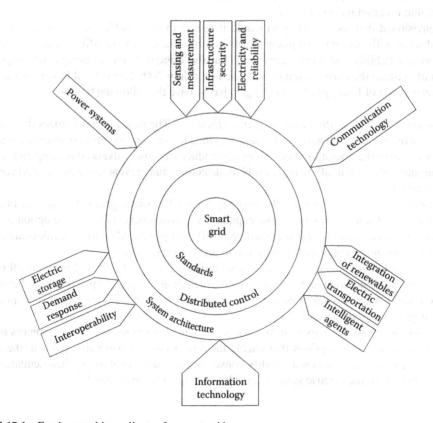

FIGURE 15.1 Fundamental ingredients of a smart grid.

TABLE 15.1

A Comparison of the Features of the Smart Grid with the Existing Grid

Smart Grid	Existing Grid
Digital	Electromechanical
Two-way communication	One-way communication
Distributed generation	Centralized generation
Network	Hierarchical
Sensors throughout	Few sensors
Self-monitoring	Blind
Self-healing	Manual restoration
Adaptive and islanding	Failures and blackouts
Old-fashioned customer metering	Intelligent customer metering
Remote checking/testing	Manual checking/testing
Pervasive control	Limited control
Many customer choices	Few customer choices

Table 15.1 provides a comparison of features of the smart power grid with the existing power grid. In summary, a smart grid uses sensors, communications, and computational ability and control in some form to enhance the overall functionality of the electric power delivery system. In other words, a dumb system becomes smart by sensing, communicating, applying intelligence, exercising control and through feedback, continually adjusting. This allows several functions that permit optimization, in combination, of the use of bulk generation, and storage, transmission, distribution, distributed resources, and consumer end uses toward goals that ensure reliability and optimize or minimize the use of energy, mitigate environmental impact, manage assets, and minimize costs. In other words, the philosophy of the smart grid is a brand-new way of looking at the electric power delivery system and its operation to achieve the optimality, and maximum efficiency and effectiveness. Presently, it is hoped and expected that a smart grid will provide the following:

1. Higher penetration of renewable resources.
2. Extensive and effective communication overlay from generation to consumers.
3. The use of advanced sensors and high-speed control to make the grid more robust.
4. It will provide higher operating efficiency.
5. It will provide a greater resiliency against attack and natural disasters.
6. It will provide effective automated metering and rapid service restoration after storms.
7. It will facilitate real-time or time-of-use pricing of the electric energy.
8. It will provide greater customer participation in generation and selling of the energy generated by using renewable resources.

In summary, a smart grid is an electricity network that can intelligently integrate the actions of all users connected to it, that is, generators, consumers, and those that do both, in order to efficiently deliver sustainable, economic, and secure electricity supplies.

A smart grid uses innovative products and services together with intelligent monitoring, control, communication, and self-healing technologies in order to

1. Better facilitate the connection and operation of generators of all sizes and technologies.
2. Permit consumers to play a part in optimizing the operation of the system.
3. Provide consumers with greater information and choice of supply.
4. Deliver increased levels of reliability and security of supply.
5. Significantly reduce the environmental impact of the whole electricity supply system.

15.2 NEED FOR ESTABLISHMENT OF SMART GRID

The move toward the smart grid is fueled by a number of needs. For example, there is the need for improved grid reliability while dealing with an aging infrastructure, there is the need for environmental compliance and energy conservation. Also, there is the need for improved operational efficiencies and customer service.

Furthermore, with an increase in regulating influence and the focus on smart-grid advanced technologies, there is a renewed interest in increasing the investment in distribution networks to defer infrastructure build-out and to reduce operating and maintenance costs through improving grid efficiency, network reliability, and asset management programs. Thus, since the roots of power system issues are usually found the electric distribution system, the point of departure for the grid overhaul is the distribution system.

As said before, most of the utilities believe that the biggest return on the investment will be investing in distribution automation which will provide them with fast increasing capability over time. Thus, "blind" and manual operations, along with electromechanical components in the electric distribution grid, will need to be transformed into a "smart grid."

Such transformation is necessary to meet environmental targets, to accommodate a greater emphasis on demand response (DR), and to support plug-in hybrid electric vehicles (PHEVs) as well as distributed generation and storage capabilities. Also, as succinctly put by Gellings [6], the attributes of the good smart grid are as follows:

1. Absolute reliability of supply
2. Optimal use of bulk power generation and storage in combination with distributed resources and controllable/dispatchable consumer loads to assure lowest cost
3. Minimal environmental impact of electricity production and delivery
4. Reduction in electricity used in the generation of electricity and an increase in the efficiency of the power delivery system and in the efficiency and effectiveness of end use
5. Resiliency of supply and delivery from physical and cyber attacks and major natural phenomena (e.g., hurricanes, earthquakes, and tsunamis)
6. Assuring optimal power quality for all consumers who require it
7. Monitoring of all critical components of the power system to enable automated maintenance and outage prevention

Furthermore, the recommended renewable portfolio standards (RPSs) mechanism generally places an obligation on the utility companies to provide a minimum percentage of their electricity from approved renewable energy sources. According to the U.S. Environmental Protection Agency, as of August 2008, 32 states plus the District of Columbia had established RPSs targets.

Together, these states represent for almost half of the electricity sales in the United States. The RPS targets presently range from a low 2% to a high of 25% of electricity generation, with California leading the pact that requires 20% of the energy supply come from renewable resources by 2010 and 33% by 2020. RPS noncompliance penalties imposed by states range from $10 to $25 per MWh.

Many states regulatory commissions have initiated proceedings or adopted policies for the implementation of advanced metering infrastructures (AMIs) to enable demand response (DR). In this ruling on October 17, 2008, the Federal Energy Regulatory Commission (FERC) established a policy aimed at eliminating barriers to the participation of demand response in the organized power markets (independent service operators [ISOs] and regional transmission organizations [RTOs] by ensuring the comparable treatment of resources. In this ruling,

FERC [5] states: possible rate of return [9]

demand response can provide competitive pressure to reduce wholesale power prices: increases awareness of energy usage; provides for more efficient operation of markets; mitigates market power; enhances reliability; and in combination with certain new technologies, can support the use of renewable energy resources, distributed generation, and advanced metering. Thus, enabling demand-side resources, improves the economic operation of electric power markets by aligning prices more closely with the value customers place on electric power.

Among other things, the order directs regional transmission operator (RTOs) and ISOs to accept bids from distributed resources (DR) for energy and ancillary services, eliminate penalties for taking less energy than scheduled, and permit aggregators to bid DR on behalf of retail customers.

It is well known that the reliable supply of electrical power plays a critical role in the economy. The new operating strategies for environmental compliance, together with our aging transmission and distribution infrastructure, constitute a great challenge to the security, reliability, and quality of the electric power supply [14].

When implemented throughout the system, intermittent energy resources, such as wind, will greatly stress transmission grid operation. The distribution grid will also be stressed with the introduction and, perhaps, rapid adaptation of on-site solar generation as well as PHEVs and plug-in electric vehicles (PEVs). Such plug-in vehicles could considerably increase the circuit loading if the charging times and schedules are not properly managed and controlled.

Major upgrades to distribution system infrastructure may flow patterns due to the integration of the distributed generation and microgrids. Therefore, the existing power delivery infrastructure can be substantially improved through automation and information management.

As succinctly put by Farhangi [4], the convergence of communication technology and information technology (IT) with power system engineering helped by a number of new approaches, technologies, and applications permits the existing grid to penetrate condition monitoring and asset management, especially on the distribution system.

It is obvious that smart distribution system applications are at the core of the energy delivery systems between the transmission system and customers, all smart distribution applications target three main objectives:

1. Improving distribution network performance
2. Improving distribution system energy efficiency by increasing distribution network transit and capacity
3. Empowering the customer by providing choices

The smart distribution applications will have to integrate power system technologies, such as distribution automation, volt/var control (VVC), advanced distribution line monitoring with telecommunications, and data management technologies. The primary purpose of VVC is to maintain acceptable voltage at all pints along the distribution feeder under all loading conditions. The purpose of using voltage regulators or transformers with *load tap changers* (LTCs) is to automatically raise or lower the voltage in response to changes in load.

It is often that there is a need that requires the use of *capacitor banks* to supply some reactive power that would otherwise be drawn from the supply substations. As penetration of intermittent renewable resource-based generating units increase in the future, high-speed dynamic load/var control will play a significant role in *maintaining power quality* and *voltage stability* on the distribution feeders.

Hence, as said before, electric utility companies believe that investing in distribution automation will provide them with increasing capabilities over time. Thus, the first step in the evolution of the smart grid starts at the distribution side, enabling new applications and operational efficiencies to be introduced into the system.

15.3   ROOTS OF THE MOTIVATION FOR THE SMART GRID

The electric power grid is now focusing on a large number of technological innovations. Utility companies around the world are incorporating new technologies in their various operations and infrastructures.

At the bottom of this transformation is the requirement to make more efficient use of present assets. Several utility companies have developed their own vision of future smart distribution systems to reach the smart-grid objectives. Figure 15.2 shows such representation in terms of a smart-grid tree. Notice that the trunk of the tree is the asset management, which is the base of smart-grid development. Based on this foundation, the utility company builds its smart-grid system by a careful overhaul of their IT infrastructure, communication, and network infrastructure. Well-designed layer of intelligence over utility assets enables the emergence of the smart-grid capabilities will be built on vertical integration of the upper-layer applications. For example, an important capability such as microgrid may not be possible without the integration of distributed generation and home area networks.

Thus, the emergence of the truly smart grid will require a drastic overhaul of the existing system. It will require the establishment of distributed control and monitoring systems within and alongside the present electric power grid. Most likely, this change will be gradual but continuous.

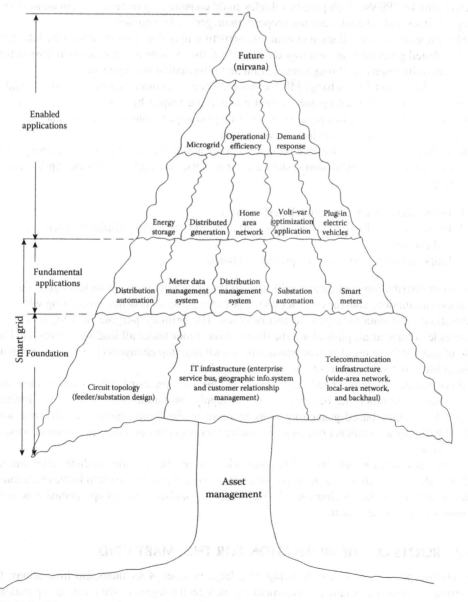

FIGURE 15.2 The representation of a smart grid as a tree.

Such smart grids will facilitate the distributed generation and cogeneration of energy as well as the integration of alternative sources of energy and the management and control of a power system's emissions and carbon footprint. Furthermore, they will help the utilities to make more efficient use of their present assets through a peak-shaving demand response and service quality control [4].

Here, the dilemma of a utility company is how to establish the smart grid to achieve the highest possible rate of return on the needed investments for such fundamental overhauls, the new architecture, protocols, and standards toward the smart grid.

However, as supply constraints continue, there will be more focus on the operational effectiveness of the distribution network in terms of cost reduction and capacity relief. Monitoring and control requirements for the distribution system will increase, and the integrated smart-grid architecture will benefit from data exchange between the distribution management system (DMS) and other project applications. The appearance of widespread distribution generation and consumer-demand response programs also introduces considerable impact on the DMS operation.

It is important to point out that smart-grid technologies will add a tremendous amount of real-time and operational data with the increase in sensors and the need for more information on the operations of the system. In addition, utility customers will be able to generate and deliver electricity to the grid or consume the electricity from the grid based on predetermined rules and schedules.

Thus, the roles of the consumers have changed; they are no longer only buyers but are sellers and/or buyers, switching back and forth from time to time. This results in two-way power flows in the grid and the need for monitoring and controlling the generation and consumption points on the distribution network.

As a result of these changes, the distribution generation will be from dissimilar sources and subject to great uncertainty. At the same time, the electricity consumption of the individual customers is also subject to a great uncertainty when they respond to the real-time pricing and rewarding policies of power utilities for economic benefits.

The conventional methods of load management (LM) and load estimation (LE) in the traditional DMS are no longer effective, causing other DMS applications ineffective or altogether useless.

However, the impact of demand response management (DRM) and consumer behaviors may be mandated and predicted, from the utility pricing rules and rewarding policies for specified time periods. The fundamental benefits of automation applied to a distribution system are as follows:

1. Released capacity
2. Reduced losses
3. Increased service reliability
4. Extension of the lives of equipment
5. Effective utilization of assets

The drivers for advanced distribution automation include the following:

1. Worldwide energy consumption is increasing due to population growth and increased energy use per capita in most developing countries.
2. Increased emphasis on system efficiency, reliability, and quality.
3. The need to serve increasing amounts of sensitive loads.
4. Need to do more with less capital expenditure.
5. Performance-based rates.
6. Increasing focus on renewable energy due to the increasing costs to extract and utilize the fossil fuels in an environmentally benign manner is becoming increasingly expensive.
7. Availability of real-time analysis tools for faster decision making.
8. Worldwide energy consumption.

In general, distribution automation and control functions include the following [10]:

1. Discretionary load switching
2. Peak load pricing
3. Load shedding
4. Cold load pickup
5. Load reconfiguration
6. Voltage regulation
7. Transformer load management (TLM)
8. Feeder load management (FLM)
9. Capacitor control
10. Dispersed storage and generation
11. Fault detection, control, and isolation
12. Load studies
13. Condition and state monitoring
14. Automatic customer meter reading
15. Remote service connect or disconnect
16. Switching operations

15.4 DISTRIBUTION AUTOMATION

IEEE definition of the distribution automation is "a system that enables an electric utility to remotely monitor, coordinate, and operate distribution companies in real-time mode from remote locations." For example, various DMS management information system (MIS) applications that are commonly used today include the following [2]:

1. Fault detection and service restoration (FDIR) is designed to improve system reliability. It detects a fault on a feeder section based on the remote measurements from the feeder terminal units (FTUs), rapidly isolates the faulted feeder section, and then restores service to the *unfaulted* feeder sections.
2. The topology processor (TP) is a background, offline processor that accurately determines the distribution network topology and connectivity for display colorization and to provide accurate network data for other DMS control applications.
3. Optimal network reconfiguration (ONR) is a module that recommends switching operations to reconfigure the distribution network to minimize network energy losses, maintain optimum voltage profile, and balance the loading conditions among the substation transformers, the distribution feeders, and the network phases.
4. Integrated voltage/var control (IVVC) has three basic objectives: reducing feeder network losses by energizing or de-energizing the feeder capacitor banks, ensuring that an optimum voltage profile is maintained along the feeder during normal operating conditions, and reducing peak load through feeder voltage reduction by controlling the transformer tap positions in substations and voltage regulators on feeder sections, as illustrated in Figure 15.8.
5. ONR is a module that recommends switching operations to reconfigure the distribution network to minimize network energy losses, maintain optimum voltage profiles, and balance the loading conditions among the substation transformers, the distribution feeders, and the network phases.
6. Switch order management (SOM) is very useful for system operators in real-time operation. It provides advanced analysis and execution features to better manage all switch operations in the system.
7. Dynamic load modeling/load estimation (LM/LE) is the base module in DMS. It uses all the available information from the distribution network to accurately estimate individual loads and aggregate bulk loads.

8. The dispatcher training simulator (DTS) is used to simulate the effects of normal and abnormal operating conditions and switching scenarios before they are applied to the real system.

9. Short-circuit analysis (SCA) is used to calculate the short-circuit current to evaluate the possible impact of a fault on the network.

10. Relay protection coordination (RPC) manages and verifies the relay settings of the distribution feeders under various operating conditions and network configurations.

11. Optimal capacitor placement/optimal voltage regulator placement (OCP/OVP) is used to determine the optimal locations for capacitor banks and voltage regulators in the distribution networks for the most effective control of the feeder volt/var profile.

15.5 ACTIVE DISTRIBUTION NETWORKS

An *active network* is a passive network that has been converted to an active one by the connection of distributed generation in terms of, for example, combined heat and power (CHP) cogens and/or other renewable-based energy-producing units such as wind turbines or solar units.

The necessary monitoring, communications, and control in terms of both preventive and corrective actions are provided for such networks. They are flexible, adaptable, most likely autonomous (such as the case with microgrids), and most likely intelligent.

Hence, *active distribution networks* are the distribution network to which renewable energy sources are connected. An active distribution network can be considered as an active network. Microgrids are *autonomous active networks*. The active network management is receiving considerable attention in the development of smart grid.

Also, in active distribution networks, the older-style voltage regulators were often designed for hardly a pure radial situation, that is, power flow is always from the same direction (from the substation). They may not work correctly, if power flow is from the opposite direction. For example, they could raise voltage during light load, creating higher voltage situation, or they could lower voltage during heavy load, creating low-voltage situation.

Therefore, it is necessary to use "bidirectional" voltage regulator controller to handle feeder reconfiguration. Feeder reconfiguration may become a more frequent occurrence due to load transferred to another feeder during service restoration, or due to having an ONR to reduce losses. In general, a distribution generation of sufficient size can reverse power flow.

15.6 VOLT/VAR CONTROL IN DISTRIBUTION NETWORKS

In general, it is agreed that the following are the three main approaches to VVC in the distribution systems:

1. Traditional approach
2. Supervisory control and data acquisition (SCADA) volt/var approach
3. IVVC optimization approach

15.6.1 TRADITIONAL APPROACH TO VOLT/VAR CONTROL IN THE DISTRIBUTION NETWORKS

In the traditional approach to VVC in the distribution networks, the process is controlled by individual, independent, and stand-alone volt/var regulating equipment, such as substation transformer's LTCs, by using line-voltage regulators, and by using fixed and switched capacitor banks.

Note that VVC is a fundamental operating requirement of all electric distribution systems. Its primary purpose is to maintain acceptable voltage at all points along the distribution feeder under all loading conditions, that is, under the full-load or light-load conditions.

Example 15.1

Using the traditional VVC approach, conceptually illustrate how to manage a typical distribution primary line (i.e., feeder) volt/var flows by using the traditional VVC approach. Use the individual, independent, stand-alone volt/var regulating equipment, such as substation transformer's LTCs, by using line-voltage regulators and fixed and switched capacitor banks.

Solution

Figure 15.3 illustrates how to manage volt/var flows related to a distribution primary line using the individual, independent, stand-alone volt/var regulating equipment under the *traditional volt/var control approach*. Note that the current/voltage sensors located on the pole tops send the "local" current/voltage measurements stand-alone controllers [19].

They, in turn, send out individual on/off control/command signal to the associated capacitor bank and/or voltage regulators, after using the substation transformer's LTCs. The process continues until appropriate voltage profile for the feeder is obtained. However, such traditional approach has the following limitations:

1. *Power factor correction/loss reduction*: Many traditional capacitor bank controllers have voltage control, that is, they switch on when voltage is low. This approach is good at maintaining acceptable voltage and has reactive power controllers, but it is expensive since it requires the addition of CTs. Also, the approach is good at power factor correction during peak-load times, but it may not come on at all during off-peak times. As a result, power factor is nearly unity during the peak-load times, but it is low during the off-peak times, causing higher electrical losses [10].
2. *Monitoring of switched capacitor bank performance*: It is well known that switch capacitor banks are often out of service due to blown-out fuses, etc. With the traditional approach, the switched capacitor bank could be out of service for extended periods without operator knowing it. This results in higher losses due to capacitor bank being out of service. As a result, it requires routine inspections which are costly.
3. *Voltage regulation problem when large distributed generation unit is connected*: As a result of connecting a large distributed generation unit, load current through voltage regulator will be reduced since the voltage regulator "thinks" the load is light on the feeder. Thus, the voltage regulator lowers its tap settings in order to avoid "light-load, high-voltage" condition. This, in turn, makes the actual "heavy-load, low-voltage" condition even worse.

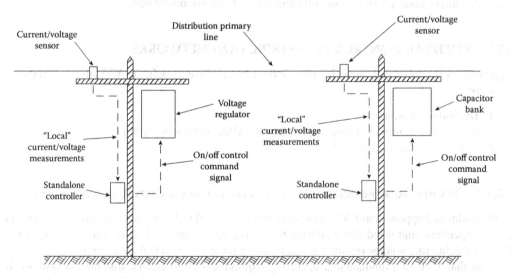

FIGURE 15.3 An illustration for how to manage volt/var flows related to a distribution primary line using the individual, independent, stand-alone volt/var regulating equipment under the *traditional* volt/var control approach.

These problems can be remedied by implementing DMS. Such system uses voltage regulators or transformers with LTCs that automatically raise or lower the voltage in response to changes in load. Also, capacitor banks are used to supply some of the reactive power that would otherwise be drawn from the supply substations.

However, today utilities are seeking to do more with VVC than just keeping voltage within the allowable limits. System optimization is an important part of the normal operating strategy under smart grid.

Especially in the future, as penetration of intermittent renewable generating resources increases, high-speed dynamic VVC will be essential in sustaining power quality and voltage stability on the distribution feeders. But the traditional approach has the aforementioned limitations.

15.6.2 SCADA Approach to Control Volt/Var in the Distribution Networks

Volt/var power apparatus is monitored and controlled by SCADA system. Such VVC is typically handled by two separate (independent) systems, that is, by var dispatch system that controls capacitor banks to improve power factor, reduce electrical losses, etc., and by *voltage control* system that controls LTCs and/or voltage regulators to reduce demand and/or energy consumption, which is also known as *conservation voltage reduction* (CVR)[1].

Operation of these systems is primarily *based on a stored set of predetermined* rules. For example, "if power factor is less than 0.95, than switch capacitor bank #1 off." The overall objective of var dispatch is to maintain power factor as close as to unity at the beginning of the feeder without causing leading factor. In summary, the *objectives of SCADA voltage control* are as follows:

1. Maintain acceptable voltage at all locations under all loading conditions.
2. Operate at as low as voltage as possible to reduce power consumption through CVR.

Example 15.2

Using the SCADA volt/var approach, conceptually illustrate how to manage a typical distribution primary line volt/var flows using var dispatch components. Use switched and fixed capacitor banks, capacitor bank control interface, communications facility, means of monitoring three-phase var flow at the substation, and master station running var dispatch software.

Solution

Figure 15.4 shows var dispatch components. Note that var dispatch processor contains rules for capacitor switching. There is a one-way communication link for capacitor bank control between the var dispatch processor and the capacitor bank controller. (Note that at the present time, the capacitor bank is de-energized by having the capacitor switch at off-position.)

Substation *remote terminal unit* (RTU) measures the real and reactive power at substation end of the feeder; accordingly, var dispatch processor sends commands to the capacitor bank controller. Var dispatch processor applies the rules to determine if the capacitor bank switching is needed. If the reactive power is below the threshold, there is no need for any action, as illustrated in Figure 15.5 (*Note that there is no communication between the radios.*).

---

[1] The U.S. utilities' average CVR factor is about 0.8. Here,

$$\text{CVR factor} = \frac{\Delta P}{\Delta V}$$

The CVR performed during peak load period can be viewed as demand (capacity) reduction. The resultant annual energy savings due to CVR can be calculated as

Annual energy savings = (Average load) (Number of hours per year) (%voltage reduction) × (CVR factor)

(Value of energy conservation) − (Loss of revenue form kWh sales)

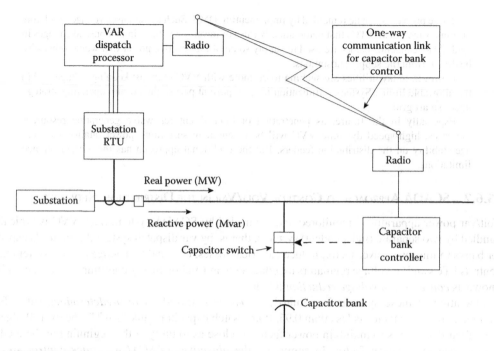

FIGURE 15.4   Var dispatch components of a SCADA system.

On the other hand, if the reactive power is above the threshold, action is required. Thus, var dispatch processor applies the rules to determine whether capacitor bank switching is needed.

The necessary communication is established between the radios and the capacitor bank controller sends a command to capacitor switch and it switches to the "on" position and the capacitor bank is energized, as illustrated in Figure 15.6.

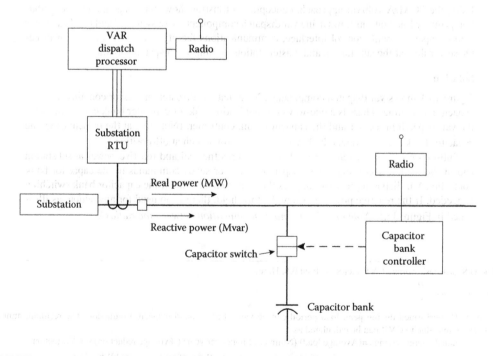

FIGURE 15.5   Var dispatch rules applied (and no action is required).

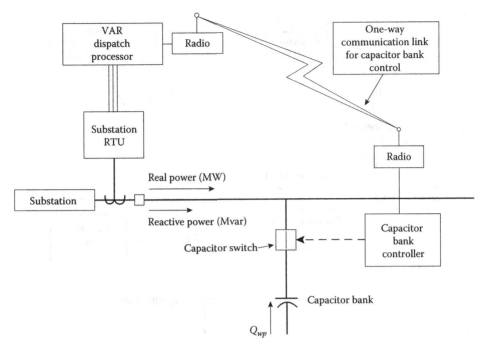

FIGURE 15.6 The capacitor bank is switched *"on."* (From Farhangi, H., The path of the smart grid, *IEEE Power & Energy Magazine*, 8(1), 18–28, ©2010 IEEE. With permission.)

Thus, the reactive power coming from the supplier is reduced. This change is detected by the substation RTUs. They measure real and reactive power at the substation end of the feeder. The var dispatch processor applies the rules for capacitor switching and sends a signal to the capacitor bank controller, which in turn de-energizes the capacitor bank by switching the capacitor switch to the "off" position, as illustrated in Figure 15.7.

At the substation end of the feeder, the var dispatch processor applies the rules for capacitor switching and sends a signal to the capacitor bank controller which, in turn, de-energized the capacitor bank by switching the capacitor switch to the "off" position, as illustrated in Figure 15.7.

Note that such application of SCADA system provides (1) self-monitoring, (2) operator over-rode capability, and (3) some improvement inefficiency. Here, the objectives of SCADA voltage control are (1) maintaining acceptable voltage at all locations under all loading conditions, and (2) operating at as low as voltage as possible to reduce power consumption, that is, CVR. However, SCADA cannot accomplish the following:

1. It does not adapt to changing feeder configurations (i.e., the rules are fixed in advance).
2. It does not adapt to varying operating needs (i.e., the rules are fixed in advance).
3. The overall efficiency with SCADA is improved with respect to the traditional approach, but it is not necessarily optimal under all conditions.
4. The operations of var/volt devices are not coordinated. It does not adapt well to the presence of modern grid devices such as distributed generation.

15.6.3 INTEGRATED VOLT/VAR CONTROL OPTIMIZATION

What is needed is an IVVC optimization approach (i.e., centralized approach) that develops and executes a coordinated "optimal" switching plan for *all* voltage control devices based on an optimal power flow program to decide the plans for action.

In the process achieves utility-specific objectives as well which include (1) minimizing power demand in terms of total customer demand and distribution system power losses, (2) minimizing

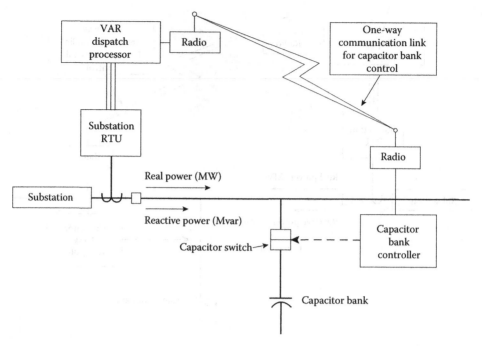

FIGURE 15.7 Change in reactive power detected, and the capacitor bank is switched "*off.*" (From Farhangi, H., The path of the smart grid, *IEEE Power & Energy Magazine*, 8(1), 18–28, ©2010 IEEE. With permission.)

"wear and tear" on the control equipment, and (3) maximizing revenue, which is the difference between energy sales and energy prime cost.

Example 15.3

Consider the system given in Example 15.2, and conceptually design an integrated volt/var optimization (IVVO) system configuration that develops a coordinated "optimal" switching plan for all voltage control equipment and executes the plan.

Solution

Figure 15.8 shows the IVVO system configuration. All the inputs from essential devices are fed to the VVC regulation coordination algorithm. They are provided through the communication links between the devices and the coordination algorithm.

Here, the volt/var regulation coordination algorithm manages tap changer settings, inverter, and rotating machine var levels, and capacitors to regulate voltage, reduce losses, conserve energy, and system resources. The AMI, meter distribution management system (MDMS), and line switch provides inputs to distribution SCADA.

The inputs of bank voltages and status, and switch control are also provided by the switched capacitor banks and line-voltage regulators. IVVC optimizing engine develops a coordinated "optimal" switching plan for *all* voltage control devices and executes the plan. IVVC regulates an accurate, up-to-date electrical model based on control of substation and feeder devices.

The AMI provides information on voltage feedback and accurate load data. The line switch provides information on switch status. The line-voltage regulator provides inputs on monitor and control tap position, and measures load voltage and load. The bank voltage and status, switch control of substation capacitor bank, as well as substation transformer with tap changer under load (TLC) are also used as inputs to the SCADA system.

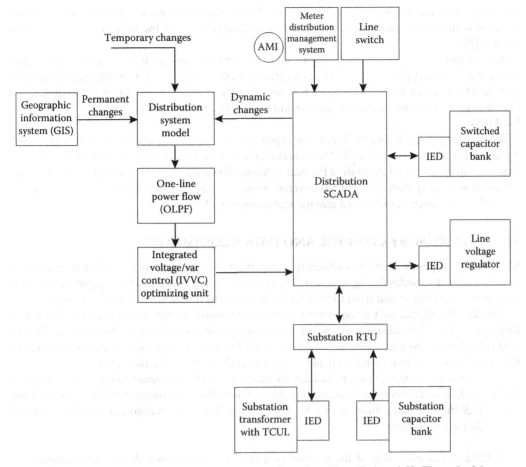

FIGURE 15.8 Volt/var optimization (VVO) system configuration. (From Farhangi, H., The path of the smart grid, *IEEE Power & Energy Magazine*, 8(1), 18–28, ©2010 IEEE. With permission.)

Also, it is required that the substation transformer (with TLC) monitor and control tap position, and measure load voltage and load. Temporary changes (i.e., cuts, jumpers, and manual switching) as well as permanent asset changes (i.e., line extension and/or reconductoring) are inputted into distribution system model. Also provided are the real-time updates of dynamic changes. The necessary power flow results are provided by OLP, which calculates losses, voltage profile, etc. The IVVC determines optimal set of control actions to achieve a desired objective and provides the optimal switching plan to SCADA. In general, the IVVC has the following benefits:

1. When the network reconfiguration takes place, the dynamic model upgrades automatically.
2. The VVC actions are coordinated.
3. The system can model the effects of distributed generation and other modern grid elements.
4. It produces the "optimal" results.
5. It accommodates varying operating objectives, depending on the present need.

15.7 EXISTING ELECTRIC POWER GRID

The present electricity grid is the result of fast urbanization and infrastructure development. However, the growth of the electric power system has been influenced by economic, political, and geographic factors that are utility specific. Nevertheless, the basic topology of the present power system has remained the same. As it can be easily observed, the basic topology is a vertical one [8].

Due to lack of proper communications and real data, the system is overengineered to meet maximum expected peak demand of its aggregated customer load. Since the peak demand is infrequent due to its nature, the system is intrinsically vulnerable due to increase in demand for electric energy and decreased investments in plant and equipment, resulting in extensive blackouts.

In order to prevent it and maintain any expensive upstream plant and equipment without any damage, the utilities have established various command-and-control functions, such as SCADA systems, as will be discussed in detail in the next sections. However, the application of the SCADA has remained not totally effective, and has covered about 15%–20% of distribution system. Primarily, the SCADA has been implemented into the transmission system.

15.8 SUPERVISORY CONTROL AND DATA ACQUISITION

SCADA is the equipment and procedures for controlling one or more remote stations from a master control station. It includes the digital control equipment, sensing and telemetry equipment, and two-way communications to and from the master stations and the remotely controlled stations.

The SCADA digital control equipment includes the control computers and terminals for data display and entry. The sensing and telemetry equipment includes the sensors, digital-to-analog and analog-to-digital converters, actuators, and relays used at the remote station to sense operating and alarm conditions and to remotely activate equipment such as circuit breakers [10].

The communications equipment includes the modems (modulator/demodulator) for transmitting the digital data, and the communications link (radio, phone line, and microwave link, or power line). Figure 15.9 shows a block diagram of a SCADA system. Typical functions that can be performed by SCADA are as follows:

1. Control and indication of the position of a two or three position device, for example, a motor driven switch or a circuit breaker.
2. State indication without control, for example, transformer fans on or off.
3. Control without indication, for example, capacitors switched in or out.
4. Set point control of remote control station, for example, nominal voltage for an automatic tap changer.
5. Alarm sensing, for example, fire or the performance of a non-commanded function.
6. Permit operators to initiate operations at remote stations from a central control station.
7. Initiation and recognition of sequences of events, for example, routing power around a bad transformer by opening and closing circuit breakers, or sectionalizing a bus with a fault on it.
8. Data acquisition from metering equipment, usually via analog/digital converter and digital communication link.

Today, in this country, all routine substation functions are remotely controlled. For example, a complete SCADA system can perform the following substation functions:

1. Automatic bus sectionalizing
2. Automatic reclosing after a fault
3. Synchronous check
4. Protection of equipment in a substation
5. Fault reporting
6. Transformer load balancing

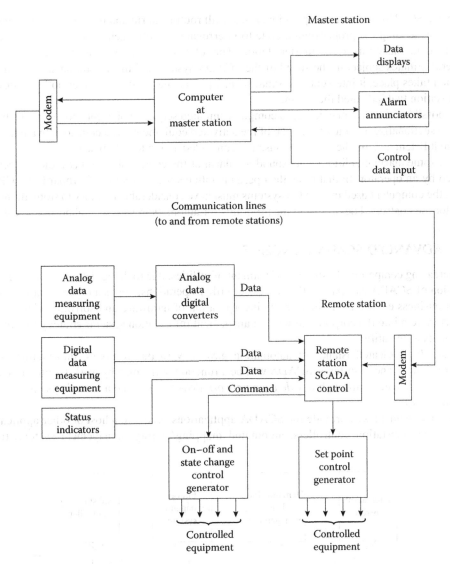

FIGURE 15.9 Supervisory control and data acquisition (SCADA). (From Gönen, T., *Electric Power Distribution System Engineering*, 2nd Ed., CRC Press, Boca Raton, FL, 2008.)

7. Voltage and reactive power control
8. Equipment condition monitoring
9. Data acquisition
10. Status monitoring
11. Data logging

All SCADA systems have two-way data and voice communication between the master and the remote stations. Modems at the sending and receiving ends modulate, that is, put information on the carrier frequency, and demodulate, that is, remove information from the carrier, respectively.

Here, digital codes are utilized for such information exchange with various error detection schemes to assure that all data are received correctly. The RTU properly codes remote station information into the proper digital form for the modem to transmit, and to convert the signals received from the master into the proper form for each piece of remote equipment.

When a SCADA system is in operation, it scans all routine alarm and monitoring functions periodically by sending the proper digital code to interrogate, or poll, each device. The polled device sends its data and status to the master station. The total scan time for a substation might be 30 s to several minutes subject to the speed of the SCADA system and the substation size. If an alarm condition takes place, it interrupts a normal scan. Upon an alarm, the computer polls the device at the substation that indicated the alarm.

It is possible for an alarm to trigger a computer-initiated sequence of events, for example, breaker action to sectionalize a faulted bus. Each of the activated equipment has a code to activate it, that is, to make it listen, and another code to cause the controlled action to take place.

Also, some alarm conditions may sound an alarm at the control station that indicates action is required by an operator. In that case, the operator initiates the action via a keyboard or a CRT. Of course, the computers used in SCADA systems must have considerable memory to store all the data, codes for the controlled devices, and the programs for automatic response to abnormal events.

15.9 ADVANCED SCADA CONCEPTS

The increasing competitive business environment of utilities, due to deregulation, is causing a reexamination of SCADA as a part of the process of utility operations, not as a process unto itself. The present business environment dictates the incorporation of hardware and software of the modern SCADA system into the corporation-wide, management information systems strategy to maximize the benefits to the utility.

Today, the dedicated islands of automation gave way to the corporate information system. Tomorrow, in advanced systems, SCADA will be a function performed by workstation-based applications, interconnected through a *wide area network* (WAN) to create a virtual system, as shown in Figure 15.10.

This arrangement will provide the SCADA applications access to a host of other applications, for example, substation controllers, automated mapping/facility management system, trouble

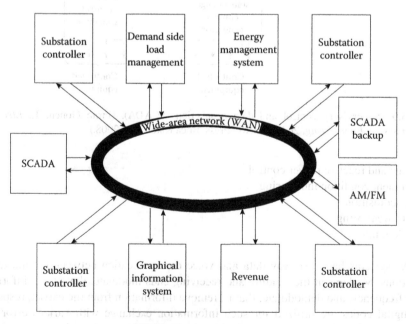

FIGURE 15.10 Supervisory control and data acquisition (SCADA) in a virtual system established by a WAN. (From Gönen, T., *Electric Power Distribution System Engineering*, 2nd Ed., CRC Press, Boca Raton, FL, 2008.)

call analysis, crew dispatching, and demand-side LM. The WAN will also provide the traditional link between the utility's energy management system (EMS) and SCADA processors. The workstation-based applications will also provide for flexible expansion and economic system reconfiguration.

Also, unlike the centralized database of most exiting SCADA systems, the advanced SCADA system database will exist in dynamic pieces that are distributed throughout the network. Modifications to any of interconnected elements will be immediately available to all users, including the SCADA system. SCADA will have to become a more involved partner in the process of economic delivery and maintained quality of service to the end user.

In most applications today, SCADA and the *energy management system* (EMS) operate only on the transmission and generation sides of the system. In the future, economic dispatch algorithms will include demand-side (load) management and voltage control/reduction solutions. The control and its hardware and software resources will cease to exist.

15.10   SUBSTATION CONTROLLERS

In the future, RTUs will not only provide station telemetry and control to the master station, but also will provide other primary functions such as system protection, local operation, *graphical user interface* (GUI), and data gathering/concentration from other subsystems.

Therefore, the future's RTUs will evolve into a class of devices that performs multiple substation control, protection, and operation functions. Besides these functions, the substation controller also develops and processes data required by the SCADA master, and it processes control commands and messages received from the SCADA master.

The substation controller will provide a gateway function to process and transmit data from the substation to the WAN. The substation controller is basically a computer system designed to operate

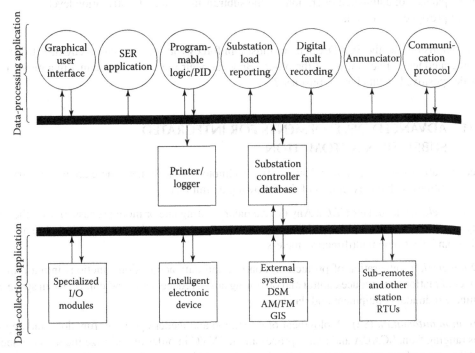

FIGURE 15.11   Substation controller. (From Gönen, T., *Electric Power Distribution System Engineering*, 2nd Ed., CRC Press, Boca Raton, FL, 2008.)

in a substation environment. As shown in Figure 15.11, it has hardware modules and software in terms of

1. *Data processing applications*: These software applications provide various users access to the data of the substation controller in order to provide instructions and programming to the substation controller, collect data from the substation controller, and perform the necessary functions.
2. *Data collection applications*: These software applications provide the access to other systems and components that have data elements necessary for the substation controller to perform its functions.
3. *Control database*: All data resides in a single location, whether from a data processing application, data collection application, or derived from the substation controller itself.

Therefore, the substation controller is a system that is made up of many different types of hardware and software components and may not even be in a single location. Here, RTU may exist only as a software application within the substation controller system. Substation controllers will make all data available on WANS. They will eliminate separate stand-alone systems and thus provide greater cost savings to the utility company.

According to Sciaca and Block [17], the SCADA planner must look beyond the traditional roles of SCADA. For example, the planner must consider the following issues:

1. Reduction of substation design and construction costs
2. Reduction of substation operating costs
3. Overall lowering of power system operating costs
4. Development of information for non-SCADA functions
5. Utilization of existing resources and company standard for hardware, software, and database generation
6. Expansion of automated operations at the subtransmission and distribution levels
7. Improved customer relations

To accomplish these, the SCADA planner must join forces with the substation engineer to become an integrated team. Each must ask the other "How can your requirements be met in a manner that provides positive benefits for my business?"

15.11 ADVANCED DEVELOPMENTS FOR INTEGRATED SUBSTATION AUTOMATION

Since the substation integration and automation technology is fairly new, there are no industry standard definitions with the exception of the following definitions:

Intelligent electronic device (IED): Any device incorporating one or more processors with the capability to receive or send data/control from or to an external source, for example, digital relays, controllers, and electronic multifunction meters.

IED integration: Integration of protection, control, and data acquisition functions into a minimal number of platforms to reduce capital and operating costs, reduce panel and control room space, and eliminate redundant equipment and databases.

Substation automation (SA): Deployment of substation and feeder operating functions and applications ranging from SCADA and alarm processing to IVVC in order to optimize the management of capital assets and enhance operation and maintenance efficiencies with minimal human intervention.

Open systems: A computer system that embodies supplier-independent standards so that software can be applied on many different platforms and can interoperate with other applications on local and remote systems.

A SA project prior to the 1990s typically involved three major functional areas: SCADA; plus station control, metering, and display; and plus protection.

In recent years, the utility industry has started using IEDs in their systems. These IEDs provided additional functions and features, including self-check and diagnostics, communication interfaces, the ability to store historical data, and integrated RTU I/O. The IED also enabled redundant equipment to be eliminated, as multiple functions were integrated into a single piece of equipment. For example, when interfaced to the potential transformers and current transformers of an individual circuit, the IED could simultaneously handle protection, metering, and remote control.

As more and more traditional SA functions become integrated into single piece of equipment, the definition of IED began to expand. The term is now applied to any microprocessor-based device with a communications port, and therefore includes protection relays, meters, RTUs, *programmable logic controllers* (PLCs), load survey and operator indicating meters, digital fault recorders, revenue meters, and power equipment controllers of various types.

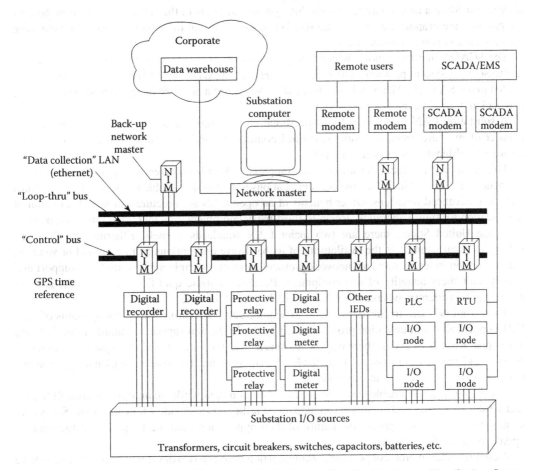

FIGURE 15.12 Configuration of SA system. (From Gönen, T., *Electric Power Distribution System Engineering*, 2nd Ed., CRC Press, Boca Raton, FL, 2008.)

The IED can thus be considered as the first level of automation integration. Additional economies of scale can be obtained by connecting all of the IEDs into a single integrated substation control system. The use of a fully integrated control system can lead to further streamlining of redundant equipment, as well as reduced costs for wiring, communications, maintenance, and operation, and improved power quality and reliability.

However, the process of implementation has been slow, largely because hardware interfaces and protocols for IEDs are not standardized. Protocols are as numerous as the vendors, and in fact more so, since products even from same end or often have different protocols. Figure 15.12 shows the configuration of an SA system.

The electric utility SA system uses a variety of devices integrated into a functional package by a communications technology for the purpose of monitoring and controlling the substation. Common communications connections include utility operations centers, finance offices, and engineering centers.

Communications for other users is usually through a bridge, gateway, or processor. A library of standard symbols should be used to represent the substation power apparatus on graphical displays. In fact, this library should be established and used in all substations and coordinated with other systems in the utility, such as distribution SCADA system, the EMS, the *geographical information system* (GIS), and the trouble call management system [7].

According to McDonald [15], the *global positioning* satellite (GPS) clock time reference is shown, providing a time reference for the SA system and IEDs in the substation. The host processor provides the graphical user interface (GUI) and the historical information system for achieving operational and nonoperational data.

The SCADA interface knows which SA system points are sent to the SCADA system, as well as the SCADA system protocol. The *local area network* (LAN) enabled IEDs can be directly connected to the SA LAN. The non-LAN-enabled IEDs require a network interface module (NIM) for protocol and physical interface conversion.

A substation LAN is typically high speed and extends into the switchyard, which speeds the transfer of measurements, indications, control commands, and configuration and historical data between intelligent devices at the site [11].

This architecture reduces the amount and complexity of cabling currently required between intelligent devices. Also, it increases the communications bandwidth available to support faster updates and more advanced functions. Other benefits of an open LAN architecture can include creation of a foundation for future upgrades, access to third-party equipment, and increased interoperability.

In the United States, there are two major LAN standards, namely, Ethernet and Profibus. Ethernet's great strength is the availability of its hardware and options from a myriad of vendors, not to mention industry-standard network-protocol support, multiple application-layer support and quality, and sheer quantity of test equipment. Because of these qualifications, Ethernet is more popular in this country, whereas Profibus is widely used in Europe.

There are interfaces to substation IEDs to acquire data, determine the operating status of each IED, support all communication protocols used by the IEDs, and support standard protocols being developed. Besides SCADA, there may be an interface to the EMS that allows system operators to monitor and control each substation and the EMS to receive data from the substation integration and automation system at different time intervals.

The data warehouse enables users to access substation data while maintaining a firewall to protect substation control and operation functions. The utility has to decide who will use the SA system data, the type of data required, the nature of their application, and the frequency of the data, or update, required for each user.

A communication protocol permits communication between two devices. The devices must have the same protocol and its version implemented. Any protocol differences will result in communication errors. The substation integration and automation architecture must permit devices from different supplies to communicate employing an industry standard protocol. The primary capability of an IED is its stand-alone capability, for example, protecting the power system for a relay IED. Its secondary capability is its integration capabilities, such as its physical interface, for example, RS-232, RS-485, and Ethernet, and its communication protocol, for example, Modbus, Modbus Plus, DNP3, UCA2, and MMS.

To get all IEDs and their heterogeneous protocols onto a common substation LAN and platform, the gateway approach is best. The gateway will act not only as an interface between the local network physical layer and the RS-232/RS-485 ports found on the IEDs, but also as a protocol converter, translating the IED's native protocol (like SEL, DNP3, or Modbus) into the protocol standard found on the substation's local network.

Two approaches can be used when using gateways to interface to the substation network. In one, a single low-cost gateway is used for each IED, and in the other, a multi-ported gateway interfaces with multiple IEDs. Which approach is more economical will depend on where the intelligent devices are located. If the IEDs are clustered in a central location, then the multi-ported gateway is certainly better.

The design of the substation integration and automation for new substations is easier than the one for existing substations. The new substation will typically have many IEDs for different functions, and the majority of operational data for the SCADA system will come from these IEDs. The IEDs will be integrated with digital two-way communications.

Typically, there are no conventional RTU in new substations. The RTU functionality is addressed using IEDs and PLCs and an integration network, using digital communications. In existing substations, there are several alternative approaches, depending on whether the substation has a conventional RTU installed.

The utility has three choices for their conventional substation RTUs: (1) integrate RTU with IEDs; (2) integrate RTU as another substation IED; and (3) retire RTU and use IEDs and PLCs, as with a new substation.

The environment of a substation is challenging for SA equipment. Substation control buildings are seldom heated or air-conditioned. Ambient temperatures can range from well below freezing to above $104°F$ ($40°C$). Metal-clad switchyard substations can reach ambient temperatures in excess of $140°F$ ($60°C$). Temperature changes stress the stability of measuring components in IEDs, RTUs, and transducers. In many environments, self-contained heating or air-conditioning may be recommended.

In summary, the integrated substation control system architecture (which is made up of IEDs, LANs, protocols, GUIs, and substation computers) is the foundation of the automated substation. However, the application building blocks consisting of operating and maintenance software are what produce the really substantial savings that can justify investment in an integrated substation control system.

15.12 EVOLUTION OF SMART GRID

As illustrated in Figure 15.13, the metering side of the distribution system has received most of the attention in terms of the recent infrastructure investments. The earlier applications in the distribution system included the automated meter reading (AMR). The AMR technology has facilitated utilities with the ability to read the customer consumption meters, alarms, and status remotely.

As shown in Figure 15.14, even though AMR technology has received a substantial attention initially, in time, it became clear that AMR is not the answer for the demand-side management, primarily due to its one-way communication nature. It simply reads the customers' meter data [3].

It does not permit the transition to the smart grid where extensive control at all levels is essential [4]. Thus, AMR technology applications become extinct. It was replaced by AMI. This system provides utility companies with a two-way communication system to the customers' meters, and the ability to modify customers' service-level parameters.

Hence, by using AMI, utilities can reach to their goals in LM and increased revenues. With AMI technology, power companies not only can collect instantaneous information about individual

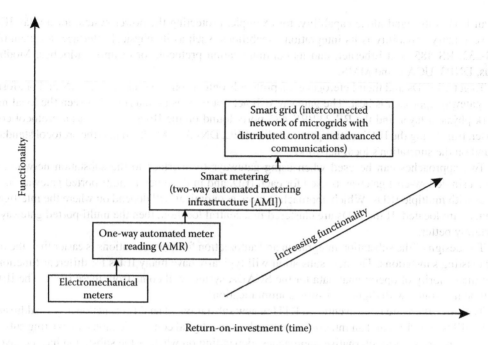

FIGURE 15.13 The evolution of smart grid as a function of return on investment versus time.

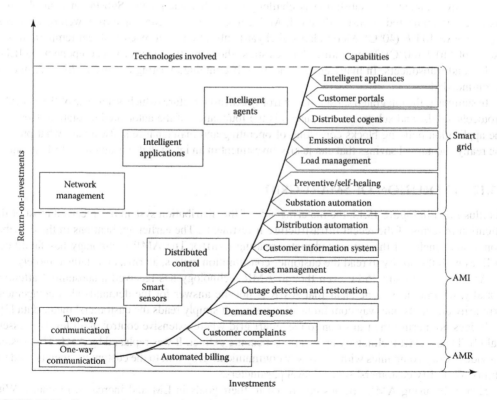

FIGURE 15.14 Return on investments for a smart grid.

and aggregated demand but can also modify the energy consumption, as well as implement their cost-cutting measures. As said by Farhangi [4], the emergence of AMI started a concerted move by stakeholders to further refine the ever-changing concepts around the smart grid.

As a next step, according to Farhangi [4], the smart grid requires to leverage the AMI infrastructure and implement its distribution command-control strategies over the AMI backbone. The penetrating control and intelligent that are properties of the smart grid has to be located across all geographic areas, as well as components, and functions of the power system. The distinguished three elements mentioned earlier, that is, geographic areas, components, and functions determine the topology of the smart grid and its components.

Again, it is important to point out that smart distribution systems are essential part of the smart grid; here are the necessary steps to establish a smart distribution system:

Step 1: Design information models based on overall requirements of a smart distribution system.

Step 2: Establish substation data integration and associated applications, such as traditional SCADA and fault location.

Step 3: Add feeder automation for selected applications, for example, automatic reconfiguration and VVC.

Step 4: Add advanced metering integration, for example, state estimation and outage management.

Step 5: Add integrated applications such as energy management optimization, risk assessment, and advanced equipment diagnostics

Step 6: Add interface to distributed energy resources as well as islanding, advanced energy management and optimization, and PQ management capabilities

The necessary next steps to achieve the volt/var optimization have been illustrated in Figure 15.15. The volt/var regulation coordination algorithm manages tap changer settings, inverter and rotating machine var levels, and capacitors to regulate voltage, reduce losses, and conserve energy and

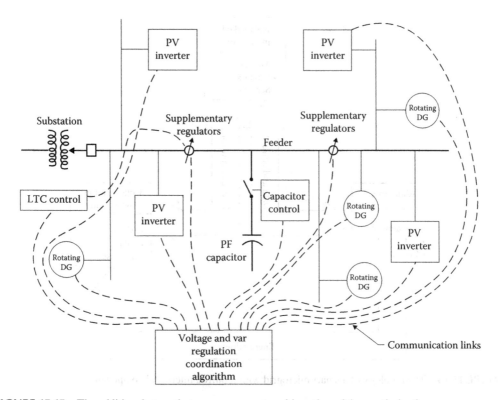

FIGURE 15.15 The additional steps that are necessary to achieve the volt/var optimization.

system resources. The necessary links between the algorithm and the individual components have been indicated in the figure[13].

15.13 SMART MICROGRIDS

As succinctly put by Farhangi [4], the smart grid is the collection of all technologies, concepts, topologies, and approaches that permit to maintain hierarchies of generation, transmission, and distribution to be replaced with an end-to-end, organically intelligent, fully integrated environment where the business processes, objectives, and needs of all stakeholders are supported by the efficient exchange of data, services, and transactions.

A smart grid, is hence, defined as a grid that accommodates a wide variety of generation options, for example, central, distributed, intermittent, and mobile. It provides customers with the ability to interact with the EMS to adjust their energy use and reduce their energy costs.

A smart grid is also a self-healing system. It foresees the forthcoming failures and takes the necessary corrective actions to avoid or mitigate system problems. It uses the IT to continuously optimize the employment of its capital assets while minimizing operational and maintenance costs [4].

However, the smart grid should not be seen as a replacement for the present electric power grid, but a complement to it. Thus, it can coexist with the present electric power grid, adding to its capabilities, functionalities, and capacities by means of evolutionary path. This dictates a topology for the smart grid that permits for organic growth, the inclusion of forward-looking technologies, and full backward compatibility with the present systems [4].

The smart grid can be also defined as the ad hoc integration of complementary components, subsystems, and functions under the extensive control of a highly intelligent and distributed command-and-control system.

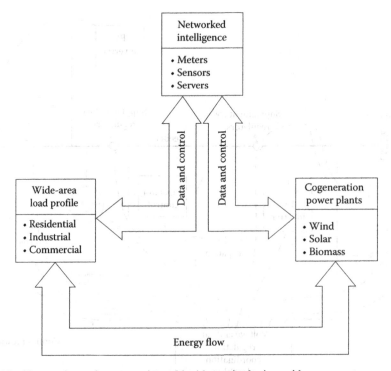

FIGURE 15.16 The topology of a smart microgrid with required microgrid components.

Furthermore, the organic growth and evolution of the smart grid is achieved by the inclusion of intelligent microgrids. Here, the microgrid is defined as interconnected networks of distributed energy systems, including loads and resources, that can function whether they are connected to or separated from the electric power grid [4].

15.14 TOPOLOGY OF A MICROGRID

As discussed in Section 11.8, a small microgrid network can operate in both grid-connected and islanded modes. The topology of a smart microgrid is illustrated in Figure 15.16. As said by Farhangi [4], a small grid integrates the following components:

1. It includes power plants capable of meeting local demand as well as feeding the unused energy back to the electric power grid. They are known as cogenerators and often use renewable resources, such as sun, wind, and biomass. Some microgrids have CHP thermal power plants that are capable of recovering the waste heat in terms of district cooling or heating in the vicinity of the power plant.
2. It employs local and distributed power-storage capability to smooth out the intermittent performance of renewable energy sources.
3. It services a variety of loads, including residential, commercial, and industrial loads.
4. It has communication infrastructure that facilitates system components to exchange information and commands reliably and securely.
5. It employs smart meters and sensors capable of measuring a number of consumption parameters (e.g., real and reactive powers, voltage, current, and demand) with acceptable accuracy.
6. It includes an intelligent core, made of integrated networking, computing, and communication infrastructure elements, that appears to users in terms of energy management applications that permit command and control on all network nodes.
7. It includes smart terminations, loads, and appliances capable of communicating their status and accepting commands to adjust and control their performance and service levels according to consumer and/or utility requirements.

15.15 TOPOLOGY OF A SMART GRID

Farhangi [4] predicts that the smart grid of the future will be interconnected through dedicated highways for power exchange, data, and commands, as shown in Figure 15.17. But, it is expected that not all microgrids will have the same capabilities and needs. It will be subject to the load diversity, geography, economics, and the mix of the primary energy resources.

The necessary AMI systems now being established will facilitate the evolution of the smart grid. However, due to the high costs involved, it is foreseeable that the new and the old grids may coexist for some time, eventually though it is expected that the new system will replace the old grid.

Thus, during the transition period, there will be a hybrid system. The new power grid will appear as a system of organically integrated collection of smart grids with extensive command-and-control functions implemented at all levels [12,13].

15.16 STANDARDS OF SMART GRIDS

It is very possible that some substantial problems to emerge when distinctively different systems, components, and functions begin to be integrated as part of a distributed command-and-control system of a smart grid.

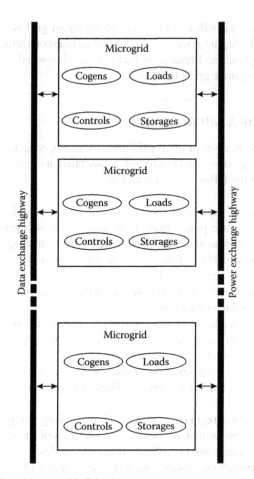

FIGURE 15.17 The envisioned smart grid of the future.

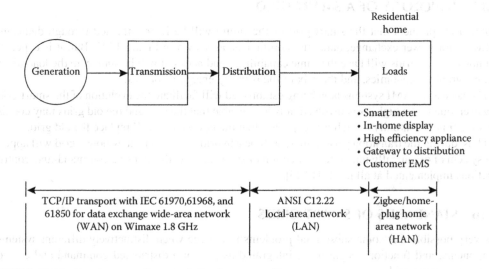

FIGURE 15.18 Development of standards for the smart grid.

A part of the problem is the fact that at the present time, there are no commonly accepted interfaces, messaging and control protocols, and standards that would be abode by to ensure a common communication vocabulary among system components of a smart grid.

In order to assist the development of the required standards, the power industry is slowly adopting different terminologies for the purpose of segmentation of the command-and-control layers of the smart grid. The examples of this include HAN, LAN, and WAN.

HAN stands for "home area network"; it is used to identify the network of communicating loads, appliances, and sensors beyond the smart meter and within the customer's property.

The LAN denotes the local area network. It is used to identify the network of integrated smart meters, field components, and gateways that constitute the logical network between distribution substations and customer's premises.

Finally, WAN is used to identify the network of upstream utility assets, including power plants, distribution storage, substations, etc. As shown in Figure 15.18, the interface between the WAN and LAN interface is provided by smart meters.

In the United States, the U.S. National Institute of Standards and Technology (NIST) is leading the effort for the standardization for smart grid [18]. In Europe and other places, similar efforts indicate the need for the development of common information model (CIM) to enable vertical and lateral integration of applications and functions within the smart grid.

Out of the proposed standards, IEC 61850 and its related standards, appear to be favorites for WAN data communication, supporting TCP/IP, among other protocols, over fiber or a 1.8 GHz WiMax.

In North America, ANSI C12.12, and its related standards, is considered as the favorite LAN standard, facilitating a new generation of smart meters capable of communicating substation gateways over a numerous wireless technologies.

Also, the European Community is pushing for the development of the AMI standard for Europe, replacing the aging DLMS/COSEM standard. Thus, it is pushing for efforts to develop a European counterpart for ANSI-C12.22.

It appears that ZigBee with Smart Energy Profile is the favorite for HAN, partially due to the lack of initiatives by the home appliance manufacturers.

15.17 EXISTING CHALLENGES TO THE APPLICATION OF THE CONCEPT OF SMART GRIDS

It is estimated that the electric power grid will make a transition from an electromechanically controlled system to an electronically controlled network within the next two decades. According to Amin and Wollenberg [1], there are some fundamental challenges to achieve this transition, namely,

1. The lack of transmission capacity to meet the substantially increasing loads
2. The difficulties of grid operation in a competitive market environment
3. The redefinition of power system planning and operation in the competitive era
4. The determination of the optimum type, mix, and placement of sensing, communication, and control hardware
5. The coordination of centralized and decentralized control

Smart grids are not really about doing things a lot differently than the way they are being done today. Instead, they are about doing *more* of what is being done, that is, sharing communication, infrastructures, filling in product gaps, and leveraging existing technologies to a greater extent while driving a higher level of integration to realize the synergies across enterprise integration.

A smart grid is not an off-the-shelf product or something that can be installed and turned on the next day. Rather, it is an integrated solution of technologies driving incremental benefits in capital expenditures, operation and maintenance expenditures, and customer and societal benefits.

A well-designed smart-grid imitative build on the existing infrastructure provides a greater level of integration at the enterprise level, and has a long-term focus. It is definitely a onetime solution but a change in how utilities look at a set of technologies that can enable both strategic and operational processes. It is the means to leverage benefits across applications and remove the barriers that are created by the past company practices [16].

REFERENCES

1. Amin, M. and Wollenberg, B. F., Toward a smart grid-power delivery for the 21st century, *IEEE Power & Energy Magazine*, 3(5), September/October 2005, 34–41.
2. Bricker, S., Rubin, L., and Gönen, T., Substation automation techniques and advantages, *IEEE Computer Applications in Power*, 14(3), July 2001, 31–37.
3. Farhangi, H., Intelligent microgrid research at BCIT, in *Proceedings of the IEEE Electric Power System Conference (EPEC'08)*, Vancouver, British Columbia, Canada, October 2008, pp. 1–7.
4. Farhangi, H., The path of the smart grid, *IEEE Power & Energy Magazine*, 8(1), January/February 2010, 18–28.
5. FERC assessment of demand response and advanced metering, 2007 Staff Report, September 2007. [On line.] Available: http://www.ferc.gov/legal/staff-reports/09-07-demand-response.pdf
6. Gellings, C. W., *The Smart Grid*, Boca Raton, FL: CRC Press, 2009.
7. Gnadt, P. A. and Lawler, J. S., *Automating Electric Utility Distribution Systems*, Englewood Cliffs, NJ: Prentice Hall, 1990.
8. Gönen, T., *Modern Power System Analysis*, New York: Wiley, 1988.
9. Gönen, T., *Engineering Economy for Engineering Managers: With Computer Applications*, New York: Wiley, 1990.
10. Gönen, T., *Electric Power Distribution System Engineering*, 2nd Ed., Boca Raton, FL: CRC Press, 2008.
11. Gönen, T., *Electric Power Transmission System Engineering*, 2nd Ed., Boca Raton, FL: CRC Press, 2009.
12. http://www.oe.energy.gov/DocumentsandMedia/SGSRMain_lowres.pdf
13. IEEE PES Power & Energy Society, *Smart Distribution Systems Tutorial*, IEEE PES General Meeting, Minneapolis, MN, July 2110.
14. Kintner-Meyer, M., Schneider, K., and Pratt, R., Impacts assessment of plug-in hybrid vehicles on electric utilities and regional U.S. power grids. [On line.] Available: http://www.pnl.gov/energy/eed/etd/pdfs/phev_feasibility_analysis_combined.pdf
15. McDonald, D. J., Substation integration and automation, in *Electric Power Substation Engineering*, Chapter 7, Boca Raton, FL: CRC Press, 2003.
16. More, D. and McDonnell, D., Smart grid meets distribution utility reality, *Electric, Light, Power*, March 2007, 1–6.
17. Sciaca, S. C. and Block, W. R., Advanced SCADA concepts, *IEEE Computer Applications in Power*, 8(1), January 1995, 23–28.
18. U.S. National institute of standards and technology, NIST framework and roadmap for smart grid interoperability standards, Release 1.0, September 2009, Washington, DC.
19. IEEE PES, "Smart Distribution Systems", a tutorial, IEEE PES General Meeting, July 20, 2010, Mineapolis, Minnesota.

Appendix A: Brief Review of Phasors

A.1 INTRODUCTION

The instantaneous value of a sinusoidally varying voltage can be expressed as

$$v(t) = V_m \sin \omega t \tag{A.1}$$

where

V_m represents amplitude (or maximum value) of the voltage
ωt represents argument
ω represents radian frequency (or angular frequency)

Since the sine wave is periodic, the function repeats itself every 2π radians. Its period T is 2π radians and its frequency f is $1/T$ hertz. Thus, the radian frequency can be expressed as $\omega = 2\pi f$. The voltage given by Equation A.1 can be expressed as a cosine wave as

$$v(t) = V_m \cos(\omega t - 90°) \tag{A.2}$$

A more general form of the sinusoid is

$$v(t) = V_m \cos(\omega t + \phi) \tag{A.3}$$

where ϕ is the phase angle.

Euler's identity states that

$$e^{j\phi} = \cos\phi + j\sin\phi \tag{A.4}$$

Therefore, Equation A.3 can be expressed as

$$
\begin{aligned}
v(t) &= \Re e[V_m \cos(\omega t + \phi) + jV_m \sin(\omega t + \phi)] \\
&= \Re e\left[V_m e^{j(\omega t + \phi)} \right] \\
&= \Re e\left[V_m e^{j(\omega t + \phi)} \right] \\
&= \sqrt{2}\,\Re e\left[V_m e^{j\omega t} \right]
\end{aligned}
\tag{A.5}
$$

Also, by using the definition of the effective (i.e., root-mean-square [rms]) value of the voltage,

$$V = \frac{V_m}{\sqrt{2}} e^{j\omega t} \tag{A.6}$$

Therefore,

$$v(t) = \sqrt{2}\Re e\left[Ve^{j\omega t}\right] \tag{A.7}$$

Thus, the complex amplitude of the sinusoid given by Equation A.3 can be expressed in exponential form as

$$V = Ve^{j\phi} \tag{A.8}$$

or in polar form as

$$V = |V|\angle\phi \tag{A.9}$$

or

$$V = V\angle\phi \tag{A.10}$$

or in rectangular form as

$$V = V(\cos\phi + j\sin\phi) \tag{A.11}$$

The complex quantity, given by Equation 1.10, is also called a *phasor* (or sometimes, erroneously called a *vector*). Anderson (1993) defines a phasor as *a complex number which is related to the time domain sinusoidal quantity by the following expression*:

$$a(t) = \Re e\left(\sqrt{2}Ae^{j\omega t}\right) \tag{A.12}$$

If we express A in terms of its magnitude $|A|$ and angle α, we have

$$A = |A|e^{j\alpha}$$

and

$$\alpha(t) = \Re e\left(\sqrt{2}|A|e^{j(\omega t+\alpha)} = \sqrt{2}|A|\cos(\omega t+\alpha)\right) \tag{A.13}$$

Thus, Equations A.12 and A.13 convert the rms phasor (complex) quantity to the actual time domain variable. For example, assume that a sinusoidal voltage of

$$v(t) = V_m\cos(\omega t + \phi) \tag{A.14}$$

is applied to a circuit with an impedance

$$Z = |Z|e^{j\theta} \tag{A.15}$$

Therefore, the current can be expressed as

$$i(t) = \frac{V_m e^{j(\omega t + \phi)}}{|Z| e^{j\theta}}$$

$$i(t) = \frac{V_m e^{j(\omega t + \phi)}}{|Z e j\theta|} \qquad = \left(\frac{V_m}{|Z|}\right) e^{j(\omega t + \phi - \theta)}$$

$$= I_m e^{j(\omega t + \phi - \theta)}$$

$$I_m e^{j(\omega t + \phi - \theta)} = \frac{V_m e^{j(\omega t + \phi)}}{|Z| e^{j\theta}} \tag{A.16}$$

Since time appears in both the current and voltage expressions, the equation is given in time domain. If the equality is multiplied by $e^{-j\omega t}$ to suppress $e^{j\omega t}$ and multiplied again by $1/\sqrt{2}$ to give the effective current and voltage values, then

$$\frac{e^{-j\omega t}}{\sqrt{2}} I_m e^{j(\omega t + \phi - \theta)} = \frac{e^{-j\omega t}}{\sqrt{2}} \frac{V_m e^{j(\omega t + \phi)}}{|Z| e^{j\theta}} \tag{A.17}$$

which gives

$$\frac{I_m}{\sqrt{2}} e^{j(\phi - \theta)} = \frac{V_m}{\sqrt{2}} \frac{e^{j\theta}}{|Z| e^{j\theta}} \tag{A.18}$$

which becomes

$$|I| \angle \phi - \theta = \frac{|V| \angle \phi}{|Z| \angle \theta} \tag{A.19}$$

or

$$I = \frac{V}{Z} \tag{A.20}$$

Equations A.18 through A.20 are in frequency domain. Equation A.18 is a *transformed* equation. I and V values without subscripts in Equation A.19 represent the effective current and voltage values. Thus, the I, V, and Z in Equation A.20 are complex quantities. Table A.1 gives a comparison and summary of the relationships between V and I in the time domain and V and I in the frequency domain for the three basic passive ideal circuit elements, R, L, and C. Figure A.1a shows the voltage and current functions in the time domain, while Figure A.1b shows the voltage and current phasors in the frequency domain.

An impedance that has a resistance R in series with a reactance X can be represented by the impedance operator

$$Z = R + jX \tag{A.21}$$

TABLE A.1

A Comparison of the Current and Voltage Relationships between the Time Domain and Frequency Domain for *R*, *L*, and *C*

Time Domain		Frequency Domain (Phasors)	
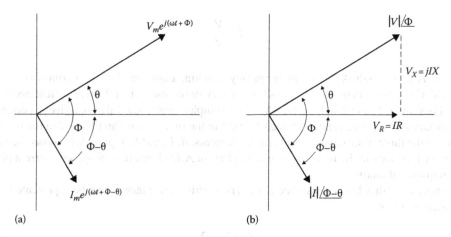 *R* circuit	$v = iR$	*R* circuit	$V = RI$
	$i = \dfrac{v}{R}$		$I = \dfrac{V}{R}$
L circuit	$v = L\dfrac{di}{dt}$	$j\omega L$ circuit	$V = j\omega LI$
	$i = \dfrac{1}{L}\displaystyle\int_{-\infty}^{t} v\,dt$		$I = -j\dfrac{V}{\omega L}$
C circuit	$v = \dfrac{1}{C}\displaystyle\int_{-\infty}^{t} i\,dt$	$-j\dfrac{1}{\omega c}$ circuit	$V = -j\dfrac{1}{\omega C}\,I$
	$i = C\dfrac{dv}{dt}$		$I = j\omega CV$

which is a complex quantity. A sinusoidal current in an impedance **Z** can be represented by a current phasor **I**, as shown in Figure A.1b. Therefore, the voltage drop across the impedance can be expressed as

$$\mathbf{V} = \mathbf{IZ}$$

$$= IR + jIX \tag{A.22}$$

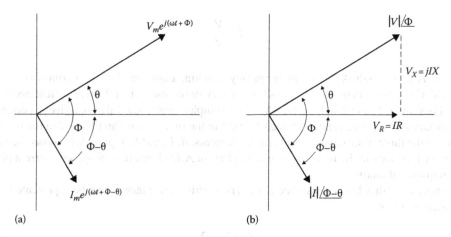

FIGURE A.1   Voltage and current functions given in: (a) the time domain and (b) the frequency domain.

where
- IR represents horizontal component of the phasor V
- IX represents vertical component of the phasor V
- θ represents phase angle

$$\theta = \tan^{-1} X/R$$

Neglecting the corresponding phase angles, the numerical values of the voltage drop and current can be expressed as

$$|V| = |IZ| = I\sqrt{R^2 + X^2} \tag{A.23}$$

$$|I| = \frac{|V|}{|Z|} = \frac{|V|}{\sqrt{R^2 + X^2}} \tag{A.24}$$

If the admittance of a network is already determined, the corresponding impedance can be expressed as

$$Z = \frac{1}{Y} \tag{A.25}$$

where Z and Y are complex quantities. Therefore,

$$Z = R + jX \tag{A.26}$$

or

$$Z = \frac{1}{G - jB} \tag{A.27}$$

On the other hand, an admittance that has a conductance G in parallel with a susceptance B can be represented by the admittance operator

$$Y = G - jB \tag{A.28}$$

When inductive and capacitive susceptances are in parallel,

$$B = \frac{L}{\omega L} - \omega C \tag{A.29}$$

When the voltage is represented by a phasor V, the total current can be expressed as

$$I = VY$$

$$= VG - jVB \tag{A.30}$$

When two branches of a network are connected in parallel, the complex quantities representing the admittances are added as

$$G - jB = G_1 - jB_1 = G_2 - jB_2 \tag{A.31}$$

Neglecting the corresponding phase angles, the numerical values of the current and voltage drop can be expressed as

$$|I| = |VY| = |V|\sqrt{G^2 + B^2} \tag{A.32}$$

$$|V| = \frac{|I|}{|Y|} = \frac{|I|}{\sqrt{G^2 + B^2}} \tag{A.33}$$

If the impedance is known, the corresponding admittance is found as

$$Y = \frac{1}{Z} \tag{A.34}$$

Therefore,

$$Y = G - jB$$

$$= \frac{1}{R + jX} \tag{A.35}$$

$$Y = \frac{R}{R^2 + X^2} - j\frac{X}{R^2 + X^2} \tag{A.36}$$

However, as shown in Equations A.35 and A.36, G is *not* equal to $1/R$ and B is *not* equal to $1/X$, as is the case when there is only a single circuit element R or X for which $G = 1/R$ and $B = 11X$, respectively.

But, it makes more sense to define the impedances due to resistor, inductance, and capacitance as

$$Z_R = R \tag{A.37}$$

$$Z_L = j\omega_L = jX_L \tag{A.38}$$

$$Z_L = -j\frac{1}{\omega C} = -jX_c \tag{A.39}$$

and by using reduction techniques apply the rules of series and parallel combinations to impedances to determine the equivalent impedance value. Therefore, the equivalent phasor impedance can be defined as the ratio of the phasor voltage to phasor current as

$$Z = \frac{V}{I} \tag{A.40}$$

The equivalent resistance and reactance values can be found as

$$R = \mathcal{R}e|Z| \tag{A.41}$$

$$X = \mathcal{F}e|Z| \tag{A.42}$$

where R and X are determined by taking only the real and imaginary portions of the impedance Z, respectively. As given in Table A.1, in *a purely resistive* circuit, current I is in phase with its voltage V. However, in *a purely inductive* circuit, current I lags its voltage V by 90°, but in *a purely capacitive* circuit, current I leads its voltage V by 90°.

PROBLEMS

A.1 The rms value of a sinusoidal ac current can be expressed as

$$I_{rms} = \sqrt{\frac{1}{T}\int_0^T i^2(t)dt}$$

where

$$i(t) = I_m \cos(\omega t - \theta)$$

$$\omega = \frac{2\pi}{T}$$

Verify that the rms value of the current is

$$I_{rms} = \frac{I_m}{\sqrt{2}}$$

A.2 The voltage and current values are given in the time domain as

$$v(t) = 162.6346\cos(10t - 50°)$$

and

$$i(t) = 14.142\sin(5t + 120°)$$

Convert them to the corresponding phasor values expressed in polar forms.

A.3 A voltage value is given in the frequency domain as

$$V = 70.711\angle - 60° \text{ kV}$$

Express this in a sine wave form.

A.4 Assume that only three currents $i_1(t)$, $i_2(t)$, and $i_3(t)$ enter a node and that the $i_1(t)$ and $i_2(t)$ are given as 14.14 cos(600t + 30°) A and 28.28 sin(600t − 50°) A, respectively. Determine the I_3 current in polar form.

A.5 Assume that only three currents enter a node and that the I_1 and I_3 are given as $19\angle-70°$ and $24\angle45°$ A, respectively. Determine the I_2 current in the time domain.

Appendix B: Per-Unit System

B.1 INTRODUCTION

Because of various advantages involved, it is customary in power system analysis calculations to use impedances, currents, voltages, and powers in per-unit values (which are scaled or normalized values) rather than in physical values of ohms, amperes, kilovolts, and megavoltamperes (or megavars, or megawatts).

A per-unit system is a means of expressing quantities for ease in comparing them. The per-unit value of any quantity is defined as the ratio of the quantity to an "*arbitrarily*" chosen base (i.e., *reference*) value having the same dimensions. Therefore, the per-unit value of any quantity can be defined as physical quantity:

$$\text{Quantity in per unit} = \frac{\text{physical quantity}}{\text{base value of quantity}} \qquad (B.1)$$

where "*physical quantity*" refers to the given value in ohms, amperes, volts, etc. The *base value* is also called unit value since in the per-unit system it has a value of 1, or unity. Therefore, a base current is also referred to as a unit current.

Since both the physical quantity and base quantity have the same dimensions, the resulting per-unit value expressed as a decimal has no dimension and therefore is simply indicated by a subscript pu. The base quantity is indicated by a subscript B. The symbol for per unit is pu, or 0/1. The percent system is obtained by multiplying the per-unit value by 100. Hence,

$$\text{Quantity in percent} = \frac{\text{physical quantity}}{\text{base value of quantity}} \times 100 \qquad (B.2)$$

However, the percent system is somewhat more difficult to work with and more subject to possible error since it must always be remembered that the quantities have been multiplied by 100.

Thus, the factor 100 has to be continually inserted or removed for reasons that may not be obvious at the time. For example, 40% reactance times 100% current is equal to 4000% voltage, which, of course, must be corrected to 40% voltage. Hence, the per-unit system is preferred in power system calculations. The advantages of using the per unit include the following:

1. Network analysis is greatly simplified since all impedances of a given equivalent circuit can directly be added together regardless of the system voltages.
2. It eliminates the $\sqrt{3}$ multiplications and divisions that are required when balanced three-phase systems are represented by per-phase systems. Therefore, the factors $\sqrt{3}$ and 3 associated with delta and wye quantities in a balanced three-phase system are directly taken into account by the base quantities.
3. Usually, the impedance of an electrical apparatus is given in percent or per unit by its manufacturer based on its nameplate ratings (e.g., its rated voltamperes and rated voltage).
4. Differences in operating characteristics of many electrical apparatus can be estimated by a comparison of their constants expressed in per units.

5. Average machine constants can easily be obtained since the parameters of similar equipment tend to fall in a relatively narrow range and therefore are comparable when expressed as per units based on rated capacity.
6. The use of per-unit quantities is more convenient in calculations involving digital computers.

B.2 SINGLE-PHASE SYSTEM

In the event that any two of the four base quantities (i.e., base voltage, base current, base voltamperes, and base impedance) are "*arbitrarily*" specified, the other two can be determined immediately.

Here, the term *arbitrarily* is slightly misleading since in practice the base values are selected so as to force the results to fall into specified ranges. For example, the base voltage is selected such that the system voltage is normally close to unity.

Similarly, the base voltampere is usually selected as the kilovoltampere or megavoltampere rating of one of the machines or transformers in the system, or a convenient round number such as 1, 10, 100, or 1000 MVA, depending on system size. As aforementioned, on determining the base voltamperes and base voltages, the other base values are fixed. For example, current base can be determined as

$$I_B = \frac{S_B}{V_B} = \frac{VA_B}{V_B} \tag{B.3}$$

where
I_B represents current base in amperes
S_B represents selected voltampere base in voltamperes
V_B represents selected voltage base in volts

Note that

$$S_B = VA_B = P_B = Q_B = V_B I_B \tag{B.4}$$

Similarly, the impedance base* can be determined as

$$Z_B = \frac{V_B}{I_B} \tag{B.5}$$

where

$$Z_B = X_B = R_B \tag{B.6}$$

Similarly,

$$Y_B = B_B = G_B = \frac{I_B}{V_B} \tag{B.7}$$

Note that by substituting Equation B.3 into Equation B.5, the impedance base can be expressed as

$$Z_B = \frac{V_B}{VA_B / V_B} = \frac{V_B^2}{VA_B} \tag{B.8}$$

---

* It is defined as that impedance across which there is a voltage drop that is equal to the base voltage if the current through it is equal to the base current.

or

$$Z_B = \frac{(kV_B)^2}{MVA_B} \tag{B.9}$$

where

kV_B is the voltage base in kilovolts

MVA_B is the voltampere base in megavoltamperes

The per-unit value of any quantity can be found by the *normalization process*, that is, by dividing the physical quantity by the base quantity of the same dimension. For example, the per-unit impedance can be expressed as

$$Z_{pu} = \frac{Z_{physical}}{Z_B} \tag{B.10a}$$

or

$$Z_{pu} = \frac{Z_{physical}}{V_B^2 / (kVA_B \times 1000)} \tag{B.10b}$$

or

$$Z_{pu} = \frac{(Z_{physical})(kVA_B)(1000)}{V_B^2} \tag{B.11}$$

or

$$Z_{pu} = \frac{(Z_{physical})(kVA_B)}{(kV_B)^2(1000)} \tag{B.12}$$

or

$$Z_{pu} = \frac{(Z_{physical})}{(kV_B)^2 / MVA_B} \tag{B.13}$$

or

$$Z_{pu} = \frac{(Z_{physical})(MVA_B)}{(kV_B)^2} \tag{B.14}$$

Similarly, the others can be expressed as

$$I_{pu} = \frac{I_{physical}}{I_B} \tag{B.15}$$

or

$$V_{pu} = \frac{V_{physical}}{V_B} \tag{B.16}$$

or

$$kV_{pu} = \frac{kV_{physical}}{kV_B} \tag{B.17}$$

or

$$VA_{pu} = \frac{VA_{physical}}{VA_B} \tag{B.18}$$

or

$$kVA_{pu} = \frac{kVA_{physical}}{kVA_B} \tag{B.19}$$

or

$$MVA_{pu} = \frac{MVA_{physical}}{MVA_B} \tag{B.20}$$

Note that the base quantity is always a real number, whereas the physical quantity can be a complex number. For example, if the actual impedance quantity is given as $Z\angle\theta\,\Omega$, it can be expressed in the per-unit system as

$$\mathbf{Z}_{pu} = \frac{Z\angle\theta}{Z_B} = Z_{pu}\angle\theta \tag{B.21}$$

that is, it is the magnitude expressed in per-unit terms.

Alternatively, if the impedance has been given in rectangular form as

$$\mathbf{Z} = R + jX \tag{B.22}$$

then

$$\mathbf{Z}_{pu} = R_{pu} + jX_{pu} \tag{B.23}$$

where

$$R_{pu} = \frac{R_{physical}}{Z_B} \tag{B.24}$$

and

$$X_{pu} = \frac{X_{physical}}{Z_B} \tag{B.25}$$

Similarly, if the complex power has been given as

$$\mathbf{S} = P + jQ \tag{B.26}$$

then

$$\mathbf{S}_{pu} = P_{pu} + jQ_{pu} \tag{B.27}$$

where

$$P_{pu} = \frac{P_{physical}}{S_B} \tag{B.28}$$

and

$$Q_{pu} = \frac{Q_{physical}}{S_B} \tag{B.29}$$

If the actual voltage and current values are given as

$$\mathbf{V} = V \angle \theta_{\mathbf{V}} \tag{B.30}$$

and

$$\mathbf{I} = I \angle \theta_{\mathbf{I}} \tag{B.31}$$

the complex power can be expressed as

$$\mathbf{S} = \mathbf{V}\mathbf{I}^* \tag{B.32}$$

or

$$S \angle \theta = (V \angle \theta_{\mathbf{V}})(I \angle -\theta_{\mathbf{I}}) \tag{B.33}$$

Therefore, dividing through by S_B,

$$\frac{S \angle \phi}{S_B} = \frac{(V \angle \theta_{\mathbf{V}})(I \angle -\theta_{\mathbf{I}})}{S_B} \tag{B.34}$$

However,

$$S_B = V_B I_B \tag{B.35}$$

Thus,

$$\frac{S\angle\theta}{S_B} = \frac{(V\angle\theta_V)(I\angle-\theta_I)}{V_B I_B} \tag{B.36}$$

or

$$S_{pu}\angle\theta = (V_{pu}\angle\theta_V)(I_{pu}\angle-\theta_I) \tag{B.37}$$

or

$$\mathbf{S}_{pu} = \mathbf{V}_{pu}\mathbf{I}_{pu}^* \tag{B.38}$$

Example B.1

A 240/120 V single-phase transformer rated 5 kVA has a high-voltage winding impedance of 0.3603 Ω. Use 240 V and 5 kVA as the base quantities and determine the following:

(a) The high-voltage side base current
(b) The high-voltage side base impedance in ohms
(c) The transformer impedance referred to the high-voltage side in per unit
(d) The transformer impedance referred to the high-voltage side in percent
(e) The turns ratio of the transformer windings
(f) The low-voltage side base current
(g) The low-voltage side base impedance
(h) The transformer impedance referred to the low-voltage side in per unit

Solution

(a) The high-voltage side base current is

$$I_{B(HV)} = \frac{S_B}{V_{B(HV)}}$$

$$= \frac{5000\,\text{VA}}{240\,\text{V}}$$

$$= 20.8333\,\text{A}$$

(b) The high-voltage side base impedance is

$$Z_{B(HV)} = \frac{V_{B(HV)}}{I_{B(HV)}}$$

$$= \frac{240\,\text{V}}{20.8333\,\text{A}}$$

$$= 11.52\,\Omega$$

(c) The transformer impedance referred to the high-voltage side is

$$Z_{pu(HV)} = \frac{Z_{HV}}{Z_{B(HV)}}$$

$$= \frac{0.3603\,\Omega}{11.51\,\Omega}$$

$$= 0.0313\,pu$$

(d) The transformer impedance referred to the high-voltage side is percent

$$\%Z_{HV} = Z_{pu(HV)} \times 100$$

$$= (0.0313\,pu)100$$

$$= 3.13\%$$

(e) The turns ratio of the transformer windings is

$$n = \frac{V_{HV}}{V_{LV}}$$

$$= \frac{240\,V}{120\,V}$$

$$= 2$$

(f) The low-voltage side base current is

$$I_{B(LV)} = \frac{S_B}{V_{B(LV)}}$$

$$= \frac{5000\,VA}{120\,V}$$

$$= 41.6667\,A$$

or

$$I_{B(LV)} = nI_{B(HV)}$$

$$= 2(20.8333\,A)$$

$$= 41.6667$$

(g) The low-voltage side base impedance is

$$Z_{B(LV)} = \frac{V_{B(LV)}}{I_{B(LV)}}$$

$$= \frac{120\,V}{41.667\,A}$$

$$= 2.88\,\Omega$$

or

$$Z_{B(HV)} = \frac{Z_{B(LV)}}{n^2}$$

$$= 2.88\,\Omega$$

(h) The transformer impedance referred to the low-voltage side is

$$Z_{LV} = \frac{Z_{HV}}{n^2}$$

$$= \frac{0.3603\,\Omega}{2^2}$$

$$= 0.0901\,\Omega$$

Therefore,

$$Z_{pu(LV)} = \frac{Z_{LV}}{Z_{B(LV)}}$$

$$= \frac{0.0901\,\Omega}{2.88\,\Omega}$$

$$= 0.0313\,\text{pu}$$

or

$$Z_{pu(LV)} = Z_{pu(HV)}$$

$$= 0.0313\,\text{pu}$$

Notice that in terms of per units the impedance of the transformer is the same whether it is referred to the high-voltage side or the low-voltage side.

Example B.2

Redo the Example B.1 using MATLAB®.

(a) Write the MATLAB program script.
(b) Give the MATLAB program output.

Solution

(a) Here is the MATLAB program script:

```
% MATLAB SCRIPT for Example B.1
clear
clc
%System Parameters
 VBhv = 240;
 VBlv = 120;
 SB = 5e3;
 Zhv = 0.3603;
%Solution for part (a)
```

```
    IBhv = SB/VBhv
%Solution for part (b)
    ZBhv = VBhv/IBhv
%Solution for part (c)
    Zpu_hv = Zhv/ZBhv
%Solution for part (d)
    percent_Zhv = Zpu_hv*100
%Solution for part (e)
    n = VBhv/VBlv
%Solution for part (f)
    IBlv = SB/VBlv
    IBlv = n*IBhv
%Solution for part (g)
    ZBlv = VBlv/IBlv
    ZBlv = ZBhv/n^2
%Solution for part (h)
    Zlv = Zhv/n^2
    Zpu_lv = Zlv/ZBlv
    Zpu_lv = Zpu_hv
```

(b) Here is the MATLAB program output:
```
IBhv =
  20.8333
ZBhv =
  11.5200
Zpu _ hv =
  0.0313
percent _ Zhv =
   3.1276
n =
   2
IBlv =
  41.6667
IBlv =
  41.6667
ZBlv =
   2.8800
ZBlv =
   2.8800
Zlv =
   0.0901
Zpu _ lv =
   0.0313
Zpu _ lv =
   0.0313
>>
```

B.3   CONVERTING FROM PER-UNIT VALUES TO PHYSICAL VALUES

The physical values (or system values) and per-unit values are related by the following relationships:

$$\mathbf{I} = \mathbf{I}_{pu} \times I_B \tag{B.39}$$

$$\mathbf{V} = \mathbf{V}_{pu} \times V_B \tag{B.40}$$

$$\mathbf{Z} = \mathbf{Z}_{pu} \times Z_B \tag{B.41}$$

$$R = R_{pu} \times Z_B \tag{B.42}$$

$$X = X_{pu} \times Z_B \tag{B.43}$$

$$VA = VA_{pu} \times VA_B \tag{B.44}$$

$$P = P_{pu} \times VA_B \tag{B.45}$$

$$Q = Q_{pu} \times VA_B \tag{B.46}$$

B.4   CHANGE OF BASE

In general, the per-unit impedance of a power apparatus is given based on its own voltampere and voltage ratings and consequently based on its own impedance base. When such an apparatus is used in a system that has its own bases, it becomes necessary to refer all the given per-unit values to the system base values. Assume that the per-unit impedance of the apparatus is given based on its nameplate ratings as

$$Z_{pu(given)} = (Z_{physical}) \frac{MVA_{B(given)}}{[kV_{B(given)}]^2} \tag{B.47}$$

and that it is necessary to refer the very same physical impedance to a new set of voltage and voltampere bases such that

$$Z_{pu(new)} = (Z_{physical}) \frac{MVA_{B(new)}}{[kV_{B(new)}]^2} \tag{B.48}$$

By dividing Equation B.47 by Equation B.48 side by side,

$$Z_{pu(new)} = Z_{pu(old)} \left[\frac{MVA_{B(old)}}{MVA_{B(given)}} \right] \left[\frac{kV_{B(given)}}{kV_{B(old)}} \right]^2 \tag{B.49}$$

In certain situations, it is more convenient to use subscripts 1 and 2 instead of subscripts "*given*" and "*new*," respectively. Then Equation B.49 can be expressed as

$$Z_{pu(2)} = Z_{pu(1)} \left[\frac{MVA_{B(2)}}{MVA_{B(1)}} \right] \left[\frac{kV_{B(1)}}{kV_{B(2)}} \right]^2 \tag{B.50}$$

In the event that the kV bases are the same but the MVA bases are different, from Equation B.49,

$$Z_{pu(new)} = Z_{pu(given)} \frac{MVA_{B(new)}}{MVA_{B(given)}} \tag{B.51}$$

Similarly, if the megavoltampere bases are the same but the kilovolt bases are different, from Equation B.49,

$$Z_{pu(new)} = Z_{pu(given)} \left[\frac{kV_{B(given)}}{kV_{B(new)}} \right]^2 \tag{B.52}$$

Equations B.49 through B.52 must only be used to convert the given per-unit impedance from the base to another but not for referring the physical value of an impedance from one side of the transformer to another.

Example B.3

Consider Example B.1 and select 300/150 V as the base voltages for the high-voltage and the low-voltage windings, respectively. Use a new base power of 10 kVA and determine the new per-unit, base, and physical impedances of the transformer referred to the high-voltage side.

Solution

By using Equation B.49, the new per-unit impedance can be found as

$$Z_{pu(new)} = Z_{pu(old)} \left[\frac{MVA_{B(old)}}{MVA_{B(given)}} \right] \left[\frac{kV_{B(given)}}{kV_{B(old)}} \right]^2$$

$$= (0.0313 \, pu) \left(\frac{10{,}000 \, VA}{300 \, V} \right) \left(\frac{240 \, V}{300 \, V} \right)^2$$

$$= 33.334 \, A$$

The new current base is

$$I_{B(HV)new} = \frac{S_B}{V_{B(HV)new}}$$

$$= \frac{10{,}000 \, VA}{300 \, V}$$

$$= 33{,}334 \, A$$

Thus,

$$Z_{B(HV)new} = \frac{V_{B(HV)new}}{I_{B(HV)new}}$$

$$= \frac{300 \, V}{33.334 \, A}$$

$$= 9 \, \Omega$$

Therefore, the physical impedance of the transformer is still

$$Z_{HV} = Z_{pu,new} \times Z_{B(HV)new}$$

$$= (0.0401\,\text{pu})(9\,\Omega)$$

$$= 0.3609\,\Omega$$

B.5 THREE-PHASE SYSTEMS

The three-phase problems involving balanced systems can be solved on a per-phase basis. In that case, the equations that are developed for single-phase systems can be used for three-phase systems as long as per-phase values are used consistently. Therefore,

$$I_B = \frac{S_{B(1\phi)}}{V_{B(L-N)}} \tag{B.53}$$

or

$$I_B = \frac{VA_{B(1\phi)}}{V_{B(L-N)}} \tag{B.54}$$

and

$$Z_B = \frac{V_{B(L-N)}}{I_B} \tag{B.55}$$

or

$$Z_B = \frac{[kV_{B(L-N)}]^2 (1000)}{kVA_{B(1\phi)}} \tag{B.56}$$

or

$$Z_B = \frac{[kV_{B(L-N)}]^2}{MVA_{B(1\phi)}} \tag{B.57}$$

where the subscripts 1ϕ and $L-N$ denote per phase and line to neutral, respectively. Note that, for a balanced system,

$$V_{B(L-N)} = \frac{V_{B(L-L)}}{\sqrt{3}} \tag{B.58}$$

and

$$S_{B(1\phi)} = \frac{S_{B(3\phi)}}{3} \tag{B.59}$$

However, it has been customary in three-phase system analysis to use line-to-line voltage and three-phase voltamperes as the base values. Therefore,

$$I_B = \frac{S_{B(3\phi)}}{\sqrt{3}V_{B(L-L)}} \tag{B.60}$$

or

$$I_B = \frac{kVA_{B(3\phi)}}{\sqrt{3}kV_{B(L-L)}} \tag{B.61}$$

and

$$Z_B = \frac{V_{B(L-L)}}{\sqrt{3}I_B} \tag{B.62}$$

$$Z_B = \frac{[kV_{B(L-L)}]^2(1000)}{kVA_{B(3\phi)}} \tag{B.63}$$

or

$$Z_B = \frac{\left[kV_{B(L-L)}\right]^2}{MVA_{B(3\phi)}} \tag{B.64}$$

where the subscripts 3ϕ and $L-L$ denote per three phase and line, respectively. Furthermore, base admittance can be expressed as

$$Y_B = \frac{1}{Z_B} \tag{B.65}$$

or

$$Y_B = \frac{MVA_{B(3\phi)}}{[kV_{B(L-L)}]^2} \tag{B.66}$$

where

$$Y_B = B_B = G_B \tag{B.67}$$

The data for transmission lines are usually given in terms of the line resistance R in ohms per mile at a given temperature, the line inductive reactance X_L in ohms per mile at 60 Hz, and the line shunt capacitive reactance X_c in megohms per mile at 60 Hz. Therefore, the line impedance and shunt susceptance in per units for 1 mile of line can be expressed as

$$\mathbf{Z}_{pu} = (\mathbf{Z}, \Omega/\text{mile}) \frac{MVA_{B(3\phi)}}{[kV_{B(L-L)}]^2} \text{ pu} \tag{B.68}$$

where

$$\mathbf{Z} = R + jX_L = Z \angle \theta \, \Omega/\text{mile}$$

and

$$B_{pu} = \frac{[kV_{B(L-L)}]^2 \times 10^{-6}}{[MVA_{B(3\phi)}][X_c, M\Omega/\text{mile}]} \tag{B.69}$$

In the event that the admittance for a transmission line is given in microsiemens per mile, the per-unit admittance can be expressed as

$$Y_{pu} = \frac{[kV_{B(L-L)}]^2 (Y, \mu S)}{[MVA_{B(3\phi)}] \times 10^6} \tag{B.70}$$

Similarly, if it is given as reciprocal admittance in megohms per mile, the per-unit admittance can be found as

$$Y_{pu} = \frac{[kV_{B(L-L)}]^2 \times 10^{-6}}{[MVA_{B(3\phi)}][Z, M\Omega/\text{mile}]} \tag{B.71}$$

Figure 4.29 shows conventional three-phase transformer connections and associated relationships between the high-voltage and low-voltage side voltages and currents. The given relationships are correct for a three-phase transformer as well as for a three-phase bank of single-phase transformers. Note that in the figure, n is the turns ratio, that is,

$$n = \frac{N_1}{N_2} = \frac{V_1}{V_2} = \frac{I_2}{I_1} \tag{B.72}$$

where the subscripts 1 and 2 are used for the primary and secondary sides. Therefore, an impedance Z_2 in the secondary circuit can be referred to the primary circuit provided that

$$Z_1 = n^2 Z_2 \tag{B.73}$$

Thus, it can be observed from Figure 4.29 that in an ideal transformer, voltages are transformed in the direct ratio of turns, currents in the inverse ratio, and impedances in the direct ratio squared; power and voltamperes are, of course, unchanged. Note that a balanced delta-connected circuit of $Z_\Delta \Omega/\text{phase}$ is equivalent to a balanced wye-connected circuit of $Z_Y \Omega/\text{phase}$ as long as

$$Z_Y = \frac{1}{3} Z_\Delta \tag{B.74}$$

The per-unit impedance of a transformer remains the same without taking into account whether it is converted from physical impedance values that are found by referring to the high-voltage side or low-voltage side of the transformer. This can be accomplished by choosing separate appropriate bases for each side of the transformer (whether or not the transformer is connected in wye–wye, delta–delta, delta–wye, or wye–delta since the transformation of voltages is the same as that made by wye–wye transformers as long as the same line-to-line voltage ratings are used). In other words, the designated per-unit impedance values of transformers are based on the coil ratings.

Since the ratings of coils cannot alter by a simple change in connection (e.g., from wye–wye to delta–wye), the per-unit impedance remains the same regardless of the three-phase connection. The line-to-line voltage for the transformer will differ. Because of the method of choosing the base in various sections of the three-phase system, the per-unit impedances calculated in various sections can be put together on one impedance diagram without paying any attention to whether the transformers are connected in wye–wye or delta–wye.

Example B.4

Assume that a 19.5 kV 120 MVA three-phase generator has a synchronous reactance of 1.5 per-unit ohms and is connected to a 150 MVA 18/230 kV delta–wye-connected three-phase transformer with a 0.1 per-unit ohm reactance. The transformer is connected to a transmission line at the 230 kV side. Use the new MVA base of 100 MVA and 240 kV base for the line and determine the following:

(a) The new reactance value for the generator in per-unit ohms
(b) The new reactance value for the transformer in per-unit ohms

Solution

(a) Using Equation B.49, the new per-unit impedance of the generator is

$$Z_{pu(new)} = Z_{pu(old)} \left[\frac{MVA_{B(new)}}{MVA_{B(old)}} \right] \left[\frac{kV_{B(old)}}{kV_{B(old)}} \right]^2$$

But, first determining the new kV base for the generator,

$$kV_{B(new)}^{gen} = (240\,kV) \left(\frac{18\,kV}{230\,kV} \right) = 18.783\,kV$$

Thus, the new and adjusted synchronous reactance of the generator is

$$X_{pu(new)}^{gen} = X_{pu(old)}^{gen} \left[\frac{MVA_{B(new)}}{MVA_{B(old)}} \right] \left[\frac{kV_{B(old)}}{kV_{B(new)}} \right]^2$$

$$= (1.5\,pu) \left[\frac{100\,MVA}{120\,MVA} \right] \left[\frac{19.5\,kV}{18.783\,kV} \right]^2$$

$$= 1.347\,pu$$

(b) The new reactance value for the transformer in per-unit ohms, referred to high-voltage side is

$$X_{pu(new)}^{trf} = (0.1\,\text{pu}) \left[\frac{100\,\text{MVA}}{150\,\text{MVA}} \right] \left[\frac{230\,\text{kV}}{240\,\text{kV}} \right]^2$$

$$= 0.061\,\text{pu}$$

And referred to the low-voltage side is

$$X_{pu(new)}^{trf} = (0.1\,\text{pu}) \left[\frac{100\,\text{MVA}}{150\,\text{MVA}} \right] \left[\frac{18\,\text{kV}}{18.783\,\text{kV}} \right]^2$$

$$= 0.061\,\text{pu}$$

Note that the transformer reactance referred to the high-voltage side or the low-voltage side is the same, as it should be!

Example B.5

A three-phase transformer has a nameplate rating of 20 MVA, 345Y/34.5Y kV with a leakage reactance of 12% and the transformer connection is wye–wye. Select a base of 20 MVA and 345 kV on the high-voltage side and determine the following:

(a) Reactance of transformer in per units.
(b) High-voltage side base impedance
(c) Low-voltage side base impedance
(d) Transformer reactance referred to high-voltage side in ohms
(e) Transformer reactance referred to low-voltage side in ohms

Solution

(a) The reactance of the transformer in per units is 12/100, or 0.12 pu. Note that it is the same whether it is referred to the high-voltage or the low-voltage sides.
(b) The high-voltage side base impedance is

$$Z_{B(HV)} = \frac{\left[kV_{B(HV)} \right]^2}{MVA_{B(3\phi)}}$$

$$= \frac{345^2}{20} = 5951.25\,\Omega$$

(c) The low-voltage side base impedance is

$$Z_{B(LV)} = \frac{\left[kV_{B(LV)} \right]^2}{MVA_{B(3\phi)}}$$

$$= \frac{34.5^2}{20} = 59.5125\,\Omega$$

(d) The reactance referred to the high-voltage side is

$$X_{(HV)} = X_{pu} \times X_{B(HV)}$$

$$= (0.12)(5951.25) = 714.15\,\Omega$$

(e) The reactance referred to the low-voltage side is

$$X_{(LV)} = X_{pu} \times X_{B(LV)}$$

$$= (0.12)(59.5125) = 7.1415\,\Omega$$

or

$$X_{(LV)} = \frac{X_{(HV)}}{n^2}$$

$$= \frac{714.15\,\Omega}{\left(345/\sqrt{3}\big/34.5/\sqrt{3}\right)^2} = 7.1415\,\Omega$$

where n is defined as the turns ratio of the windings.

Example B.6

A three-phase transformer has a nameplate rating of 20 MVA, and the voltage ratings of 345Y/34.5Δ kV with a leakage reactance of 125 and the transformer connection is wye–delta. Select a base of 20 MVA and 345 kV on the high-voltage side and determine the following:

(a) Turns ratio of windings
(b) Transformer reactance referred to low-voltage side in ohms
(c) Transformer reactance referred to low-voltage side in per units

Solution

(a) The turns ratio of the windings is

$$n = \frac{345/\sqrt{3}}{34.5} = 5.7735$$

(b) Since the high-voltage side impedance base is

$$Z_{B(HV)} = \frac{\left[kV_{B(HV)}\right]^2}{MVA_{B(3\phi)}}$$

$$= \frac{345^2}{20} = 5951.25\,\Omega$$

and

$$X_{(HV)} = X_{pu} \times X_{B(HV)}$$

$$= (0.12)(5951.25) = 714.15\,\Omega$$

Thus, the transformer reactance referred to the delta-connected low-voltage side is

$$X_{(LV)} = \frac{X_{(HV)}}{n^2}$$

$$= \frac{714.14\,\Omega}{5.7735^2} = 21.4245\,\Omega$$

(c) From Table 2.1, the reactance of the equivalent wye connection is

$$Z_Y = \frac{Z_\Delta}{3}$$

$$= \frac{21.4245\,\Omega}{3} = 7.1415\,\Omega$$

Similarly,

$$Z_{B(LV)} = \frac{\left[kV_{B(LV)}\right]^2}{MVA_{B(3\phi)}}$$

$$= \frac{34.5^2}{20} = 59.5125\,\Omega$$

Thus,

$$X_{pu} = \frac{7.1415\,\Omega}{Z_{B(LV)}}$$

$$= \frac{7.1415\,\Omega}{59.5125\,\Omega} = 0.12\,\text{pu}$$

Alternatively, if the line-to-line voltages are used,

$$X_{(LV)} = \frac{X_{(HV)}}{n^2}$$

$$= \frac{714.14\,\Omega}{(345/34.5)^2} = 7.1415\,\Omega$$

and therefore,

$$X_{pu} = \frac{X_{(LV)}}{Z_{B(LV)}}$$

$$= \frac{7.1415\,\Omega}{59.5125\,\Omega} = 0.12\,\text{pu}$$

as before.

Example B.7

Consider a three-phase system which has a generator connected to a 2.4/24 kV, wye–wye-connected, three-phase step-up transformer T_1. Suppose that the transformer is connected to three-phase power line. The receiving end of the line is connected to a second, wye–wye-connected, three-phase 24/12 kV step-down transformer T_2. Assume that the line length between the two transformers is negligible and the three-phase generator is rated 4160 kVA, 2.4 kV, and 1000 A and that it supplies a purely inductive load of $I_{pu} = 2.08\angle - 90°$ pu. The three-phase transformer T_1 is rated 6000 kVA, 2.4Y–24Y kV, with leakage reactance of

0.04 pu. Transformer T_2 is made up of three single-phase transformers and is rated 4000 kVA, 24Y–12Y kV, with leakage reactance of 0.04 pu. Determine the following for all three circuits, 2.4, 24, and 12 kV circuits:

(a) Base kilovoltampere values
(b) Base line-to-line kilovolt values
(c) Base impedance values
(d) Base current values
(e) Physical current values (neglect magnetizing currents in transformers and charging currents in lines)
(f) Per-unit current values
(g) New transformer reactances based on their new bases
(h) Per-unit voltage values at buses 1, 2, and 4
(i) Per-unit apparent power values at buses 1, 2, and 4
(j) Summarize results in a table

Solution

(a) The kilovoltampere base for all three circuits is arbitrarily selected as 2080 kVA
(b) The base voltage for the 2.4 kV circuit is arbitrarily selected as 2.5 kV. Since the turns ratios for transformers T_1 and T_2 are

$$\frac{N_1}{N_2} = 10 \quad \text{or} \quad \frac{N_2}{N_1} = 0.10$$

and

$$\frac{N_1'}{N_2'} = 2$$

The base voltages for the 24 and 12 kV circuits are determined to be 25 and 12.5 kV, respectively.

(c) The base impedance values can be found as

$$Z_B = \frac{[kV_{B(L-L)}]^2(1000)}{kVA_{B(3\phi)}}$$

$$= \frac{[2.5 \text{ kV}]^2 1000}{2080 \text{ kVA}} = 3.005 \ \Omega$$

and

$$Z_B = \frac{[25 \text{ kV}]^2 1000}{2080 \text{ kVA}} = 300.5 \ \Omega$$

and

$$Z_B = \frac{[12.5 \text{ kV}]^2 1000}{2080 \text{ kVA}} = 75.1 \ \Omega$$

(d) The base current values can be determined as

$$I_B = \frac{kVA_{B(3\phi)}}{\sqrt{3}kV_{B(L-L)}}$$

$$= \frac{2080\,kVA}{\sqrt{3}(2.5\,kV)} = 480\,A$$

and

$$I_B = \frac{2080\,kVA}{\sqrt{3}(25\,kV)} = 48\,A$$

and

$$I_B = \frac{2080\,kVA}{\sqrt{3}(12.5\,kV)} = 96\,A$$

(e) The physical current values can be found based on the turns ratios as

$$I = 1000\,A$$

$$I = \left(\frac{N_2}{N_1}\right)(1000\,A) = 100\,A$$

$$I = \left(\frac{N_1'}{N_2'}\right)(100\,A) = 200\,A$$

(f) The per-unit current values are the same, 2.08 pu, for all three circuits
(g) The given transformer reactances can be converted based on their new bases using

$$Z_{pu(new)} = Z_{pu(given)}\left[\frac{kVA_{B(new)}}{kVA_{B(given)}}\right]\left[\frac{kV_{B(given)}}{kV_{B(new)}}\right]^2$$

Therefore, the new reactances of the two transformers can be found as

$$Z_{pu(T_1)} = j0.04\left[\frac{2080\,kVA}{6000\,kVA}\right]\left[\frac{2.4\,kV}{2.5\,kV}\right]^2 = j0.0128\,pu$$

and

$$Z_{pu(T_2)} = j0.04\left[\frac{2080\,kVA}{4000\,kVA}\right]\left[\frac{12\,kV}{12.5\,kV}\right]^2 = j0.0192\,pu$$

(h) Therefore, the per-unit voltage values at buses 1, 2, and 4 can be calculated as

$$V_1 = \frac{2.4\,kV\angle0°}{2.5\,kV} = 0.96\angle0°\,pu$$

TABLE B.1
Results of Example B.7

Quantity	2.4 kV Circuit	24 kV Circuit	12 kV Circuit
$kVA_{B(3\phi)}$	2080 kVA	2080 kVA	2080 kVA
$kV_{B(L-L)}$	2.5 kV	25 kV	12.5 kV
Z_B	3005 Ω	300.5 Ω	75.1 Ω
I_B	480 A	48 A	96 A
$I_{physical}$	1000 A	100 A	200 A
I_{pu}	2.08 pu	2.08 pu	2.08 pu
V_{pu}	0.96 pu	0.9334 pu	0.8935 pu
S_{pu}	2.00 pu	1.9415 pu	1.8585 pu

$$V_2 = V_1 - I_{pu}Z_{pu(T_1)}$$

$$= 0.96\angle 0° - (2.08\angle - 90°)(0.0128\angle 90°) = 0.9334\angle 0° \, pu$$

$$V_4 = V_2 - I_{pu}Z_{pu(T_2)}$$

$$= 0.9334\angle 0° - (2.08\angle - 90°)(0.0192\angle 90°) = 0.8935\angle 0° \, pu$$

(i) Thus, the per-unit apparent power values at buses 1, 2, and 4 are

$$S_1 = 2.00 \, pu$$

$$S_2 = V_2 I_{pu} = (0.9334)(2.08) = 1.9415 \, pu$$

$$S_4 = V_4 I_{pu} = (0.8935)(2.08) = 1.8585 \, pu$$

(j) The results are summarized in Table B.1.

PROBLEMS

B.1 Solve Example B.1 for a transformer rated 100 kVA and 2400/240 V that has a high-voltage winding impedance of 0.911.

B.2 Consider the results of Problem B.1 and use 3000/300 V as new base voltages for the high-voltage and low-voltage windings, respectively. Use a new base power of 200 kVA and determine the new per-unit, base, and physical impedances of the transformer referred to the high-voltage side.

B.3 A 240/120 V single-phase transformer rated 25 kVA has a high-voltage winding impedance of 0.65 Q. If 240 V and 25 kVA are used as the base quantities, determine the following:
   (a) The high-voltage side base current
   (b) The high-voltage side base impedance in Q
   (c) The transformer impedance referred to the high-voltage side in per unit
   (d) The transformer impedance referred to the high-voltage side in percent
   (e) The turns ratio of the transformer windings
   (f) The low-voltage side base current
   (g) The low-voltage side base impedance
   (h) The transformer impedance referred to the low-voltage side in per unit

B.4 A 240/120 V single-phase transformer is rated 25 kVA and has a high-voltage winding impedance referred to its high-voltage side that is 0.2821 pu based on 240 V and 25 kVA. Select 230/115 V as the base voltages for the high-voltage and low-voltage windings, respectively. Use a new base power of 50 kVA and determine the new per-unit base, and physical impedances of the transformer referred to the high-voltage side.

B.5 After changing the S base from 5 to 10 MVA, redo the Example B.1 by using MATLAB.

(a) Write the MATLAB program script.

(b) Give the MATLAB program output.

Appendix C: Salient-Pole Synchronous Machines

C.1 INTRODUCTION

Almost always low-speed synchronous machines with four or more poles have salient-pole rotors. Such low-speed alternators are normally driven by water turbines and are often used as hydroelectric alternators, as shown in Figure 7.2.

In cylindrical-rotor synchronous machines there is no saliency. Therefore, the reluctance and thereby the reactances are constants due to the existence of a uniform air gap; but in the salient-pole rotor the area midway between the poles (quadrature or q-axis) has a larger air gap than the area between the pole centers (direct or d-axis). As a result, the armature mmf produces more flux along the d-axis than along the q-axis. Because of this saliency, the reactance measured at the terminals of a salient-rotor machine will vary as a function of the rotor position. The reactance associated with the d-axis is called the *direct-axis synchronous reactance*, X_d. The one associated with the q-axis is called the *quadrature-axis synchronous reactance*, X_q. They can be expressed as

$$X_d = X_a + X_{\phi d} \tag{C.1}$$

$$X_q = X_a + X_{\phi q} \tag{C.2}$$

where X_a is the armature self-reactance and is independent of the rotor angle. The direct- and quadrature-axis magnetizing reactances $X_{\phi d}$ and $X_{\phi q}$ represent the inductive effects of the direct- and quadrature-axis armature-reaction flux waves produced by the d-axis current I_d and the q-axis current I_q, respectively. The reactance X_q is less than the reactance X_d due to the existence of greater reluctance in the air gap of the quadrature axis. The X_q is often between 0.6 and 0.7 of X_d. Therefore, the armature reaction resulting from the armature current I_a can be taken into account by resolving it into two components along the d-axis and the q-axis, namely, I_d and I_q. Thus, as shown in Figure C.1a, the total voltage in the stator can be expressed as

$$V_\phi = E_a + E_d + E_q - I_a R_a \tag{C.3}$$

where E_d and E_q represent the d-axis component and the q-axis component of armature reaction voltage, respectively. Hence,

$$E_d = -j I_d X_d \tag{C.4}$$

$$E_q = -j I_q X_q \tag{C.5}$$

Therefore,

$$V_\phi = E_a - j I_d X_d - j I_q X_q - I_a R_a \tag{C.6}$$

where the term represents the voltage drop due to the armature resistance. The armature current is

$$I_a = I_d + I_q \tag{C.7}$$

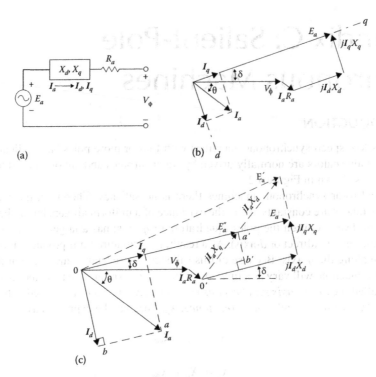

FIGURE C.1 Equivalent circuit and phasor diagrams of a salient-pole synchronous generator: (a) equivalent circuit; (b) phasor diagram; and (c) drawing the phasor diagram without knowing the power angle 8.

If the armature resistance is negligibly small, then

$$V_\phi = E_a - jI_dX_d - I_qX_q \tag{C.8}$$

or

$$E_a = V_\phi + jI_dX_d + jI_qX_q \tag{C.9}$$

The equivalent circuit of a salient-pole synchronous generator is shown in Figure C.1a. The corresponding phasor diagram is shown in Figure C.1b.

Note that this phasor diagram is based on the assumption that the angle between E_a and I_a (i.e., $\theta + \delta$) is known in advance. Normally, however, the internal power factor angle $\theta + \delta$ is not known beforehand, only the power-factor angle θ is known. Figure C.1c shows how to draw the phasor diagram when the power angle δ is not known in advance. Observe that the phasor E''_a is

$$E''_a = V_\phi + R_aI_a + jI_aX_q \tag{C.10}$$

The phasor E''_a and the angle δ can be determined by drawing the phasor I_aR_a parallel to the current I_a and from its tip the phasor I_aX_q perpendicular to the phasor I_a, and finally by drawing a line connecting the phasor origin to the tip of the phasor I_aX_q. Once the angle δ is known, then the internal generated voltage can be readily found, as shown in Figure C.1c.

Also note that if the armature resistance R_a is ignored, then the angle δ can be found from

$$\tan \delta = \frac{I_aX_q \cos \theta}{V_\phi + I_aX_q \sin \theta} \tag{C.11}$$

However, if the armature resistance R_a is not neglected, then the angle δ can be determined from

$$\tan \delta = \frac{I_a X_q \cos\theta - I_a R_a \sin\theta}{V_\phi + I_a X_q \sin\theta + I_a R_a \cos\theta} \tag{C.12}$$

The power output of a three-phase synchronous generator is

$$P = 3V_\phi I_a \cos\theta \tag{C.13a}$$

$$= 3V_\phi I_d \cos(90° - \delta) + 3V_\phi I_q \cos\delta \tag{C.13b}$$

$$= 3V_\phi I_d \sin\delta + 3V_\phi I_q \cos\delta \tag{C.13c}$$

Since

$$V_\phi \sin\delta = I_q X_q$$

or

$$I_q = \frac{V_\phi \sin\delta}{X_q} \tag{C.14}$$

and

$$V_\phi \cos\theta = E_a - I_d X_d$$

or

$$I_d = \frac{E_a - V_\phi \cos\delta}{X_d} \tag{C.15}$$

Substituting for I_d and I_q in Equation 7.68c,

$$P = 3V_\phi \left(\frac{E_a - V_\phi \cos\delta}{X_d} \right) \sin\delta + 3V_\phi \left(\frac{V_\phi \sin\delta}{X_q} \right) \cos\delta \tag{C.16}$$

or

$$P = \frac{3V_\phi E_a}{X_d} \sin\delta + 3V_\phi^2 \left(\frac{1}{X_q} - \frac{1}{X_d} \right) \sin\delta \cos\delta \tag{C.17}$$

Since

$$\sin\delta \cos\delta = \frac{(\sin 2\delta)}{2}$$

$$P = \frac{3V_\phi E_a}{X_d} \sin\delta + \frac{3V_\phi^2}{2} \left(\frac{1}{X_q} - \frac{1}{X_d} \right) \sin 2\delta \tag{C.18}$$

The first term is the same as the power in a cylindrical rotor synchronous machine, and the second term is the additional power due to the reluctance torque in the machine. Note that the second term depends on the saliency of the machine. Here,

$$\text{Saliency} = \frac{1}{X_q} - \frac{1}{X_d} \tag{C.19}$$

It disappears when $X_d = X_q$, which is the case for a cylindrical rotor. The developed torque of a synchronous machine can be expressed as

$$T_d = \frac{3V_\phi E_a}{\omega_s X_d} \sin\delta + \frac{3V_\phi^2}{2\omega_s}\left(\frac{1}{X_q} - \frac{1}{X_d}\right)\sin 2\delta \tag{C.20}$$

The torque developed by a salient-pole generator as a function of torque angle δ is plotted in Figure C.2. Note that the maximum reluctance torque occurs when δ is 45° and is about 20% of the available maximum torque. Also note that this type of machine is "*stiffer*" than a cylindrical rotor machine since its torque increases faster for small values of 3. The maximum power takes place at a value of δ less than 90°.

For a salient-pole machine, it is necessary to know the values of both X_d and X_q. These reactances are determined by the slip test. The slip test is performed by applying reduced three-phase voltage to the stator while the rotor is driven slightly above or below synchronous speed with the field winding open-circuited (i.e., unexcited). Oscillograms are taken of the armature terminal voltage, the armature current, and the induced voltage across the open-field winding. Figure C.3 shows such oscillograms. Approximate values of X_d and X_q can be found from the voltmeter and ammeter readings. For example, when armature resistance is ignored, the X_d and X_q can be determined from

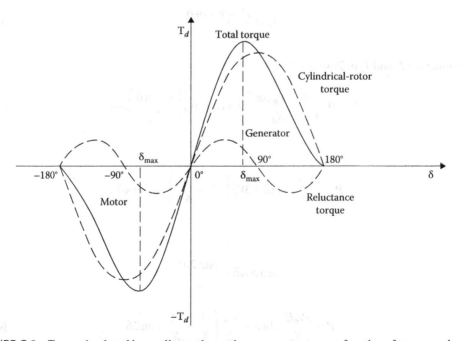

FIGURE C.2 Torque developed by a salient-pole synchronous generator as a function of torque angle.

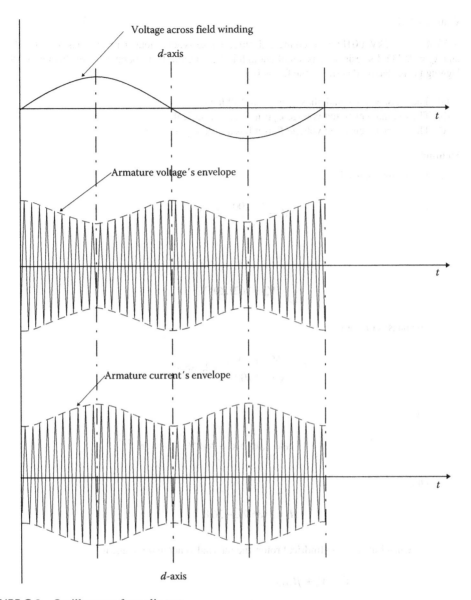

FIGURE C.3 Oscillograms from slip test.

$$X_d = \frac{V_{max}}{I_{min}} \tag{C.21}$$

and

$$X_q = \frac{V_{min}}{I_{max}} \tag{C.22}$$

Also, the ratio of maximum to minimum armature current gives the ratio of X_d/X_q. For example, assume that from the oscillograms it is found that X_d/X_q is 1.7. If X_d is known from the open- and short-circuit tests for the cylindrical-rotor machine, then X_q can be determined.

Example C.1

A 50 MVA, 13.2 kV, 60 Hz, wye-connected, three-phase synchronous generator has $X_d = 1.52\,\Omega$ and $X_q = 0.91\,\Omega$. Its resistance is small enough to be ignored. If it operates at full load at a 0.8 lagging power factor, determine the following:

(a) The phase voltage and phase current at full load
(b) The internal generated voltage E_a if it has a cylindrical rotor
(c) The internal generated voltage E_a if it has a salient-pole rotor

Solution

(a) The phase voltage is

$$V_\phi = \frac{13,200}{\sqrt{3}} = 7,621\,\text{V}$$

or

$$V_\phi = 7621\angle 0^\circ\,\text{V}$$

The phase current is

$$I_a = \frac{50 \times 10^6\,\text{VA}}{\sqrt{3}\,(13,200\,\text{V})} = 2,186.9\,\text{A}$$

and

$$\theta = \cos^1(0.8) = 36.87^\circ$$

Thus,

$$I_a = 2186.9\angle -36.87^\circ\,\text{A}$$

(b) If the machine has a cylindrical rotor, the internal generated voltage is

$$E_a = V_\phi + jI_aX_d$$

$$= 7621\angle 0^\circ + j\,1.52\,(2186.9\angle -36.87^\circ)$$

$$= 7621\angle 0^\circ + (1.52\angle 90^\circ)(2186.9\angle -36.87^\circ)$$

$$= 9976.41\angle 15.46^\circ\,\text{V}$$

(c) If the machine has a salient-pole rotor,

$$E_a = V_\phi + jI_aX_q$$

$$= 7621\angle 0^\circ + j(2186.9\angle -36.87^\circ)(0.91)$$

$$= 8815.14\angle 10.24^\circ\,\text{V}$$

Thus, the direction of E_a is $\delta = 10.24^\circ$. Hence, the magnitude of the d-axis component of current is

$$I_d = I_a \sin(\theta + \delta)$$

$$= 2186.9 \sin (36.87° + 10.24°)$$

$$= 1602.26 \, \text{A}$$

and the magnitude of the q-axis component of current is

$$I_q = I_a \cos(\theta + \delta)$$

$$= 2186.9 \cos(36.87° + 10.24°)$$

$$= 1488.39 \, \text{A}$$

By combining magnitudes and angles,

$$I_d = 1602.26 \angle - 79.76° \, \text{A}$$

$$I_q = 1488.39 \angle 10.24° \, \text{A}$$

and

$$\mathbf{E}_a = \mathbf{V}_\phi + j\mathbf{I}_d X_d + j\mathbf{I}_q X_q$$

$$= 7621 \angle 0° + j(1602.26 \angle - 79.76°)(1.52) + j(1488.39 \angle 10.24°)0.91$$

$$= 9935 \angle 10.24° \, \text{V}$$

Note that the angle δ for the salient-pole rotor is smaller than the one for the cylindrical rotor. Therefore, a salient-pole machine is said to be *stiffer* than one with a cylindrical rotor.

Appendix D: Unit Conversions from the English System to SI System

The following are useful when converting from the English system to SI system:

Length	1 in. = 2.54 cm = 0.0245 m
	1 ft = 30.5 cm = 0.305 m
	1 mile = 1609 m
Area	1 square mile = $2.59 \times 106\,\text{m}^2$
	$1\,\text{in.}^2 = 0.000645\,\text{m}^2$
	$1\,\text{in.}^2 = 6.45\,\text{cm}^2$
Volume	$1\,\text{ft}^3 = 0.0283\,\text{m}^3$
Linear speed	1 ft/s = 0.305 m/s = 30.3 cm/s
	1 mph = 0.447 m/s
	1 in./s = 0.0254 m/s = 2.54 cm/s
Rotational speed	1 rev/min = 0.105 rad/s = 6 deg/s
Force	1 lb = 4.45 N
Power	1 hp = 746 W = 0.746 kW
Torque	1 ft-lb = 1.356 N-m
Magnetic flux	1 line = 1 maxwell = 10^{-8} Wb
	1 kiloline = 1000 maxwells = 10^{-5} Wb
Magnetic flux density	$1\,\text{line/in.}^2 = 15.5{\times}10^{-6}$ T
	100 kilolines/in.^{2} = 1.55 T = 1.55 Wb/m^2
Magnetomotive force	1 ampere-turn = 1 A
Magnetic field intensity	1 A·turn/in. = 39.37 A/m

Appendix E: Unit Conversions from the SI System to English System

The following are useful when converting from the SI system to English system:

Length	$1\,m = 100\,cm = 39.37\,in.$
	$1\,m = 3.28\,ft$
	$1\,m = 6.22 \times 10^{-4}\,mile$
Area	$1\,m^2 = 0.386 \times 10^{-6}\,mile$
	$1\,m^2 = 1550\,in.^2$
	$1\,cm^2 = 0.155\,in.^2$
Volume	$1\,m^3 = 35.3\,ft^3$
Linear speed	$1\,m/s = 100\,cm/s = 3.28\,ft/s$
	$1\,m/s = 2.237\,mph$
	$1\,m/s = 39.37\,in./s$
Rotational speed	$1\,rad/s = 9.55\,rev/min = 57.3°/s$
Force	$1\,N = 0.225\,lb$
Power	$1\,kW = 1000\,W = 1.34\,hp$
Torque	$1\,N\text{-}m = 0.737\,ft\text{-}lb$
Magnetic flux	$1\,Wb = 10^8\,lines = 10^8\,maxwells$
	$1\,Wb = 10^5\,kilolines$
Magnetic flux density	$1\,T = 6.45 \times 10^4\,lines/in.^2$
	$1\,T = 1\,Wb/m^2$
Magnetomotive force	$1\,A = 1\,A\cdot turn$
Magnetic field intensity	$1\,A/m = 0.0254\,A\cdot turn/in.$

Appendix F: Stator Windings

F.1 INTRODUCTION

The stator (i.e., armature) winding of an ac generator is the source of the induced voltage. The coils of these windings are distributed in space, as shown in Figure F.1, so that the voltages induced in the coils are not in phase, but are displaced from each other by the *slot angle* α.

The coils may have *full pitch or fractional pitch*. A full-pitch* coil spans 180 electrical degrees (i.e., the peripheral distance from the center line of a north pole to the center line of an adjacent south pole) whereas a fractional-pitch coil spans less than 180°, but seldom less than 120°. Polyphase stator windings are usually *double-layer*, with the two layers arranged one above the other in the slot. In such an arrangement, the number of coils equals the number of slots. Thus, if there are n slots per pole per phase, each pole has n coils. The *single-layer* polyphase winding is employed in small induction motors. The main advantage of the two-layer winding is that of accommodating fractional-pitch coils, which have lower resistance turns or end connections than full-pitch coils. They also have a lower resistance without a corresponding decrease in their flux linkage. The windings can be of the *lap or wave* type. Wave windings are usually used in induction motors. Figure F.1 shows the coil interconnection of a three-phase, six-pole single-layer stator winding. Here, the armature has three slots per pole, which corresponds to one slot per phase per pole for a three-phase winding. The three-phase winding results from the addition of two more sets of armature coils displaced by 120° and 240° (in electrical degrees) from the first phase to produce a system of three voltages equal in magnitude and displaced from each other by 120°.

As previously stated, the armature winding coils are distributed over several slots in space.[†] Thus, the induced voltages in the coils are not in phase but are displaced from each other by the slot angle α. Hence, the winding voltage is the phasor sum of the coil voltages, as shown in Figure F.2. The *distribution* factor[‡] k_d is defined as

$$k_d = \frac{\text{phasor sum of coil voltages}}{\text{arithmetic sum of coil voltages}} \tag{F.1}$$

$$K_d = \frac{E_g}{nE_c} \tag{F.2}$$

where

E_g represents resultant group emf given by the phasor sum of coil voltages
E_c represents individual coil voltage
n represents number of slots per pole per phase

Therefore, from the geometry shown in Figure 7.2, the distribution factor can be reexpressed as

$$k_d = \frac{\sin(n\alpha/2)}{n\sin(\alpha/2)} \tag{F.3}$$

---

[*] It is also called **pole** pitch and is $2n/p$ radians, where p is the number of poles. Thus, the pole pitch is always 7r electrical radians or 180 electrical degrees, regardless of the number of poles on the machine.
[†] Such a winding is called a *distributed* winding.
[‡] It is also called the *breadth* factor.

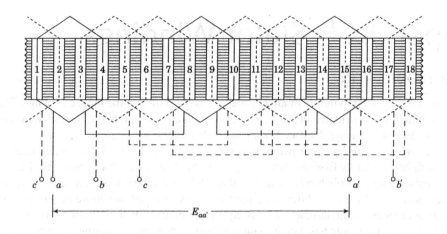

FIGURE F.1   The coil interconnection of a three-phase, six-pole single-layer armature (stator) winding.

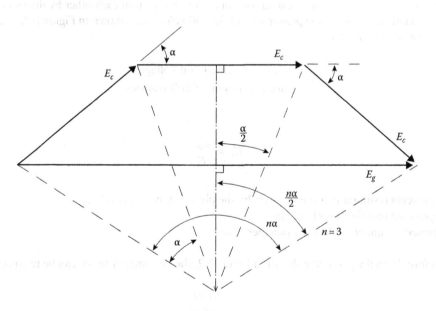

FIGURE F.2   Phasor addition of individual coil voltages to determine the distribution factor.

Note that the slot angle α is also called the slot *pitch* and is given in electrical degrees. Therefore, if q is the number of stator phases and s is the number of slots that exist in the stator, then the slot angle can be determined from

$$\alpha = \frac{180°}{nq} \qquad (F.4)$$

$$n = \frac{s}{pq} \qquad (F.5)$$

Since there are p coil groups, the phase voltages are

$$E_a = pE_g \qquad (F.6)$$

Therefore, the groups are interconnected in such a way that all of the group voltages add in phase. If there are n slots per pole per phase, then each pole group has n coils.

If the individual coils of a winding are short-pitched (i.e., if the coil span is less than a pole pitch), the induced voltage is less than the voltage that would be induced if the coil span were a full pole pitch. The *pitch factor* k_p of the coil is defined as

$$k_p = \frac{\text{voltage induced in short-pitch coil}}{\text{voltage induced in full-pitch coil}} \qquad (F.7)$$

or

$$k_p = \sin\frac{\rho}{2} \qquad (F.8a)$$

$$= \sin\left\{\frac{\text{coil span in electrical degrees}}{2}\right\} \qquad (F.8b)$$

where ρ is the *coil pitch or coil span* in electrical degrees.

If the coils of a winding are distributed in several slots and the coils are short-pitched, the voltage induced in the winding will be affected by both factors k_p and k_d. They can be combined into a single winding factor k_w. The winding factor of a stator is given by

$$k_w = k_p \times k_d \qquad (F.9)$$

Example F.1

A three-phase, six-pole, synchronous machine has a stator with 36 slots. Each coil in this double-layer winding has a span of five slots. Determine the following:

(a) The number of slots per pole per phase
(b) The slot pitch (or slot angle) in electrical degrees
(c) The coil pitch
(d) The pitch factor
(e) The number of coils in a phase group
(f) The distribution factor
(g) The winding factor

Solution

   (a) The number of slots per pole per phase is

$$n = \frac{s}{pq} = \frac{36}{(6)(3)} = 2$$

   (b) The slot pitch or slot angle is

$$\alpha = \frac{180°}{nq} = \frac{180°}{(2)(3)} = 30°$$

   (c) The coil pitch (i.e., coil span) is

$$\rho = (5)(30°) = 150°$$

   (d) The pitch factor is

$$k_p = \sin\frac{\rho}{2}$$

$$= \sin\frac{150°}{2}$$

$$= 0.9659$$

   (e) The number of coils in a phase group in such a double-layer winding is equal to the number of slots per pole per phase, which is 2.

   (f) The distribution factor is

$$k_d = \frac{\sin(n\alpha/2)}{n\sin(\alpha/2)}$$

$$= \frac{\sin\left[2(30°)/2\right]}{2\sin(30°/2)}$$

$$= 0.9659$$

   (g) The winding factor is

$$k_w = k_p k_d$$

$$= (0.9659)(0.9659)$$

$$= 0.933$$

PROBLEMS

F.1   A three-phase two-pole synchronous generator has stator windings with a double layer. There are 12 coils per phase with 20 turns in each coil. The windings have an electrical pitch of 150°. The shaft speed is 3600 rev/min. Assume that the flux per pole in the alternator is 0.0203 Wb, and determine, after reading this appendix, the following:

   (a) The slot pitch of the armature in electrical degrees
   (b) The coil span in terms of slots
   (c) The magnitude of the phase voltage
   (d) The terminal voltage of this alternator if the stator windings are connected in wye

Appendix G: Glossary for Electrical Machines Terminology

Some of the most commonly used terms, both in this book and in general usage, are defined in the following pages. Most of the definitions given in this glossary are based on Acarnley (1984), AIEE Standards Committee report (1946), Alger (1970), Anderson (1981, 1993), Bergseth (1987), Bewley (1949), and Blume et al. (1951).

ac: See **Alternating current**

Acceptance angle: The total range of sun positions from which sunlight can be collected by a system.

Activation polarization: The polarization resulting from the rate-determining step of the electrode reaction. See also **Polarization**.

Active distribution network: It is a distribution network that has been integrated with distributed generation.

Active filter: Any of a number of sophisticated power electronic devices for eliminating harmonic distortion.

Active material: The constituents of a cell that participate in the electrochemical charge–discharge reaction.

Additive polarity: If in a given transformer, terminals H_1 and X_1 are diagonally opposite, the transformer is said to have additive polarity. In an additive polarity, the mmfs are added to each other. Per ANSI rule, additive polarities are required for high-voltage power transformers that are greater than 200 kVA.

Admittance: The ratio of the phasor equivalent of the steady-state sine-wave current to the phasor equivalent of the corresponding voltage.

Adverse weather: Weather conditions which cause an abnormally high rate of forced outages for exposed components during the periods such conditions persist, but which do not qualify as major storm disasters. Adverse weather conditions can be defined for a particular system by selecting the proper values and combinations of conditions reported by the weather bureau: thunderstorms, tornadoes, wind velocities, precipitation, temperature, etc.

Ah: See **Ampere-hours**.

Air-blast transformer: A transformer cooled by forced circulation of air through its core and coils.

Air circuit breaker: A circuit breaker in which the interruption occurs in air.

Air-core transformer: A transformer having no ferromagnetic material but only air is present in its core.

Air gap: The space that is located between the outside of the rotor and the inside of the stator.

Air mass (AM): The length of the path through the earth's atmosphere traversed by direct radiation, expressed as a multiple of path length with the sun at zenith (overhead).

Air mass 0 (AM0): The amount of sunlight falling on the surface in outer space just outside the earth's atmosphere (1.4kW/m^2).

Air mass 1 (AM1): The amount of sunlight falling on the earth at sea level when the sun is shining straight down through a dry clean atmosphere. (A close approximation is Sahara Desert at high noon.) The sunlight intensity is very close to 1 kilowatt per square meter (1kW/m^2).

Air mass 2 (AM2): A closer approximation to usual sunlight conditions; may be simulated by ELH projector bulb. The illumination is 800W/m^2.

Air switch: A switch in which the interruptions of the circuit occur in air. Al: Symbol for aluminum.

Alternating current: The electrical current which reverses its direction of flow. Sixty cycles per second is standard current used by utilities in the United States.

Alternator: It is an ac generator.

Ambient condition: The conditions of the surroundings.

Ambient temperature: The prevailing temperature outside a building.

Ampacity: Current rating in amperes, as of a conductor.

Ampere's circuital law: The magnetic potential drop around a closed path is balanced by the mmf, giving rise to the field.

Ampere-hours (Ah): A current of one ampere running 1 h.

Ampere-hour efficiency: See **Efficiency, ampere-hour**.

Ampere's right-hand rule: See **Right-hand rule**.

Ampere-turns (of a coil): The product of the current in a coil in amperes and the number of turns of the coil. It represents the amount of magnetomotive force which in turn represents the coil's ability to produce flux.

Ångstrom (abbreviation, Å) A unit of length, 10^{-10} m.

Anode: The electrode in an electrochemical cell at which chemical oxidation takes place. During charge, the positive terminal of the cell is the anode. During discharge, the situation reverses and the negative terminal of the cell is the anode. However, battery manufacturers sometimes refer to the negative terminal as the anode during both charge and discharge. See also **Cathode**.

Anolyte: The electrolyte associated with the anode in an electrochemical cell. In a secondary battery, the anolyte is associated with the negative terminal.

ANSI: Abbreviation for American National Standards Institute.

Armature winding: In synchronous machine, it is located on the stator and has an ac current in its winding whereas, in a dc machine, it is located on the rotor and has also an ac current in its winding.

Askarel: A generic term for a group of nonflammable synthetic chlorinated hydrocarbons used as electrical insulating media. Askarels of various compositional types are used. Under arcing conditions the gases produced, while consisting predominantly of noncombustible hydrogen chloride, can include varying amounts of combustible gases depending upon the askarel type. Because of environmental concerns, it is not used in new installations anymore.

Automatic substations: Those in which switching operations are so controlled by relays that transformers or converting equipment are brought into or taken out of service as variations in load may require, and feeder circuit breakers are closed and reclosed after being opened by overload relays.

Autotransformer: A transformer in which at least two windings have a common section. An autotransformer has a single winding, part of which is common to both the primary and the secondary simultaneously. Hence, in an autotransformer, there is no electrical isolation between the input side and the output side. As a result, the power is transferred from the primary to the secondary through both induction and conduction. They can be used as a step-down or step-up transformer. The *common winding* is the winding between the low-voltage terminals while the remainder of the winding belonging exclusively to the high-voltage circuit is called the *series winding*. This combined with the common winding forms the series-common winding between the high voltage terminals. In a sense, an autotransformer is just a normal two-winding transformer connected in a special way. But, the series winding must have extra insulation in order to be just as strong as the one on the common winding. Autotransformers are increasingly used to interconnect two high-voltage transmission lines operating at different voltages. They can be used as step-down or step-up transformers.

Average penetration: It is the ratio of annual energy from renewable energy–powered generators in kWh to total annual energy delivered to loads in kWh. It is used when fuel or CO_2-emission savings are being considered.

AWG: Abbreviation for American Wire Gauge. It is also sometimes called the Brown & Sharpe Wire Gauge.

Backey-Robinson machines: They are brushless synchronous machines that operate based on rotor magnetic structure with changing reluctance. They are mainly used in aerospace applications.

Balanced load: When the load impedances of all three-phase are identical, the load is a balanced load.

Balanced system: A three-phase power system that is made of a balanced three-phase source is connected to a balanced three-phase load over three-phase balanced lines.

Balanced voltages (or currents): Three-phase voltages (or currents) which are equal in magnitude and are 120° out of phase with respect to each other and form a symmetrical three-phase set.

Balance-of-systems (BOS): All the equipment needed to make a complete working power system, except the solar cell modules.

Base load: The minimum load over a given period of time.

Basic impulse insulation level: See **BIL**.

Battery: Two or more electrochemical cells electrically interconnected in an appropriate series or parallel arrangement to provide the required operating voltage and current levels. Under common usage, the term battery also applies to a single cell if it constitutes the entire electrochemical storage system.

Battery, marine: A deep-discharge battery used on boats; capable of discharging small amounts of electricity over a long period of time.

Battery, motive-power: A large-capacity deep-discharge battery designed for long life when used in electric vehicles.

Battery, secondary: See **Battery, storage**.

Battery, stationary: For use in emergency standby power systems, a battery with long life but poor deep-discharge capabilities.

Battery, storage: A secondary battery; rechargeable electric storage unit that operates on the principle of charging electric energy into chemical energy by means of a reversible chemical reaction. The lead-acid automobile battery is the most familiar of this type.

BIL: Abbreviation for basic impulse insulation levels, which are reference levels expressed in impulse-crest voltage with a standard wave not longer than $1.5 \times 50\,\mu s$. The impulse waves are defined by a combination of two numbers. The first number is the time from the start of the wave to the instant crest value; the second number is the time from the start to the instant of half-crest value on the tail of the wave.

Braking mode (of operation): A three-phase induction motor running at a steady-state speed can be brought to a quick stop by interchanging two of its leads. By doing this, the phase sequence, and therefore the direction of rotation of the magnetic field, is suddenly reversed, the motor comes to a stop under the influence of torque and is immediately disconnected from the line before it can start in other direction. This is also known as the *plugging operation*.

Branch circuit: A set of conductors that extend beyond the last overcurrent device in the low-voltage system of a given building. A branch circuit usually supplies a small portion of the total load.

Breakdown: Also termed puncture, denoting a disruptive discharge through insulation.

Breaker, primary-feeder: A breaker located at the supply end of a primary feeder which opens on a primary-feeder fault if the fault current is of sufficient magnitude.

British thermal unit (Btu): A unit of energy which is equal to the amount of heat required to raise the temperature of 1 lb of water (at 4°C) 1°F.

Built-in potential: The electrical potential that develops across the junction when two dissimilar solids are joined. The open-circuit voltage of photovoltaic cell will approach but always less than this potential.

Bus: A conductor or group of conductors that serves as a common connection for two or more circuits in a switchgear assembly.

Bushing: An insulating structure including a through conductor, or providing a passageway for such a conductor, with provision for mounting on a barrier, conductor or otherwise, for the purpose of insulating the conductor from the barrier and conducting from one side of the barrier to the other.

BX cable: A cable with galvanized interlocked steel spiral armor. It is known as ac cable and used in a damp or wet location in buildings at low voltage.

Cable: Either a standard conductor (single-conductor cable) or a combination of conductors insulated from one another (multiple-conductor cable).

Cable fault: A partial or total load failure in the insulation or continuity of the conductor.

Capability: The maximum load-carrying ability expressed in kilovoltamperes or kilowatts of generating equipment or other electric apparatus under specified conditions for a given time interval.

Capability, net: The maximum generation expressed in kilowatt-hours per hour which a generating unit, station, power source, or system can be expected to supply under optimum operating conditions.

Capacitor bank: An assembly at one location of capacitors and all necessary accessories (switching equipment, protective equipment, controls, etc.) required for a complete operating installation.

Capacity: The rated load-carrying ability expressed in kilovoltamperes or kilowatts of generating equipment or other electric apparatus.

Capacity, energy: The total number of watt-hours (kilowatt-hours) that can be withdrawn from a fully charged cell or battery. The energy capacity of a given cell varies with temperature, rate, age, and cutoff voltage. This term is more common to system designers than it is to the battery industry where the capacity usually refers to ampere-hours. Generally, the total number of ampere-hours that can be withdrawn from a fully charged cell or battery.

Capacity factor: The ratio of the average load on a machine or equipment for the period of time considered to the capacity of the machine or equipment.

Capacity, installed: The total number of ampere-hours that can be withdrawn from a new cell or battery when discharged to the system-specified cutoff voltage at the system design rate and temperature (i.e., discharge to the system design–specified maximum depth of discharge).

Capacity, rated: The manufacturer's conservative estimate of the total number of ampere-hours that can be withdrawn from a new cell or battery for a specific discharge rate, temperature, and cutoff voltage.

Cascade cells: See **tandem solar cells**.

Cathode: The electrode in an electrochemical cell at which chemical reduction takes place. During charge, the negative terminal of the cell is the cathode. During discharge, the situation reverses and the positive terminal of the cell is the cathode. However, battery manufacturers sometimes refer to the positive terminal as the cathode during both discharge and charge. See also **Anode**.

Catholyte: The electrolyte associated with the cathode of an electrochemical cell in a secondary battery. The catholyte is usually associated with the positive terminal.

Cell: The basic electrochemical unit used to store electrical energy.

Cell capacity: Expressed in ampere-hours, the total amount of electricity that can be drawn from a fully charged battery until it is discharged to a specific voltage.

Characteristic quantity: The quantity or the value of which characterizes the operation of the relay.

Characteristics (of a relay in steady state): The locus of the pickup or reset when drawn on a graph.

Charge: The amount paid for a service rendered or facilities used or made available for use.

Charging: The conversion of electrical energy into chemical potential energy within a cell by the passage of a direct current in the direction opposite to that of discharge.

Charge rate: The current applied to a cell or battery to restore its available capacity. This rate is commonly normalized with respect to the rated capacity of the cell or battery. For example, the 10 h charge rate of a 500 A h cell or battery is expressed as

$$\text{Rated capacity/Charge time} = (500 \text{ A h})/10 \text{ h} = C/10 \text{ rate}$$

For the same cell or battery, a 5 h charge rate would be designated the C/5 rate and would result in a charging current of 100 A. Unfortunately, the ampere-hour efficiency of many secondary batteries is less than 100% depth discharge, is often longer than indicated by so-called hour rate.

CHP systems: Combined heat and power systems. (*Cogeneration* systems.)

Chopped-wave insulation level: It is determined by test using waves of the same shape to determine the BIL, with exception that the wave is chopped after about 3 μs.

CIGRÉ: It is the international conference of large high-voltage electric systems. It is recognized as a permanent nongovernmental and nonprofit international association based in France. It focuses on issues related to the planning and operation of power systems, as well as the design, construction, maintenance, and disposal of high-voltage equipment and plants.

Circuit breaker: A device that interrupts a circuit without injury to itself so that it can be reset and reused over again.

Circuit, earth (ground) return: An electric circuit in which the earth serves to complete a path for current.

Circular mil: A unit of area equal to $\pi/4$ of a square mil (= 0.7854 square mil). The cross-sectional area of a circle in circular mils is therefore equal to the square of its diameter in mils. A circular inch is equal to 1 million circular mils. A mil is one-thousandth of an inch. There are 1974 circular mils in a square millimeter. Abbreviated *cmil*.

CL: Abbreviation for current-limiting (fuse).

cmil: Abbreviation for circular mil.

Collector: Any device gathering the sun's radiation and converting it to a useful energy form.

Collector efficiency: The ratio of the energy collected by a solar collector to the radiant energy incident on the collector.

Collector tilt angle: The angle at which the collector aperture is slanted upward from the horizontal plane.

Complex (or phasor) power: It is a phasor power that is equal to $P + j Q$.

Component: A piece of equipment, a line, a section of a line, or a group of items which is viewed as an entity.

Computer usage:

    Off-line usage: It includes research, routine calculations of the system performance, and data assimilations and retrieval.

    On-line usage: It includes data-logging and the monitoring of the system state, including switching, safe interlocking, plant loading, post-fault control, and load shedding.

Concentration ratio (concentration factor): The ratio of radiant energy intensity at the hot spot of a focusing collector to the intensity of unconcentrated direct sunshine at the collector site.

Concentrator: A solar collector that focuses or funnels solar radiation onto a relatively small absorber at a focal point or line.

Condenser: Also termed capacitor; a device whose primary purpose is to introduce capacitance into an electric circuit. The term condenser is deprecated.

Conduction: The heat transfer through matter by exchange of kinetic energy from particle to particle.

Conductor: A substance which has free electrons or other charge carriers that permit charge flow when an emf is applied across the substance.

Connection charge: The amount paid by a customer for connecting the customer's facilities to the supplier's facilities.

Contactor: An electric power switch not operated manually and designed for frequent operation.

Contract demand: The demand that the supplier of electric service agrees to have available for delivery.

Core losses: See **Magnetic core losses**.

Cress factor: A value which is displayed on many power quality monitoring instruments representing the ratio of the crest value of the measured waveform to the rms value of the waveform. For example, the cress factor of a sinusoidal wave is 1.414.

CT: Abbreviation for current transformers.

Cu: Symbol for copper.

Current transformers (CTs): Current transformers are connected in series with the line. They are used to step down currents to measurable levels. The secondaries of both voltage and current transformers are normally grounded. They are usually rated on the basis of 5 A secondary current and used to reduce primary current to usable levels for transformer-rated meters and to insulate and isolate meters from high-voltage circuits. *During the operation of a current transformer, its secondary terminals must never be open-circuited.* When the secondary is open-circuited, the primary mmf is not balanced by corresponding secondary mmf (in other words, there will be no secondary mmf to oppose the primary mmf.) Thus, all of the primary current becomes excitation current. Consequently, a very high flux density is produced in the core, causing a very high voltage to be induced in the secondary. In addition to endangering the user, it may damage the transformer insulation and also cause overheating due to excessive core losses. Thus, *if the ammeter needs to be removed, the proper procedure is to close the shorting switch first.*

Current transformer burdens: CT burdens are normally expressed in ohms impedance such as B-0.1, B-0.2, B-0.5, B-0.9, or B-1.8. Corresponding volt-ampere values are 2.5, 5.0, 12.5, 22.5, and 45.

Current transformer ratio: CT ratio is the ratio of primary to secondary current. For current transformer rated 200:5, the ratio is 200:5 or 40:1.

Cutoff voltage: The cell or battery voltage at which discharge is terminated. The cutoff voltage is specified by the cell manufacturer and is generally a function of discharge rate. Discharge beyond the specified cutoff voltage usually results in increasingly rapid decreases in cell voltage and energy output; it may permanently damage a cell and it may void a manufacturer's warranty (also called *final voltage*).

Cycle: One discharge–charge sequence to a specified depth of discharge.

Cycle life: The number of cycles, to a specified depth of discharge, that a cell or battery can undergo before failing to meet its specified capacity or efficiency performance criteria.

Cycle service: A duty cycle characterized by frequent and usually deep discharge–charge sequences, such as motive power applications.

dc: direct current; electric current that always flows in the same direction, positive to negative. Photovoltaic cells and batteries are all dc devices.

Declination: The angular position of the sun north (+) or south (−) of the plane of the earth's equator; it is a function of the time of year.

Deep-discharge: The withdrawal of a significant percentage of rated capacity (50% or more).

Deep-discharge cycles: Cycles in which a battery is nearly completely discharged.

Delay angle (α): The time, expressed in electrical degrees, by which the starting point of commutation is delayed. It cannot exceed 180°. It is also called *ignition angle* or *firing angle*.

Delta–delta connection: In such connection, there is no phase shift and no problem with unbalanced loads or harmonics. If one transformer fails in service, the remaining two transformers in the bank can be operated as an open-delta or (V–V) connection at about 58% of the original capacity of the bank. However, in a complete delta–delta bank, transformers tend to share the load inversely to their internal impedances, and therefore identical transformers have to be used.

Depth of discharge (DOD): The ampere-hours removed from a fully charged cell or battery, expressed as a percentage of rated capacity. For example, the removal of 25 A h from a fully charged 100 A h rated cell results in a 25% depth of discharge. Under certain conditions, such as discharge rates lower than that used to rate the cell, the depth of discharge can exceed 100%.

Delta–wye connection: There is no problem with third-harmonic components in its voltages. It has the same advantages and the same phase shift as the wye–delta connection. The secondary voltage lags the primary voltage by 30°, as is the case for the wye–delta connection. In general, when a wye–delta or delta–wye connection is used, the wye is preferably on the high-voltage side, and the neutral is grounded. Hence, the transformer insulation can be manufactured to withstand $\sqrt{3}$ *times* the line voltage, instead of the total line voltage.

Demand: The load at the receiving terminals averaged over a specified interval of time.

Demand factor: The ratio of the maximum coincident demand of a system, or part of a system, to the total connected load of the system, or part of the system, under consideration.

Demand, instantaneous: The load at any instant.

Demand, integrated: The demand integrated over a specified period.

Demand interval: The period of time during which the electric energy flow is integrated *in determining demand*.

Depreciation: The component which represents an approximation of the value of the portion of plant consumed or "used up" in a given period by a utility.

DER: See **Distributed energy resources**.

Developed torque: The mechanical torque developed by the electromagnetic energy conversion process. It can be found by dividing the developed power by the shaft speed. Further, it can be found by dividing the air-gap power by the angular synchronous speed. Because of this, the air-gap power is also called *the torque in synchronous watts*.

Diffuse insulation: The sunlight scattered by atmospheric particulates that arrives from a direction other than the direction of direct sunlight. The blue color of sky is an example of diffuse solar radiation.

Direct axis: It is the *stator (pole) axis*. It is also called *d-axis*.

Direct current: See **dc**.

Direct conversion: The conversion of sunlight directly into electric power, instead of collecting sunlight as heat and using the heat to produce power. Solar cells are direct conversion devices.

Direct radiation: The radiation from sunlight directly into electric power. Solar cells are direct conversion devices.

Discharge: The process of withdrawing current from a cell or battery by the conversion of chemical energy into electrical energy.

Discharge rate: The current removed from a cell or battery. This rate can be expressed in amperes, but more commonly, it is normalized to rated capacity C and expressed as C/X. For example, drawing 290 A from a cell with a rated capacity of 100 A h is referred to as $C/5$ discharge rate (100 A h/20 A). Similarly, discharge currents of 5, 10, and 33.3 A would be designated as the $C/20$, $C/10$, and $C/3$ rates respectively. See also **Hour rate**.

Disconnecting or isolating switch: A mechanical switching device used for changing the connections in a circuit or for isolating a circuit or equipment from the source of power.

Disconnector: A switch that is intended to open a circuit only after the load has been thrown off by other means. Manual switches designed for opening loaded circuits are usually installed in a circuit with disconnectors to provide a safe means for opening the circuit under load.

Distributed energy resources (DERs): The new energy resources that are connected to the utility distribution network.

Distributed generation: Small generators are mostly connected to the distribution system network. Such generation is known as distributed generation (DG) or *dispersed generation*. It is also called *embedded generation* since it is embedded in the distribution network. Power in such power systems may flow from point to point within the distribution network.

Distributed generation (DG): It is a new approach for generating power using nonconventional/renewable energy sources such as natural gas, wind power, solar photovoltaic cells, biogas, cogenerators, Stirling engines, and microturbines.

Distribution automation: It is a system that enables an electric utility to remotely monitor, coordinate, and operate distribution companies in real-time mode from remote locations.

Distribution center: A point of installation for automatic overload protective devices connected to buses where an electric supply is subdivided into feeders and/or branch circuits.

Distribution switchboard: A power switchboard used for the distribution of electric energy at the voltages common for such distribution within a building.

Distribution system: That portion of an electric system which delivers electric energy from transformation points in the transmission, or bulk power system, to the consumers.

Distribution transformer: A transformer for transferring electric energy from a primary distribution circuit to a secondary distribution circuit or consumer's service circuit; it is usually rated in the order of 5–500 kVA.

Doping: The deliberate addition of a known impurity (dopant) to a pure semiconductor to produce the desired electric properties.

Dot convention: It is a convention that implies that (1) currents entering at the dotted terminals will result in mmfs that will produce fluxes in the same direction, and (2) voltages from the dotted to undotted terminals have the same sign.

Dynamo: It is also called a dc generator.

Eddy currents: Because iron is a conductor, time-varying magnetic fluxes induce opposing voltages and currents eddy currents that circulate within the iron that tend to flow perpendicular to the flux and in a direction that opposes any change in the magnetic field due to Lenz's law. The induced eddy currents tend to establish a flux that opposes the original change imposed by the source.

Eddy-current loss: It is a power loss due to the existence of eddy currents.

Effectively grounded: Grounded by means of a ground connection of sufficiently low impedance that fault grounds which may occur cannot build up voltages dangerous to connected equipment.

Efficiency: The ratio of the measure of a desired effect to the measure of the input causing the effect, both expressed in the same units of measure.

Efficiency, ampere-hour (*Columbic*): The ratio of ampere-hours removed from a cell or battery during discharge to the ampere-hours required to restore the initial capacity.

Efficiency, energy (*watt-hour*): The ratio of the energy delivered by a cell or battery during a discharge to total energy required to restore the initial state of charge. The watt-hour efficiency is approximately equal to the product of the voltage and ampere-hour efficiencies. This is sometimes referred to as the round-trip efficiency. Round-trip efficiencies usually do not include energy losses resulting from self-discharge, auxiliary equipment (parasitic losses), or battery equalization.

Efficiency, voltage: The ratio of the average discharge voltage of a cell or battery to the average charge voltage during the subsequent restoration of an equivalent capacity.

EHV: Abbreviation for extra high voltage.

Electric rate schedule: A statement of an electric rate and the terms and conditions governing its application.

Electric system loss: Total electric energy loss in the electric system. It consists of transmission, transformation, and distribution losses between sources of supply and points of delivery.

Electrical reserve: The capability in excess of that required to carry the system load.

Electrolyte: A liquid conductor of electricity. The medium which provides the ion transport mechanism between the positive and negative electrodes of a cell. In some cells, such as the lead-acid type, the electrolyte may also participate directly in electrochemical charge–discharge reactions.

Electronic-grade silicon: Silicon, electronic-grade.

Emergency rating: Capability of installed equipment for a short time interval.

Emittance: The ratio of radiation emitted by real surface to the radiation emitted by a perfect radiator at the same temperature. Normal emittance is the value measured at 90° to the plane of the sample; hemispheric emittance is total amount emitted in all directions.

End-of-charge voltage: The cell or battery voltage at which the finishing charge is normally terminated by the charging source.

Energy: That which does work or is capable of doing work. As used by electric utilities, it is generally a reference to electric energy and is measured in kilowatt-hours.

Energy conversion principle: Energy is neither created nor destroyed. It is simply changed in form. The role of electromagnetic (or electromechanical) machines is to transmit energy from one type of energy to another.

Energy density: The ratio of rated energy available from a cell or battery, normalized to its weight or volume.

Energy loss: The difference between energy input and output as a result of transfer of energy between two points.

Energy management system (EMS): A computer system that monitors, controls, and optimizes the transmission and generation facilities with advanced applications. A SCADA system is a subject of an EMS.

Equivalent commutating resistance (R_c): The ratio of drop of direct voltage to direct current. However, it does not consume any power.

Express feeder: A feeder which serves the most distant networks and which must traverse the systems closest to the bulk power source.

Extra high voltage: A term applied to voltage levels higher than 230 kV. Abbreviated EHV.

FA: Self-cooled.

Facilities charge: The amount paid by the customer as a lump sum, or periodically, as reimbursement for facilities furnished. The charge may include operation and maintenance as well as fixed costs.

Faraday's law of induction: Whenever a flux passes through a turn of a coil, a voltage (i.e., an electromotive force [*emf*]) is induced in each turn of that coil, which is directly proportional to the rate of change in the flux with respect to time.

Feeder: A set of conductors originating at a main distribution center and supplying one or more secondary distribution centers, one or more branch-circuit distribution centers, or any combination of these two types of load.

Feeder, multiple: Two or more feeders connected in parallel.

Fermi level: Energy level at which the probability of finding an electron is one-half. In a metal, the Fermi level is very near the top of the filled levels in the partially filled valance band. In a semiconductor, the Fermi level is in the band gap.

Ferromagnetic materials: They include certain forms of iron and its alloys in combination with cobalt, nickel, aluminum, and tungsten. They are relatively easy to magnetize since they have a high value of relative permeability. They also include ferrites and therefore made up of iron oxides.

Field winding: It is the excitation or exiting winding on a machine. Its location changes depending on the machine type. In a synchronous machine, it is located on the rotor. But, in a dc machine, it is located on the stator.

First-contingency outage: The outage of one primary feeder.

Fixed-capacitor bank: A capacitor bank with fixed, not switchable, capacitors.

Flash: A term encompassing the entire electrical discharge from cloud to stricken object.

Flashover: An electrical discharge completed from an energized conductor to a grounded support. It may clear itself and trip a circuit breaker.

Flexible ac transmission systems (FACTS): They are the converter stations for ac transmission. It is an application of power electronics for control of the ac system to improve the power flow, operation, and control of the ac system.

Flicker: Impression of unsteadiness of visual sensation induced by a light stimulus whose luminance or spectral distribution fluctuates with time.

Flicker factor: A factor used to quantify the load impact of electric arc furnaces on the power system.

Flux linkage (λ): It is the product of the turns N of a winding and the flux Φ that links its turns. It is in Weber (Wb).

FOA: Forced oil cooled.

Forced interruption: An interruption caused by a forced outage.

Forced outage: An outage that results from emergency conditions directly associated with a component, requiring that it be taken out of service immediately, either automatically or as soon as switching operations can be performed or an outage caused by improper operation of equipment or by human error.

Frequency deviation: An increase or decrease in the power frequency. Its duration varies from few cycles to several hours.

Fringing: If there is an air gap in a magnetic circuit, there is a tendency for the flux to bulge out outward (or spread out) along the edges of the air gap rather than to flow straight through the air gap parallel to the edges of the core. This phenomenon is called fringing and is taken into account by assuming a larger (effective) air-gap cross-sectional area. The common practice is to use an effective air-gap area by adding the air-gap length to each of the two dimensions which make up the cross-sectional area.

Fuse: An overcurrent protective device with a circuit-opening fusible part that is heated and severed by the passage of overcurrent through it.

Generation embedded distribution network: See **Active distribution network**.

Green power: It is used in a narrow sense to mean electric products that generated from renewable sources that are environmentally and socially acceptable. Most renewable energy forms are readily converted to electricity.

Grid: It is a term that is usually used to describe the totality of the electric power network.

Grid-connected: It means connected to any part of the network.

Grid-connected generator: The renewable energy generator supplies power to a large interconnected grid which is also supplied power by the other generators. The power supplied by the generator is only a small portion of the total power supplied to the grid.

Ground: Also termed earth; a conductor connected between a circuit and the soil; an accidental ground occurs due to cable insulation faults, an insulator defect, etc.

Ground protective relay: Relay which functions on the failure of insulation of a machine, transformer, or other apparatus to ground.

Ground wire: A conductor having grounding connections at intervals, which is suspended usually above but not necessarily over the line conductor to provide a degree of protection against lightning discharges.

Grounding: It is the connection of a conductor or frame of a device to the main body of the earth. Thus it must be done in a way to keep the resistance between the item and the earth is under the limits. It is often that the burial of large assemblies of conducting rods in the earth, and the use of connectors in large cross diameters are needed.

Han: It stands for "home area network"; it is used to identify the network of communicating loads, appliances, and sensors beyond the smart meter and within the customer's property.

Harmonics: Sinusoidal voltages or currents having frequencies that are an integer multiples of the fundamental frequency at which the supply system is designed to operate.

Harmonic distortion: Periodic distortion of the sign wave.

Harmonic resonance: A condition in which the power system is resonating near one of the major harmonics being produced by nonlinear elements in the system, hence increasing the harmonic distortion.

Hazardous open circulating (in *CTs*): The operation of the CTs with the secondary winding open can result in a high voltage across the secondary terminals, which may be dangerous to personnel or equipment. Therefore, the secondary terminals should always be short-circuited before a meter is removed from service.

High-speed relay: A relay that operates in less than a specified time. The specified time in present practice is 50 ms (i.e., 3 cycles on a 60 Hz system).

Holes: An energy level that could be an electron, but currently is not. Holes act like charged particles, with energy and momentum, and are capable of carrying an electric current.

Hour rate: The discharge rate of a cell or battery expressed in terms of the length of time that a new fully charged cell or battery can be discharged at a specific current before reaching a specified cutoff voltage:

$$\text{hour rate} = C/I$$

where
  C is the rated capacity of the cell or battery
  i is the specified discharge current

For example, if a new fully charged cell rated at 100 A h can be discharged at 20 A for a period of 5 h before reaching the cutoff voltage, discharge of the battery at 20 A is referred to as the 5 h rate voltage, discharge of the battery of 20 A is referred to as 5 h rate (C/I = 100 A h/20 A). Unfortunately, for most cells, the available capacity is not constant with discharge rate. Therefore, the discharge current cannot always be calculated by dividing discharge time into rated capacity. For example, a lead-acid battery with a rated capacity of 100 A h at the 5 h rate may only deliver 88 A h at the 3 h rate before reaching the cutoff voltage. The discharge current at the 3 h rate would therefore be 29.3 A rather than 33 A as might otherwise be expected and would be designed as the C/3.4 rate. See also **Discharge rate**.

Hours of capacity: The total number for hours for which a fully charged cell or battery is capable of supplying the system load demand before reaching the specified maximum depth of discharge.

HV: Abbreviation for high voltage.

Hybrid system: A system that produces both usable heat as well as electricity.

Hydrogen economy: A system in which hydrogen is substituted for fossil fuels. The hydrogen is basically a means of transporting and storing renewable energy.

Hysteresis machines: It is similar to a reluctance machine with a solid cylindrical rotor made of a permanent magnet material that needs only one electrical input. It is used in phonograph turntables and in other constant-speed applications, such as electrical clocks.

Ideal transformer: A transformer having no losses and its magnetic core material does not saturate.

IED: Any device incorporating one or more processors with the capability to receive or send data or control from or to an external source (e.g., electronic multifunction meters, digital relays, or controllers).

IED integration: Integration of protection, control, and data acquisition functions into a minimal number of platforms to reduce capital and operating costs, reduce panel and control room space, and eliminate redundant equipment and database.

Impedance: The ratio of the phasor equivalent of a steady-state sine-wave voltage to the phasor equivalent of a steady-state sine-wave current.

Impedance matching: It is used to determine the *maximum power transfer* from a source with an internal impedance Z_s to a load impedance Z_L. It is necessary to select the turns ratio for maximum transfer so that

$$Z'_L = \left(\frac{N_1}{N_2} \right)^2 Z_L = a^2 Z = Z_s$$

Impedance ratio: It is the square of the ratio of the number of turns in primary winding to the number of turns in the secondary winding.

Impedance (*referring of*): Transferring an impedance from one side of the transformer to the other side. It is also called as *reflecting, transferring,* or *scaling* the impedance.

Incremental energy costs: The additional cost of producing or transmitting electric energy above some base cost.

Index of reliability: A ratio of cumulative customer minutes that service was available during a year to total customer minutes demanded; can be used by the utility for feeder reliability comparisons.

Indium oxide: A wide band gap semiconductor that can be heavily doped with tin to make a highly conductive transparent thin film. Often used as a front contact or one component of a heterojunction solar cell.

Indoor transformer: A transformer that must be protected from the weather.

Inductance (*L*) **of a coil**: It is the flux linkage per ampere of current in the coil and is measured in H.

Induction machine: Even though the induction machine with a wound rotor can be used as a generator, its performance characteristics (especially in comparison to asynchronous generator) have not been satisfactory in most applications. However, induction generators are occasionally used at hydroelectric power plants. Recently, they are also used in wind turbines. They, with a wound rotor, can be used as a *frequency changer.*

Inductor and flux-switch machines: They have inductor flux-switch configurations based on a changeable-reluctance principle similar to the reluctance machines, and a function of rotor position accomplished by the rotor design. They can be used as brushless synchronous motors and generators in aerospace and traction applications.

Input kVA: It is slightly more than the output kVA due to the losses involved. It is equal to the rated primary voltage which is multiplied by the rated primary current.

Inrush current: Occasionally upon energizing a power transformer, a transient phenomenon (due to magnetizing current characteristics) takes place even if there is no load connected to its secondary. As a result, its magnetizing current peak may be several times (about eight times) the rated transformer current, or it may be practically unnoticeable. Because of losses in the excited winding and magnetic circuit, this current ultimately decreases to the normal value of the excitation current.) Such transient event is known as the current inrush phenomenon. It may cause (1) a momentary dip in the voltage if the impedance of the excitation source is significant; (2) undue stress in the transformer windings; or (3) improper operation of protective devices (e.g., tripping overload or common differential relays). The magnitude of such an inrush current depends on the magnitude, polarity, and rate of change in applied voltage at the time of switching.

Installed capacity: See **Capacity, installed**.

Installed reserve: The reserve capability installed on a system.

Instantaneous penetration: It is the ratio of power from renewable energy powered generators in kW to the total powered delivered to loads in kW. (The maximum instantaneous penetration is much greater than the average penetration.)

Instrument transformers: They are used in ac power circuits to provide safety for the operator and equipment from high voltage; they permit proper insulation levels and current-carrying capacity in relays, meter, and other equipments. They are of two types: *current transformers* (CRs) and *voltage transformers* (VTs) which were formerly called *potential transformers* (PTs). In the United States, the standard instruments and relays are rated at 5 A and/or 120 V, 60 Hz. Regardless of the type of instrument transformer in use, the external load applied to its secondary is referred to as its *burden*. The burden usually describes the impedance connected to the transformer's secondary winding, but may specify the voltamperes supplied to the load. For example, a transformer supplying 5 A to a resistive burden of 0.5 Ω may also be said to have a burden of 12.5 VA at 5 A.

Integrated: It means the physical connection of the generator to the network with due attention to the secure and safe operation of the system and the control of the generator to achieve optimality in terms of the energy resource usage.

Integration: It means the physical connection of the generator to the network with due attention to the secure and safe operation of the system and the control of the generator to achieve optimality in terms of the energy resource usage.

Interconnections: See tie lines.

International Electrotechnical Commission (IEC): An international organization whose mission is to prepare and publish standards for all electrical, electronic, and related technologies.

Interpole axis: It is also called *quadrature axis*, or simply *q-axis*.

Interruptible load: A load which can be interrupted as defined by contract.

Interruption: It is the loss of service to one or more consumers or other facilities and is the result of one or more component outages, depending on system configuration. An interruption is the result of one or more component outages.

Interruption duration: The period from the initiation of an interruption to a consumer until service has been restored to that consumer.

Inverse time-delay relay: A dependent time-delay relay having an operating time that is an inverse function of the electrical characteristic quantity.

Inverse time-delay relay with definite minimum: A relay in which the time delay varies inversely with the characteristic quantity up to a certain value, after which the time delay becomes substantially independent.

Inverter: A converter for changing direct current to alternating current.

Investment-related charges: Those certain charges incurred by a utility which are directly related to the capital investment of the utility.

Iron losses: See **Magnetic core losses**.

I_{sc}: See **Short-circuit current**.

ISO: Independent system operator.

Isokeraunic level: The average number of thunder-days per year at that locality (i.e., the average number of thunder will be heard during a 24 h period).

Isokeraunic map: A map showing mean annual days of thunderstorm activity within the continental United States.

Isolated ground: It originates at an isolated ground-type receptacle or equipment input terminal block and terminates at the point where neutral and ground are bonded at the power source. Its conductor is insulated from the metallic raceway and all ground points throughout its length.

Isolating transformers: They are used to electrically isolate electric circuits from each other or to block dc signals while maintaining ac continuity between the circuits, and to eliminate electromagnetic noise in many types of circuits.

Junction diode: A semiconductor device with a junction and a built-in potential that passes current in one direction than the other. All solar cells are junction diodes.

kcmil: Abbreviation for 1000 circular mils.

K-factor: A factor used to quantify the load impact of electric arc furnaces on the power system.

Knee-point emf: That sinusoidal electromotive force (emf) applied to the secondary terminals of a current transformer, which, when increased by 10%, causes the exciting current to increase by 50%.

Knot: A measure of wind speed; equal to q nautical mil h^{-1}.

kVA rating (of a transformer): It is the output kVA measured at the secondary (load) terminals. It is slightly less than its input kVA due to the losses involved.

Lag: Denotes that a given sine wave passes through its peak at a later time than a reference time wave.

Lambda: The incremental operating cost at the load center, commonly expressed in mils per kilowatt-hour.

Lan: It denotes the local area network. It is used to identify then network of integrated smart meters, field components, and gateways that constitute the logical network between distribution substations and customer's premises.

Langley: A unit of solar radiation intensity equivalent to 1.0 g cal cm^2.

Latitude: The angular distance north (+) or south (−) of the equator, measured in degrees.

Lenz's law of induction: If the coil ends were connected together, the voltage built up would produce a current that would create a new flux opposing the original flux change.

Life: The period during which a cell or battery is capable of operating above a specified capacity or efficiency performance level. For example, with lead-acid batteries, the end of life is generally taken as the point in time when a fully charged cell can only deliver 80% of its rated capacity. Beyond this state of aging, deterioration and loss of capacity begins to accelerate rapidly. Life may be measured in cycles and/or years, depending on the type of service for which the cell or battery is intended.

Lightning arrestor: A device that reduces the voltage of a surge applied to its terminals and restores itself to its original operating condition.

Line: A component part of a system extending between adjacent stations or from a station to an adjacent interconnection point. A line may consist of one or more circuits.

Linear concentrator: A solar concentrator which focuses sunlight along a line, such as the parabolic trough concentrator and the fixed-mirror concentrator.

Line currents: In a delta connection, the line currents are 30° behind the corresponding phase currents. The magnitudes of line currents lag the line-to-neutral voltages. The line currents also lag the line-to-neutral voltages by phase-impedance angle, regardless of whether or not the circuit is delta or wye.

Line loss: Energy loss on a transmission or distribution line.

Line voltages: In a wye-connected three-phase load, the line voltages are 30° ahead of associated phase voltages. The magnitudes of line voltages are square root of three times of those phase voltages. It is also called *line-to-line voltages* or *phase-to-phase voltages*.

L–L: Abbreviation for line to line.

L–N: Abbreviation for line to neutral.

Load center: A point at which the load of a given area is assumed to be concentrated.

Load diversity: The difference between the sum of the maxima of two or more individual loads and the coincident or combined maximum load, usually measured in kilowatts over a specified period of time.

Load duration curve: A curve of loads, plotted in descending order of magnitude, against time intervals for a specified period.

Load factor: The ratio of the average load over a designated period of time to the peak load occurring in that period.

Load, interruptible: A load which can be interrupted as defined by contract.

Load losses, transformer: Those losses which are incident to the carrying of a specified load. They include I^2R loss in the winding due to load and eddy currents, stray loss due to leakage fluxes in the windings, etc., and the loss due to circulating currents in parallel windings.

Load tap changer: A selector switch device applied to power transformers to maintain a constant low side or secondary voltage with a variable primary voltage supply, or to hold a constant voltage out along the feeders on the low-voltage side for varying load conditions on the low-voltage side. Abbreviated LTC.

Load-tap-changing transformer: A transformer used to vary the voltage, or phase angle, or both, of a regulated circuit in steps by means of a device that connects different taps of tapped winding(s) without interrupting the load.

Local backup: Those relays that do not suffer from the same difficulties as remote backup, but they are installed in the same substation and use some of the same elements as the primary protection.

Locked-rotor test: The rotor of the machine is blocked to prevent it from moving. A reduced voltage is applied to the machine so that the rated current flows through the stator windings. This input power, voltage, and current suggest a blocked rotor test frequency of 25% of the rated frequency. If full voltage at the rated frequency were applied, the current would be five to eight or more times the rated value. Because of this, blocked-rotor test are not done at full voltage except for small motors. Even then, such tests are made as rapidly as possible to prevent overheating of the windings.

Loss factor: The ratio of the average power loss to the peak-load power loss during a specified period of time.

Low-side surges: The current surge that appears to be injected into the transformer secondary terminals upon a lighting strike to grounded conductors in the vicinity.

LTC: Abbreviation for load tap changer.

LV: Abbreviation for low voltage.

Magnetic core losses: When a magnetic material is subjected to a time-varying flux, there is some energy loss in the material in the form of magnetic losses. They are also called **iron** or **core losses**.

Magnetic fields: Such fields exist whenever current flows in a conductor. They are not voltage dependent. When a conductor carries an electric current I, a magnetic field is created around the conductor. The direction of magnetic field (or lines of force) is determined using Ampere's right-hand rule. See **Ampere's right-hand rule**.

Magnetic field direction (of a coil): If the coil is held in the right hand with fingers pointing in the direction of the current in the coil, the thumb then points toward the north pole of the coil.

Magnetic flux density (B): It is produced by the magnetic field intensity H and thus $B = \mu H$ where μ represents the permeability of the medium and has the unit of H/m. The unit of magnetic flux density is Wb/m^2 or Tesla (T).

Magnetic skin effect: At very high frequencies, the interior of the magnetic core is practically unused because of the large and circulating eddy currents induced and their inhibiting effect.

Magnetomotive force (mmf): The ability of a coil to produce flux.

Main distribution center: A distribution center supplied directly by mains.

Maintenance expenses: The expense required to keep the system or plant in proper operating repair.

Maximum demand: The largest of a particular type of demand occurring within a specified period.

Maximum efficiency (of a transformer): It is achieved when the core loss is equal to the copper loss. The efficiency of transformers at a rated load is very high and increases with their ratings.

Maximum torque: It is also called *pull-out torque* or the *maximum internal torque,* or *breakdown torque.*

MC: Abbreviation for metal-clad (cable).

Megawatt (MW): A million watts of electric power.

Microgrid: It is basically an *active distribution network* and is made up of collection of distributed generation systems and various loads at distribution voltage level. They are generally small low-voltage combined heat loads of a small community.

Minimum demand: The smallest of a particular type of demand occurring within a specified period.

Momentary interruption: An interruption of duration limited to the period required to restore service by automatic or supervisory-controlled switching operations or by manual switching at locations where an operator is immediately available.

Monthly peak duration curve: A curve showing the total number of days within the month during which the net 60 min clock-hour integrated peak demand equals or exceeds the percent of monthly peak values shown.

Nadyne Rice Machines: See **Backey-Robinson machines**.

Nameplate (of a transformer): A nameplate that is attached to the tank of a transformer providing information about its apparent power in kVA or MVA, voltage ratings, impedance, name, and address of its manufacturer.

National grid: It usually means the EHV transmission network.

N.C.: Abbreviation for normally closed.

NEC: Abbreviation for National Electric Code.

NEMA design classes: They are also simply called *motor design classes*. Each of them provides different torque-speed curves for motors.

NESC: Abbreviation for National Electrical Safety Code.

Net system energy: Energy requirements of a system, including losses, defined as (1) net generation of the system, plus (2) energy received from others, less (3) energy delivered to other systems.

Network distribution system: A distribution system which has more than one simultaneous path of power flow to the load.

Neutral current: It is the current in the neutral wire of a wye-connected power system.

N.O.: Abbreviation for normally open.

Noise: An unwanted electrical signal with a less than 200 kHz superimposed upon the power system voltage or current in phase conductors, or found on neutral conductors or signal lines. It is not a harmonic distortion or transient. It disturbs microcomputers and programmable controllers.

No-load current: The current demand of a transformer primary when no current demand is made on the secondary.

No-load loss: Energy losses in an electric facility when energized at rated voltage and frequency but not carrying load.

Nominal operating voltage: The average terminal voltage of a cell or battery discharging at a specified rate and at a specified temperature.

Noncoincident demand: The sum of the individual maximum demands regardless of time of occurrence within a specified period.

Nonlinear load: An electrical load which draws current discontinuously or whose impedances varies throughout the cycle of the input ac voltage waveform.

Normal rating: Capacity of installed equipment.

Normal weather: All weather not designated as adverse or major storm disaster.

Normally closed: Denotes the automatic closure of contacts in a relay when deenergized. Abbreviated N.C.

Normally open: Denotes the automatic opening of contacts in a relay when deenergized. Abbreviated N.O.

Notch: A switching (or other) disturbance of the normal power voltage waveform, lasting less than a half-cycle; which is initially of opposite polarity than the waveform. It includes complete loss of voltage for up to a 0.5 cycle.

Notching: A periodic disturbance caused by normal operation of a power electronic device, when its current is commutated from one phase to another.

OA: Oil-immersed.

Off-peak energy: Energy supplied during designated periods of relatively low system demands.

OH: Abbreviation for overhead.

Ohm's law (of the magnetic circuit): It says that the magnetomotive force ($\mathcal{F}$) of a magnetic circuit is equal to its flux Φ in the magnetic circuit multiplied by the reluctance $\mathcal{R}$ of the magnetic circuit.

Ohmic contacts: Contacts that do not impede the flow of current into or out of the semiconductor.

One-line diagram: When a circuit composed of three or more conductors is pictorially represented by a single line with standard symbols for transformers, switchgear, and other system components.

On-peak energy: Energy supplied during designated periods of relatively high system *demands*.

Open-circuit test: It is done by applying rated voltage to one of the windings of a transformer, with the other winding (or windings) open circuited. It is done to determine the excitation admittance of the transformer-equivalent circuit. The input power, current, and voltage are measured. (For reasons of safety and convenience, usually the high-voltage winding is open-circuited and the test conducted by placing the instruments on the low-voltage side of the transformer. It is also called *core-loss* test, *iron-loss* test, and the *no-load* test, or the *magnetization* test.

Open-circuit voltage: An equilibrium voltage, reached when the number of carriers drifting back across the junction is equal to the number being generated by the incoming light.

Open-circuit voltage (of a battery): The terminal voltage of a cell or battery at a specified state of charge and temperature under no-load conditions.

Operating expenses: The labor and material costs for operating the plant involved.

Outage: The state of a component when it is not available to perform its intended function due to some event directly associated with that component. An outage may or may not cause an interruption of service to consumers depending upon the system configuration.

Outage duration: The period from the initiation of an outage until the affected component or its replacement once again becomes available to perform its intended function.

Outage rate: For a particular classification of outage and type of component, the mean number of outages per unit exposure time per component.

Oscillatory transient: A sudden and non-power frequency change in the steady-state condition of voltage or current that includes both positive and negative polarity values.

Overcharge: The forcing of current through a cell after all the active material has been converted to the state. In other words, charging (overcharging) continued after 100% state of charge is achieved. Overcharging does not increase the energy stored in a cell and usually results in gassing and/or excessive heat generation, both of which reduce battery life.

Overhead expenses: The costs which in addition to direct labor and material are incurred by all utilities.

Overload: Loading in excess of normal rating of equipment.

Overload protection: Interruption or reduction of current under conditions of excessive demand, provided by a protective device.

Overvoltage: A voltage that has a value at least 10% above the nominal voltage for a period of time greater than 1 min.

Panelboard: A distribution point where an incoming set of wires branches into various other circuits.

Parallel: In module construction; to increase current output, cells are wired with the back contact of one cell connected to the back contact of the next. The total current is the sum of the individual current outputs of the cells. Cells are usually wired in series to form an array and arrays are wired in parallel to obtain desired current.

Passivate: To chemically react a substance with the surface of a solid to tie up or remove the reactive atoms on the surface. For example, the air oxidation of a fresh aluminum surface to form a thin layer of aluminum oxide.

Passive filter: A combination of inductors, capacitors, and resistors designed to eliminate one or more harmonics. The most common variety is simply an inductor in series with a shunt capacitor, which short-circuits the major distorting harmonic component from the system.

PE: An abbreviation used for polyethylene (cable insulation).

Peak current: The maximum value (crest value) of an alternating current.

Peak voltage: The maximum value (crest value) of an alternating voltage.

Peaking station: A generating station which is normally operated to provide power only during maximum load periods.

Peak-to-peak value: The value of an ac waveform from its positive peak to its negative peak. In the case of a sine wave, the peak-to-peak value is double the peak value.

Pedestal: A bottom support or base of a pillar, statue, etc.

Penetration (*of renewable energy*): The penetration of electrical energy or power being supplied from wind turbines or from other renewable energy sources is usually referred as the penetration. It is usually given in percentage.

Percent regulation: See **Percent voltage drop**.

Percent voltage drop: The ratio of voltage drop in a circuit to voltage delivered by the circuit, multiplied by 100 to convert to percent.

Permanent forced outage: An outage whose cause is not immediately self-clearing but must be corrected by eliminating the hazard or by repairing or replacing the component before it can be returned to service. An example of a permanent forced outage is a lightning flashover which shatters an insulator, thereby disabling the component until repair or replacement can be made.

Permanent forced outage duration: The period from the initiation of the outage until the component is replaced or repaired.

Permanents magnet machines: They are ordinary synchronous machines with the field excitation provided by a permanent magnet. They have a very high efficiency since there are no field losses.

Permeance: It is the easiness that a magnetic circuit exhibits to the flow of flux in it. It is the reciprocal of the reluctance of a magnetic circuit.

Phase: The time of occurrence of the peak value of an ac waveform with respect to the time of occurrence of the peak value of a reference waveform.

Phase angle: An angular expression of phase difference.

Phase shift: The displacement in time of one voltage waveform relative to other voltage waveform(s).

Phase voltage: It is also called the line-to-neutral or line-to-ground voltage.

Photovoltaic cell (*solar cell*): A semiconducting device that converts sunlight directly into electric power. The conversion process is called the *photovoltaic effect*.

Point of common coupling (PCC): It is the connection point to the network.

Polarization: The deviation from open-circuit voltage caused by the flow of current in an electrochemical cell.

Pole: A column of wood or steel, or some other material, supporting overhead conductors, usually by means of arms or brackets.

Pole fixture: A structure installed in lieu of a single pole to increase the strength of a pole line or to provide better support for attachments than would be provided by a single pole. Examples are A fixtures, H fixtures.

Polyphase source: A power source that has more than one phase.

Potential energy: The energy stored in matter because of its position or because of the arrangement of its parts, e.g., water stored behind a dam.

Power: The rate at which work is done. The rate (in kilowatts) of generating, transferring, or using energy.

Power, active: The product of the rms value of the voltage and the rms value of the in-phase component of the current.

Power, apparent: The product of the rms value of the voltage and the rms value of the current.

Power density: The ratio of rated power available from a cell or battery, normalized to its weight or volume.

Power factor: The ratio of active power to apparent power.

Power-factor adjustment clause: A clause in a rate schedule that provides for an adjustment in the billing if the customer's power factor varies from a specified reference.

Power, instantaneous: The product of the instantaneous voltage multiplied by the instantaneous current.

Power pool: A group of power systems operating as an interconnected system and pooling their resources.

Power, reactive: The product of the rms value of the voltage and the rms value of the quadrature component of the current.

Power tower: A device that generates electric power from sunlight, consisting of a field of small mirrors tracking the sun to focus the light onto a tower-mounted boiler. The steam produced runs a conventional turbine generator.

Power triangle: A triangle that shows the relationships between the real, reactive, and apparent power.

Power transformer: A transformer which transfers electric energy in any part of the circuit between the generator and the distribution primary circuits.

Power system stability: The ability of an electric power system, for a given initial operating condition, to regain a state of operating equilibrium after being subjected to a physical disturbance.

Present value: The value of a future cash flow discounted to the present based on the premise that a dollar worth today is worth more than a dollar received in the future by virtue of the amount of interest (or return) that it earns.

Primary cell or battery: A cell or battery whose initial capacity cannot be significantly restored by charging and is therefore limited to a single discharge.

Primary disconnecting devices: Self-coupling separable contacts provided to connect and disconnect the main circuits between the removable element and the housing.

Primary distribution feeder: A feeder operating at primary voltage supplying a distribution circuit.

Primary distribution network: A network consisting of primary distribution mains.

Primary distribution system: A system of ac distribution for supplying the primaries of distribution transformers from the generating station or substation distribution buses.

Primary distribution trunk line: A line acting as a main source of supply to a distribution system.

Primary feeder: That portion of the primary conductors between the substation or point of supply and the center of distribution.

Primary main feeder: The higher-capacity portion of a primary distribution feeder that acts as a main source of supply to primary laterals or direct connected distribution transformers and primary loads.

Primary network: A network supplying the primaries of transformers whose secondaries may be independent or connected to a secondary network.

Primary transmission feeder: A feeder connected to a primary transmission circuit.

Primary unit substation: A unit substation in which the low-voltage section is rated above 1000 V.

Protective relay: A device whose function is to detect defective lines or apparatus or other power-system conditions of an abnormal or dangerous nature and to initiate appropriate control circuit action.

PT: Abbreviation for potential transformers.

pu: Abbreviation for per unit.

Pull-out torque: It is the maximum torque for motor operation takes place. Any load that requires a torque greater than the maximum torque causes an unstable operation of the machine; the machine pulls out of synchronism and comes to a standstill.

Radial distribution system: A distribution system which has a single simultaneous path of power flow to the load.

Radial service: A service which consists of a single distribution transformer supplied by a single primary circuit.

Radial system, complete: A radial system which consists of a radial subtransmission circuit, a single substation, and a radial primary feeder with several distribution transformers each supplying radial secondaries; has the lowest degrees of service continuity.

Rate base: The net plant investment or valuation base specified by a regulatory authority upon which a utility is permitted to earn a specified rate of return.

Rated capacity: See **Capacity, rated**.

Rated voltage: It is the full load voltage (of secondary of a transformer). At no load, the secondary terminal voltage may change from the rated voltage value due to the effect of the impedance of the transformer. It is also called the *nameplate voltage*, *nominal transformer voltage*, or *full-load voltage*.

Reactive power: It is the reactive portion of the complex power and it is equal to $V I \sin \theta$.

Reactor: An inductive reactor between the dc output of the converter and the load. It is used to smooth the ripple in the direct current adequately, to reduce harmonic voltages and currents in the dc line, and to limit the magnitude of fault current. It is also called a *smoothing reactor*.

Real power: It is the real portion of the complex power. It is the average or active power. It is calculated by $I V \cos \theta$.

Reclosing device: A control device which initiates the reclosing of a circuit after it has been opened by a protective relay.

Reclosing fuse: A combination of two or more fuse holders, fuse units, or fuse links mounted on a fuse support(s), mechanically or electrically interlocked, so that one fuse can be connected into the circuit at a time and the functioning of that fuse automatically connects the next fuse into the circuit, thereby permitting one or more service restorations without replacement of fuse links, refill units, or fuse units.

Reclosing relay: A programming relay whose function is to initiate the automatic reclosing of a circuit breaker.

Reclosure: The automatic closing of a circuit-interrupting device following automatic tripping. Reclosing may be programmed for any combination of instantaneous, time-delay, single-shot, multiple-shot, synchronism-check, deadline-live-bus, or dead-bus-live-line operation.

Recovery voltage: The voltage that occurs across the terminals of a pole of a circuit-interrupting device upon interruption of the current.

Rectifier: A device that passes current in one direction only.

Reference phase: When the knowledge of variables in one phase provides information about all phases due to the compete symmetry that exists between phases.

Reluctance machine: A synchronous machine without the dc excitation. It is used in timers, electrical clocks, and recording appliances.

Reluctance (of a magnetic circuit): It is the opposition of the magnetic material to the flow of the flux.

Renewable energy generator: It can be defined as stand-alone or grid-connected. A stand-alone renewable energy generator provides for the greater part of the demand with or without other generators or storage. On the other hand, in a grid-connected system, the renewable energy generator supplies power to a large interconnected network which is also supplied power by other generators.

Required reserve: The system planned reserve capability needed to ensure a specified standard of service.

Resistance: The real part of impedance.

Return on capital: The requirement which is necessary to pay for the cost of investment funds used by the utility.

Reversal: The continued discharge of a cell to the point that the cell's electrical terminals change polarity.

Reversible potential: The open circuit (zero-current) potential of a reversible chemical reaction.

Right-hand rule: If the conductor is held in the right hand with the thumb pointing in the direction of the current flow, the fingers then point in the direction of magnetic field around the conductor.

Risk: The probability that a particular threat will exploit a particular vulnerability of an equipment, plant, or system.

Risk management: Decisions to accept exposure or to reduce vulnerabilities by either mitigating the risks or applying cost-effective controls.

Rotating rectifiers: They have similar performance as regular synchronous machines except that field excitation is provided by an auxiliary generator and by rectifiers that are stationed on the rotor.

Rotor: An inner (rotating) part of an electrical machine. The rotor is centered within the stator. The rotor is supported by a steel rod that is called a shaft. It has winding located in the slots around the rotor.

SAG of a conductor (at any point in a span): The distance measured vertically from the particular point in the conductor to a straight line between its two points of support.

Sag: The distance measured vertically from a conductor to the straight line joining its two points of support. Unless otherwise stated, the sag referred to is the sag at the midpoint of the span, or a decrease between 0.1 and 0.9 pu in rms voltage and current at the power frequency for a duration of 0.5 cycles to 1 min.

Sag, final unloaded: The sag of a conductor after it has been subjected for an appreciable period to the loading prescribed for the loading district in which it is situated, or equivalent loading, and the loading removed. Final unloaded sag includes the effect of inelastic deformation.

Sag, initial unloaded: The sag of a conductor prior to the application of any external load.

Sag section: The section of line between snub structures. More than one sag section may be required to properly sag the actual length of conductor which has been strung.

Sag span: A span selected within a sag section and used as a control to determine the proper sag of the conductor, thus establishing the proper conductor level and tension. A minimum of two, but normally three, sag spans are required within a sag section to sag properly. In mountainous terrain or where span lengths vary radically, more than three sag spans could be required within a sag section.

SCADA: Deployment of substation and feeder operating functions and applications ranging from supervisory control and data acquisition (SCADA) and alarm processing to integrated volt/var control in order to optimize the management of capital assets and enhance operational and maintenance efficiencies with minimal human intervention.

SCADA communication line: It is the communication link between the utility's control center and the RTU at the substation.

Scheduled interruption: An interruption caused by a scheduled outage.

Scheduled maintenance (generation): Capability which has been scheduled to be out of service for maintenance.

Scheduled outage: An outage that results when a component is deliberately taken out of service at a selected time, usually for purposes of construction, preventive maintenance, or repair.

Scheduled outage duration: The period from the initiation of the outage until construction, preventive maintenance, or repair work is completed.

Seasonal diversity: Load diversity between two (or more) electric systems which occurs when their peak loads are in different seasons of the year.

Secondary battery: See **Battery, secondary**.

Secondary cell or battery: A cell or battery that is capable of being charged repeatedly.

Secondary current rating: The secondary current existing when the transformer is delivering rated kilovoltamperes at rated secondary voltage.

Secondary disconnecting devices: Self-coupling separable contacts provided to connect and disconnect the auxiliary and control circuits between the removable element and the housing.

Secondary distributed network: A service consisting of a number of network transformer units at a number of locations in an urban load area connected to an extensive secondary cable grid system.

Secondary distribution feeder: A feeder operating at secondary voltage supplying a distribution circuit.

Secondary distribution mains: The conductors connected to the secondaries of distribution transformers from which consumers' services are supplied.

Secondary distribution network: A network consisting of secondary distribution mains.

Secondary distribution system: A low-voltage ac system that connects the secondaries of distribution transformers to the consumers' services.

Secondary fuse: A fuse used on the secondary-side circuits, restricted for use on a low-voltage secondary distribution system that connects the secondaries of distribution transformers to consumers' services.

Secondary mains: Those which operate at utilization voltage and serve as the local distribution main. In radial systems secondary mains that supply general lighting and small power are usually separate from mains that supply three-phase power because of the dip in voltage caused by starting motors. This dip in voltage, if sufficiently large, causes an objectionable lamp flicker.

Secondary network: It consists of two or more network transformer units connected to a common secondary system and operating continuously in parallel.

Secondary network service: A service which consists of two or more network transformer units connected to a common secondary system and operating continuously in parallel.

Secondary, radial: A secondary supplied from either a conventional or completely self-protected (type CSP) distribution transformer.

Secondary selective service: A service which consists of two distribution transformers, each supplied by an independent primary circuit, and with secondary main and tie breakers.

Secondary spot network: A network which consists of at least two and as many as six network-transformer units located in the same vault and connected to a common secondary service bus. Each transformer is supplied by an independent primary circuit.

Secondary system, banked: A system which consists of several transformers supplied from a single primary feeder, with the low-voltage terminals connected together through the secondary mains.

Secondary unit substation: A unit substation whose low-voltage section is rated 1000 V and below.

Secondary voltage regulation: A voltage drop caused by the secondary system, it includes the drop in the transformer and in the secondary and service cables.

Second-contingency outage: The outage of a secondary primary feeder in addition to the first one.

Sectionalizer: A device which resembles an oil circuit recloser but lacks the interrupting capability.

Self-discharge: The process by which the available capacity of a cell is reduced by internal chemical reactions (local action).

Self-discharge rate: The rate at which a battery will discharge on standing; affected by temperature and battery design.

Series: In array construction, connecting cells by overlapping the front edge of one cell to the front contact of the next to obtain a higher voltage.

Service area: Territory in which a utility system is required or has the right to supply or make available electric service to ultimate consumers.

Service availability index: See **Index of reliability**.

Service entrance: All components between the point of termination of the overhead service drop or underground service lateral and the building main disconnecting device, with the exception of the utility company's metering equipment.

Service entrance conductors: The conductors between the point of termination of the overhead service drop or underground service lateral and the main disconnecting device in the building.

Service entrance equipment: Equipment located at the service entrance of a given building that provides overcurrent protection to the feeder and service conductors, provides a means of disconnecting the feeders from energized service conductors, and provides a means of measuring the energy used by the use of metering equipment.

Service lateral: The underground conductors, through which electric service is supplied, between the utility company's distribution facilities and the first point of their connection to the building or area service facilities located at the building or other support used for the purpose.

SF$_6$: Formula for sulfur hexafluoride (gas).

Shaft: A steel rod that supports the rotor. The shaft, itself in turn, is supported by bearings so that the rotor can turn freely.

Short circuit current (I_{SC}): The maximum current a cell can deliver into a short circuit; directly proportional to the area of the cell and the light intensity, or a fault current due to a short circuit.

Short-circuit test: It is used to determine the equivalent resistance and reactance of the transformer under rated conditions. It is done by short-circuiting one winding (usually low-voltage winding) to the other winding. The reduced voltage is adjusted until the current in the shorted winding is equal to its rated value. The input voltage, current, and power are measured as before. The applied voltage V_{SC} is only a small percentage of the rated voltage and is sufficient to circulate rated current in the windings of the transformer. It is about 2%–12% of the rated voltage. Hence, the excitation current is small enough to be ignored. It is also called the *impedance test* or the *copper-loss test*.

Skin effect: The phenomenon by which alternative current tends to flow in the outer layer of a conductor. It is a function of conductor size, frequency, and the relative resistance of the conductor material.

Slip: It describes the relative motion between synchronous speed of the magnetic field and rotor's mechanical speed in per unit or in percent.

$$s = \frac{n_s - n_m}{n_s} \text{ pu} \quad s = \frac{n_s - n_m}{n_s} \times 100$$

Slip-ring motors: See **Wound-rotor motors**.

Slip speed: It is also called the *slip rpm*. It is defined as the difference between synchronous speed and rotor speed and indicates how much the rotor slips behind the synchronous speed.

Solar array: A set of modules assembled for a specific application; may consist of modules in series for increased voltage or in parallel for increased current, or a collector combination of both. Also see **Solar module**.

Solar cell: A device that converts sunlight directly into electricity.

Solar constant: The intensity of solar radiation beyond the earth's atmosphere, at the average earth–sun distance, on a surface perpendicular to the sun's rays. The value for the solar constant is 1353 W/m^2 (within + or −1.6%).

Solar furnace: A solar concentrator for producing very high temperatures. (Installations in France, Russia, and Japan produce temperatures as high as 7000°F.)

Solar module: A string of 32–36 cells, producing an open-circuit voltage in bright sunlight of about 18 bolts, or 16 volts when producing maximum power. Total current output of a series string is the same as a single cell. Also see solar array.

Solar ponds: A pond of stratified water that collects and retains heat. Convection, normally present in ponds, is suppressed by imposing a stable density gradient of dissolved salts.

Solar power farms: Large centralized photovoltaic array systems that generate electricity to power the utility grid.

Solar radiation: The radiant energy received from the sun both directly as a beam component and diffusely by scattering from the sky and reflection from the ground.

Solar-thermal electric conversion: The conversion of solar energy to thermal energy, which powers turboelectric generators.

Specific heat capacity: The amount of heat required to raise the temperature of unit mass of material one degree, usually measured in British thermal units per pound per degree Fahrenheit or joules per kilogram per Kelvin.

Specular reflection: A mirror-like reflection in which incident and reflected angles are equal.

Squirrel-cage-rotor induction motor: Instead of a winding, the slots in the squirrel-cage rotor have bars of copper or aluminum, known as rotor bars, which are short-circuited with two end rings of the same material. There is one ring at each end of the stack of rotor laminations. The solid bars are placed parallel, or approximately parallel, to the shaft and embedded in the surface of the core; the conductors are not insulated from the core, since the rotor currents naturally flow the path of least resistance through the rotor conductors.

St: Abbreviation for steel.

Stability: The quality whereby a protective system remains operative under all conditions other than those for which it is specifically designed to operate.

Standard method of marking terminals: For single-phase transformer windings, the terminals on the high-voltage side are labeled H_1 and H_2, while those on the low-voltage side are identified as X_1 and X_2. Here, in this convention the terminal with subscript 1 is equivalent to the dotted terminal in the dot-polarity notation.

STATCOM: It is a static compensator. It provides variable lagging or leading reactive powers without using inductors or capacitors for var generation.

State of charge (SOC): The available capacity in a cell or battery expressed as a percentage of rated capacity. For example, id 25 A h have been removed from a fully charged 100 A h cell, the new state of charge is 75%.

Static var system: It is a static var compensator that can also control mechanical switching of shunt capacitor banks or reactors.

Stationary cell or battery: A cell or battery that is generally intended for float service usually in a fixed location.

Stationary-reflector tracking-absorber concentrating collector: A spherical dish with movable absorber rod that tracks the movement of the sun.

Stator: The outside (stationary) part of an electrical machine. It has windings located in the slots of the inner surface of the stator.

Step-down transformer: In a given transformer, if the number of turns of the primary winding is greater than those of the secondary winding.

Step-up transformer: In a given transformer, if the number of turns of the secondary winding is greater than those of the primary winding.

Stirling engines: It is a closed-cycle piston heat engine where the working gas is permanently contained within the cylinder. Its traditionally classified as an external combustion engine, though heat can also be supplied by noncombustible sources like solar, geothermal, chemical, and nuclear energy. It uses an external heat source and an external heat sink. Each is maintained within a limited temperature range and has a sufficiently large temperature difference between them.

Strand: One of the wires, or groups of wires, of any stranded conductor.

Stranded conductor: A conductor composed of a group of wires, or of any combination of groups of wires. Usually, the wires are twisted together.

Submersible transformer: A transformer so constructed has to be successfully operable when submerged in water under predetermined conditions of pressure and time.

Substation: An assemblage of equipment for purposes other than generation or utilization, through which electric energy in bulk is passed for the purpose of switching or modifying its characteristics.

Substation LAN: It is a communications network, typically high speed, and within the substation and extending into the switchyard.

Substation local area network (LAN): A technology that is used in a substation environment and facilitate interfacing to process-level equipment (IEDs and PLCs) while providing immunity and isolation to substation noise. It is basically a communication network, typically high speed, and within the substation and extending into the switchyard.

Substation voltage regulation: The regulation of the substation voltage by means of the voltage regulation equipment which can be LTC (load-tap-changing) mechanisms in the substation transformer, a separate regulator between the transformer and low-voltage bus, switched capacitors at the low-voltage bus, or separate regulators located in each individual feeder in the substation.

Subsynchronous: Electrical and mechanical quantities associated with frequencies below the synchronous frequency of a power system.

Subsynchronous oscillation: The exchange of energy between the electric network and the mechanical spring-mass system of the turbine generator at subsynchronous frequencies.

Subsynchronous resonance: It is an electric power system condition where the electric power network exchanges energy with a turbine generator at one or more of the natural frequencies of the combined system below the synchronous frequency of the system.

Subtractive polarity: In a given transformer, if the current i_1 flows into the dotted end of the primary winding and the current i_2 flows out of the dotted end of the secondary winding, the mmfs will be subtracted from each other. Such transformer can be said having a subtractive polarity. Here, current i_2 flowing in the direction of the induced current, according to Lenz's law. Small transformers have subtractive polarities in order to reduce voltage stress between adjacent leads.

Subtransmission: That part of the distribution system between bulk power source(s) (generating stations or power substations) and the distribution substation.

Sulfation (or Sulphation): A condition which afflicts unused and discharged batteries; large crystals of lead sulfate grow on the plate of a lead-acid battery, instead of the usual tiny crystals, making the battery extremely difficult to recharge. The large crystals are more difficult to reduce by charging current than are the smaller crystals that result from normal and self-discharge reactions. It can be caused by leaving the battery in a discharged state for long periods of time.

Sun time: The time measured on a basis of sun's virtual motion.

Super paramagnetic materials: They are made up of powdered iron (or other magnetic material) particles that are mixed in a nonferrous epoxy or other plastic material.

Supply security: Provision must be made to ensure continuity of supply to consumers even with certain items of plant out of action. Usually two circuits in parallel are used and a system is said to be secure when continuity is assured. It is the prerequisite in design and operation.

Susceptance: The imaginary part of admittance.

Sustained interruption: The complete loss of voltage (<0.1 pu) on one or more phase conductors for a time greater than 1 min.

SVC: Static var compensator.

Swell: An increase to between 1.1 and 1.8 pu in rms voltage or current at the power frequency for durations from 0.5 cycle to 1 min.

Switch: A device for opening and closing or for changing connections in a circuit.

Switchboard: A large single panel, frame, or assembly of panels on which are mounted (on the face, or back, or both) switches, fuses, buses, and usually instruments.

Switched-capacitor bank: A capacitor bank with switchable capacitors.

Switchgear: A general term covering switching or interrupting devices and their combination with associated control, instrumentation, metering, protective, and regulating devices; also assemblies of these devices with associated interconnections, accessories, and supporting structures.

Switching time: The period from the time a switching operation is required due to a forced outage until that switching operation is performed.

Switch, isolating: An auxiliary switch for isolating an electric circuit from its source of power; it is operated only after the circuit has been opened by other means.

Switch, limit: A switch that is operated by some part or motion of a power-driven machine or equipment to alter the electric circuit associated with the machine or equipment.

System: A group of components connected together in some fashion to provide flow of power from one point or points to another point or points.

Systems: It is used to describe the complete electrical network, generators, loads, and prime movers.

System interruption duration index: The ratio of the sum of all customer interruption durations per year to the number of customers served. It gives the number of minutes out per customer per year.

Tandem solar cells: Two photovoltaic cells are constructed on top of each other such that the light passes through the wide band gap cell to reach the narrow band gap cell. These cells very efficiently utilize the sunlight giving a greater output for a given area. Also called *cascade cells*.

Tertiary winding: If the three-phase transformer has three windings, the third winding is always connected in delta. It is called tertiary winding and it is in addition to the primary and secondary windings. The third-harmonic voltages are suppressed by trapping the third-harmonic (circulating) currents within the delta tertiary winding. However, if a neutral is not provided in such transformer, the phase voltages become drastically unbalanced when the load is unbalanced. This causes neutral instability that makes unbalanced loading impractical, even though the line-to-line voltages remain normal. There also problems with third harmonics. Thus, any attempt to operate a wye–wye connection of transformers without the presence of a primary neutral connection will lead to difficulty and potential failure.

Thermal conductivity: The amount of heat that can be transferred by conduction through a material of unit area and thickness per unit temperature difference.

Thermocouple: A thermoelectric device consisting of two dissimilar wires with their ends connected together. A small voltage is generated when two junctions are at different temperatures; if one junction is kept at a reference temperature, the voltage generated in the other is a measure of the temperature of the other junction above the reference.

Three-phase transformers: It has a one magnetic core and three primary windings and three secondary windings are wound on the legs of the core. The core itself can be made of using a core-type design or shell-type design. In the core-type design, both the primary and secondary windings of each phase are placed on one leg of each transformer. The core-type design has two windows and three legs in it whereas the shell-type design has six windows and the primary and secondary winding. Here, the center phase windings are wound in the opposite direction of the other two phases. The no-load losses are less than those in a core-type transformer, the coils are put onto the legs in the middle of the core. There are four possible connections for a three-phase transformer, namely, wye–wye, delta–delta, wye–delta, and delta–wye.

Three-winding transformers: They are usually used in bulk power (transmission) substations to lower the transmission voltage to the subtransmission voltage level. They are also increasingly used at distribution substations. While the primaries and secondaries are usually connected in wye–wye, the tertiary windings of a three-phase and three-winding transformer bank are connected in delta. The tertiaries are used for (1) providing a path for the third harmonics and their multiples in the excitation and the zero-sequence currents (the zero-sequence currents are trapped and circulate in the delta connection); (2) in-plant power distribution; and (3) the application of power factor correcting capacitors or reactors.

Torque: Twisting action on the cylinder. It is defined as force times the radial distance at which it is applied, measured from the axis of rotation. Thus, the torque T is a function of the magnitude of the applied force F, and the distance between the axis of rotation and the line of action of the force. $T = F r \sin \theta$.

Torque in synchronous watts: It is also called *air-gap power*.

Total demand distortion (TDD): The ratio of the root-mean-square (rms) of the harmonic current to the rms value of the rated or maximum demand fundamental current, expressed as a percent.

Total energy system: Hybrid system producing both usable heat and electricity. A system for providing all energy requirements, including heat, air-conditioning, and electric power.

Total harmonic distortion (THD): The ratio of the root-mean-square of the harmonic content to the root-mean-square value of the fundamental quantity, expressed as a percent of the fundamental.

Tracking collector: A collector that can rotate about one or two axes to face the sun; usually restricted to high-temperature concentrating collectors because of the complexity and cost of a tracking system.

Tracking system, one-axis: A mount pointing capable of pointing in one axis only: reoriented seasonally by hand and used with linear concentrators or flat plates.

Tracking system, two-axis: A mount capable of pivoting both daily and seasonally to follow the sun.

Transformer: It is a static machine (i.e., it has no moving parts) that is used for changing the voltage and current levels in a given electrical system, establishing electrical isolation, impedance matching, and measuring instruments.

Triplen harmonics: A term frequency used to refer to the odd multiples of the third harmonic, which deserve special attention because of their natural tendency to be zero sequence.

True power factor (TPF): The ratio of the active power of the fundamental wave, in watts, to the apparent power of the fundamental wave, in root-mean-square voltamperes (including the harmonic components).

Turns ratio: It is the ratio of the number of turns in the primary winding to the number of turns in the secondary winding. It is also equal to the ratio of primary voltage to the secondary voltage. It is also equal to the ratio of the secondary current to the primary current.

Ultracapacitors: Experimental carbon-based electrolytes used to deliver large current to accelerate large motors.

Ultra high-speed: It is a term that is not included in the relay standards but is commonly considered to be operation in 4 ms or less.

Ultraviolet radiation: A radiation having wavelengths longer than those of x-rays.

Ultraviolet (UV) wavelengths: Wavelengths shorter than 400 nm; the energetic rays of the sun, invisible but responsible for suntans and sunburns. Our atmosphere filters out most of UV rays.

Underground distribution system: That portion of a primary or secondary distribution system which is constructed below the earth's surface. Transformers and equipment enclosures for such a system may be located either above or below the surface as long as the served and serving conductors are located underground.

Undervoltage: A voltage that has a value at least 10% below the nominal voltage for a period of time greater than 1 min.

Undervoltage relay: A relay that functions on a given value single-phase ac under voltage.

Unit: A self-contained relay unit that in conjunction with one or more other relay units performs a complex relay function.

Unit substation: A substation consisting primarily of one or more transformers which are mechanically and electrically connected to and coordinated in design with one or more switchgear or motor control assemblies or combinations thereof.

Unreach: The tendency of the relay to restrain at impedances larger than its setting. That is, it is due to error in relay measurement resulting in wrong operation.

URD: Abbreviation for underground residential distribution.

Utilization factor: The ratio of the maximum demand of a system to the rated capacity of the system.

Vacuum evaporation: Method of depositing thin coatings of a substance by heating it in a vacuum system.

VD: Abbreviation for voltage drop.

VDIP: Abbreviation for voltage dip.

V_{oc}: See open-circuit voltage.

Voltage, base: A reference value which is a common denominator to the nominal voltage ratings of transmission and distribution lines, transmission and distribution equipment, and utilization equipment.

Voltage dip: A voltage change resulting from a motor starting.

Voltage drop: The difference between the voltage at the transmitting and receiving ends of a feeder, main or service.

Voltage flicker: Voltage fluctuation caused by utilization equipment resulting in lamp flicker, i.e., in a lamp illumination change.

Voltage fluctuation: A series of voltage changes or a cyclical variation of the voltage envelope.

Voltage imbalance (or *unbalance*): The maximum deviation from the average of the three-phase voltages or currents, divided by the average of the three-phase voltages or currents, expressed in percent.

Voltage interruption: Disappearance of the supply voltage on one or more phases. It can be momentary, temporary, or sustained.

Voltage magnification: The magnification of capacitor switching oscillatory transient voltage on the primary side by capacitors on the secondary side of a transformer.

Voltage, maximum: The greatest 5 min average or mean voltage.

Voltage, minimum: The least 5 min average or mean voltage.

Voltage, nominal: A nominal value assigned to a circuit or system of a given voltage class for the purpose of convenient designation.

Voltage, open circuit: See **Open-circuit voltage**.

Voltage regulation: The percent voltage drop of a line with reference to the receiving-end voltage.

$$\% \text{ regulation} = \frac{\left|\bar{E}_s\right| - \left|\bar{E}_r\right|}{\left|\bar{E}_r\right|} \times 100$$

where

$\left|\bar{E}_s\right|$ is the magnitude of the sending-end voltage

$\left|\bar{E}_r\right|$ is the magnitude of the receiving-end voltage

Voltage regulator: An induction device having one or more windings in shunt with, and excited from, the primary circuit, and having one or more windings in series between the primary circuit and the regulated circuit, all suitably adapted and arranged for the control of the voltage, or of the phase angle, or of both, of the regulated circuit.

Voltage, service: Voltage measured at the terminals of the service entrance equipment.

Voltage, short circuit: See short-circuit voltage.

Voltage spread: The difference between maximum and minimum voltages.

Voltage stability: It is the ability of a power system to maintain steady voltages at all buses in the system after being subjected to a disturbance from a given initial operational condition. It can be either fast (short term, with voltage collapse in the order of fractions of a few seconds), or slow (long term, with voltage collapse in minutes or hours).

Voltage stability problems: It is manifested by low system voltage profiles, heavy reactive line flows, inadequate reactive support, and heavy-loaded power systems.

Voltage transformation: It is done by substation power transformers by raising or lowering the voltage.

Voltage transformer (VTs): It is also called *potential transformer* (*PT*). The transformer that is connected across the points at which the voltage is to be measured. They are single-phase transformers that are use to step down the voltage to be measured to a safe value.

Voltage transformer burdens: The VT burdens are normally expressed as voltamperes at a designated power factor. It may be a *W*, *X*, *M*, *Y*, or *Z* where *W* is 12.5 VA at 0.10 power factor, *X* is 25 VA at 0.70 power factor, *M* is 35 VA at 0.20 power factor, *Y* is 75 VA at 0.85 power factor, and *Z* is 200 VA at 0.85 power factor. The complete expression for a current transformer accuracy classification might be 0.3 at B-0.1, B-0.2, and B-0.5 while the potential transformer might be 0.3 at *W*, *X*, *M*, and *Y*.

Voltage transformer ratio: It is also called *VT ratio*. It is the ratio of primary to secondary voltage. For a voltage transformer rated 480:120, the ratio is 4:1 and for a voltage transformer rated 7200:120, it is 60:1.

Voltage, utilization: Voltage measured at the terminals of the machine or device.

VRR: Abbreviation for voltage-regulating relay.

Wan: It is used to identify the network of upstream utility assets, including power plants, distribution storage, substations, etc.

Watt (W): The energy rate of 1 J/s.

Wattmeter: A device that has a potential coil and a current coil, which are designed and connected in such a way that its pointers deflection is proportional to $V I \cos \theta$.

Waveform distortion: A steady-state deviation from an ideal sine wave of power frequency principally characterized by the special content of the deviation.

Wind energy: The kinetic energy resulting from air motion caused by the interaction of earth's rotation and solar-driven atmospheric currents.

Work: The energy transfer causing a force to move through a distance with measure equal to the dot product of the force and displacement vectors.

Wound-rotor motors: They are also called *slip-ring motors.*

Wye–delta connection: This connection can be used with unbalanced loads. There is no problem with third-harmonic components in its voltages, since they are absorbed in a circulating current on the delta side. In high-voltage transmission systems, the high-voltage side is connected in delta and the low-voltage-side is connected in wye. Due to delta connection, the secondary voltages are shifted 30° with respect to the corresponding primary voltages. In the United States, it is a standard practice to make the secondary voltage (i.e., the lower voltage) lag the primary voltage (i.e., the higher voltage) by 30°. This connection is basically used to step down a high voltage to a lower voltage.

Wye–wye connection: Both windings are connected in wye. There is no phase displacement between the primary and secondary line-to-line voltages, even though it is possible to shift the secondary voltages by 180° by reversing all three secondary windings. The use of a wye–wye connection creates no problem as long as it has *solidly grounded neutrals* (especially the neutral on the primary side). Here, the addition of a primary neutral connection makes each transformer independent of the other two.

REFERENCES

1. IEEE Committee Report, Proposed definitions of terms for reporting and analyzing outages of electrical transmission and distribution facilities and interruptions, *IEEE Transactions on Power Apparatus and System*, PAS-87(5), May 1968, 1318–1323.
2. IEEE Committee Report, Guidelines for use in developing a specific underground distribution system design standard, *IEEE Transactions on Power Apparatus and System*, PAS-97(3), May/June 1978, 810–827.
3. *IEEE Standard Definitions in Power Operations Terminology*, IEEE Standard 346-1973, November 2, 1973.
4. *Proposed Standard Definitions of General Electrical and Electronics Terms*, IEEE Standard 270, 1966.
5. Pender, H. and W. A. Del Mar, *Electrical Engineers' Handbook—Electrical Power*, 4th Ed., New York: Wiley, 1962.
6. *National Electrical Safety Code*, 1977 ed., ANSI C2, New York: IEEE, November, 1977.
7. Fink, D. G. and J. M. Carroll (eds.), *Standard Handbook for Electrical Engineers*, 10th Ed., New York: McGraw-Hill, 1969.
8. *IEEE Standard Dictionary of Electrical and Electronics Terms*, New York: IEEE, 1972.

Answers to Selected Problems

Chapter 1

1.1 $P_{tot} = 18$ kW

1.2 $C = 378.93$ μF

1.4 (a) $S_{tot} = 309,232.9\angle -14.04°$ VA; (b) $I_{tot} = 64.42 \angle 14.04°$ A; (c) PF = 0.97 leading

Chapter 2

2.2 (a) $I_L = 75.4$ A; (b) $\theta = 31.8°$; (c) $V_{R(\phi)} = 66,474$ V; (d) $I_L Z = 8,474.4\angle 31.6°$ V; (e) $V_{S(L)} = 127,692.9$ V; (f) $P_{3\phi} = 852.77$ kW

2.5 (a) $S_{12} = 146,715.5 - j138,741.9$ VA; (b) $P_{12} = 146,715.5$ W; (c) $Q_{12} = -138,741.9$ var

2.8 (a) $P_1 = 8,860.5$ W, $P_2 = 15,495.8$ W; (b) $P_{tot} = 20,393.3$ W

2.10 $I_a = 19.9\angle -35°$ A, $I_b = 17.13\angle 168°$ A

2.12 (a) $Z = 2.2412\angle 51.66°\Omega$; (b) $I_a = 53.5427 \angle -51.66°$ A; (c) $S_{tot} = 19,275.4\angle 51.66°$ VA

2.14 (a) $I_L = 73.83$ A; (b) $C = 0.608$ μF

2.16 (a) $I_b = 566.1\angle -75°$ A; (b) $P_{tot} = 2884.3$ MW

2.18 (a) $I_a = 1,044.14\angle 54.9°$ A, $I_b = 626.3 \angle -26.6°$ A, $I_c = 1,0294.55\angle -153.69°$ A; (b) V_{DE} 1944.3 $\angle -36.6°$ V; (c) $P_{tot} = 6.656$ MW

2.21 (a) $S_{tot} = 4 + j3$ MVA; (b) $Z_\phi = 13.84\angle 36.87°$ Ω

2.23 (a) $I_a = 100\angle -30°$ A, $I_b = 100\angle -150°$ A, $I_c = 100 \angle 90°$ A; (b) $I_n = 0$

2.26 (a) $Z_{eq} = 2.5473\angle 55.84°$ Ω; (b) $I_a = 108.79\angle -55.84°$ A; (c) $S_{tot} = 90,449\angle 55.84°$ VA

2.30 (a) $V_1 = 479.9513 \angle 30°$ V, $V_2 = 479.9513\angle 150°$ V, $V_3 = 479.9513\angle -90°$ V
(b) $I_1 = 17.3205 \angle 70°$ A, $I_2 = 17.3205\angle 190°$ A, $I_3 = 17.3205\angle -50°$ A;
(c) $I_{L1} = 30\angle 40°$ A, $I_{L2} = 30\angle -80°$ A, $I_{L3} = 30\angle 160°$ A

2.33 (a) $P_1 = 8,860.53$ W, $P_2 = 4,897.51$ W; (b) $P_{tot} = 13,758.04$ W

Chapter 3

3.1 (a) $l = 0.4084$ m, $A = 0.7854$ m^2; (b) $B = 1.09$ T, (c) $\mathcal{F} = 408.4$ A · turns; (d) $I = 1.6336$ A

3.3 $\mathcal{F}_{coil} = 5260.986$ A · turns

3.5 (a) $\Phi = 0.0018$ Wb; (b) $\Phi = 0.0009$ Wb

3.7 (a) $I = 5.6$ A; (b) $\mu_r = 426.5$; (c) $\mathcal{R}_c = 93,333$ A · turns/Wb

3.10 (a) $I = 1.2031$ A; (b) $\mu_r = 723.4$; (c) $\mathcal{R}_c = 76,998.4$ A · turns/Wb

3.13 (a) $I = 7.2$ A; (b) $\mu_r = 221$; (c) $\mathcal{R}_c = 144,000$ A · turns/Wb

3.17 $\mathcal{F}_{coil} = 5092.6$ A · turns

3.20 $\Phi_{tot} = 0.026$ Wb

3.22 $\Phi_{tot} = 7.9 \times 10^{-2}$ Wb

3.26 $B = 0.634$ Wb/m^2

3.30 $B = 2.44$ Wb/m^2

3.34 $\Phi_{tot} = 2.0196 \times 10^{-3}$ Wb

Chapter 4

4.1 (a) $a = 5.75$; (b) $I_1 = 7.2464$ A, $I_2 = 41.6667$ A; (c) $Z'_L = 1904.4\Omega$; (d) $Z_L = 57.6$ Ω

4.4 (a) $I_1 = 20.8333$ A, $I_2 = 208.3333$ A; (b) $a = 10$, $N_1 = 500$ turns; (c) $\Phi_m = 0.018$ Wb

4.7 (a) $a = 6.32$; (b) $I_1 = 0.75$ mA, $I_2 = 0.00474$ A, $V_2 = 0.475$ V, $P_2 = 2.25$ mW

4.11 $I_{load} = 20.2973\angle -73.05°$ Ω, $V_{load} = 211.91\angle 0.25°$ V, $P_{line\,loss} = 65.92$ W

4.15 $I_{load} = 19.7815\angle -73.38°$ Ω, $V_{load} = 159.5\angle -13.13°$ V, $P_{line\,loss} = 1.96$ W

4.21 (a) $R_c = 88.617 \text{k}\Omega$, $X_m = 3.696 \text{k}\Omega$; (b) $Z_{eq,1} = 0.69 + j2.05\,\Omega$; (c) $R_{eq,1} = 0.69\,\Omega$, $X_{eq,1} = 2.05\,\Omega$

4.25 $Z_{th} = 0.002 + j0.009\,\Omega$

4.31 (a) $VReg = 5.15\%$; (b) $VReg = 1.9\%$; (c) $VReg = -1.73\%$

4.35 $kVA_{rated} = 125$ kVA, $I_{se} = 52.1$ A, $I_L = 62.5$ A, $I_c = 10.4$ A

4.38 (a) $\eta = 97.1\%$; (b) $\eta = 97.38\%$; (c) $\eta = 97.1\%$

Chapter 5

5.1 $\omega = 377$ rad/s

5.4 (a) $P = 75,398.2$ W; (b) $P = 101.07$ hp

5.8 (a) $F = 0.1125$ N; (b) to the left; (c) to the right

Chapter 6

6.2 (a) $n_s = 2$ poles; (b) $s_{FL} = 5\%$; (c) 3 Hz; (d) $\mathbf{n}_{rr} = 180$ rpm; (e) $n_{rs} = 3600$ rpm; (f) $n_{rsf} = 0$ rpm

6.5 (a) $\mathbf{n}_s = 3600$ rpm; (b) $n = 3420$ rpm; (c) $f_2 = 3$ Hz; (d) $T_{shaft} = 52.07$ N·m or $T_{shaft} = 38.39$ lb.ft

6.8 (a) $P_g = 34,834$ W; (b) $P_d = 33,334$ W; (c) $P_{shaft} = 32,834$ W= 44 hp; (d) $\eta = 92.9\%$

6.16 (a) $n = 1740$ rpm; (b) $I_1 = 116.33\angle 26.05°$ A; $PF = 0.9$; (c) $PF = 0.9$; (d) $P_{out} = 77,125$ W; (e) $\eta = 89\%$; (f) $T_d = 439.44$ N.m, $T_{load} = 423.25$ N.m

6.21 (a) $s = 5.5\%$; (b) $T_d = 308.76$ lb.ft; (c) $n_m = 1602$ rpm; (d) 140,516.4 W

6.25 (a) $s_{max} = 0.1858$; (b) $n_m = 814.19$ rpm; (c) $T_{max} = 1411.17$ N·m; (d) $T_{start} = 574.8$ N·m; (e) $s_{max} = 0.3716$ (f) $R_2 = 0.14\,\Omega$; (g) $T_{start} = 984.49$ N·m

6.30 (a) $s = 0.001107$; (b) $I_1 = 38.426\angle -83.7°$ A; (c) $P_{in} = 3221$ W

Chapter 7

7.2 (a) $\omega = 188.5$ rad/s; (b) $E_{max} = 391.723$ V; (c) 277 V; (d) 480 V

7.6 $E_\phi = 2731.3\angle 8.59°$ V (a) $V_a = const.$ For a $PF = 0.85$, $I_a = 117.65 A$; (b) The same; (c) Bus voltage normally increases with leading power factor

7.8 (a) $P_{in} = 20,271.74$ W; (b) $S_{in} = 25,339.67$ VA

7.9 (a) $I_a = 1,924.5\angle -31.8°$ A; (b) $E_a = 8,288\angle 14.6°$ V; (c) $\delta = 14.5°$; (d) $P_s = 154,422$ W/electrical rad; (e) $P_s = 5,408,828$ W/mehanical rad; (f) $T_s = 28,695$ N.m/mechanical degree

7.12 (a) $E_a = 227.36\angle -40.52°$ V; (b) $\delta = 40.52°$; (c) $\eta = 90.23\%$; (d) $P_S = 76,692$ W, $T_s = 407$ N·m

7.13 (a) $X_{s,un} = 0.668\Omega$; (b) $X_s = 0.61\,\Omega$; (c) SCR = 0.75

7.17 (a) $I_f = 4.7$ A; (b) $I_f = 5.9$ A; (c) $\eta = 92.73\%$; (d) $E_a = 534.77$ V; (e) $I_f = 4.25$ A

7.21 (a) at 0.8 lagging PF: $V_t = 423$ V, *at* 1.0 PF: $V_t = 472$ V, at 0.8 leading PF: $V_t = 527$ V; (b) $\eta = 94.7\%$; (c) $T_{app} = 328.5$ N·m; $T_{ind} = 311$ N·m; (d) at 0.8 lagging PF: $VReg = 13.5\%$, at 1.0 PF: $VReg = 1.7\%$, at 0.8 leading *PF*: $VReg = -8.9\%$

7.26 (a) $I_a = 29.6\angle 36.87°$ A, $E_a = 360.7\angle -15.21°$ V; (b) $P_{out} = 55.95$ kW; (c) $I_L = 129.6$ A, PF = 0.976 lagging

7.29 $S_{tot} = 665.87\angle 233.39°$ kVA

7.30 (a) $I_a = 2,165\angle -36.87°$ A; (b) $E_a = 14,731\angle 14.3°$ V; (c) $E_a = 14,633\angle 7.7°$ V

Chapter 8

8.2 (a) $Z = 500\,\Omega$; (b) $A_p = 0.1979$ m^2, $\Phi_d = 0.1385$ Wb; (c) $a = 6$ lap, $K_a = 79.58$; (d) $E_a = 1040.1$ V; (e) $I_{coil} = 33.33$ A

8.4 (a) $E_a = 264.9$ V; (b) $E_a = 235.1$ V

8.7 (a) 217.7 N·m; (b) 305 N·m

8.11 (a) $R_a = 0.03\,\Omega$; (b) $I_a = 403$ W; (c) total losses = 11,928 W; (d) $\eta = 89.3\%$

8.14 (a) $E_a = 262.3$ V; (b) $T_d = 171.16$ N·m; (c) $\eta = 88.9\%$

8.17 $N_{se} = 7$ turns

8.21 (a) $n_{100} = 1061$ rpm; (b) $T_{d(100)} = 206.9$ N-in

8.26 (a) $R_a B = 0.3556$ Ω; (b) $R_a = 0.0178$ Ω; (c) $I_{a, start} = 13,500$ A

8.30 $R_a = 01202$ Ω, $R_f = 78.13$ Ω

Chapter 9

9.2 (a) $Z_f = 27.893 + j36.65$ Ω, $Z_b = 1.83 + j1.9$ Ω; (b) $I_1 = 4.35\angle -51.05°$ A; (c) $P_m = 328.2$ W; (d) $P_g = 246.6$ W; (e) $T_d = 1.31$ N·m; (f) $P_d = 234.3$ W; (g) $P_{out} = 190.3$ W; (h) $T_{out} = 1.06$ N·m; (i) $\eta = 58\%$

9.6 (a) $I_1 = 7.533\angle -51.05°$ A, $P_g = 739.5$ W, $T_d = 3.92$ N·m; (b) $I_1 = 8.69\angle -51.05°$ A, $P_g = 984.2$ W, $T_d = 5.22$ N·m

9.9 $C = 187.9$ μF

9.12 (a) $n_m = 476.6$ rpm; (b) $T_d = 23.57$ N·m

Chapter 10

10.1 $V_t = E_a = 2704(1.0 - \exp - 2t)u(t)$

10.4 (a) $E_a(t) = 250(1 - \exp - 5t)$; (b) $E_a(\infty) = 250$ V; (c) $t = 0.92$ s

10.6 (a) $i_{ac}(t) = \sqrt{2}\left[8748 + 29,288\exp\left(-\dfrac{t}{1.15}\right) + 34,861\exp\left(-\dfrac{t}{0.0355}\right)\sin \omega t\right]$;

(b) $I_{tot} = 103,092$ A; (c) $39,019$ A; (d) $55,181$ A; (e) $103,091.9$ A

Chapter 12

12.1 73 kW

12.2 (a) At point a, 22.2 ft

(b) At point b, 31.0.2 ft

12.4 (a) 97.71%

(b) 22.83%

(c) 1999.9%

12.6 19.15%

12.8 19.4%

12.10 23.55

Chapter 13

13.1 21.5 langleys

13.2 120 kW/day

13.5 8%

13.6 12%

13.8 (a) 24 modules

(b) 6 in series with 4 parallel paths

(c) 28.7 A; 31.8 V; 912.7 W; 1.11 Ω

Appendix A

A.2 $I = 10\angle 30°$ A

A.4 $I_1 = 10\angle 30°$ A, $I_2 = 20\angle -140°$ A, $I_3 = 10.30\angle 49.7°$ A

Appendix B

B.2 $Z_{pu,new} = 0.02$ pu, $I_{B(HV)new} = 66.67$ A, $Z_{B(HV)new} = 45$ Ω

Bibliography

Acarnley, R. P., *Stepping Motors*, 2nd Ed., London, U.K.: Peter Peregrinus Ltd., 1984.

AIEE Standards Committee report, *Electrical Engineering*, 65(11), 1946, 512–516.

Alger, P., *Induction Machines*, 2nd Ed., New York: Gordon and Breach, 1970.

Anderson, L. R., *Electric Machines and Transformers*, Reston, VA: Reston Publishing Company, 1981.

Anderson, P. M., *Analysis of Faulted Power Systems*, New York: IEEE Press, 1993.

Bergseth, F. R. and Venkata, S. S., *Introduction to Electric Energy Devices*, Englewood Cliffs, NJ: Prentice-Hall, Inc., 1987.

Bewley, L. V., *Alternating Current Machinery*, New York: Macmillan Publishing Company, 1949.

Blume, L. F., Boyajian, A., Camilli, G., Lennox, T. C., Minneci, S., Montsinger, V. M., *Transformer Engineering. A Treatise on the Theory, Operation, and Application of Transformers*, 2nd Ed., New York: John Wiley & Sons, Inc., 1951.

Bose, B. K., *Power Electronics and AC Drives*, Englewood Cliffs, NJ: Prentice-Hall Inc., 1951.

Brown, D. and Hamilton III, E. P., *Electromechanical Energy Conversion*, New York: Macmillan Publishing Company, 1984.

Bumby, J. R., *Superconducting Rotating Electrical Machines*, Oxford, U.K.: Clarendon Press, 1983.

Carr, L., *The Testing of Electric Machines*, London, U.K.: MacDonald Company Ltd., 1960.

Carry, C. C., *Electric Machinery. A Coordinated Presentation of AC and DC Machines*, New York: John Wiley & Sons, Inc., 1958.

Chapman, S. J., *Electric Machinery Fundamentals*, New York: McGraw-Hill Book Company, 1985.

Concordia, C. D., *Synchronous Machines Theory and Performance*, Schenectady, NY: General Electric Company, 1951.

Crosno, C. D., *Fundamentals of Electromechanical Conversion*, New York: Harcourt, Brace & World, Inc., 1968.

Daniels, A. R., *The Performance of Electrical Machines*, London, U.K.: McGraw-Hill Publishing Company, 1968.

Del Toro, V., *Electric Machinery and Power Systems*, Englewood Cliffs, NJ: Prentice-Hall, Inc., 1985.

Del Toro, V., *Electromechanical Devices for Energy Conversion and Control Systems*, Englewood Cliffs, NJ: Prentice-Hall, Inc., 1968.

Dudley, A. M. and Henderson, S. F., *Connecting Induction Motors: Operation and Practice*, 4th Ed., New York: McGraw-Hill Book Company, 1960.

Electro-Craft Corporation, *DC Motors, Speed Controls, Servo Systems*, 3rd Ed., Oxford, U.K.: Pergamon Press Ltd., 1977.

Elgerd, O., *Basic Electric Power Engineering*, Reading, MA: Addison Wesley Publishing Company, 1977.

Elgerd, O., *Electric Energy System Theory*, 2nd Ed., New York: McGraw-Hill Book Company, 1982.

El-Hawary, M. E., *Principles of Electric Machines and Power Electronic Applications*, Reston, VA: Reston Publishing Company, Inc., 1966.

Emanuel, P., *Motors, Generators, Transformers, and Energy*, Englewood Cliffs, NJ, Prentice Hall, Inc., 1985.

Ergeneli, A., *Elektroteknik*, Vol. 1, Istanbul, Turkey: Teknik Okulu Yayinlari, 1962.

Ergeneli, A., *Elektroteknik*, Vol. 2, Istanbul, Turkey: Teknik Okulu Yayinlari, 1962.

Feinberg, R., *Modern Power Transformer Practice*, New York: John Wiley & Sons, Inc., 1979.

Fitzgerald, A. E., Kingsley, Jr. C., Umans, S. D., *Electric Machinery*, 4th Ed., New York: McGraw-Hill Book Company, 1983.

Garik, M. L. and Whipple, C. C., *Alternating-Current Machines*, 2nd Ed., New York: John Wiley & Sons, Inc., 1986.

Gehmlich, D. K. and Hammond, S. B., *Electromechanical Systems*, New York: McGraw-Hill Book Company, 1967.

Gönen, T., *Engineering Economy for Engineering Managers*, New York: John Wiley & Sons Inc., 1990.

Gönen, T., *Electric Power Distribution System Engineering*, 2nd Ed., Boca Raton, FL: CRC Press, 2008.

Gönen, T., *Electric Power Transmission System Engineering: Analysis and Design*, 2nd Ed., Boca Raton, FL: CRC Press, 2009.

Gönen, T., *Modern Power System Analysis*, New York: John Wiley & Sons, Inc., 1988.

Gönen, T. and Haj-mohamadi, M. S., Electromagnetic unbalances of six-phase transmission lines, *Electrical Power & Energy Systems*, 11(2), 1989, 78–84.

Guru, B. S. and Hiziroglu, H. R., *Electric Machinery and Transformers*, Orlando, FL: Harcourt Brace Jovanocich, Inc., 1988.

Hancock, N. N., *Electric Power Utilization*, London, U.K.: Sir Isaac Pitman & Sons, Ltd. 1967.

Hancock, N. N., *Matrix Analysis of Electrical Machinery*, 2nd Ed., Oxford, U.K.: Pergamon Press Ltd., 1974.

Hindmarsh, J., *Electrical Machines*, Oxford, U.K.: Pergamon Press Ltd., 1965.

Hubert, C. I., *Preventive Maintenance of Electrical Equipment*, 2nd Ed., New York: McGraw-Hill Book Company, 1969.

IEEE, Std, 112–1978, *IEEE Standard Test Procedure for Polyphase Induction Motors and Generators*, New York: IEEE, Inc., 1984.

Kenjo, T., *Stepping Motors and Their Microprocessor Controls*, Oxford, U.K.: Clarendon Press, 1984.

Kosow, I. L., *Control of Electric Motors*, Englewood Cliffs, NJ: Prentice-Hall, Inc., 1972.

Kosow, I. L., *Electric Machinery and Transformers*, Englewood Cliffs, NJ: Prentice-Hall, Inc., 1972.

Kostenko, M. and Piotrovsky, L., *Electrical Machines*, Vol. 1, Moscow, Russia: Mir Publishers, 1974.

Kostenko, M. and Piotrovsky, L., *Electrical Machines*, Vol. 2, Moscow, Russia: Mir Publishers, 1974.

Krause, P. C. and Wasynczuk, O., *Electromechanical Motion Devices*, New York: McGraw-Hill Book Company, 1989.

Kuhlmann, J. H., *Design of Electric Apparatus*, 2nd Ed., New York: John Wiley & Sons, Inc., 1940.

Kuo, B. C., *Step Motors*, St. Paul, MN: West Publishing Company, 1974.

Lawrence, R. R. and Richards, H. E., *Principles of Alternating Current Machinery*, 4th Ed., New York: McGraw-Hill Book Company, 1953.

Leonhard, W., *Control of Electric Drives*, Berlin, Germany: Springer-Verlag, 1985.

Lindsay, J. F. and Rashid, M. H., *Electromechanics and Electrical Machinery*, Englewood Cliffs, NJ: Prentice-Hall, Inc., 1986.

Loew, E. A. and Bergseth, F. R., *Direct and Alternating Currents: Theory and Machinery*, 4th Ed., New York: McGraw-Hill Book Company, 1954.

Mablekos, V. E., *Electric Machine Theory for Power Engineers*, New York: Harper & Row Publishers, 1980.

Majmudar, H., *Electromechanical Energy Converters*, Boston, MA: Allyn and Bacon, Inc., 1965.

Matsch, L. W. and Morgan, J. D., *Electromagnetic and Electromechanical Machines*, 3rd Ed., New York: Harper & Row Publishers, 1986.

McIntyre, R. L., *Electric Motor Control Fundamentals*, 3rd Ed., New York: McGraw-Hill Book Company, 1974.

McLaren, P. G., *Elementary Electric Power and Machines*, Chichester, U.K.: Ellis Horwood Ltd., 1984.

McLyman, Wm. T., *Transformer and Inductor Design Handbook*, New York: Marcel Dekker, Inc., 1978.

McPherson, G., *An Introduction to Electrical Machines and Transformers*, New York: John Wiley & Sons, Inc., 1981.

Meisel, J., *Principles of Electromechanical Energy Conversion*, New York: McGraw-Hill Book Company, 1966.

Millermaster, R., *Harwood's Control of Electric Motors*, 4th Ed., New York: John Wiley & Sons, Inc., 1970.

MIT Staff, *Magnetic Circuits and Transformers*, New York: John Wiley & Sons, Inc., 1943.

Molloy, E. (ed.), *Small Motors and Transformers: Design and Construction*, London, U.K.: George Newnes Ltd., 1953.

Nasar, S. A. and Boldea, I., *Linear Motion Electric Machines*, New York: John Wiley & Sons, Inc., 1976.

Nasar, S. A. (ed.), *Handbook of Electric Machines*, New York: McGraw-Hill Book Company, 1987.

Nasar, S. A., *Electric Energy Conversion and Transmission*, New York: Macmillan Publishing Company, 1985.

Nasar, S. A., *Electric Machines and Electromechanics, Outline Series in Engineering*, New York: McGraw-Hill Book Company, 1981.

Nasar, S. A., *Electric Machines and Transformers*, New York: Macmillan Publishing Company, 1984.

Nasar, S. A., *Electromechanics and Electric Machines*, New York: John Wiley & Sons, Inc., 1979.

National Electric Manufacturing Association, *Motors and Generators*, Publication No. MGI-1972, New York, 1972.

Nilsson, J. W., *Introduction to Circuits, Instruments, and Electronics*, New York: Harcourt, Brace & World, Inc., 1968.

Patrick, D. R. and Fardo, S. W., *Rotating Electrical Machines and Power Systems*, Englewood Cliffs, NJ: Prentice-Hall, Inc., 1985.

Pearman, R. A., *Power Electronics: Solid State Motor Control*, Reston, VA: Reston Publishing, 1980.

Puschstein, A. F., Lloyd, T. C., Conrad, A. G., *Alternating Current Machines*, 3rd Ed., New York: John Wiley & Sons, Inc., 1954.

Ramshaw, R. and Van Heeswijk, R. G., *Energy Conversion: Electric Motors and Generators*, Orlando, FL: Saunders College Publishing, 1990.

Richardson, D. V, *Handbook of Rotating Electric Machinery*, Reston, VA: Reston Publishing Company, Inc., 1980.

Robertson, B. L. and Black, L. J., *Electric Circuits and Machines*, New York: D. Van Nostrand Company, Inc., 1949.

Sarma, M. S., *Electric Machines: Steady-State Theory and Dynamic Performance*, Dubuque, IA: Wm. C. Brown Publishers, 1985.

Say, M. G., *Electric Engineering Design Manual*, 3rd Ed., London, U.K.: Chapman and Hall Ltd., 1962.

Say, M. G., *Introduction to the Unified Theory of Electromagnetic Machines*, London, U.K.: Pitman Publishing, 1971.

Seely, S., *Electromechanical Energy Conversion*, New York: McGraw-Hill Book Company, 1962.

Sen, P. C., *Principles of Electric Machines and Power Electronics*, New York: John Wiley & Sons, Inc., 1989.

Shultz, R. D. and Smith, R. A., *Introduction to Electric Power Engineering*, New York: Harper & Row Publishers, 1985.

Siskind, C. S., *Direct Current Machinery*, New York: McGraw-Hill Book Company, 1952.

Siskind, C. S., *Electrical Control Systems in Industry*, New York: McGraw-Hill Book Company, 1963.

Siskind, C. S., *Electrical Machines, Direct and Alternating Current*, 2nd Ed., New York: McGraw-Hill Book Company, 1959.

Skilling, H. H., *Electrical Engineering Circuits*, 2nd Ed., New York: John Wiley & Sons, Inc., 1966.

Skilling, H. H., *Electromechanics*, New York: John Wiley & Sons, Inc., 1962.

Skrotzki, B. G. A. and Vopat, W. A., *Power Station Engineering and Economy*, New York: McGraw-Hill Book Company, 1960.

Slemon, G. R. and Straughen, A., *Electric Machines*, Reading, MA: Addison-Wesley Publishing Company, 1980.

Smeatson, R. W. (ed.), *Motor Application and Maintenance Handbook*, New York: McGraw-Hill Book Company, 1969.

Smith, R. T., *Analysis of Electric Machines*, New York: Pergamon Press, 1982.

Stein, R. and Hunt, W. T., *Electric Power System Components: Transformers and Rotating Machines*, New York: Van Nostrand Reinhold Company, 1979.

Stevenson, W. D., *Elements of Power System Analysis*, 4th Ed., New York: McGraw-Hill Book Company, 1981.

Stigant, S. A. and Franklin, A. C., *J&P Transformer Book*, 10th Ed., London, U.K.: Newnes-Butterworth, 1973.

Thaler, G. J. and Wilcox, M. L., *Electric Machines: Dynamics and Steady State*, New York: John Wiley & Sons, Inc., 1966.

Veinott, C. G. and Martin, J. E., *Fractional- and Subfractional-Horsepower Electric Motors*, 4th Ed., New York: McGraw-Hill Book Company, 1986.

Veinott, C. G., *Theory and Design of Small Induction Motors*, New York: McGraw-Hill Book Company, 1959.

Wenick, E. H. (ed.), *Electric Motor Handbook*, London, U.K.: McGraw-Hill Book Company, 1978.

Westinghouse Staff, *Electrical Transmission and Distribution Reference Book*, East Pittsburgh, PA: Westinghouse Electric Corporation, 1964.

Zorbas, D., *Electric Machines: Principles, Applications, and Control Schematics*, St. Paul, MN: West Publishing Company, 1989.

Index

A

Active distribution network, 430–431, 517
Additive polarity, 104–105
Air-core transformers, 94
Air-gap, 66, 68, 74, 76, 82–83, 178, 184, 189, 225–226, 282–283, 285, 287, 302–303, 315, 374–376, 390, 402
Air-gap power, 230, 233, 236, 238, 243–244, 257, 290, 375, 378–379
Alignment principle, 191
All-day efficiency, 136
Alternator, 8, 18, 273, 276, 294, 442–443, 569
Ampere's circuital law, 56
Ampere turns, 57, 59, 74, 154, 328, 330
Angular velocity, 165–166, 169–170, 190, 198, 219, 221, 319, 349, 420–422
Apparent power, 9–10, 12, 26–28, 30, 103, 107, 124, 148, 252, 567
Armature, 18, 55, 72, 177, 183–187, 273, 276, 282–286, 288, 290, 292, 295–297, 299, 301–305, 313–323, 335, 340–341, 358, 360, 363–364, 387–389, 395–396, 402–408, 410–411, 413–416, 442, 569–573, 581–582
Armature reactance, 283, 286
Armature reaction, 283–284, 323–325, 327–328, 330–331, 401, 412, 569
Armature reaction voltage, 285, 333–334, 569
Armature voltage, 319–323, 329, 333, 342, 356–357, 395, 442
Armature winding, *see* Rotor winding
Asynchronous machine, 8
Asynchronous speed, 199, 221
Autotransformer, 94, 146–152
Autotransformer starting, 253–254, 257–258

B

Back emf, *see* Internal generated voltage
Balanced three-phase loads, 12, 25–26, 28, 30
B–H curve, 66, 70, 75, 82, 85, 87
Blocked-rotor test, 260, 262–266, 272
Braking mode, 209
Breakdown torque, *see* Maximum internal torque
Breathing field, 371
Brush-contact voltage drop, 321, 333, 338, 346
Brushless excitation, 276–277

C

Cage rotor, 211–212
Capability curve, 305–306
Capacitors, 4, 41, 130, 152, 299, 382–386, 427–428, 497, 533
Coenergy, 185–189, 195
Coercive field intensity, 71

Coercive force, *see* Coercive field intensity
Coil pitch, 583–584
Coil span, 583–584
Collector rings, *see* Slip rings
Commutating poles, 314–315, 326–327
Commutation, 325–328
Commutator, 259, 313, 315, 317–319, 321, 325–327, 350, 442
Commutator winding, 316
Compensating windings, 315, 321, 328
Composite structure, 66–67
Compound generator:
    cumulative compound, 324, 336
    differential compound, 324, 336
Constant volts per hertz, 260
Construction of synchronous machines, 273–276
Copper losses, 97, 109, 123, 135, 221, 227, 230–231, 235, 238, 250, 261–262, 265, 343–344, 376
Core losses:
    induction motor, 72–81, 230–231, 234
    transformer, 101, 113, 135
Counter emf, 100, 115, 169, 225, 351, 358–359, 361, 364–365
Counter torque, 169, 290–291, 358, 364
Critical field resistance, 335
CTs, *see* Current transformers
Cumulative compound, 323–324, 331, 336, 350–351
Current inrush, 154–156
Current limiting starting, 251–252
Current transformers (CTs), 94, 153, 155, 529
Cycloconverter, 260, 273
Cylindrical machines, 197–200

D

Damper winding, 274–275, 298–299, 415
d-axis, *see* Direct axis
dc motor braking, 364–366
dc motor control:
    armature resistance control, 355–356
    armature voltage control, 355–357
    field control, 355–356
dc motor starting, 358–360
dc test, 260, 262–264, 272, 301–303
Delta-connected source, 14
Delta-delta connection, 140–145, 561
Delta-wye connection, 140–143, 145, 561
Developed power
    of dc motor, 341–342
    of induction motor, 232–233, 236, 238, 257
Developed torque, of induction motor, 232, 236, 239–240, 243–244, 249, 257, 260
DG, *see* Distributed generation
Differential compound, 323–324, 331, 336
Direct axis, 188–189, 274, 315
Direct-axis inductance, 191, 193

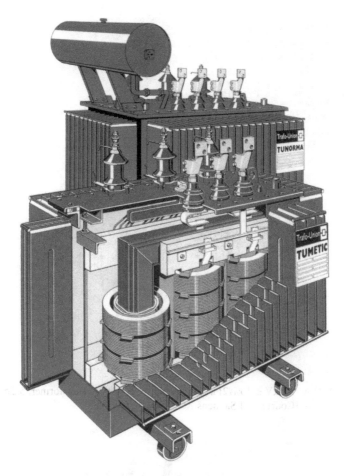

FIGURE 3.7 Oil distribution transformers: Cutaway of a TUMETIC transformer with an oil expansion tank shown in the foreground, and a TUNORMA with an oil expansion tank shown in the background. (Courtesy of Siemens AG, Munich, Germany.)

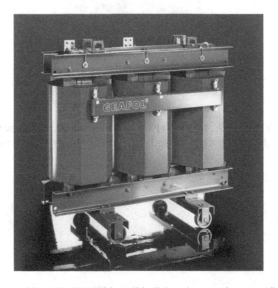

FIGURE 3.8 A 630 kVA, 10/0.4 kV GEAFOL solid dielectric transformer. (Courtesy of Siemens AG, Munich, Germany.)

FIGURE 4.25 A 40 MVA, 110 kV ± 16%/21 kV, three-phase, core-type transformer, 5.2 m high, 9.4 m long, 3 m wide, weighing 80 tons. (Courtesy of Siemens AG, Munich, Germany.)

FIGURE 4.26 A 850/950/1100 XWA, 415 kV ± 11%/27 kV, three-phase, shell-type transformer, 11.3 in high, 14 in long, 5.7 in wide, weighing (without cooling oil) 552 tons. (Courtesy of Siemens AG, Munich, Germany.)

FIGURE 4.27 10 MVA and 50 kVA, core-type, three-phase transformers with GEAFOL solid dielectric cores. (Courtesy of Siemens AG, Munich, Germany.)

FIGURE 4.28 Atypical core and coil assembly of a three-phase, core-type, power transformer. (Courtesy of North American Transformer, Milpitas, CA.)

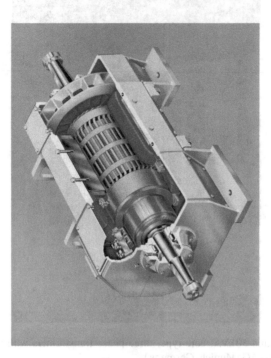

FIGURE 8.4   Cutaway view of a mill duty dc motor. (Courtesy of General Electric, Fairfield, CT.)

FIGURE 8.7   The armature of a 500 V, 150 hp dc motor. (Courtesy of General Electric, Fairfield, CT.)